ANATOMY OF THE MONOCOTYLEDONS

VII

HELOBIAE
(ALISMATIDAE)

ANATOMY OF THE MONOCOTYLEDONS

EDITED BY C. R. METCALFE

*Honorary Research Associate of the Royal Botanic Gardens, Kew, and
formerly Keeper of the Jodrell Laboratory*

VII. HELOBIAE *(ALISMATIDAE)*

(including the seagrasses)

P.B. TOMLINSON

*Professor of Botany,
Harvard University, Harvard Forest, Petersham, Mass., U.S.A.
(Research Collaborator, Fairchild Tropical Garden,
10901 Old Cutler Road, Miami, Fla. 33156, USA)*

With the Assistance of:
Priscilla Fawcett (Miami), Mary Gregory (Kew), Usher Posluszny (Guelph),
and Margaret Stant (Kew)

CLARENDON PRESS • OXFORD
1982

Oxford University Press, Walton Street, Oxford OX2 6DP

OXFORD   LONDON   GLASGOW
NEW YORK   TORONTO   MELBOURNE   WELLINGTON
IBADAN   NAIROBI   DAR ES SALAAM   LUSAKA   CAPE TOWN
KUALA LUMPUR   SINGAPORE   JAKARTA   HONG KONG   TOKYO
DELHI   BOMBAY   CALCUTTA   MADRAS   KARACHI

Published in the United States of America by Oxford University Press, New York

British Library Cataloguing in Publication Data

Tomlinson, P. B.
    Anatomy of the monocotyledons
    Vol. 7: Helobiae (Alismatidae)
    1. Monocotyledons—Anatomy
    2. Botany—Anatomy
    1. Title   II. Metcalfe, C. R.
    584'.04   QK643.M7

ISBN 0–19–854502–9

Printed in the United States of America.

EDITOR'S PREFACE

THE author of this volume, Professor P. B. Tomlinson, now at Harvard University, needs no introduction to readers of this series of volumes entitled *Anatomy of the Monocotyledons,* for he has already contributed Vol. II on the Palmae, published in 1961 and Vol. III, published in 1969, covering the families Commelinales to Zingiberales in Hutchinson's system of classification. Now, rather more than a decade later, we are indebted to Professor Tomlinson for the present volume on the Helobiae (Alismatidae). The plants described in this volume are nearly all aquatic and some of them marine. Their morphology, structure, and biology are therefore very different from those of the plants described in the author's two previous volumes. This also provides evidence of the author's versatility in facing the problems of investigating plants that range in size from arboreal giants such as some of the palms to the smallest members of the Helobiae.

The production of this volume is a notable and timely event, partly because there has hitherto been no equally comprehensive account of the morphology and anatomy of the Helobiae, and partly because of an increasing interest in the group at the present time. The reasons for this burgeoning interest in the Helobiae are stated very clearly in the author's preface, so there is no need to repeat them here.

One of the problems in studying the anatomy of the Helobiae is that it has, until now, been almost impossible to obtain living or well preserved material of some of the taxa. Herbarium specimens of aquatic plants are seldom used for anatomical investigation except as a last resort, for it is very difficult to 'revive' the tissues sufficiently to avoid errors of structural interpretation. Certain members of the group are well known and either easily accessible in the wild, or available as cultivated specimens in botanic gardens. Many of these have been repeatedly examined over the years, but the less accessible taxa have received little or no attention. The result has been a lack of balance in the treatment of the group and this has led to some previous misinterpretations of the morphology and anatomy of its members and consequently of their classification. The author has taken untold trouble to obtain well preserved, accurately named specimens of as many of the taxa as possible. He has collected many of them himself and it is indeed fortunate that other investigations in which he has been engaged have taken him to localities in which members of the Helobiae are to be found. Nevertheless there have been some taxa that he has not been able to examine, and for descriptions of some of these he has had to rely on publications by other botanists.

A second problem in studying the anatomy of the Helobiae from the taxonomic standpoint arises from the very fact that their aquatic mode of life ensures that many of the taxa show adaptive structural similarities even when they are not closely related. Nevertheless the author has shown great skill in selecting characters that have allowed him to interpret the morphology and anatomy of the plants in a sense that will be helpful to taxonomists. However, to obtain a satisfactory interpretation of what he has seen he has given attention to the morphology of the floral as well as of the vegetative parts of the plant. In this respect the present volume differs from others which have preceded it in the present series. The result has been rewarding, and perhaps this is a direction that we should take again in the future.

Professor Tomlinson acknowledges with due gratitude the assistance he has received from other botanists. Indeed as editor I know that he has been fortunate. In particular the collaboration of Priscilla Fawcett who produced most of the habit and morphological illustrations adds very considerably to the interest of the book. Without considerable assistance from Miss Mary Gregory at the Jodrell Laboratory the extensive bibliography would not have been so complete. Dr Margaret Stant, also at the Jodrell Laboratory, had already published accounts of the Alismataceae, Butomaceae, and Limnocharitaceae which served as starting points for Professor Tomlinson's descriptions of these families. Dr Stant also provided drafts of the families of Scheuchzeriaceae, Petrosaviaceae, and Lilaeaceae of which the author has made full use.

It is indeed a notable achievement to have produced this volume while the author has been very fully occupied with other projects as well as with teaching. In consequence it has taken a decade to complete the volume, but it is no less welcome on account of this delay. Indeed its value has been enhanced because it has been possible to refer to a number of recent important publications which would not have been available if the volume had been finished sooner.

The volume should have a wide appeal because, besides attracting the interest of those who are concerned with systematic, developmental and ecological anatomy, the subject matter covers taxonomy, and the author speculates on the phylogeny of the 16 families that are described. Many interesting facts about the biology of the plants are also presented. In short the volume should serve as an important work of reference that is likely to remain in vogue for many years to come. It will doubtless also be used in courses of instruction for students.

Royal Botanic Gardens,   C.R. Metcalfe
  Kew
February 1981

AUTHOR'S PREFACE

WHEN Asa Gray referred in his correspondence to *Potamogeton* as a 'vile set of weeds' it was an opinion about water plants that many of his contemporaries might well have shared. The group is difficult taxonomically because of its extreme morphological plasticity and *Potamogeton*, like other wetland plants, had no obvious commercial value at the time. The taxonomic difficulties still remain, but we now have a much greater appreciation of the direct value of aquatic plants in wetland and aquatic communities. Modern biologists find their study a fascinating and rewarding occupation so that there is continual need for systematic evaluation.

A measure of the importance which is awarded water-plants is the considerable amount of scientific effort now devoted to them. There are organizations which promote interchange of information about them (e.g. the Association of Aquatic Vascular Plant Biologists). There are entire journals devoted to aquatic plants (e.g. *Aquatic Botany*). International meetings to discuss them have been convened (e.g. The International Seagrass Workshop, Leiden, Netherlands, October, 1973). Numerous regional floras dealing exclusively with these plants on an ecological basis have been produced (e.g. Aston 1973—Australia; Correll and Correll 1975—Southwestern United States; Godfrey and Wooten 1979—Southeastern United States; Mason 1957—California; Subramanyam 1962—India. The scope is limited entirely by a compiler's concept of 'aquatic plant'. Freshwater aquatics have been dealt with at the general systematic level on a world-wide scale (Fasset 1957; Cook et al., 1974). There are major governmental agencies which have a special interest in aquatic communities (e.g. the United States Fish and Wildlife Service, Department of the Interior). There is even a special horticultural interest in aquatic plants, especially those which can be grown in aquaria, and this has produced its own popular literature. More specialized treatments have dealt with river plants (Haslam 1978), and the taxonomy or the structure of water plants of limited geographic areas (e.g. Haynes 1978; Ogden 1974).

It is against this background of burgeoning interest that this volume has to be placed. Although it deals only with a taxonomic order of monocotyledons, their predisposition for wet sites is indicated in classic names for the group—'Helobiae' or 'Fluviales'. Nevertheless they occupy a wide range of habitats. In particular, all those flowering plants which occur in strictly marine communites (the 'seagrasses') are included.

The family descriptions in this book are more complete than those of other volumes in the series since they attempt to deal with aspects of reproductive morphology and anatomy to the extent that a fairly comprehensive taxonomic account of each family is included. There is a separate taxonomic discussion of each group and the introductory sections provide overviews of major topics, discuss structural details, and summarize recent research.

The survey of the literature has also been broad, but clearly it is selective since emphasis is placed on anatomical and morphological publications. It is no longer possible to deal with all aspects of the biology of any one major group of plants in a limited space, although the book by Sculthorpe (1967)

The Biology of Aquatic Vascular Plants comes as close to this for water plants generally as seems possible at the hands of a single author. This volume by Sculthorpe continues the tradition established by Arber (1920) in her classical *Water Plants.*

Who will find this book useful? Taxonomists, of course, since the presentation is set within a systematic framework and presents evidence upon which taxonomic conclusions may be drawn. Anatomists will find the extensive summary of information useful and the layout is intended to lead directly to the more detailed published sources of information whose broadly scattered distribution is suggested by the fact that papers in about 170 different journals have been cited! Developmental morphologists will find an attempt to bring a dynamic approach to the descriptive writing, since I believe that much structural detail can only be interpreted on a developmental basis, no matter for what ultimate purpose the structural information is used (Tomlinson 1970). It is unfortunate that the telegraphic style of presentation, so necessary for conciseness, does not permit a more extended developmental discussion.

Evolutionary morphologists will find the concise presentation of information useful for comparative purposes. The Helobiae have long occupied an important place in discussions of the evolution of angiosperms, particularly those referring to the evolutionary relationships between dicotyledons and monocotyledons, although the discussions have frequently been based on very simplistic morphological interpretations. One fact which is clearly established by studying this order in detail is the remarkable amount of adaptive radiation its members show, with frequent evidence for convergence. At the same time unidirectional biological trends are mirrored in lines of morphological development, notably the transition from pollination above water to pollination at the air–water interface and finally to pollination below water. I think that the view of the Helobiae as an essentially 'primitive' group is based on too superficial an appreciation of their morphological diversity. I cannot accept them as ancestral to other monocotyledonous groups, a point of view increasingly supported by recent morphological study (e.g. Sattler and Singh 1978).

It is to be hoped that the ecologist will extract useful information from this volume since it must be emphasized continually that the structure and form of a plant is the clearest expression of its adaptive success. An appreciation of this success depends on a precise understanding of morphology and anatomy. The fact that some correlations are easily understood should lead to greater effort in interpreting those correlations which are not. It is easy to account for linear or strap-shaped leaves in plants which have intercalary leaf meristems, growing in flowing waters. But why should the Helobiae show a marked propensity for apical bifurcation? What is the function of squamules? How important in internal transport are protoxylem lacunae? Does aerenchyma have significance in internal gas exchange? Why do the submerged or even emergent leaves of some aquatic plants develop an 'apical pore' and others not? Obviously we enter the realm of plant physiology at this point.

If, having been promised a feast, the reader finds lean picking, some attempt should be made to anticipate criticism, because I am painfully aware of much that has been left out and there are specialized topics where studies are moving rapidly.

The treatment of embryological and other details of sexual reproduction is perfunctory. This is because information up to 1966 has been summarized in the compilation by Davis (1966); some more recent papers have been cited where they have direct taxonomic relevance. Pollen morphology is scarcely touched upon, largely because there are few detailed studies of this group in what might be termed the 'post-Erdtman era' (cf. Erdtman 1943, 1952). However, the field is a developing one in view of the current appreciation of biochemical problems imposed upon the pollen–stigma recognition system in hydrophilous (underwater) pollination (e.g., Pettitt 1976, 1980; Pettitt and Jermy 1975). Cytology is little mentioned, which is clearly almost a misdemeanour in view of the probable resolution of some long-standing taxonomic problems by the cytotaxonomic approach (e.g. Reese 1967).

The chemistry of secondary metabolites is briefly included in certain sections where the information is sufficiently conclusive, as in the distribution of triglochinin (Eyjólfsson 1970; Ruijgrok 1974), but in most instances the work being done is still preliminary, though obviously of ultimate great importance.

Floral biology cannot be dealt with in any detail even though comparative analysis offers enormous scope for an understanding of breeding mechanisms. A simple analysis of the distribution of dioecism, monoecism, and hermaphroditism has not been attempted, even though this has a potential for generating broad evolutionary interpretations. The simple question—What is the relation between the breeding mechanism of *Potamogeton* and its high incidence of putative hybridism?—remains unanswered.

Therefore the value of this present work must be judged very much by the individual reader. My ultimate defence is that I started out to survey the systematic anatomy of a small group of monocotyledons. Unfortunately biological investigation cannot be so constrained—'vile weeds' offer too many challenges.

Harvard Forest,   P.B. Tomlinson
Petersham, Mass.
June 1981

ACKNOWLEDGEMENTS

MUCH of the material used in this study has been assembled by myself, since I have had the opportunity to collect aquatic monocotyledons incidental to my research on other topics, mainly in the tropics. The close juxtaposition of both mangrove and seagrass communities, in which I have retained a special research interest, has facilitated this collecting. Financial research support has therefore been both direct and indirect. The Maria Moors Cabot Foundation and Atkins Garden Funds of Harvard University have provided primary support. Supplementary support in the early stages came from grants from the National Science Foundation, Washington, DC (GB-31844-X) and the American Philosophical Society. In later stages grants from the US–Australia and US–New Zealand Scientific Collaborative Programs of the Office of International Programs of the National Science Foundation were useful, as was a visiting Erskine Fellowship at the University of Canterbury, New Zealand in 1977. A grant from the Bentham–Moxon Trust permitted a period of study at the Royal Botanic Gardens, Kew in 1979.

The Keeper of the Herbarium and the Keeper of the Jodrell Laboratory, Royal Botanic Gardens, Kew; the Directors of the Gray Herbarium and Arnold Arboretum of Harvard University; and of Fairchild Tropical Garden are thanked for continued access to specimens and libraries in their care.

The several editors and publishers gave permission to reproduce illustrations which first appeared in the pages of the following journals: *American Journal of Botany, Botanical Journal of the Linnean Society of London, Aquaculture, Canadian Journal of Botany, Bulletin of Marine Science* (Miami), *Botanical Gazette,* and the *Handbook of Seagrass Biology,* edited by R. C. Phillips and C. P. McRoy.

In addition I have relied upon the help of numerous individuals who have supplied either assistance in the field or fluid-preserved material. Most of these collections are cited in the lists of 'Material Examined' at the end of each family or generic description. I fear that it is so extensive that in the following list I may yet fail, inadvertently, to acknowledge all sources of help:

Australia: H. I. Aston, W. R. Birch, J. S. Bunt, M. L. Cambridge, B. C. Clough, S. C. Ducker, H. I. Kirkman, J. Kuo, I. A. Staff.

New Zealand: E. Edgar, E. J. Godley, L. B. Moore.

Papua New Guinea: E. E. Henty, J. S. Womersley.

Fiji: J. W. Parham, A. C. Smith.

United States: I. A. Abbott, G. N. Avery, D. W. Bierhorst, V. I. Cheadle, W. R. Countryman, J. B. Fisher, R. C. Phillips, V. I. Sullivan, D. B. Ward, G. J. Wilder.

Canada: U. Posluszny, R. Sattler.

Africa: J. B. Hall (Ghana), S. Moorjani (Kenya).

French Guiana: R. A. A. Oldeman.

France: R. Delépine, F. Hallé, M. Vivien.

Other individuals supplied specific assistance. W. A. Charlton, R. Jagels, U. Posluszny, R. Sattler, and V. Singh supplied published and unpublished illustrations from their own research. Natalie Uhl made available a copy of her unpublished Ph.D. thesis (1947) which served as a primer in the treatment of floral morphology. The account of the Hydrocharitaceae is based to a large extent on the published paper by E. Ancibor (1979), with access to the slides and material she used in the Jodrell Slide Collection. Several people made constructive comments and corrections on preliminary drafts of certain families, i.e. C. D. K. Cook (Zurich) on Hydrocharitaceae, W. A. Charlton (Manchester) on Alismataceae, U. Posluszny (Guelph) on Lilaeaceae, Scheuchzeriaceae.

A special debt is owed to four people whose names, I feel, should be included on the title page:

Priscilla Fawcett, Botanical Illustrator, Fairchild Tropical Garden, Miami, Florida, has made a contribution which is most obvious, since her illustrations grace the following pages. Her task was rendered difficult since many of the plates in later stages were prepared in the absence of my direct supervision.

Mary Gregory, Bibliographer, Jodrell Laboratory, Royal Botanic Gardens, Kew, allowed access to its extensive card index of anatomical references. Without her supervision the literature coverage would have been much less complete. She also carried out many important editorial duties.

Usher Posluszny, University of Guelph, Canada, provided extensive information about floral development during his stay at Harvard Forest as a Cabot Foundation Fellow, and subsequently. His manipulative skill is indicated among other things by the dissections of the minute and complex floral parts of Zannichelliaceae.

Margaret Y. Stant, Jodrell Laboratory, Royal Botanic Gardens, Kew, provided drafts of the families Lilaeaceae, Scheuchzeriaceae, and Petrosaviaceae, which together with some of her illustrations have been incorporated into the existing account. The accounts of Alismataceae, Butomaceae, and Limnocharitaceae are also based on her published reports, together with some of her published illustrations.

The bulk of the typing and retyping of manuscript was carried out with cheerful efficiency by Valerie Horwill (Kew) and Dotty Smith (Petersham). By now hieroglyphics should be well within their grasp. Preparation of microscopic slides and assistance with assembling illustrations has been provided at various times by Silvia Arroyo, Karen Esseichick, Anne Faulkner, Peter Gassen, Monika Mattmuller, Bill Ormerod, Jenny Richards, Anita Schulman, and Regi Zimmermann.

Finally, as deadlines came and went, with almost seasonal regularity, I am grateful to the editorial staff of the Clarendon Press for their patience, faith, and forebearance and the constant acceptance of the pressure of other academic duties as an excuse for delays. Dr C. R. Metcalfe, former Keeper, Jodrell Laboratory and now an Honorary Research Associate, Royal Botanic Gardens, Kew, who initiated this project, maintained his enthusiasm and gave continual encouragement, contributed much in his editorial capacity, particularly improving style and presentation in later stages. If he will accept credit for any excellence in the result, I will accept responsibility for all deficiencies!

CONTENTS

CONTENTS

II. DETAILED DESCRIPTIONS OF FAMILIES 55

LIST OF PLATES *(follows Index)*

NOTES ON TECHNICAL METHODS

Anatomical techniques.

IN a survey of the present scope simple techniques are favoured since they lead to rapid examination of large quantities of material. Aquatic monocotyledons, with little lignified tissue, lend themselves to this approach. Much of the work has been done with temporary sections cut free-hand with a double-edge razor blade. Fresh material which retains its succulent firm texture is easiest to work with. For very small thread-like organs sectioning can be done under the dissecting microscope, using the blade as a guillotine, the sections then being transferred to a slide by a Pasteur pipette. Simple histological stains provide basic information about the distribution of lignin, cutin, and suberin. For larger organs a Reichert 'OME' sliding microtome has been used. Catling (1968) has suggested that a preliminary partial air-drying of fluid-preserved material improves sectioning, the sections recovering their normal dimensions on wetting. With patience and skill excellent permanent preparations can be made from these delicate free-hand sections, as the slides prepared by Mr. F. R. Richardson and Miss D. M. Catling in the Jodrell slide collection demonstrate.

For more complete morphological and developmental analysis, serial sections of paraffin-embedded material have been employed, but the method is only suited to meristematic material because shrinkage or dehydration distorts the appearance of mature tissues considerably. No special effort has been made to study types of tracheary element, but some incidental observations made on material macerated in hot 10 per cent KOH followed by 20 percent chromic acid have been useful.

Developmental studies of floral parts have used the dipping cone technique of Sattler (1968, 1973) since this provides precise morphological information for comparative purposes.

Drawings

The black-and-white anatomical illustrations were mostly made from free-hand sections using a Wild M-20 microscope with a drawing tube attachment. Certain diagrammatic conventions are used fairly consistently. Thinner cell walls are drawn as a single line. Internal airspaces and protoxylem lacunae are delimited with a dotted outline and usually with a central x. To emphasize the difference between cell lumina and intercellular spaces the outlines of cells are often conventionally dotted, or the position of chloroplasts somewhat symbolically represented. Enlargement of one part of a low-power diagram is frequently indicated by an arrow, to facilitate location and comparison of illustrations.

The following symbols are commonly used:
Sclerenchyma—solid black
Sieve-tubes—stippled
Tracheary elements—solid black wall with a central x
Tannin cells—cross hatched.

Morphological illustrations

These have been prepared from either fresh or fluid-preserved material by Priscilla Fawcett. Floral diagrams are conventionalized, without any attempt to show the precise overlapping of tepals.

Sexuality of flowers

In the text the names male and female have been applied to staminate and pistillate (or carpellate) flowers respectively. This is a familiar descriptive convention although it is appreciated that the flowers, being sporophytic, themselves cannot have sexuality.

Conventional abbreviations used in the descriptive text

±   more or less
vb(vbs)   vascular bundle(s)
v.   very
sp(spp)   species (plural)
TS   transverse section
LS   longitudinal section

PREVIOUS STUDIES AND LITERATURE—AN OVERVIEW

THE introductory sections are not intended as an extended summary of the anatomy and morphology of the aquatic plants dealt with in this volume, since such summaries already exist in several excellent publications. On the other hand, a brief discussion of certain features which require some comparative assessment or more detailed explanation seems necessary. The prime objective throughout, however, is to lead the reader to the relevant literature, especially those papers which themselves constitute reviews of a particular topic.

GENERAL LITERATURE

There are a number of extended treatments of the taxonomy, morphology, anatomy, and biology of aquatic plants which have formed a background to this book and provide landmarks in the voluminous literature. The early surveys of Schenck (1886a, b) are classical and established the major constructional feature of aquatic plants. Arber (1920) in her *Water plants—a study of aquatic angiosperms* surveys the older morphological and anatomical literature in so complete a fashion that her work is indispensable even to the modern student. The book is so fully indexed that it is a straightforward matter to extract information dealing with individual taxa, up to 1920. A recent reprint (1963) includes some additions to the bibliography. However, the more recent literature, especially that relating to the ecology and biology of herbaceous aquatic plants has been summarized by Sculthorpe (1967) in *The biology of aquatic vascular plants* which is now widely recognized as the most authoritative treatment and continues the Arber tradition.

Furthermore we now have a taxonomic overview of fresh-water aquatic plants in the volume *Water plants of the world—a manual for the identification of the genera of freshwater macrophytes* by C. D. K. Cook and collaborators (Cook *et al.* 1974).

These three books must form the nucleus of a library for anyone studying aquatic vascular plants.

LITERATURE ON THE HELOBIAE

The literature dealing specifically with the Helobiae is extensive and scattered, but a few high points may be mentioned, although this may do a disservice to the numerous excellent older studies which are not referred to in this brief survey. An early example of work primarily systematic in orientation, but foundational in its influence on subsequent workers is the treatment by Richard (1815) of the Alismataceae (in its broader sense). His influence is seen in later works such as those of Buchenau (1857, 1882) leading up to treatments in Engler's *Das Pflanzenreich* (e.g. Buchenau 1903a–c).

A notable feature of much of this early work is that although primarily systematic in orientation it took into consideration vegetative anatomy and

3

developmental morphology, disciplines which tend nowadays to be artificially dissociated from systematics, much to its detriment.

French authors were particularly active in the nineteenth century. The work of Chatin (1855, 1856) is often cited but was severely criticized by his contemporaries (e.g. Caspary 1857; Rohrbach 1871) for its numerous inaccuracies, and later by Costantin (1884–6). Caspary (1858) himself set a high standard, notably in his treatment of the Hydrilleae (Hydrocharitaceae). Also significant are the beautifully illustrated monographs, by Bornet (1864) on *Cymodocea nodosa* (as *Phucagrostis major*), Cymodoceaceae—the illustrations in which continue to be reproduced over 100 years later (e.g. den Hartog 1970*a*)—and by Prillieux (1864) on *Althenia* (Zannichelliaceae). Unmatched standards of excellence in description and illustration occur in the extended series of papers by Camille Sauvageau (1887–1894). These are a model of descriptive science which I use myself as exercises in lucid French for science students. Sauvageau was primarily a phycologist, and his interest in the Helobiae developed in part from studies of marine angiosperms.

German authors figure prominently in this period. We can cite the systematic surveys of Ascherson and Gürke (1889) and Ascherson and Graebner (1907) which incorporated many morphological and anatomical data. Earlier studies include those by Irmisch (e.g. 1851, 1858*a, b*) who provided the first comprehensive descriptions of reproductive morphology in Potamogetonales and of squamules, and of Magnus (e.g. 1869, 1870) whose work on *Najas* is outstanding. Solereder and Meyer (1913, 1933 onwards) made extensive surveys of the systematic anatomy of Hydrocharitaceae and Alismataceae, and also provided a summary of existing anatomical information for a number of helobian families in *Systematische Anatomie der Monokotyledonen* (Solereder and Meyer 1933). This work on the Alismataceae continued into the 1940s through his associate Mayr (1943), presumably the same author who had surveyed the distribution and structure of hydropoten (Mayr 1915). Kirchner, Loew, and Schröter (1908) provided an extended summary of the Helobiae in the treatments by several authors for their *Lebensgeschichte der Blütenpflanzen Mitteleuropas*, but the work is largely derivative and the frequently redrawn illustrations do no credit to the original sources.

In Denmark, Raunkiaer made extensive comparative studies of the Helobiae and provided considerable insight into the way in which anatomical information could be applied to the systematics of the difficult genus *Potamogeton* (Raunkiaer 1895–9, 1903). This tradition was maintained in the monograph by the Swedish botanist Hagström (1916) and continued subsequently (e.g. Ogden 1943).

In Japan the work of Miki (1932–37) is prominent. He made many new taxa known to modern science, complete with anatomical descriptions, and contributed to the discussion of interpretation of floral morphology (e.g. Miki 1937).

In England, Arber's own studies are extensive (1914–40), and in part lead up to her subsequent *Water plants*, but her observation is too often clouded by her preoccupation with the phyllode theory (e.g. Arber 1918*b*).

More recently Charlton (1968 *et seq.*) has made an extensive morphological investigation of the Alismataceae using the information as the basis for experimental study. Surveys of floral vasculature in the Helobiae have been provided by Singh (1963–69). A notable development has been the series of studies on floral organogenesis by Sattler and his associates which have important phylogenetic implications (see publications by Sattler, Lieu, Posluszny, Singh in the bibliography and on p. 47). The work of Kaul (1966 to date) is complementary and balances the approach.

The taxonomic treatment of the seagrasses by den Hartog (1970a) includes much morphological information and underlies the extended work on seagrass ecology which developed in the 1960s and 1970s (e.g. Phillips and McRoy 1980). Şerbănescu-Jitariu (1964 et seq.) in Roumania has published an extended series of studies which investigate the extent to which syncarpy occurs in the Helobiae.

In an overview of recent literature one notes the extent to which new ground is being broken in several fields. Ultrastructural studies are important in understanding unusual metabolic processes in submerged plants, especially those which grow in the sea. Pollen morphology has been studied in relation to hydrophily, where pollen–stigma recognition systems may be unusual. Experimental morphology (e.g. Charlton 1979a, b) has reflected the ease with which aquatic plants can be cultured clonally and manipulated axenically. *Elodea* and related genera are familiar classroom subjects in introductory plant physiology; they have also been a favourite of research workers.

Two considerations stand out: first the extent to which studies are restricted to temperate taxa (notably those of Europe) whereas tropical groups remain relatively little studied (e.g. the tropical Hydrocharitaceae). Second, there is a dearth of developmental work. For example, the *only* developmental study of the widely distributed seagrass *Halophila* is by Balfour (1879). Developmental studies on other seagrasses are few and incomplete. There is very little developmental work on the vegetative organs of the cosmopolitan genus *Potamogeton*. The accounts by Chrysler (1907) and Monoyer (1927) of the vascular system of the Helobiae contain little developmental information, although mature construction cannot be understood without this. In contrast Sattler's work has revealed the significance of developmental study in interpreting floral morphology. The work of Charlton (e.g. 1970) and Wilder (1975) on vegetative morphology is complementary to this work on reproductive morphology. Clearly the scope for continued comparative morphological and anatomical study of a classical kind in this group of monocotyledons remains very broad, especially as modern technological methods begin to be applied to them.

SYSTEMATIC INTRODUCTION

This volume deals with a series of monocotyledonous families [Butomaceae (family 343) to Najadaceae (family 357) in Hutchinson's system (1959)] which are traditionally referred to as an order Helobiae (Eckardt 1964—the Alismatidae of Cronquist, 1968; the Alismidae of Takhtajan, 1966). Their members are typically inhabitants of wet places; in many instances the life-cycle is carried out entirely under water. In the most extremely specialized examples the plants are marine (seagrasses). All marine angiosperms fall within this group.

This volume also includes descriptions of two saprophytic families which clearly have no relationship with the aquatic members. For convenience, they are included here as an order, Triuridales, largely because they are apocarpous. The extended introductory discussion refers to them little, if at all.

The volume departs appreciably from the method of treatment in earlier books in this series (Metcalfe 1960; Tomlinson 1961, 1969; Cutler 1969; Metcalfe 1971; Ayensu 1972) in that, in addition to the standard summary of features of vegetative anatomy, the reproductive anatomy is outlined, together with a more complete description of both vegetative and reproductive morphology. This expanded treatment is justified by the rather uniform anatomy of these water plants. Members of the group are for the most part so specialized in relation to the distinctive requirement of their habitat that considerable parallels in anatomical construction can be found between plants which, on the basis of the structure of their reproductive organs, are clearly not closely related. This was a principle established early in the study of aquatic plants (e.g. Schenck 1886b). A description of the microscopic morphology of the vegetative parts alone therefore provides limited diagnostic anatomical information. It is only when diagnostic morphological evidence from all parts of the plant is drawn upon that a reasonable assessment of systematic interrelationships between families becomes possible. For this reason information relates to the whole organism, although the major emphasis is still on the vegetative parts.

The comprehensive approach allows a number of problems dealing with the systematics and phylogeny of the whole group to be addressed with a major part of the evidence at hand. Initially a systematic overview is necessary, but the presentation under the first two headings depends on assumptions established under the last two headings.

How Should the Helobiae and Triurideae be Subdivided?

Most of the major groups considered in this book have been long recognized as fairly discrete entities. Differences between major schemes of classification largely reflect the hierarchical level to which a particular assemblage has been raised and the extent to which it should be associated with other groups. With increase in knowledge, there has been, inevitably, progressive taxonomic inflation as groups of higher rank have been created, but the result certainly has produced more sharply circumscribed families and eliminated unnatural assemblages.

For a comparison which illustrates these features, five major schemes have been summarized in the following pages, i.e. those of Bentham and Hooker (1883), Engler (1904), Hutchinson (1959), Eckardt (1964), and Takhtajan (1966). Table 1 summarizes the groupings these schemes incorporate. However, if we ignore rank and deal with taxonomic units which are equivalent, as is done in the last column, the disparity between the systems is seen to be slight.

TABLE 1

Comparison of hierarchical rankings in five systems

System	Order	Family	Equivalent unit*
Bentham and Hooker (1883)	2	4	12
Engler (1904)	2	8	13
Eckardt (1964)	2	10	14
Hutchinson (1959)	7	15†	14
Takhtajan (1966)	4	16	16
This book	6	16†	15

* Excluding Petrosaviaceae.
† Including Petrosaviaceae.

The arrangement in this volume is essentially that of Takhtajan (1966) except that here Ruppiaceae is not recognized as a separate family but is included in Potamogetonaceae, while Petrosaviaceae is treated as a family, together with Triuridaceae, in a group Triuridae, whereas Takhtajan included *Petrosavia* in Liliaceae. This recognition of 16 families seems to be the subdivision which most accurately represents the sum of similarities and differences between the groups. The most directly comparable unit is the family, even though it probably had a much broader connotation in the earlier schemes than now exists. On the other hand, there is a quite variable interpretation of the concept of order, which may represent a very constrained assemblage (as in Hutchinson), but has been used in a very comprehensive way, e.g. in the German systems. The usage of names can be misleading, so that Potamogetonales, Potamogetonineae, Potamogetonaceae, Potameae can represent very different assemblages.

Having established a reasonable set of clearly circumscribed families, the task of assembling a totally acceptable arrangement of them into higher categories is now attempted, but seems impossible.

BENTHAM AND HOOKER (1883). *Genera Plantarum.* Vol. 3, pt. II.

Series I. Microspermae
   Family Hydrocharideae
Series VI. Apocarpae
  Family Triurideae
  (*Petrosavia* included in Liliaceae, tribe Tofieldiae)
  Family Alismaceae
    Tribe I. Alismeae
    Tribe II. Butomeae
        (*Butomus, Butomopsis, Hydrocleis, Limnocharis*)
  Family Naiadaceae
    Tribe I.   Juncagineae
        (*Triglochin, Scheuchzeria, Tetroncium, Lilaea*)
    Tribe II.   Aponogetoneae (*Aponogeton*)
    Tribe III. Potameae (*Potamogeton, Ruppia*)
    Tribe IV. Posidoniae (*Posidonia, Amphibolis*)

Tribe V.   Zannichellieae (*Zannichellia, Althenia, Lepilaena*)
Tribe VI.   Zostereae (*Zostera, Phyllospadix*)
Tribe VII.   Naiadeae (*Naias*)
Tribe VIII.   Cymodoceae (*Cymodocea*)
[In summary, two orders, four families, with ten tribes, excluding Hydrocharideae, Triurideae and *Petrosavia*]

ENGLER (1904). *Syllabus der Pflanzenfamilien* ed 4.

ORDER 2. HELOBIAE
  Suborder Potamogetonineae
    Family Potamogetonaceae
      Tribe I.   Zostereae (*Zostera, Phyllospadix*)
      Tribe II.   Posidonieae (*Posidonia*)
      Tribe III.   Potamogetoneae (*Potamogeton, Ruppia*)
      Tribe IV.   Cymodoceeae (*Cymodocea, Halodule*)
      Tribe V.   Zannichellieae (*Zannichellia, Althenia*)
    Family Najadaceae
    Family Aponogetonaceae
    Family Juncaginaceae
      Tribe Triglochineae (*Triglochin, Scheuchzeria, Tetroncium*)
      Tribe Lilaeeae (*Lilaea*)
  Suborder Alismatineae
    Family Alismataceae
  Suborder Butomineae
    Family Butomaceae
    Family Hydrocharitaceae
ORDER 3. TRIURIDALES
    Family Triuridaceae
[*Petrosavia* included in subfamily Melanthoideae, tribe Tofieldieae of Liliaceae]
[In summary, two orders, three suborders, eight families excluding *Petrosavia*]

HUTCHINSON (1959) *Families of Flowering Plants II. Monocotyledons*

Order 83. *Butomales*
    Family 343. Butomaceae
    Family 344. Hydrocharitaceae
Order 84. *Alismatales*
    Family 345. Alismataceae
    Family 346. Scheuchzeriaceae
    Family 347. Petrosaviaceae
Order 85. *Triuridales*
    Family 348. Triuridaceae
Order 86. *Juncaginales*
    Family 349. Juncaginaceae
    Family 350. Lilaeaceae
    Family 351. Posidoniaceae
Order 87. *Aponogetonales*
    Family 352. Aponogetonaceae
    Family 353. Zosteraceae

Order 88. *Potamogetonales*
    Family 354. Potamogetonaceae
    Family 355. Ruppiaceae
Order 89. *Najadales*
    Family 356. Zannichelliaceae
    Family 357. Najadaceae
[In summary seven orders; 15 families including *Petrosavia*]

ECKARDT, T. (1964). In *Syllabus der Pflanzenfamilien*

ORDER HELOBIAE (ALISMATALES)
  1. Suborder **Alismatineae**
      Family Alismataceae
      Family Butomaceae
        Subfamily Butomoideae
        Subfamily Limnocharitoideae
  2. Suborder **Hydrocharitineae**
      Family Hydrocharitaceae
        Subfamily Hydrocharitoideae
          Tribe Ottelieae
          Tribe Stratioteae
          Tribe Hydrochariteae
          Tribe Enhaleae
        Subfamily Vallisnerioideae
          Tribe Blyxeae
          Tribe Vallisnerieae
          Tribe Hydrilleae
        Subfamily Thalassioideae
        Subfamily Halophiloideae
  3. Suborder **Scheuchzeriineae**
      Family Scheuchzeriaceae
  4. Suborder **Potamogetonineae**
      Family Aponogetonaceae
      Family Juncaginaceae
      Family Potamogetonaceae
        Tribe Potamogetoneae
        Tribe Posidonieae
        Tribe Zostereae
      Family Zannichelliaceae
        Tribe Zannichellieae
        Tribe Cymodoceeae
      Family Najadaceae
ORDER TRIURIDALES
      Family Triuridaceae
        Tribe Sciaphileae
        Tribe Triurideae
[*Petrosavia* and *Protolirion* are included in the tribe Petrosavieae (subfamily Melanthioideae) of the family Liliaceae (order Liliiflorae)]
[In summary, two orders, four suborders, ten families excluding *Petrosavia*]

TAKHTAJAN, A. (1966). *Systema et Phylogenia Magnoliophytorum* (Liliatae-
Monocotyledones)

Subclass Alismidae
 Superorder Alismanae
  Order 76. **Alismales**
   Family Butomaceae
   Family Limnocharitaceae
   Family Alismaceae
  Order 77. **Hydrocharitales**
   Family Hydrocharitaceae
  Order 78. **Potamogetonales**
   Family Scheuchzeriaceae
   Family Juncaginaceae
   Family Lilaeaceae
   Family Aponogetonaceae
   Family Zosteraceae
   Family Posidoniaceae
   Family Potamogetonaceae
   Family Ruppiaceae
   Family Zannichelliaceae
   Family Cymodoceaceae
   Family Najadaceae
  Order 79. **Triuridales**
   Family Triuridaceae
[*Petrosavia* is included in the subclass Liliidae, superorder Lilianae, order 80. Liliales,
family Liliaceae]
[In summary, four orders, 16 families excluding *Petrosavia*]

TOMLINSON, P. B. (1981). Arrangement in this volume (see synopsis on
p. 12). Genera of monotypic families included.

A. **Helobiae (Alismatidae)**
 Order Alismatales
  Family 1. Alismataceae
  Family 2. Limnocharitaceae
  Family 3. Butomaceae (*Butomus*)
 Order Hydrocharitales
  Family 4. Hydrocharitaceae
 Order Aponogetonales
  Family 5. Aponogetonaceae (*Aponogeton*)
 Order Scheuchzeriales
  Family 6. Scheuchzeriaceae (*Scheuchzeria*)
  Family 7. Juncaginaceae
  Family 8. Lilaeaceae (*Lilaea*)
 Order Potamogetonales
  Family 9. Potamogetonaceae (Potamogetonoideae & Ruppioideae)
  Family 10. Zannichelliaceae
  Family 11. Posidoniaceae (*Posidonia*)
  Family 12. Cymodoceaceae
  Family 13. Zosteraceae
  Family 14. Najadaceae (*Najas*)

B. **Triuridae (Triuridales)**
    Family 15. Triuridaceae
    Family 16. Petrosaviaceae (*Petrosavia?*)

IN WHAT SEQUENCE ARE THE HELOBIAN FAMILIES TO BE ARRANGED?

Accepting the Helobiae as a natural order (p. 15) implies that they are monophyletic. Their subdivision into a precise number of families is relatively easily achieved, as we have seen. However, it is not possible to arrange them in a linear sequence which reflects phyletic trends in view of the marked adaptive radiation they exhibit. At the most, if it is conceived that they have diverged from a common ancestor, as is shown below to be reasonable, some members may have diverged least in their morphology from the hypothetical ancestor, others have diverged more, but the divergence has not taken place in a monopodial manner. All sequences incorporated into various taxonomic schemes, even those that claim to be evolutionary, are essentially typological since they distribute families in a linear sequence beginning with those that have flowers most like a hypothetical type, and ending with those that have highly specialized flowers least like a hypothetical type. The type may have pleiomerous flowers, as in the generally accepted interpretations of which Takhtajan's is representative, or simple flowers as in the Englerian scheme. Floral morphology in the most organizationally simple members of the group is such that the distribution of parts in relation to a type flower is disputable, as is discussed later (see also Markgraf 1936; Uhl 1947). This trend of simplification by and large follows a trend towards hydrophily which reaches its highest elaboration in the marine angiosperms. However, the fact that the seagrasses themselves have a polyphyletic origin seems clear from their taxonomic distribution within a diversity of families (cf. for example Hydrocharitaceae with Zosteraceae). That there is no single pathway to submarine pollination in the monocotyledons itself suggests a microcosmic model of the overall adaptive radiation of these plants.

The following sequence of families agrees closely with that provided by several systematists in that it illustrates a trend from relatively simple amphibious or terrestrial plants, with flowers suited to aerial and usually insect pollination (e.g. Alismataceae, Butomaceae) to highly specialized submerged aquatics in which the whole life cycle is completed beneath either fresh or salt water (e.g. Cymodoceaceae, Najadaceae, Posidoniaceae, Zannichelliaceae, Zosteraceae). The arrangement of families in a particular linear sequence is in no way intended to show evolutionary progress within the Helobiae and the sequence is essentially arbitrary.

In particular it is impossible to find a satisfactory place for such highly specialized families as Hydrocharitaceae and Aponogetonaceae in a linear sequence, since any juxtaposition which associates them with other particular families is certainly misleading.

The following synopsis thus largely illustrates the major diagnostic differences between the families. The Triuridales are included for completeness. We have commented already that this group is included in this volume largely as a matter of traditional convenience and it is doubtful if they are at all closely related to the Helobiae.

SEQUENCE, ARRANGEMENT, AND DIAGNOSES OF ORDERS AND FAMILIES IN THIS BOOK

A. *Helobiae (Alismatidae)* and B. *Triuridae (Triuridales)*

Herbaceous monocotyledons, usually with 1—several—numerous **free carpels;** either **mycotrophic** or more or less **aquatic;** flowers often with radial symmetry; ovary inferior only in the Hydrocharitaceae.

A. *Helobiae (Alismatidae)*

Plants chlorophyllous, **autotrophic** (never with mycorrhizal roots); tropical or temperate; usually aquatic and so often either facultatively or obligately submerged. **Seeds without endosperm. Squamules** almost always present.

ALISMATALES

Perianth generally differentiated into evident sepals and petals; carpels several to v. numerous; flowers bracteate, **hypogynous,** inflorescences often in complex cymes but without a pair of spathe-like enveloping bracts.

1. *Alismataceae*

Plants with **secretory canals** (laticifers); adult leaves usually petiolate and with a broad blade; carpels often v. numerous and in 2 or more whorls; **ovules single** (rarely more) at the base of each carpel; **fruit usually indehiscent** (achenous). **Embryo curved.** Emergent or facultatively (rarely obligately) submerged freshwater aquatics.

2. *Limnocharitaceae*

Plants with **secretory canals** (laticifers); adult leaves petiolate and with a broad blade; carpels not numerous, usually in 1 whorl; **ovules numerous,** scattered over the inner surface of each carpel; **fruit usually dehiscent** (follicular). **Embryo curved.** Emergent or floating freshwater aquatics.

3. *Butomaceae*

Plants without secretory canals; **leaves linear, triquetrous,** and without a differentiated blade; perianth wholly petaloid, carpels (4-)6; **ovules numerous,** scattered over the inner surface of the carpel; **fruit dehiscent. Embryo straight.** Emergent freshwater aquatics.

HYDROCHARITALES

4. *Hydrocharitaceae*

Leaves **eligulate.** Flowers **epigynous,** floral envelope various; inflorescence cymose with an enveloping pair of bracts often fused to form a **spathe.** Carpels

2–numerous, with **laminar** or ± **parietal placentation.** Emergent or floating or facultatively or obligately submerged aquatics of both fresh and salt water (i.e. some seagrasses).

APONOGETONALES

5. *Aponogetonaceae*

Laticifer-like elements present. Plants with simple or branched **spicate inflorescences** with an enveloping spathe, flowers usually bisexual, often with petaloid perianth segments, basically trimerous but often with several carpels; **ovules** usually **2 or more** at the base of each carpel; fruit follicular. Emergent, facultatively or obligately submerged freshwater aquatics.

SCHEUCHZERIALES

Leaves usually **linear, ligulate,** distichous; inflorescences spicate, flowers bracteate or ebracteate, without distinction between petals and sepals. **Embryo straight. Triglochinin** often present.

6. *Scheuchzeriaceae*

Leaves linear, **ligulate,** with a large **apical pore;** inflorescences spicate, flowers **bracteate,** with a biseriate perianth. Carpels 3 or more, **ovules** usually **2,** basal in each carpel. Fruit follicular. **Squamules absent,** replaced by axillary hairs. Emergent rhizomatous bog-plants of high latitudes in northern hemisphere.

7. *Juncaginaceae*

Leaves linear, flattened or equitant. Spikes **ebracteate,** perianth biseriate. Carpels few, sometimes sterile; **ovules single,** basal in each carpel. Achenes separating. Terrestrial or emergent freshwater or salt-marsh plants, rarely facultatively submerged.

8. *Lilaeaceae*

Spikes **bracteate,** but without a biseriate perianth; carpel **single, uniovulate.** Fruit an achene. **Laticifer**-like secretory cells present. Emergent, ephemeral freshwater aquatics of seasonally flooded sites.

POTAMOGETONALES

Submerged aquatics (sometimes with floating leaves and emergent inflorescences) with 1–4 (rarely more) **uniovulate** carpels per flower; flowers usually without tepal-like floral envelopes; leaves **ligulate** (except Najadaceae). Fruit usually achenous.

9. *Potamogetonaceae*

Leaves commonly with a broad blade. Axis **sympodially branched** below the **spicate inflorescence.** Carpels usually 4 (sometimes more) per flower;

tepals **adnate to stamens** or absent (*Ruppia*). Cotyledon usually **curved.** Usually submerged aquatics (commonly with floating leaves and emergent inflorescences) growing in fresh or saline water, but never truly marine.

10. *Zannichelliaceae*

Stem and **leaves** linear, **thread-like;** flowers unisexual; **tepals absent** or represented by scales or cupules; male flower a **single stamen** (often plurisporangiate) associated with one or more carpels. Pollen globose. Stomata virtually absent. Wholly submerged plants in fresh or brackish water; never marine.

11. *Posidoniaceae*

Inflorescence racemose, flowers naked, each with 3 stamens and a **single uniovulate carpel,** the stamen **connective expanded** adaxially. **Pollen filamentous. Fruit** with a spongy pericarp but **dehiscent.** Stomata absent. Submerged **marine** plants of **temperate** or subtropical waters (not tropical).

12. *Cymodoceaceae*

Inflorescence cymose, flowers naked but enclosed by leaf-like bracts; male flowers with **2 tetrasporangiate stamens** fused back-to-back; female flowers with 2 **free, unilocular, uniovulate ovaries** each with 1–3 stigmas. **Pollen filamentous.** Fruit indehiscent (sometimes viviparous). Stomata absent. Submerged **marine** plants mainly of **tropical** waters.

13. *Zosteraceae*

Ultimate inflorescence unit a **flattened spike-like axis** with naked flowers on one side, associated with scales, the whole enveloped by a single leaf-like bract. Male flower of **2 free bilocular anthers;** female flower an **ellipsoidal carpel** with a short style and 2 stigmas. **Pollen filamentous.** Fruit indehiscent. Stomata absent. Submerged **marine** plants mainly of **temperate** waters.

14. *Najadaceae*

Axes filamentous. Leaves short, **pseudowhorled, linear, undifferentiated,** eligulate with a **single vein.** Flowers unisexual naked or with 1–2 cupulate envelopes; male flower a 1- or 4-sporangiate anther, female flower a sessile flask-shaped carpel with a solitary **axis-borne ovule** and 2–4 styles. Pollen globose. Fruit indehiscent. Stomata absent. Submerged plants of fresh or brackish water, cosmopolitan.

B. *Triuridae* (*Triuridales*)

Plants without chlorophyll, saprophytic and with **mycorrhizal roots;** tropical **terrestrial, seeds without endosperm.** Squamules absent.

TRIURIDALES

15. *Triuridaceae*

Carpels usually **numerous,** each with a **single basal ovule;** flowers usually **unisexual,** tepals 3 or more, not often in 2 distinct series. Fruit dehiscent. **Stomata absent.**

16. *Petrosaviaceae*

Carpels **3**, each with **several marginal ovules; flowers perfect,** tepals 6 in 2 distinct series of 3 each; fruit a follicle. **Stomata present.**

DO THE HELOBIAN FAMILIES CONSTITUTE A NATURAL (I.E. PRESUMED MONOPHYLETIC) GROUP?

One reason why members of this order have been placed together is because they have a strong preference for aquatic environments and are frequently wholly submerged. Therefore, it could be suggested that the Helobiae are an unnatural assemblage based on a convergence of characters which results from a similar response to a constant set of environmental features. Evidence in support of this point of view would come from the existence of numerous other aquatic monocotyledons whose terrestrial relatives are clearly evident from similar features of reproductive morphology and whose juxtaposition in any systematic scheme would clearly be artificial. Examples are provided in Table 2.

From this standpoint the Helobiae could then be regarded as a group of reduced aquatics, whose immediate and diverse terrestrial ancestors have been lost, so that their polyphyletic status has been obscured. Evidence in favour of this would be provided if the plants shared a diversity in fundamental

TABLE 2

Aquatic or marsh groups of monocotyledons (other than Helobiae) which have some well established affinity with a terrestrial family

Aquatic taxon	Terrestrial relative
Crinum natans	*Crinum* spp
Cryptocoryne, Jasarum	Araceae
Eriocaulon spp	Eriocaulaceae
Pistia	Araceae
Hanguanaceae	(Traditionally *Flagellaria,* but isolated)
Lemnaceae	Araceae
Mayacaceae	Commelinaceae
Pontederiaceae (floating or emergent, *Hydrothrix* is submerged)	Liliaceae?
Rapateaceae (sometimes marsh plants)	Farinosae
Sparganiaceae	Pandanaceae?
Typhaceae	Pandanaceae?
Xyridaceae (often emergent aquatics)	Farinosae

Some or even numerous taxa in the following families grow as emergent aquatics: Araceae, Cyperaceae, Gramineae, Haemodoraceae, Hypoxidaceae, Iridaceae, Juncaceae, Marantaceae, Philydraceae, Pontederiaceae.

Certain taxa have amphibious or submerged forms, e.g. *Sparganium* (Sparganiaceae), *Scirpus* (Cyperaceae), *Glyceria* (Gramineae).

features which could not be interpreted as hydrophytic adaptations. Evidence to the contrary would be the demonstration of a suite of characters which the members have in common and which are unlikely to be direct adaptations to an aquatic environment. At first sight this common set of characters is scarcely recognizable, and indeed descriptions of the order give an impression of a bewildering array of characters (Cronquist 1968; Eckardt 1964). More careful assessment leads to a consensus that the members are indeed a natural assemblage, but with a remarkable array of morphologically specialized features. The order in fact serves as an example of extreme adaptive radiation.

Table 3 lists some of the more constant features found in members of the Helobiae. None of these characters is unique to the order and a more detailed discussion of some of them is given later.

To some extent negative characters can be used to define the order. Calcium oxalate in the form of raphide crystals in raphide cells does not occur although raphides are common in many families of monocotyledons. Likewise silica does not occur. The stomatal complex in its development apparently never has the type of division described as oblique (Tomlinson 1974b), although studies on the subject are few. Vessels never occur in any organ other than the root, in the majority of families they are completely absent.

Carpels are never fused in such a way that unilocular ovaries with axile or parietal placentas are produced—the nearest approach to this is the inferior ovary of the Hydrocharitaceae which has a unique construction, and can be equated with an aggregation of numerous free carpels with distinctive laminar placentation (e.g. Troll 1931b). The divergence from an unspecialized

TABLE 3

Commonly expressed features within the order Helobiae

Feature	Exceptions
1. Herbaceous habit	None
2. Leaves simple	None
3. Shoot apical bifurcation	Many
4. Hydropoten*	Many
5. Squamules	*Scheuchzeria* (1 sp)
6. Floral trimery	Many
7. Apocarpy	Hydrocharitaceae; monocarpellate families?
8. Pollen trinucleate	Few
9. Stomata (where present) paracytic	None
10. Well-developed aerenchyma*	Few
11. Seed endosperm absent (Germination epigeal)	V. few (?)
12. Sieve-plates simple	None?
13. Vessels absent	Some
14. Carpel development	Few
15. Ovule bitegmic	V. few (*Aponogeton* spp.)
16. Hypocotyl enlarged	Many
17. Helobial endosperm	Some
18. Protoxylem lacunae*	Few

* Features characteristic of aquatic plants generally.

apocarpous gynoecium is never considerable, although this assumes that solitary carpels in many families are derived by reduction from a pluricarpous condition.

On the other hand, members of the group vary enormously in major features. Morphological architecture is variable, although often precise and highly organized in a given species. Floral organization shows a wide spectrum, and methods of pollination vary from animal (mostly insects) and wind pollination of aerial flowers, to pollination at the water surface, to pollination below the water surface, with a degree of specialization distinctive within the angiosperms; the total range is scarcely exceeded by any other comparable angiosperm group. Flowers range from examples with a perianth, nectaries, and numerous free parts, to complex assemblages of reduced structures in which the male flower is represented by a single stamen and the female flower by a single carpel. The flower of *Najas* is one of the simplest known in the angiosperms. The ovary may consist of very many carpels, or commonly only a single carpel; placentation varies from the peculiar laminar placentation of Butomaceae, Limnocharitaceae, and Hydrocharitaceae, to unilocular ovaries with a single basal ovule. The fruit may be an aggregation of follicles, or become capsule-like. Achenous fruits are common. Even submerged fruits are sometimes dehiscent—underwater capsules! The embryo is diverse in its morphology and frequently has an enlarged hypocotyl. However, seedling development is most usually epigeal as far as the limited information indicates (e.g. Boyd 1932; Kaul 1978). It is because of this very diversity that the Helobiae have attracted the attention of evolutionists since many of the trends they show can be read as phyletic series. Recognition of the Helobiae as a monophyletic group is, of course, central to this attention.

Most authors claim to detect lines of reduction within the order. This has been done for vegetative architecture (Wilder 1975); vascular systems (e.g. Chrysler 1907; Monoyer 1927); floral morphology (e.g. Uhl 1947; Posluszny and Sattler 1976b); embryo structure (e.g. Yamashita 1976b). Other authors have been content to represent the diversity as a good example of adaptive radiation (as in the studies of floral development by Sattler and Singh 1978). A completely opposite point of view is taken by Burger (1978) who hypothesizes a trend of elaboration from *Lilaea*-like ancestors.

DO THE HELOBIAE REPRESENT A PRIMITIVE AND EVEN ANCESTRAL GROUP OF MONOCOTYLEDONS?

The classic view is that the monocotyledons are derived from aquatic dicotyledons, best represented by the Nymphaeales, and that certain members of the Helobiae (notably the Alismatales) are progenitors of the whole of the monocotyledons. This view is implied or expressed in numerous systems of classification, e.g. Cronquist (1968), Eckardt (1964), Hutchinson (1959), Takhtajan (1966) or in the writing of evolutionists (e.g. Henslow 1911). Evidence for this is believed to be the similar floral construction, based on the apocarpous condition and certain assumed common features of morphology, e.g. a similar vascular system, monocotyledony, and similar ecological preferences.

More recently acquired evidence, much of it summarized in this volume, casts serious doubt on this interpretation. The following comments seem pertinent:

(i) The Helobiae have trinucleate pollen and the seeds lack endosperm, both regarded as specialized features in the angiosperms.

(ii) They have specialized sieve-tubes with simple sieve-plates, hardly a primitive feature.

(iii) Floral morphology is not primitively shoot-like because numerous parts appear to be derived by secondary proliferation of a basic trimerous pattern; there is no evidence for spirally-arranged floral parts in this order (Sattler and Singh 1978).

(iv) Vegetative morphology is frequently very precisely organized, and not the type one would expect in the ancestor of a highly diversified group. It is particularly difficult to derive the more generalized types of architecture found in most terrestrial monocotyledons from these specialized aquatics. How does one produce a palm from a water plantain?

(v) Similarities of vascular organization of Nymphaeales and monocotyledons are spurious and based on superficial comparison. There appear to be no fundamental features of the vascular system that characterizes the monocotyledons (Zimmermann and Tomlinson 1972) in the very specialized rhizomes of Nymphaeaceae (Weidlich 1976a, b).

As evidence of this type continues to accumulate, it seems that the classic view of the archaic status of the Helobiae will progressively appear less substantial. The existence of apocarpy in the order has deluded workers into simplistic evolutionary concepts.

A rather different point of view is presented by Burger (1978) who seeks to discover the common ancestor of monocotyledons in plants with a piperalean type of construction so that *Lilaea* is regarded as a primitive type of monocotyledon, the usual trimerous monocotyledonous flower being derived from the whole condensed inflorescence of *Lilaea*. Here the pseudanthial interpretation of floral organization in the Potamogetonales, favoured by a number of authors (e.g. Miki 1937; Markgraf 1936; Uhl 1947) finds particularly appropriate expression (see p. 48). However, Burger's concept of the inflorescence of *Lilaea* is rather simplistic, since the distribution of appendages in actual flowers does not correspond to that shown in his stylized illustration. Also vegetative morphology in *Lilaea* is quite specialized. On the other hand, Burger's article is realistic in its appreciation of the limitations and assumptions involved in this hypothetical approach. The following quotation from Sculthorpe seems appropriate at this point:

So far as hydrophytes are concerned, much might be gained by admitting, at the outset, that the natural affinities and ancestry of many families are profoundly enigmatic. Botanists are still seriously ignorant, or deeply divided in their interpretation, of many aspects of the morphology, cytogenetics and geological history of aquatic taxa. In this present state of inadequate knowledge, phylogenetic speculation is sometimes provocative and may inspire the pursuit of some hitherto untrodden path of inquiry, but all too often it seems no more than wild intellectual diversion, flourishing precociously on the very lack of factual data and unequivocal evidence. (Sculthorpe 1967, p. 14.)

GENERAL INTRODUCTION TO THE MORPHOLOGY AND ANATOMY OF THE HELOBIAE

Endosperm development

The Helobiae are an unusual group of monocotyledons because mature seeds are typically without endosperm; nevertheless the type of development which is characteristic of the endosperm, which does appear but is resorbed by the time of embryo maturation, has been used as a diagnostic character. It is usually referred to as helobial endosperm, since it was described in principle first by Hofmeister (1861) in *Scheuchzeria* and has subsequently been shown to occur widely in other members of the order, but is neither universally present, nor restricted to them (Swamy and Parameswaran 1962*b*), as Hofmeister himself demonstrated.

Developmentally, helobial endosperm is somewhat intermediate between the more usual but contrasted cellular and nuclear types (but only in a typological sense). It involves separation of the two nuclei formed by the first mitosis of the endosperm nucleus, which occupies a chalazal position, by a transverse wall so that the embryo sac as a whole is divided into a chalazal cell and a micropylar cell. The latter is always larger and subsequently much more prolific, the former having a presumed haustorial function. Free nuclear divisions occur in either the micropylar cell alone, or in both cells, leading to a coenocytic stage. Cell formation, if any, takes place during much later stages.

Helobial endosperm has been reported for the following families of Helobiae (see summaries by Davis 1966; Sculthorpe 1967; but especially Swamy and Parameswaran, 1962*b* where the literature is surveyed in full):

Aponogetonaceae, Butomaceae, Hydrocharitaceae, Limnocharitaceae, Potamogetonaceae, Scheuchzeriaceae, Zannichelliaceae; it has been reported in some genera of Alismataceae (*Echinodorus, Limnophyton, Sagittaria*) whereas other genera may have nuclear endosperm (*Alisma, Damasonium, Luronium, Machaerocarpus*). *Ruppia* is somewhat controversial but may have helobial endosperm. The Najadaceae are reported to have a free nuclear stage but with the possibility of a very late cellular stage; they are included under the nuclear type by Swamy and Parameswaran (1962*b*). On the other hand, nuclear endosperm has been recorded in Juncaginaceae, Lilaeaceae, Zosteraceae, and also for Triuridaceae.

Helobial endosperm is by no means restricted to the Helobiae; it has been reported for 11 other families of monocotyledons. There is some controversy concerning the similarity of endosperm development in certain Nymphaeales and Helobiae and, if such similarity exists, it would be of considerable phyletic interest. The condition has been reported for a number of other dicotyledons, but Swamy and Parameswaran (1962*b*) regard the similarities as spurious and consider helobial endosperm to be a uniquely monocotyledonous feature, deserving the same order of importance as those better known morphological features in which monocotyledons and dicotyledons differ from each other. Helobial endosperm seems to be strongly correlated with an aquatic habitat, but its adaptive significance is quite obscure.

In view of the highly specialized nature of studies of this aspect of the life cycle of a flowering plant no attempt has been made to incorporate detailed information in this volume, except where recent papers, not included in the above summary articles, seem particularly relevant to systematic discussion. Because of the very precise cytological analysis which is needed in this kind of study, information is likely to accumulate slowly but it should be sought since a reliable assessment of the situation can only be based on extensive data.

The embryo

Early stages of mega- and microsporogenesis, together with embryo development, are not considered in this account, again because of the requirement of specialized expertise and the present incomplete state of our knowledge. Information has been summarized by Davis (1966) in which a diversity of types are established. This is not surprising in view of the diverse life forms shown by helobial families. Information about Triuridales is too limited to allow useful comparison.

On the other hand, the embryo itself, in its mature form, provides diagnostic information. Very commonly the embryo is poorly differentiated and it has been concluded that this is a primitive condition (e.g. Eames 1961). Equally, however, in many families the embryo is distinctive and specialized in ways that are probably related to the problem of establishment in subaerial environments (Arber 1920; Sculthorpe 1967). Furthermore, the helobial embryo has been studied extensively in relation to a controversy concerning the topographical relationship between the cotyledon and the embryonic shoot apex. Because the embryo is poorly differentiated the cotyledon has been regarded as terminal, with the plumular shoot apex lateral to it (Haccius 1952; Souèges 1940, 1954; Swamy 1963; Swamy and Lakshmanan 1962*a*; Swamy and Parameswaran 1962*a*). Most recent studies show that the positional relationships of shoot apex and hypocotyl are essentially the same as in the embryos of other angiosperms but that the cotyledon is precocious in its growth and forces the plumular apex to one side (e.g. Yamashita 1976*b*; Ly Thi Ba *et al.* 1973, 1976, 1978). In this respect the same kind of displacement may occur in other monocotyledons in the mature vegetative phase, the massive leaf buttress seeming to push the truly terminal shoot apex to one side in each plastochrone cycle, as in *Acorus* (Kaplan 1970) and *Nypa* (Tomlinson 1971).

The embryos in the Helobiae may be categorized in a very general way as follows:

(i) Straight (e.g. Aponogetonaceae, Butomaceae, Limnocharitaceae, Scheuchzeriaceae, Juncaginaceae, Lilaeaceae).

(ii) Curved (e.g. Alismataceae, Potamogetonaceae, Zannichelliaceae). The cotyledon may be characteristically rolled longitudinally as in *Groenlandia*, *Zannichellia*.

(iii) Curved, with a marked enlargement of the hypocotyl (e.g. Zosteraceae, *Ruppia*).

In the Hydrocharitaceae embryo morphology is diverse, as would be expected (Haccius 1952). Among other families, a sequence of progressive change from the more or less symmetrical embryos of the less specialized families (e.g. Scheuchzeriaceae) to the very highly specialized families with very asymmetrical embryos can be recognized. This peculiar configuration has led to the suggestion that the true radicle is lacking in the more specialized embryos, the first root representing an adventitious root since it does not develop immediately opposite the suspensor. However, Yamashita (1976b), on the basis of a comparative embryological study, which demonstrates the trend shown in Fig. 1, concludes that the first root is homologous in all these embryos, the displacement resulting from the shift in the position of the suspensor and from the increasing curvature in these embryos. On the other hand, in the embryo of *Aponogeton crispus*, which has an enlarged hypocotyl, the radicle actually seems to abort (Yamashita 1976b). The embryo structure in the Aponogetonaceae thus seems genuinely variable.

The most remarkable structural feature which the more specialized embryos show is the enlarged hypocotyl, which reaches its largest proportions in some of the seagrasses (e.g. *Thalassia*, Fig. 4.9 J,K) where it seems to serve as a stabilizing device, maintaining a constant seedling orientation, as well as providing a massive food storage reserve.

Seedling morphology

Some of the rather limited early information on seedling morphology in the Helobiae is summarized by Boyd (1932) (see also Andersson 1888; Muenscher 1936). Epigeal germination is very common and may prove to be a consistent feature; its association with the non-endospermous condition is to be expected.

Kaul (1978) has recently compared seedling development in a few members of Alismataceae, Hydrocharitaceae, and Limnocharitaceae (see also Lieu 1979b,c). His observation that seedlings in the first family either lack a developed radicle, or retain only its vestiges (a common feature in aquatic plants), whereas in the other two families the radicle is well developed, is of considerable biological and systematic significance. In all groups a characteristic circlet (collet) of hairs is present at the base of the hypocotyl; this aids in attachment. A number of these distinctive morphological features occur in unrelated families of aquatic plants (Arber 1920; Sculthorpe 1967), suggesting that they are of direct adaptive significance. The peculiar grappling-apparatus of the *Amphibolis* seedling has attracted attention (Black 1913; Ducker *et al.* 1977). It seems that the subject of embryo and seedling structure needs attacking more extensively on a comparative basis. The study of *Zostera* by Taylor (1957) is exemplary in this respect (cf. Yamashita 1973). The emphasis should be biological, not phylogenetic; the evolutionary derivation of one embryo or seedling type from another is, at best, speculative, whereas there is the possibility of establishing sound ecological and testable hypotheses on the basis of functional morphology leading to intellectually more acceptable conclusions. The systematic significance of seedling morphology could then be assessed reliably.

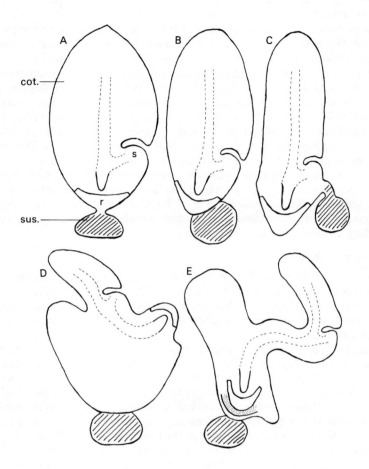

FIG. 1. Embryo structure in various Helobiae to show different degrees of modification of embryonic axis and position of origin of radicle. (After Yamashita 1976*b*.)

cot.—cotyledon; s—plumule (shoot apex); r—radicle (root apex); sus—suspensor (hatched).

A. *Scheuchzeria.*

B. *Aponogeton,* e.g. *A. madagascariensis.*

C. *Lilaea* and *Triglochin.*

D. *Ruppia.*

E. *Zostera.*

It should be emphasized that this configuration only represents the embryo at a particular developmental stage. In B, for example, the root later straightens and the suspensor is detached from the embryo base obscuring the earlier obliquity. For *Zostera* see also Taylor (1957).

GROWTH HABIT AND ARCHITECTURE

Range of growth forms

The diversity of growth forms in the Helobiae covers almost completely the range which Arber (1920, p. 5) has included in her biological classification of hydrophytes. They vary from rooted terrestrial plants (as in some *Triglochin* spp), to plants with only the basal parts submerged, to wholly submerged but rooted plants and finally to free-floating plants. The only growth habit not represented is essentially the rootless floating or submerged condition found in some Lemnaceae and Ceratophyllaceae. Such is the plasticity of many individual species in this group that a given population may be capable of a wide diversity of growth forms, depending on the extent of flooding and the rate of flow of water.

Within this broad range, the habit is largely determined by the type and arrangement of leaves in relation to the organization of the axis, and the extent to which the internodes are elongated. Commonly the leaves are in a terminal rosette, either spirally, or distichously, or spirodistichously arranged, the axis then being either cormous or rhizomatous. Proliferation by stoloniferous offsets is a very common feature. Rhizomatous plants may be monomorphic but are more usually dimorphic, when there is a sharp differentiation between horizontal and erect shoots, as in a number of seagrasses (Tomlinson 1974a).

Plants with extended herbaceous stems include many *Potamogeton* species, with the distal leaves broad and floating, the leaves solitary at a node and widely spaced, whereas in *Najas* or *Elodea* and related genera the plants are wholly submerged and the short leaves in whorls or pseudowhorls. Free-floating rosettes occur in *Stratiotes* and *Hydrocharis*, both spreading by stolons.

None of these growth habits is unique to the Helobiae, and there is a considerable indication of parallel evolution within the order. The full range of growth forms for the whole order is expressed in the single family Hydrocharitaceae. Forms found in the Helobiae also occur in quite unrelated orders. Thus *Najas* and *Hydrilla* may be compared with *Mayaca* (Commelinales); *Hydrocharis* and *Stratiotes* with *Ceratopteris, Eichhornia* and *Pistia*.

Branching of shoots

In several of the more detailed and precise studies of shoot organization in the Helobiae (see summary in Wilder 1975 and, more recently, Lieu 1979a,b,c) it has been shown frequently that there is an equal division of the shoot apical meristem, this often makes it difficult to establish whether growth is monopodial or sympodial. This bifurcation has now been reported for members of the Alismataceae, Butomaceae, Limnocharitaceae, Hydrocharitaceae, Juncaginaceae, Lilaeaceae, Najadaceae, and it may possibly occur in the Aponogetonaceae. The division of the apical meristem may occur on the main axis itself and lead to an equal division of a rhizome or stolon or it may occur as the first division in a lateral meristem, as in *Najas*, some Hydrocharitaceae. Usually the division of a main axis produces axes of two kinds, one remaining vegetative while the other becomes the inflorescence, e.g. in *Butomus, Alisma, Sagittaria*. Bifurcation (dichotomy) of the vegetative axis has been reported in a number of unrelated monocotyledons that are not closely related to each other (e.g. Flagellariaceae—*Flagellaria*, Tomlinson and Posluszny 1977; Palmae—*Chamaedorea*, Fisher 1974; —*Nypa*, Tomlinson 1971; Strelitziaceae—*Strelitzia*, Fisher 1976). Bifurcation is regarded here as a derived feature.

It can be argued that where a similar dichotomous-like forking occurs in the Helobiae the condition is polyphyletic, because leaf arrangement in such plants is usually distichous. This allows one to interpret the forking simply as precocious development of an axillary shoot (e.g. Wilder 1975). However, the biological significance of this precocious branching is not clear, although it seems to be part of the syndrome of characters which makes shoot morphology highly organized and predictable in the Helobiae. A good example of the way in which normal morphological rules are transcended is provided by *Thalassia*, in which lateral meristems are precocious, but leaf opposed (Tomlinson and Bailey 1972).

Branching of the vegetative parts in all plants is of two kinds: (i) *deterministic*, where the branch meristems are precisely related to and have a strong influence on the overall shoot morphology; (ii) *opportunistic*, where the branch meristems (always axillary) only grow out as an environmental response, usually when the plant is damaged. This categorization corresponds essentially to the distinction between sequential branching and reiterative branching made for the architecture of trees (Hallé *et al.* 1978).

In the herbaceous plants considered here, the distinction in the type of branching is usually made evident because deterministic branching produces no proliferation of new shoots (e.g. if one of the axes produced by branching is determinate, like an inflorescence) whereas opportunistic branching will either proliferate the shoot system (i.e. "cause branching" in common usage), or at least repair it. This is the basis for the distinction made by Tomlinson (1974a) between *regenerative* branching and *proliferative* branching. In regenerative branching (which is usually associated with sympodial construction) the products of axis division are differentiated and only one continues vegetative development so that there is no apparent multiplication of axes. In proliferative branching the vegetative axis is multiplied. This distinction was made with reference to seagrasses, but has general application and important ecological implications (e.g. Bell & Tomlinson, 1980). In specialized instances proliferative branching only occurs if the rhizome system is damaged and is therefore a random event.

It is clear on this basis that dichotomous branching in the Helobiae is normally of the deterministic or regenerative kind. It is under strong genetic control and therefore directly moulded by adaptive forces. Nevertheless, an acceptable biological interpretation for the high incidence of shoot bifurcation (dichotomy or precocious axillary branching) in the Helobiae has yet to be provided. The simplistic explanation is that it reflects the highly programmed shoot morphology of these plants; dichotomy cannot therefore be a primitive feature.

Turions

A number of hydrophytes, both dicotyledons and monocotyledons, produce specialized swollen or tuberous shoots, usually as a response to cold or unfavorable environmental conditions (Arber 1920, pp. 217–25). The name winter-bud or turion has been widely applied to them while the term hibernaculae exists in the older literature, with some implication of their function. In the Helobiae turions are particularly characteristic of *Potamogeton* species, but they also occur in a few Hydrocharitaceae and Alismataceae. These structures are clearly over-wintering organs but they can also serve for multiplication and dispersal.

These specialized shoots are not circumscribed very sharply in their morphology from other kinds of modified shoots such as tubers or stolons and they bear comparison with bulbils and other kinds of pseudo-viviparous shoots which can develop in place of flowers, notably in the Alismataceae and Hydrocharitaceae.

Turions are defined by Arber as specialized shoots which are stored with food material, and protected externally in some way. The protective structures are usually scale-leaves, which also may be swollen and store food; in *Potamogeton crispus* the turion is largely made up of rigid, reduced leaves. Such is the diversity in the construction of these modified shoots that there is no purpose in applying rigid definitions to them, as Sculthorpe (1967) points out, but it is unfortunate that turion has been used very imprecisely to describe rhizomatous axes in certain aquatics (e.g. *Thalassia*), which does not develop specialized over-wintering shoots, or even for the axis and associated leaf, as by Edgecumbe (1980) for *Zostera*. Such usage indicates a lack of appreciation of the diversity of shoot morphology in aquatic plants which is, in fact, frequently highly organized and with complex polymorphism of axes (cf. Tomlinson 1974a).

Information about the conditions which determine the development of turions and their subsequent germination has been summarized by Sculthorpe (1967). Mention should be made of the study by Richards and Blakemore (1975) on the factors affecting germination of the familiar turions of *Hydrocharis morsus-ranae*. They describe formation of these structures as a modification of the terminal buds of stolons which remain undeveloped with bud-scales tightly clasping the embryonic new rosette. These authors found that high water temperatures promoted germination with an additional light requirement independent of photoperiod.

In the Helobiae the most extensive accounts of turion morphology and anatomy relate to *Potamogeton* and can be found in the writings of Hagström (1916), Monoyer (1927), Raunkiaer (1903), and especially Sauvageau (1894).

LEAF MORPHOLOGY AND ANATOMY

Leaf organization

The leaf morphology of Helobiae shows a wide range of shape. Examples include (a) the linear emergent leaves of *Butomus, Scheuchzeria,* and *Triglo-*

chin species; (b) the petiolate and bladed leaves of *Alisma, Sagittaria* species and *Limnophyton*; (c) the ovate floating leaves of *Hydrocleys* and *Hydrocharis*; (d) the broad-bladed leaves of *Potamogeton* species; (e) the linear, filamentous leaves of Zannichelliaceae and *Ruppia*; (f) the strap-shaped submerged leaves of *Vallisneria, Thalassia, Enhalus*; (g) the narrow ovate single-veined leaves of *Najas, Elodea,* and *Hydrilla.* The fenestrate leaves of some *Aponogeton* species are unique. The Hydrocharitaceae show the widest range of leaf forms within one family (cf. *Blyxa, Elodea, Halophila, Limnobium, Ottelia, Thalassia*). A wide range is also shown in the Juncaginaceae, where leaves may be strap-shaped (*Cycnogeton*), terete (*Triglochin*), or even equitant (*Tetroncium*).

Heterophylly and leaf plasticity. This topic has been discussed extensively by numerous authors (e.g. Arber 1919, 1920; Costantin 1886*b*; Glück 1924; Goebel 1880, 1896; Sauvageau 1890*a*; Streitberg 1954) and is not considered in detail here. The subject has been reviewed most recently by Sculthorpe (1967, pp. 218–47).

Morphological plasticity is common and its range may depend either on the circumstances under which the plant grows (cf. *Sagittaria sagittifolia* as a rooted, terrestrial plant and the same species in running water), or the plasticity may represent ontogenetic stages (cf. juvenile and adult stages in many *Potamogeton* species). This diversity and plasticity obviously influences the leaf anatomy considerably and reduces the likelihood of there being consistent systematic diagnostic features in plants with marked heterophylly. However, there may be consistent diagnostic features in leaves of one particular stage. Certainly one can find diagnostic differences between taxa with morphologically similar adult leaves, as shown by Sauvageau (1890*a–d*) and by Tomlinson (1980) for seagrasses.

Vernation. In the Helobiae the configuration of the developing unexpanded leaf blade (its vernation is a consistent diagnostic feature for a given taxon at the family, generic, or specific level. The feature is best seen as the leaf blade expands from the bud. The range of possibilities in the order is very limited and is summarized in Fig. 2, p. 32. The leaf blade may be either flat (*adplicate*) or rolled, and if rolled it may be either *involute*, when the two halves of the blade are rolled separately against the upper surface of the midrib (as in Commelinaceae), or *convolute*, when one half of the blade is rolled entirely within the other rolled half (as in the banana leaf). The adplicate condition refers strictly to plants with distichous phyllotaxis and ribbon-like leaves, but it may be applied more generally to any linear leaf, even though the blade may be thickened or angled (as for example in Butomaceae, Lilaeaceae, and Scheuchzeriaceae). In Najadaceae, for example, the leaves are rather broad and somewhat curved in bud, but still never rolled.

All three conditions occur in different species of *Potamogeton*, as indicated by Raunkiaer (1895–9), whose nomenclature is used (Fig. 2). Differences in vernation seem mainly related to the width of the blade. Otherwise the feature seems constant at the generic or family level. In heterophyllous species, juvenile or ribbon-like leaves are adplicate and may contrast with rolled adult or emergent leaves. However, convolute and involute leaves are never found in the same species.

The following lists summarize tentatively the known information, but exceptions may still be found:

(a) *Leaves with adplicate vernation*
 1. In the strict sense, the phyllotaxis distichous and the leaf blade more or less flat:
 Cymodoceaceae, Posidoniaceae, Zannichelliaceae, Zosteraceae; *Enhalus, Thalassia, Vallisneria* (Hydrocharitaceae), *Groenlandia, Ruppia, Potamogeton* spp (Potamogetonaceae).
 2. In a less strict sense, the phyllotaxis distichous, spirodistichous or spiral:
 Butomaceae, Lilaeaceae, Juncaginaceae, Najadaceae, Scheuchzeriaceae, *Halophila,* Anachariteae (Hydrocharitaceae).
(b) *Leaves with rolled vernation*
 1. Involute:
 Alismataceae, Aponogetonaceae, Limnocharitaceae, *Hydrocharis, Ottelia* spp, *Stratiotes* (Hydrocharitaceae), some *Potamogeton* spp (Potamogetonaceae).
 2. Convolute:
 Limnobium (Hydrocharitaceae), some *Potamogeton* spp (Potamogetonaceae).

Apical pore. Sauvageau (1890c, 1891a, 1893) drew attention to the existence of an apical opening which develops in the submerged or even floating leaves of a number of Helobiae by dissolution of apical tissues so that the group of tracheids at the end of the median vein may be in contact with the surrounding medium. Such an apical pore occurs in a wide variety of aquatic plants (Arber 1920). These pores were investigated experimentally by Minden (1899), Weinrowsky (1899), and Goffart (1900), and it has been generally concluded that they function much like hydathodes, i.e. they exude water. In submerged leaves this exudation is in keeping with the concept of a transpiration stream functioning under weak positive pressure.

However, apical pores are of limited taxonomic distribution in aquatic plants. In seagrasses, for example, they are only recorded for Zosteraceae (Sauvageau 1890d) and *Halodule* (Cymodoceaceae—Sauvageau 1890b). Frequently they do occur in plants with emergent leaves (e.g. Alismataceae, Butomaceae, Limnocharitaceae). The apical pore of *Scheuchzeria* is particularly distinctive and diagnostic.

Since the apical pore, where it occurs, is often a feature which is visible only late in the ontogeny of the leaf, its presence may not always be recorded. In *Halodule* (Cymodoceaceae) the leaf apex is used as a major specific diagnostic feature (den Hartog 1970a) but because of the considerable developmental variability it is often difficult to apply.

Ligule, stipule, and sheath. An extension or outgrowth of the leaf sheath beyond the insertion of the blade to produce a tongue-like structure (ligule) is diagnostic for many Alismatidae. However, a ligule is never developed in the Hydrocharitaceae. In its most precise development, the ligule is a narrow plate 2–3 cells wide which extends across the junction of blade and sheath, in plants with ribbon-like leaves, e.g. in Cymodoceaceae, Posidoniaceae, Zosteraceae. The ligule develops as an outgrowth of the adaxial surface of the leaf primordia, relatively late in development as in *Posidonia* (Weber 1956) and *Zostera* (Sauvageau 1890*a*). Otherwise the ligule is a direct extension of the open, tubular leaf sheath, as in *Scheuchzeria*. The ligule can be a useful diagnostic feature, e.g. in seagrasses it distinguishes the broader blades of Cymodoceaceae and Zosteraceae from those of *Enhalus* and *Thalassia* (Hydrocharitaceae), which have no ligule.

In some *Potamogeton* species (Sauvageau 1894; Raunkiaer 1895–9) the ligular extension of the leaf sheath is very pronounced and the whole sheath is free of the leaf axis. It has therefore been described as a stipule (Cosson 1860) and has even been homologized with the stipule pair of dicotyledons (Colomb 1887; Glück 1901). However, the homology seems mistaken since it is doubtful if true stipules occur in monocotyledons.

The leaf insertion in the Helobiae is normally broad and encircles the stem, with the production of a sheathing base. This is normally open, i.e. split vertically even though opposite margins overwrap. A closed, tubular leaf sheath is uncommon but occurs in *Zostera* subgenus *Zostera* and some *Potamogeton* species. Leaves with a narrow insertion and without a sheathing base occur in Najadaceae and, in Hydrocharitaceae, in the Anachariteae and many species of *Halophila*.

Leaf anatomy

Information concerning the anatomy of leaves in the Helobiae is extensive but widely scattered. An attempt has been made to summarize this in the family descriptions but a few brief general comments seem appropriate. It should be emphasized that, as in all submerged leaves, the epidermis becomes the main photosynthetic tissue of the leaves and this topic has received some recent attention at the hands of anatomists using ultrastructural techniques (p. 35).

Stomata. These are absent from many families with wholly submerged shoots (e.g. Cymodoceaceae, Najadaceae, Posidoniaceae, Zannichelliaceae, and Zosteraceae). They may remain in a vestigial form, as in the leaf apex of *Zannichellia* where they are diagnostically useful for certain cytological races. Otherwise they are absent from some taxa, but present in other taxa within a given family, depending on the extent to which parts remain submerged. Presence or absence of stomata is a very plastic feature of species which are markedly heterophyllous. A frequently noted feature is the restriction of stomata to the upper surface of floating leaves (e.g. Kaul 1976*b*).

There are too few studies of stomatal development to establish whether there are consistent and distinctive patterns of cell division in the ontogeny of the individual stomatal complex (cf. Paliwal 1976; Paliwal and Lavania 1978, 1979; Tomlinson 1974*b*) in those plants where stomata are normally developed. At maturity the stomatal complex may be described as paracytic or tetracytic, the present limited information suggests a rather uniform developmental pattern.

Mesophyll structure. There are extensive reviews of the variability of the

A

B   C

FIG. 2. Leaf vernation in Helobiae. Diagrammatic TS of leaf in bud showing the three main types observed.

A. Adplicate.

B. Involute

C. Convolute

(Nomenclature after Raunkiaer 1895–9.)

internal anatomy of leaves in relation to leaf plasticity (Costantin 1886b; Sauvageau 1891a; Arber 1920; Solereder and Meyer 1933; Sculthorpe 1967; Kaul 1976b). One may briefly note the marked development of aerenchyma in the mesophyll of emergent and floating leaves; the frequent inversion of normal mesophyll differentiation in floating leaves, i.e. with the adaxial layers becoming less palisade-like and more like a spongy mesophyll, and the very simple and regular differentiation of the mesophyll in strap-shaped leaves. Despite this marked influence of environment on internal leaf form, diagnostic characters continue to be revealed in the distribution of vascular bundles, mechanical fibres, and development of transverse diaphragms. In many instances it is possible to identify a member of the Helobiae from its leaf anatomy alone, at least to the level of genus. In doing so inherent variability has to be taken into account. Most Helobiae are wide-ranging and very small population samples have been surveyed so far.

Inverted vascular bundles. A useful diagnostic feature in the leaves of Helobiae is the presence of vascular bundles with inverse orientation in the blade and petiole, i.e. with the xylem situated abaxially and not adaxially. The developmental significance of this orientation is unexplained, but the structure provided evidence for the phyllode theory of the origin of the monocotyledonous leaf (Arber 1918b, 1920, 1922b, 1925a). The inverted bundles were considered an indication of a previously ancestral petiole rather than a fully developed ancestral laminate leaf. Arber's approach was, however, very eclectic and, apart from the difficulty of refuting such a phylogenetic theory in the absence of fossil evidence it did leave unanswered the problematical question of the rather scattered distribution of inverted bundles in the Helobiae. For example *Enhalus* and *Thalassia* (Hydrocharitaceae) have similar strap-shaped leaves and both are wholly submerged marine. The first has inverted bundles the second not, and this disparity is not easily explained in terms of phyllodic anatomy. Closely related taxa may or may not have inverted vascular bundles, as in the Juncaginaceae, where the differences are not always explicable by differences in leaf morphology. A morphogenetic analysis of this phenomenon would be helpful and the Helobiae provide plentiful examples for comparative study.

Epidermal ultrastructure in submerged leaves

The leaf blade epidermis in submerged aquatic plants is unusual in that it is the main site of photosynthesis, with a high concentration of chloroplasts. Chloroplasts occur elsewhere in the mesophyll but are much more diffusely distributed. Recent ultrastructural studies have revealed cytoplasmic peculiarities of the epidermal cells, which recall those distinguishing transfer cells from their neighbours in a wide variety of plants (Cutter 1978; Gunning and Pate 1969; Pate and Gunning 1969). The most extensive and detailed studies are those on seagrasses—Jagels (1973) on *Thalassia*; Birch (1974) on *Halophila*; Benedict and Scott (1976) on *Thalassia*; Doohan and Newcomb (1976) on *Cymodocea* spp; Barnabas *et al.* (1977) on *Zostera*; Ducker *et al.* (1977) on *Amphibolis*; and Kuo (1978), Cambridge and Kuo (1979) on *Posidonia*. Although the leaves are submerged, a thin cuticle is always present and it may have distinctive properties associated with the specialized exchange function of submerged photosynthetic epidermal cells, especially in salt water. Gessner (1968) observed that the cuticle of the leaf epidermis of *Thalassia testudinum* is perforated by clusters of small pores 0.1 μm in diameter. Similarly Kuo (1978) describes the cuticle of *Posidonia australis* as being fibrillar or porous in appearance. Doohan and Newcomb (1976) have commented on this structure in relation to epiphytic bacteria.

The most distinctive cytoplasmic feature in all of the seagrasses examined except *Posidonia* is the presence of convoluted plasmalemma, which is also characteristic of transfer cells, and is frequently also associated with wall ingrowths from the inner cell wall. The cytoplasm is dense, rich in organelles, especially chloroplasts and mitochondria, and the nucleus is large (Plate 1A, B). There is said to be no protoplasmic continuity between the epidermal and underlying mesophyll cells, which implies that nutrient movement is apoplastic (i.e. along the free space within and between cell walls). *Thalassia* (Plate 1C) illustrates these features well and also shows an enlarged apoplastic region between the cell wall and the plasmalemma. In *Zostera* (Plate 2B) these structures are less pronounced.

The epidermis of *Posidonia* is distinctly different (Kuo 1978), since the wall ingrowths and convoluted plasmalemma are absent, the vacuole contains polyphenolic substances (tannins) and a well-developed system of plasmodesmata connect the epidermal and adjacent mesophyll cells suggesting that symplastic communication is possible (within a protoplasmic continuum). These differences may be significant phyletically. It should be noted that *Posidonia* also has a different mesophyll structure from that of most other seagrasses, since a multiseriate mesophyll is laid down in the young leaf primordium. This contrasts with the condition in other seagrasses in which the initially uniseriate mesophyll becomes multiseriate when mature by regular segmentation processes. Precisely the same segmentation process occurs in seemingly unrelated seagrasses, e.g. *Thalassia* (Tomlinson 1969a) (see Fig. 4.11); *Zostera* (Roth 1961); *Enhalus* (Troll 1931a) all with similar strap-shaped leaves.

Comparative study within a single leaf can also be informative. Kuo (1978) describes the epidermis of the colourless leaf sheath of *Posidonia* as lacking chloroplasts, having a non-porous cuticle and a regular microfibrillar packing of the cell wall. At the same time the vacuole is well-developed but only sometimes includes polyphenolic materials. In the leaf blade, chloroplasts in the mesophyll cells often contain starch grains, which are not usually developed in epidermal cells.

Other investigators have commented on the similar structure of epidermal cells in some seagrasses and of osmoregulatory cells in salt glands (e.g. Jagels 1973). Birch (1974) described peculiar wall properties of the epidermal cells of *Halophila ovalis* in which an annulus with an affinity for silver salts could be demonstrated, a feature first commented on by Solereder (1913). However, not all *Halophila* leaves show this property.

It is clear that continued ultrastructural study of the leaves of seagrasses will lead to the recognition of further characteristics relative to their function;

the information has to be placed in a developmental, systematic, and ecological as well as physiological context. Geographical considerations cannot be neglected; the genera of seagrasses are restricted to particular temperature regimes, with a fairly clear cut difference between tropical (e.g. *Enhalus, Halodule, Halophila, Syringodium, Thalassia, Thalassodendron*) and temperate or at most subtropical (e.g. *Amphibolis, Posidonia,* Zosteraceae) ranges (den Hartog 1970*a*). Furthermore the range of salinities they tolerate is variable. Some brackish-water species which never occur in marine situations may tolerate hypersaline conditions (e.g. *Ruppia*; Davis and Tomlinson 1974). *Ruppia* (Plate 1D) lacks the convoluted plasmalemma. Our present understanding of leaf ultrastructure in Helobiae thus at present permits no functional generalizations.

STEM VASCULAR CONSTRUCTION

Erect shoots

Since the pioneering studies of the course of vascular bundles in monocotyledonous stems, e.g. the study by de Bary (1884) and the less reliable investigation by Falkenberg (1876), there have been only two major studies of shoot vasculature in the Helobiae. This is rather surprising in view of the relative accessibility and simplicity of these shoots, but not surprising in view of their frequent nodal complexity. Chrysler (1907) made a comparative study of the adult shoots of a number of Potamogetonales, while Monoyer (1927) made more extensive comparisons with the shoots of other families of monocotyledons which grow in water. Both authors were impressed by the way in which their observations could be interpreted as revealing a progressive reduction in the vascular cylinder. Monoyer showed that there is a progressive loss of the cortical system, while at the same time the central vascular strands become progressively more aggregated. The maximum reduction of the vascular system of the stem is represented for example by *Najas, Ruppia,* and *Zannichellia* which each have a protostelic central cylinder, with a central xylem lacuna surrounded by a cylinder of phloem. The central cylinder is enclosed by an endodermis, while the cortical system is either reduced to a few vestigial bundles or it may even be eliminated. In intermediate stages, Monoyer illustrates independent vascular strands in the central cylinder (sometimes represented only by the phloem) which are typically confluent with other strands at the node and form a regular pattern. In some families, notably the Potamogetonaceae (p. 291), these patterns are highly characteristic and may be diagnostically useful at the specific level. However, the analyses by Chrysler and Monoyer are evidently subjective and presented in a somewhat stylized manner. They do not depend at all on developmental investigation. This is one area where an excellent, but limited, existing achievement could readily be elaborated by modern study now that thin-sectioning and microcinephotographic techniques are available. In particular the extension of this work to the more complex vascular system of those plants with large axes (e.g. Alismatales) is needed.

Rhizome anatomy

Chrysler and Monoyer largely studied plants with extended internodes, where vascular systems were not congested. In the condensed rhizomatous or cormous axes of many rosette-bearing Helobiae, the vascular system is organized in such a way that analyses of the course of vascular bundles are scarcely possible. Accounts of the general anatomy of condensed axes remain rather superficial, both in this present volume as well as in earlier accounts. Nevertheless, a generally similar organization does exist, as can be seen in the series of illustrations which form Plates 3–5. The common pattern is a broad, often lacunose, cortex, with or without a discrete cortical vascular system, together with a central cylinder delimited by a distinct endo-dermis and a series of irregularly distributed strands which are continuous with the leaf traces. This pattern is consistent with our general understanding of the construction of monocotyledonous vascular systems as described by Zimmermann and Tomlinson (1972) but the pattern is rendered imprecise by the virtual absence of internodes.

Stolons and inflorescences

These axes may contrast strongly with rhizomes, since they are usually extended, and sometimes have only a single internode. The ground tissue is always markedly lacunose and the vascular bundles are commonly few or scattered towards the periphery of the organ. Plate 2, showing the peduncle of a large species of *Ottelia* and of *Cycnogeton*, is typical. Comparative develop-mental and morphological study suggests that in some Helobiae stolons are modified inflorescences and a particularly clear series of intermediate stages is shown by several Alismataceae, Limnocharitaceae, and Hydrocharitaceae (Wilder 1975). The basically similar anatomy of the two types of organs is thus readily explicable.

Root Morphology and Anatomy

Anatomical reduction

The root system in Helobiae, as in all monocotyledons, is entirely adventi-tious, but shows many specialized features, usually of reduction, related to hydromorphy, (cf. Arber 1920; Bristow 1975; Sculthorpe 1967).

Amongst the anatomical features associated with the more hydrophytic taxa the following may be emphasized:

(i) Root-hairs specialized, e.g. with an endocytic base, occurring in certain Hydrocharitaceae.

(ii) Exodermis narrow (1–2-layered), the cells commonly with slightly thickened and suberized walls.

(iii) Cortex aerenchymatous, usually developed by collapse of alternate radial plates of cells.

(iv) Transverse cortical diaphragms frequently developed (their functional significance is discussed elsewhere, p. 45).

(v) Stele narrow, the xylem frequently poorly differentiated.

(vi) Sieve-tubes pericyclic.

Some general morphological and anatomical features of roots in Helobiae are outlined below.

Root distribution

In the seedling the radicle is poorly developed or aborts early. In the Alismataceae the radicle fails completely to elongate (Kaul, 1978; Lieu 1979*b,c*). Attachment is primarily by a ring (collet) of hairs at the base of the hypocotyl, a common condition in aquatic plants (Sculthorpe 1967). In certain taxa with highly specialized embryos it has been suggested that the true radicle aborts completely, the first root to appear being equivalent morphologically to the first adventitious root. However, embryological study (see Fig. 1, pp. 24) shows that the radicle is displaced to a lateral position and does not occur opposite the suspensor, as is usual in straight embryos. Nevertheless, even though a radicle is present, it may remain undeveloped (Kaul, 1978).

Roots otherwise develop on buried axes. They may be restricted to the lower surface in dorsiventral shoots (e.g. *Butomus, Enhalus, Posidonia, Potamogeton,* Zosteraceae), but in radially-symmetrical axes they occur on all sides. In shoot systems with appreciable dimorphism, roots may be absent from or infrequent on erect shoots (e.g. Cymodoceaceae). Roots may differentiate as primordia but remain suppressed (e.g. Zosteraceae). A special association between roots and branches is noticeable, e.g. in Najadaceae, Zannichelliaceae, and Anachariteae (Hydrocharitaceae) where roots may only occur at nodes that also support a branch.

A common feature, found throughout the Helobiae, is the close association between root initials and the shoot apical meristem so that roots are a feature of the primary shoot organization; they do not appear as belated appendages after primary differentiation is complete. This is part of the biological phenomenon of meristem dependence, previously described for seagrasses (Tomlinson 1974*a*), but obviously a general process in the whole order. Roots are commonly restricted to nodes and their vascular traces may have a precise topological relation to the shoot vasculature, e.g. in Zosteraceae, where they always occupy a characteristic position in two groups opposite and immediately below the leaf (Plate 6). There may be from 2–12 root primordia in each group, but commonly many of the primordia remain undeveloped.

A characteristic site for the origin of roots is in the cortical tissues close to the shoot apex (e.g. *Thalassia*). In part, one can account for both the precocious development and nodal restriction to the marked development of axial cortical aerenchyma in the first situation, and intercalary growth in the second situation so that the ground tissue itself becomes specialized early in development (e.g. *Zostera,* Plate 6). Essentially there is no undifferentiated parenchyma available for late root initiation. The same considerations apply in other aquatic monocotyledons (e.g. Pontederiaceae, notably *Eichhornia,* and in *Pistia*).

Root dimorphism

A few Helobiae show some dimorphism of roots. In *Limnobium* and *Stratiotes* for example the first-developed roots associated with each leaf rosette are thicker and with fewer root-hairs than those developed later (Wilder 1974*b*), apparently a primary genotypic expression. A similar dimorphism is known for some Pontederiaceae (e.g. *Heteranthera*, Hildebrand 1885). The presence or absence of root-hairs is itself under a degree of phenotypic control, e.g. a substrate with a high carbon dioxide concentration promotes development of root-hairs in *Elodea* (Dale 1951). Variation in the abundance of root-hairs is often observed in aquatics, root-hairs commonly develop only when roots penetrate the substrate (D'Almeida 1942).

Root traces

Although roots may originate in the cortical tissues, rather than at the surface of the central vascular cylinder, as is usual in monocotyledons, a vascular connection to the stele is made at an early stage. This root trace always connects to the peripheral vascular tissue of the axial cylinder, a feature which may be quite striking in the more specialized rhizomes of submerged plants (e.g. *Thalassia*, Plate 5). In plants with marked intercalary growth and internodal elongation the root trace is a complicating factor in the nodal vasculature.

In the Zosteraceae, by virtue of the distinctive method of development of the adventitious root complexes, the individual traces may unite with each other before they are attached to the stele, so that essentially two compound root traces form the connection with the stem vascular system.

Branching of roots

Roots in the Helobiae may be branched or unbranched, the latter type having a restricted taxonomic distribution. Unbranched roots characterize the families Zannichelliaceae, Potamogetonaceae, Zosteraceae, Najadaceae. Branched roots occur in all other families, except the genus *Halodule* in Cymodoceaceae and a number of Hydrocharitaceae (*Thalassia*, *Vallisneria*, Anachariteae). The character may be used to distinguish otherwise similar aquarium plants e.g. *Sagittaria* from *Vallisneria* (Laessle 1953).

The adaptive significance of unbranched roots is not known, but they seem characteristic of taxa which are highly specialized, i.e. submerged aquatics, often with submerged pollination. Morphogenetically, absence of branch root primordia seems strongly (but not absolutely) correlated with the presence of sieve-tubes in the pericycle and particularly with poorly differentiated tracheary elements. Further comparative and experimental examination of this phenomenon may throw light on the physiology of root branching generally.

Root apical organization

No attempt has been made to review the literature on apical organization of the root, since this is adequately summarized by von Guttenberg (1968) (see also Guttenberg and Jankuszeit 1957; Lüpnitz 1969). The root apices of aquatic plants provide experimental subjects since whole plants can provide relatively accessible roots and the apical organization and pattern of differentiation may be relatively simple (e.g. Tomlinson 1969*a*). Kadej (1966) showed that apical organization in *Elodea* spp changes several times during the growth of a single root.

Reduced vascular tissue

Xylem. A trend in the reduction of the vascular tissues of aquatic plants roughly parallels that of increasing specialization. Consequently the roots of emergent or amphibious hydrophytes are structurally most like those of terrestrial plants, while wholly submerged aquatics have the most reduced root anatomy. This trend is commented upon by numerous authors who have written about the anatomy of roots of aquatic plants (e.g. Sculthorpe 1967).

A series of examples illustrates this trend. In the larger Alismataceae, Butomaceae, and Limnocharitaceae, a normal polyarch vascular system, uninterrupted pericycle and thick-walled, lignified tracheary elements (usually vessels in the late metaxylem) are found. However, the cortex is aerenchymatous and the stele relatively narrow.

Reduction in rooted hydrophytes with floating leaves involves a further narrowing of the stele, and there are fewer metaxylem elements; commonly the metaxylem includes a single central vessel or file of tracheids, the walls often with poorly developed sculpturing and little lignification. In a still more advanced stage the central elements are missing, or represented by a xylem lacuna so that the protoxylem becomes the sole differentiated xylem.

In the final stages the xylem is represented by indistinctly differentiated cells. These either occupy the position of erstwhile protoxylem poles next to the pericycle (e.g. Cymodoceaceae, Posidoniaceae, *Thalassia*), or they are represented by a group of thin-walled central cells (as in the Zosteraceae), or even by a central xylem lacuna, as in *Vallisneria*.

Phloem. As pointed out by Sauvageau (1889*b*) and illustrated by Schenck (1886*b*), sieve-tubes commonly occupy an unusual pericyclic position in aquatic plants, especially the more highly specialized Helobiae. The arrangement involves a single series of sieve-tubes alternating with the protoxylem elements (or their reduced remains). The sieve-tube is always next to the endodermis, while its associated companion cell is always situated towards the centre of the stele. Usually the sieve-tube has a characteristic pentagonal outline in transverse section.

The ultimate in stelar reduction is probably achieved by *Halodule* (Cymodoceaceae) in which the thread-like unbranched roots have a stele scarcely 50 μm in diameter, including two opposed pericyclic sieve-tubes and a few central thin-walled cells (which may be vestigial tracheary elements) surrounding a minute central lacuna. Roots of *Elodea*, *Najas*, and *Vallisneria* may be similar, but with more sieve-tubes and a central xylem lacuna.

THE CONDUCTING TISSUES

Tracheary elements

Aquatic plants generally are characterized by reduced xylem tissue (Schenck 1886b; Sculthorpe 1967). This reduction is interpreted as a result of the loss of functional need (mechanical and conductive) in plants with a constant water supply and often supported by the aqueous medium.

Type and distribution

Vessels occur in the following helobial families where they are restricted to the roots (Cheadle 1943a,b; Ancibor 1979):

Alismataceae, Butomaceae, Hydrocharitaceae, Juncaginaceae, Lilaeaceae, Limnocharitaceae, Scheuchzeriaceae.

Cheadle recorded vessels as absent from Hydrocharitaceae, presumably because of his small sample (two species in two genera) whereas Ancibor reports them for the genera *Blyxa, Enhalus, Hydrocharis, Limnobium, Ottelia, Stratiotes.*

The perforation plates are normally scalariform, usually with numerous thickening bars, but simple perforation plates are recorded for some Alismataceae and Butomaceae. Vessels are essentially absent and the tracheary elements most reduced in entirely submerged taxa, e.g. Cymodoceaceae, Najadaceae, Posidoniaceae, Zannichelliaceae, Zosteraceae, and certain Hydrocharitaceae, e.g. Anacharitae, *Halophila, Thalassia,* where they may be represented by weakly lignified elements with annular or spiral wall thickening, the amount of secondary wall being minimal. In the roots of some submerged taxa reduction is extreme since the position of an incipient tracheary element can be recognized, but the wall sculpturing is never completed (e.g. *Thalassia*—Tomlinson 1969a).

In the stems of plants with reduced tracheary tissue xylem is always developed in the nodal vascular complex and there are even records of perforated elements (by definition vessel elements) in this position, e.g. in *Althenia* (Prillieux 1864), and in *Najas, Ruppia,* and *Zannichellia* (Chrysler 1907) but these elements seem not to be aggregated into true vessels.

In the leaves of submerged Helobiae the longitudinal vascular strands typically lack complete tracheary elements when functionally mature since the protoxylem elements are extended passively and not replaced by metaxylem elements. On the other hand transverse veins may retain a single series of narrow tracheids. In emergent aquatics, late-differentiated metaxylem is usually developed, but in the largest bundles it is always subordinate to the protoxylem lacuna.

Protoxylem lacunae

The apparent elimination of tracheary tissue during the ontogeny (and presumably phylogeny) of aquatic plants has led to the suggestion that there is no transpiration stream because (a) it is unnecessary, (b) there is no structural pathway for it. Similarly, the roots which are universally present in the plants considered here, are said to have solely an anchoring function.

This orthodox position is essentially adopted by Sculthorpe (1967) in his review of the topic. Structural and physiological considerations suggest that this view is fundamentally wrong (cf. Denny, 1980).

In all Helobiae that show close association with water the protoxylem is stretched passively by organ elongation. The space it originally occupies does not become engulfed by surrounding parenchyma as is usual for protoxylem in terrestrial plants but rather remains and even enlarges so that a cavity (protoxylem lacuna) is developed. Usually the remains of the wall thickening of protoxylem elements can be found in this cavity, adhering to the wall. The cavity may be very broad (e.g. over 300 μm wide in the leaves of emergent Alismatales). The walls of cells adjacent to the cavity often become thickened as is common in submerged Helobiae. This thickening can be easily seen and is frequently illustrated. Since the cavities of anastomosing vascular bundles may be continuous, or at least continuous with uncollapsed tracheary elements (e.g. of nodal plexi), there is a transport channel for the movement of water in aquatic plants. However, it may well function either at weak *positive* pressures in submerged plants via root pressure, or at *low negative* pressures in emergent plants which transpire. The apical pore (p. 30) seems to cater for water loss in plants with submerged leaves and no apparent opportunity for transpirational water loss.

Consequently the protoxylem lacuna constitutes a possible water-transport pathway which, however, does not function at high negative pressures as in terrestrial plants and therefore does not require rigid thickened walls. The protoxylem lacuna is a minimally strengthened tube providing the necessary conduit.

Failure to demonstrate a transpiration stream in water plants may result from the techniques used and the mistaken analogy with the water-transport mechanism of terrestrial plants. In the latter, since water movement is mostly under *negative* pressure coloured dyes can be drawn into the xylem of a cut shoot, provided embolisms are eliminated; in aquatic plants water movement may be mostly under weak *positive* pressure so that coloured dyes cannot be drawn into the xylem. More elaborate methods are needed to insert a tracer without making the protoxylem lacunar system leak.

The structure of the root itself also suggests an absorptive function as has already been discussed (p. 38). In summary, roots of aquatic plants usually have root-hairs, especially when they are anchored in some firm substrate. Roots may be dimorphic in this respect (i.e. with or without root-hairs). There is frequently some specialized relation between root-hairs and contiguous cells. Roots in the Helobiae are frequently septate, developing transverse diaphragms which could facilitate transport across the cortex. The rhizosphere of certain aquatic Helobiae (notably seagrasses) has been shown to be rich in nitrogen-fixing microorganisms and it is suggested that this may be a major source of nitrogen for whole ecosystems, absorbed initially by the plant through its root and subsequently made available for food-chains by decomposition (Patriquin 1972; Patriquin and Knowles 1972). Carignan and Kalff (1980) have shown that the sediment in which aquatic plants grow is the source of all their phosphorus. This was done by growing plants rooted in ^{32}P-labelled substrate but with the leafy shoots in open (unlabelled) water. The plants studied included species of *Elodea, Najas, Potamogeton,* and *Vallisneria.*

Consequently the existence of a water-conducting channel, however modified, is seemingly necessary to the normal functioning of aquatic plants.

Phloem

Sieve-tubes in all Helobiae have simple sieve-plates, as far as has been observed. It has been suggested (p. 18) that this fact can be used as evidence that the Helobiae are a specialized and not a primitive group, since primitive groups are considered to be categorized by compound sieve-plates.

MISCELLANEOUS MORPHOLOGICAL AND ANATOMICAL FEATURES

Aerenchyma

In this volume the term *aerenchyma* is used to describe lacunose ground tissue generally, which is the commonly accepted usage (Esau 1977). However, the original intention of Schenck (1889b) who coined the term, was that it should be restricted to the secondary tissue with large intercellular spaces produced by a cork cambium, as in the roots of *Ludwigia* (*Jussiaea*), in contrast to lacunose tissue of primary origin. This restricted usage has been employed by recent authors (e.g. Eames and MacDaniels 1951, and the work of Schenck reviewed accurately by Troll 1943, and Guttenberg 1968). Applying the term more generally is justifiable biologically, if not morphologically.

The presence of aerenchyma, in the wide sense, is a feature of aquatic plants, particularly those which are submerged (but is by no means confined to them). Aerenchyma is characteristic of most Helobiae, it occurs in all vegetative parts and its ubiquity tends to obscure systematic anatomical features.

Development. Aerenchyma in the Helobiae is almost always developed schizogenously, i.e. by the enlargement and expansion of existing intercellular spaces, although in the radial cortical lacunae of many roots it is often supplemented by partial collapse of radial plates of cortical cells. The pattern of development of the aerenchyma may be very simple and consists in the enlargement of spaces between cells laid down in a regular pattern close behind the shoot apex, as in most stems. Where internodal elongation is pronounced the lacunae may extend from one node to the next as in *Elodea*; but the system may form a longitudinal reticulum, as in the cortex of many non-septate roots. More highly specialized segmentation patterns exist in leaves. The most specialized condition is found where the lacunae are divided at regular intervals by transverse septa or diaphragms, often of highly specialized cells.

Transverse septa in leaves. These are usually uniseriate, the cells may be thin-walled, but are commonly unevenly thick-walled and quite elaborate in shape (e.g. in many Hydrocharitaceae). The cytoplasm may be dense and

the nucleus, though small, is conspicuous. The compactness of these diaphragm cells varies considerably, but they are commonly stellate with minute
but regular intercellular spaces. In most leaves and petioles certain transverse
septa include transverse veins (Duval-Jouve 1873b; LeBlanc 1912), but the
regular pattern of alternation of different kinds of transverse septa described
by Kaul (1972, 1973) for *Sparganium* has not been reported for the Helobiae.

Studies in the method of development of the transverse septa are few,
but a very regular and complex segmentation pattern has been described
for certain strap-shaped leaves (e.g. *Thalassia*, Fig. 4.12) in which the transverse septa represent the initial mesophyll cells of the young leaf primordium
(Tomlinson 1972). Curiously, precisely the same pattern occurs in *Enhalus*
(Troll 1931a) and *Zostera* (Roth 1961). Whether this similarity reflects extreme parallelism or indicates phyletic relationships remains unanswered.

Transverse septa in stems. Diaphragms similar to those of the leaves frequently occur in the aerenchyma of stems, although where the leaf diaphragm
cells are highly elaborated, as in many Hydrocharitaceae, the stem diaphragm
cells are usually simpler. In filamentous stems the nodal vascular plexus
may be the only interruption of the lacuna (as in *Najas*, *Hydrilla*, etc.).
Detachment above the node may facilitate breakage and vegetative dispersal,
as in *Egeria* (Jacobs 1946).

Transverse septa in roots. The roots of a number of Helobiae are characteristically septate, with transverse diaphragms at regular intervals. The diaphragm
cells may be elaborated, often much more so than in the leaf of the same
plant (as in *Thalassia*). Here the diaphragm cells develop from plates of
cells at regular intervals which remain unextended (Tomlinson 1969a). One
can recognize something of a transition towards this in the Alismataceae
and Limnocharitaceae, where some taxa have a root cortex with irregularly
distributed short cells which are often contiguous laterally but scarcely form
diaphragms. A similar kind of transitional series occurs in Eriocaulaceae
(Tomlinson 1969b).

Function of aerenchyma and septa. It is generally assumed that aerenchyma
is developed to increase the internal gas space of aquatic plants since they
may grow in an oxygen-poor substrate (e.g. in submerged plants, or the
subterranean parts of plants rooted in wet soil). The gas space is said to be
continuous and to facilitate internal diffusion, e.g. of oxygen, especially to
roots or rhizomes.

Williams and Barber (1961) suggest, on the other hand, that the advantages
are partly mechanical, and represent economy of material, the largest plant
size being achieved with a minimum expenditure of biomass, this being possible in organisms where mechanical requirements are minimized. These considerations certainly apply to some large terrestrial plants with well-developed
aerenchyma, notably in many Zingiberales (Tomlinson 1969b), where there
would seem to be no need for an extensive internal atmosphere.

The diaphragms themselves serve three important functions:

(i) They may resist compression forces, particularly where they are thick-
walled, and so prevent collapse of air-canals.

(ii) They may prevent extensive flooding where an organ is broken. In this respect the very regular size of the intercellular pores of many of these diaphragms may be functionally significant. The pores are sufficiently wide to permit movement of gas molecules, but are too narrow to permit the movement of a gas–water interphase. At a pressure difference of 0.1 atmosphere (i.e. in an organ 1 m deep) a perforation up to 3 μm diameter could confine a water–air interphase bubble. These are the same considerations which determine the limit to which pores in the pit membranes of tracheary elements can contain embolisms.*

(iii) They provide lateral transport across the cortex, i.e. from the root surface to the stele (or vice versa). This may be important where the cortex is lacunose and the radial plates of cells are partially collapsed. However, Drew *et al.* (1980) have shown in corn roots subjected to oxygen stress that the resulting collapse of cortical parenchyma cells, producing extensive aerenchyma, did not influence the ability of the roots to translocate certain ions. It appeared that strands of wall residues bridging the cortex could be involved in conduction across the cortex. The close association between root-hairs and pitted exodermal cells immediately below root-hairs, the exodermis and the transverse diaphragms, and finally the transverse diaphragms and inner cortex suggest a continuous symplastic pathway. Such a pathway exists, for example, in *Thalassia* (Tomlinson 1969*a*). This pathway may be important where there is a high concentration of nitrogen-fixing microorganisms in the rhizosphere. It has been suggested that this association is one of mutualism, which may be a major source of nitrogen for the entire ecosystem, a suggestion made for a seagrass (*Thalassia*) (e.g. Patriquin 1972; Patriquin and Knowles 1972). Nitrogen (and possibly phosphorus) is made available to other organisms via decomposition of the *Thalassia*. Further discussion of transport is included under conducting tissues (p. 42).

Squamules

Distribution and structure. Irmisch (1858*a*) first drew attention to axillary scales, which he called squamulae intravaginales, which are almost ubiquitous in the Helobiae, having been found in all taxa that have been investigated, with the exception of *Scheuchzeria*, where they are replaced by a weft of uniseriate, filamentous hairs. Squamules are typically restricted to the axil of foliage and scale leaves (including prophylls and cotyledons) but may also be associated with bracts. They rarely occur in association with floral appendages. Squamules do occur in some other monocotyledonous families (e.g. certain Araceae, Pontederiaceae), most typically in taxa which can be described generally as 'aquatic'. Information about the frequency and distribution of squamules in the Helobiae has been summarized by Wilder (1975), but see also Gibson (1905) and Arber (1923, 1925*b*).

---

* From the equation:

$$\text{Height of a water column in a capillary} = \frac{2T}{rgs}$$

where T = surface tension of water, against air; r = capillary radius; s = specific density of water; and g = acceleration, due to gravity. (See Zimmermann and Brown 1971, p. 207.)

Squamules originate early, often at the second or third plastochrone and may briefly exceed associated appendage primordia. However, their appearance is delayed in *Elodea* as reported by Dale (1957), probably because of the unusual cylindrical shoot apex. Arber (1923) suggested that squamules develop on the back of a leaf primordium, rather than from internode tissue, but examination of a diversity of examples does not support this interpretation (e.g. Dale 1957; Wilder 1974c).

Squamules mature as non-vasculated, bi- or more usually multiseriate structures, the constituent cells of which are non-vacuolate and have densely-staining cytoplasm and a conspicuous nucleus. They are usually scale-like, but may be stalked, have dentate margins and may include tannin cells. It is frequently remarked that the shape of individual squamules is diagnostic for a given taxon (e.g. Ancibor 1979, for certain Hyrocharitaceae). The number of squamules at a node is also fairly constant for a given taxon, but tends to be correlated with the size of the axis. It varies from very numerous (as in Alismataceae, Butomaceae, Limnocharitaceae, Lilaeaceae, some Hydrocharitaceae, some *Potamogeton* spp) to two in more diminutive forms (as in Najadaceae, Zannichelliaceae, *Ruppia, Groenlandia*, small spp of *Potamogeton*, Anacharitae of Hydrocharitaceae). Where numerous, the squamules form a kind of palisade in the leaf axil, adjacent scales sometimes being partly fused at their bases. Where there are two these are always lateral. In *Vallisneria* Wilder (1974c) showed that two lateral primary squamules develop prior to others formed later on either side.

Function of squamules. Two main views exist:

(i) They are close-packing organs which fill the space between developing appendages. This theory does not account particularly well for their association with aquatic plants.

(ii) They secrete materials (e.g. mucilage) which may discourage or even prevent the growth of microorganisms which are otherwise likely to be favoured by the wet conditions (Schilling 1894). The mucilaginous secretion is itself easily demonstrated since it often persists in histological preparations, but there is little published evidence to show that the squamules actually produce it, the only definitive study is that by Rougier (1972).

It should be noted that the only helobial family which lacks squamules, Scheuchzeriaceae, is the one that grows at highest latitudes.

Hydropoten

Localized regions of irregular flat cells on either leaf surface, or on the petiole, usually with a thin or discontinuous cuticle, were named *Hydropoten* (lit. 'water-drinkers') by Mayr (1915). The subject has been reviewed recently by Lyr and Streitberg (1955) and for dicotyledons by Wilkinson (1980). The cells have a distinct affinity for certain dyes (e.g. Drawert 1938). Two functions have been ascribed to them: (a) they are regions of localized salt uptake or at least ion exchange which may involve active transport, (b) they represent regions which facilitate either water loss or water uptake.

Recent studies in Nymphaeales and some other aquatic plants, including *Sagittaria*, have clarified certain structural and physiological aspects of hydropoten and it seems certain they have a salt-transporting function, e.g. Lüttge (1964), Lüttge and Krapf (1969), Kristen (1969), Lüttge et al. (1971).

Hydropoten are common in certain Alismatideae, e.g. on the under surface of floating leaves or on submerged leaves in Alismataceae, Aponogetonaceae, Hydrocharitaceae, Potamogetonaceae, but they seem absent from other families, e.g. Zannichelliaceae and all seagrasses.

FLORAL ORGANIZATION

Two separate problems present themselves in the interpretation of floral organization in the Alismatidae.

(i) Is the pleiomerous, apocarpous, and polycarpous condition ancestral? The presumed primitive apocarpy and polycarpy is one of the main pieces of evidence sustaining the concept of the Alismatidae as a primitive group.

(ii) In plants with a simple reproductive organization, as in the Potamogetonales, to what extent is the flower morphologically equivalent to either a reduced branch axis (i.e. a pseudanthium), or a reduced floral axis?

Recent work on both topics has provided new evidence which relates to both these problems. This is briefly reviewed here.

Apocarpy and polycarpy

Apocarpy is generally recognized in Alismatidae with polycarpic gynoecia, although there is some tendency both towards basal fusion of otherwise separate carpels and towards epigyny (Kaul 1976a; Şerbănescu-Jitariu 1973a,b). Epigyny is, of course, most pronounced in and characteristic of the Hydrocharitaceae. The same method of initiation of carpels as separate horseshoe-shaped primordia which may become partly adnate below occurs in many Helobiae. In the Hydrocharitaceae, the carpels become enveloped by the surrounding hypanthium and Troll (1931b) considers this a direct modification of the apocarpous condition.

More controversial has been the interpretation of the flower in the Alismataceae as primitive, on the basis of spiral organization of the frequently numerous appendages, a condition considered to resemble most precisely the hypothetical ancestral shoot (e.g. Salisbury 1926).

A series of developmental studies (e.g. Leins and Stadler 1973) and in particular by Sattler and Singh using epi-illumination techniques on whole stained flower primordia, has revealed little or no evidence for spiral initiation of parts. There is, on the other hand, considerable evidence as summarized by Sattler and Singh (1978) for a basically trimerous condition, somewhat contradicting the observations of numerous early authors (e.g. Buchenau 1857; Eichler 1875; Ronte 1891; Salisbury 1926). The interpretation by Sattler and Singh is based on the recognition of three primary organizing centres (CA primordia) in the early stages of appendage initiation. In most taxa, each of these primary primordia gives rise to a petal and two adjacent stamens. CA primordium is used because each primordium initiates a part of the corolla (C) and androecium (A). In Limnocharis the comparable primary primordium initiates three stamens. In subsequent development there is the production of additional stamens, sometimes in a centrifugal direction, so that the original trimery is obscured in the mature flower. The observations are summarized diagrammatically in Fig. 3 (p. 50) (taken from Sattler and Singh 1978). There is some limited evidence for primary gynoecial primordia, each of which initiates one or more carpels in Alismataceae (e.g. *Alisma*) but this condition does not seem general.

In these developmental studies it is concluded that floral parts are intrinsically whorled in origin; no evidence for obvious spiral organization has been found. This developmental evidence can be used to suggest that polycarpy and the large number of floral parts in putatively 'primitive' Alismatidae is a recent condition and that these flowers actually represent a derivation from the trimerous organization basic to monocotyledonous flowers. Similar conclusions have been drawn for the pleiomerous state in certain palms (Uhl 1976).

Pseudanthia

That it is possible to interpret the 'flowers' of certain biologically more specialized Alismatidae (particularly the Potamogetonales) as condensed inflorescences comes from the distinctive organization of many of them, in which there are many structural peculiarities, which makes it difficult to recognize a radially-symmetrical flower. The following features may be noted:

(i) The perianth is never petaloid.
(ii) The perianth parts are few or absent.
(iii) The unusual position of perianth parts. Thus in *Potamogeton* the four tepals are opposite the stamens and are attached to the connective. Perianth parts stand opposite the stamens in other taxa (e.g. Juncaginaceae, Scheuchzeriaceae).
(iv) The variation in numbers of parts. For example, in *Lilaea*, flowers may include a single stamen, a single carpel, or a stamen and associated carpel. Variation in numbers of parts and the kind of envelope associated with them is particularly noticeable in Zannichelliaceae.
(v) Unusual associations of floral parts. This is, for example, pronounced in Zosteraceae, where the resemblance of a normal flower is particularly remote. The parts are borne on a flattened axis with the appendages all on one side. The carpels are in two rows; each carpel is associated with a pair of stamens and usually also a scale (retinacule). Different workers have recognized different associations of floral appendages in their search for interpretative homologies (e.g. Uhl 1947; Markgraf 1936), so that just about every possible combination of appendages has been equated with an archetypical flower.

Furthermore, the individual units (ultimate aggregation of several parts) may be bractless, as in Juncaginaceae, Lilaeaceae, some *Potamogeton* species. However, the bractless condition makes it difficult to interpret the reduced units as lateral axes, i.e. they are not subtended by a leaf. Also bracteate and bractless inflorescences occur in related families, e.g. Juncaginaceae and Scheuchzeriaceae.

No attempt is made here to review the extensive literature and diversity of interpretations, which are referred to under the descriptions of individual families. It should be noted that no one interpretative approach is wholly conclusive, whether this is based on evidence from comparative morphology, floral vasculature or developmental morphology. Summaries of contrasted viewpoints are presented in Posluszny and Sattler (1976a,b).

Developmental evidence is often very informative. The floral envelope in *Potamogeton* appears at maturity to be an outgrowth of the connective; if it is accepted as such, as by Markgraf (1936), interpretation of the flower becomes difficult. However, that the tepal in *Potamogeton* is equivalent to a tepal in a normal flower is made clear in its early development, it becomes displaced by growth of the common tissue supporting tepal and opposed stamens (Sattler 1973). In *Triglochin* the existence of two outer tepal whorls outside the stamens, as in a normal flower, has been revealed by Lieu (1979a). This simplifies the acceptance of this as a flower rather than a condensed branch system. Anatomical evidence, relied upon extensively by certain workers (e.g. Uhl 1947), has been less informative. It may depend on the level at which traces to individual appendages diverge from the common vascular supply of the flower and the relationships between traces to associated appendages. These relationships are primarily morphogenetic and phyletic use of them seems very subjective.

The failure of comparative study to reveal completely the true 'nature' of the reproductive parts in the specialized Alismatidae in terms of either a typological or an ancestral 'flower' may be frustrating to comparative morphologists and evolutionists, but is itself a further reflection of the diversity of the members of this group and their highly specialized reproductive biology.

BIOCHEMICAL DIVERSITY

The study of secondary metabolites is still in a developing phase, but a number of recent observations on the systematic significance of secondary constituents in Helobiae deserve mention. General summaries are provided by Hegnauer (1963), Gibbs (1974), and for aquatic angiosperms by McClure (1970).

Carbohydrates

The distribution of apiose, a pentose with a branched C chain, has been surveyed in the Helobiae by van Beusekom (1967). Its distribution shows a correlation with seagrasses, since it is universally present in all taxa examined (*Cymodocea, Phyllospadix, Posidonia, Thalassia,* and *Zostera*) and some submerged plants of brackish water, e.g. in *Potamogeton pectinatus, Ruppia spiralis,* but not in *Najas marina* and *Zannichellia palustris.* It is absent from freshwater Helobiae *(Aponogeton, Butomus, Sagittaria, Scheuchzeria)* as well as *Sparganium.* The distribution is thus in part ecological, in part taxonomic.

Sakai and Hayashi (1973) include a number of Helobiae in their survey of those monocotyledonous plants that accumulate either sugar or starch in their leaves (saccharophylly versus amylophylly). Within the order there is a fairly clear-cut distinction between starchy-leaved taxa in members of

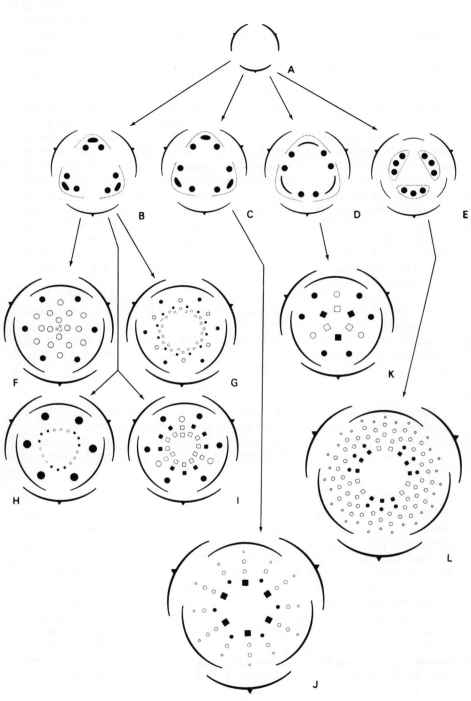

FIG. 3. Alismatales. Floral development. (After Sattler and Singh 1978.)

Diagrams of developmental stages (A–E) and mature flowers (F–L) of several taxa to illustrate basic trimerous organization of the flower, as revealed by developmental study (cf. Plates 7–9).

Sepals—curved line with a central triangle;
Petals—curved line, without a central triangle;
Stamens—large circles;
Staminodes—small circles (in J and L);
Carpels—large squares;
Carpellodes—small squares (in F);
Solid black—first formed stamens or staminodes and carpels;
Dotted lines—primary primordia.

A. Stage after sepal inception, common to all taxa.

B–E. Stage in different taxa after inception of petals and stamen (or staminode) pairs.

B. Common pattern for four different taxa. C. *Hydrocleys nymphoides*.

D. *Butomus umbellatus*. E. *Limnocharis flava*.

F–I. Mature flowers derived from the earlier stages represented by B–E.

F. Male flower of *Sagittaria cuneata*. G. Female flower of *Sagittaria cuneata* (only three of the numerous carpel whorls are drawn). H. *Alisma triviale*. I. *Echinodorus amazonicus*.

J–L. Mature flowers.

J. *Hydrocleys nymphoides*. K. *Butomus umbellatus*. L. *Limnocharis flava*.

In B and C obvious CA primordia are formed, the developmental relationships between petals and first-formed stamens subsequently obscured.

the Hydrocharitaceae (four spp in three genera) and Alismataceae (four spp in two genera) and non-starchy leaves in Potamogetonales (three spp of *Potamogeton* and two spp of *Najas*). *Ottelia japonica* is anomalous in the first group in having non-starchy leaves. It is suggested by these authors that this test, which is technically simple, may provide a useful chemotaxonomic tool.

Flavones and flavonols

In their survey of 23 species of Helobiae, Boutard *et al.* (1973) suggest that the order may be divided into two groups, the first represented by Alismataceae, Aponogetonaceae, Butomaceae, Hydrocharitaceae, and Juncaginaceae in which flavonols are common, but flavones uncommon. In the second group represented by the Potamogetonaceae (including Ruppiaceae), Zannichelliaceae, and Zosteraceae flavones are present (the most common being luteolin) while flavonols are absent.

These observations were confirmed and extended by Harborne and Williams (1976; see also Bate-Smith 1968) in a survey of 50 species in which the distribution of flavone sulphates was examined. Apart from the general division between the orders Alismatales, Hydrocharitales, Scheuchzeriales (and possibly Aponogetonales) on the one hand, and the Potamogetonales, on the other, a strong association between a halophytic habitat and the ability to develop sulphated and conjugated flavones was noted. The most clearcut observation was the complete absence of caffeyl and sulphate conjugates from *Potamogeton*.

Tannin and lignin

Variation in the amount and particularly distribution of polyphenolic substances (tannins) is a visual feature which has obvious taxonomic implications. In the very preliminary and subjective analysis which this present survey has permitted it is clear that there is a distinction between tannin-rich and tannin-poor families. Tannin-poor families are represented by the Potamogetonaceae, Najadaceae, and Zannichelliaceae whereas particularly tannin-rich families are the Cymodoceaceae, Posidoniaceae, and Zosteraceae. Some taxa within the Hydrocharitaceae are tannin-rich (e.g. *Thalassia*, *Enhalus*).

Of particular interest are those families in which tannin cells are structurally differentiated, as with the large tannin idioblasts in the transverse diaphragms of the air-lacunae in many Hydrocharitaceae. Cymodoceaceae are generally characterized by large epidermal tannin cells, which have a narrow surface exposure but expand into hypodermal tissues, notably in *Syringodium*. In *Posidonia* the transverse veins of the leaves are characteristically accompanied by tannin cells and Kuo (1978) has commented on the abundance of polyphenols in mature leaf epidermal cells. A quantitative, objective survey of the distribution of tannins in helobial families would be systematically rewarding, especially in conjugation with other biochemical analyses.

It should be noted that although all the Helobiae are capable of generating lignin (as in the Casparian thickenings of the endodermal cells), many representatives of the group otherwise have a very limited ability to do so in other cells. Tracheary elements may remain little lignified or completely unlignified. In Zosteraceae for example, the numerous leaf fibres remain unlignified.

Laticifers

Metcalfe (1967), in surveying the distribution of latex in the plant kingdom, records it as uncommon in monocotyledons (in five families). Laticifer-like elements (actually latex-secreting schizogenous lacunae) occur in the Helobiae only in Alismataceae and Limnocharitaceae, but not Butomaceae. Laticifer-like elements distinguish *Lilaea* and *Aponogeton*. However, there are few developmental observations on these elements in these plants and detailed investigations of their contents are necessary before their taxonomic usefulness can be assessed.

This brief overview indicates the extended scope for structural and chemical analysis of the Helobiae in relation to their secondary metabolic constituents.

Isozymes (a note added in proof)

The recent work of C. McMillan and his associates on isozymes in seagrasses especially *Phyllospadix,* appeared too late to be described in detail (see McMillan, C. and Phillips, R. C. 1981. *Canad. J. Bot.* **59:** 1494–1500). The results of examination of leaf extracts by standard gel electrophoresis showed that isozyme differences exist in populations referrable to different species of *Phyllospadix* on morphological grounds. Plants of *P. scouleri* and *P. torreyi* are more easily distinguished from each other in northern sites than in southern sites and some plants from the Monterey Peninsula showed intermediate characteristics which were interpreted by these authors as suggestive of hybridization between these two species (*cf.* identification of material in Fig. 13.5, p. 447).

This work is still at a preliminary stage but might well be applied to the numerous groups within the Helobiae where taxonomic uncertainty exists.

II. DETAILED DESCRIPTIONS OF FAMILIES

Family 1
ALISMATACEAE (Ventenat 1799)
(Figs. 1.1–1.8; Plates 7 and 8)

SUMMARY

THIS family consists of about ten broadly-defined (14 more narrowly-defined) genera and *c*. 80 species. The family is almost cosmopolitan, with a concentration of species in the northern hemisphere and especially North America, but its distribution is often somewhat localized. The plants occur in wet habitats, and they are phenotypically often very plastic; some are annual. The plants have sympodially-branched cormous (Fig. 1.1 A), stoloniferous, or rhizomatous (Fig. 1.3 A, B) rootstocks with distichous (Fig. 1.3 D, E), spirodistichous or spirally-arranged (Fig. 1.1 C) leaves, the spiral often initiated by the plumular apex. The axes sometimes also become stoloniferous by modification of the terminal inflorescences (pseudostolons of den Hartog 1957) or by extension of stoloniferous laterals (e.g. *Sagittaria* L. spp, Lieu 1979*b*). The plants may be wholly terrestrial but the rootstock is commonly submerged. The leaves are usually erect and emergent (Figs. 1.1 B and 1.4 A), but sometimes floating (e.g. *Luronium* Rafin.) or even submerged and ribbon-like. Fertile inflorescences are always emergent. The leaves are either linear, or more usually, petiolate with a broad lanceolate (Figs. 1.1 D and 1.3 F) or hastate (*Limnophyton* Miq., *Sagittaria*) blade, which is usually involute in bud. The form of the leaf within a single species is much influenced by variations in habitat. The terminal inflorescence arises by ± equal bifurcation of the vegetative meristem, often at regular intervals, and the continuing vegetative meristem can be interpreted as a precociously developed axillary meristem. The ground plan of the inflorescence is uniform (Charlton 1973). It is usually an erect panicle (Fig. 1.1 B), of which the ultimate units (e.g. Fig. 1.1 G) are typically helicoid cincinni (bostrycoid cymes). The inflorescence is sometimes spicate (Fig. 1.3 B). The inflorescences in smaller species are few-flowered by reduction with a tendency for vegetative and often stoloniferous proliferation leading to development of monopodial pseudostolons.

The flowers are usually bisexual (Fig. 1.2), or have become unisexual by abortion (Fig. 1.4). The plants are monoecious (e.g. *Sagittaria* spp), polygamous (e.g. *Limnophyton, Sagittaria* spp) or rarely dioecious (*Burnatia* M. Mich.). Wooten (1971) and Kaul (1979) note that the sex ratios are constant in monoecious and dioecious populations of *Sagittaria* spp. The basically trimerous flowers have a biseriate perianth differentiated into a distinct calyx and a deciduous or ephemeral, white, pink, or purple corolla, but the corolla is rarely reduced or absent *(Burnatia)*. There are three (e.g. *Wiesneria* M. Mich.), or six (e.g. *Alisma* L., *Baldellia* Parl., *Limnophyton, Luronium*) stamens but these may become more or very numerous, apparently by secondary proliferation (e.g. *Sagittaria*). The carpels are usually free but basally connate (e.g. in *Damasonium* Mill.). There are *c*. six carpels in *Luronium*, but they are usually numerous, and arranged in a single whorl (e.g. in *Alisma*) or very numerous (hundreds) and attached to an expanded *(Sagittaria)* or somewhat extended (*Echinodorus* L.C. Rich.) receptacle. Carpels are often compressed laterally. The style is short; rarely it is beaked (e.g. in *Damasonium* (including *Machaerocarpus* Small). The ovules are usually solitary and ad-axial, but there are two or more in *Damasonium* (e.g. *D. polyspermum*). The fruit is usually indehiscent (dehiscent at maturity in *Damasonium*), the individual carpels becoming achenes which are dispersed separately. The achenes are sometimes rigid or sculptured. The non-endospermous seeds are sometimes, and the embryo is always, curved. Germination is epigeal, and the radicle is suppressed (Kaul 1978).

Anatomically the family is characterized by abundant squamules (Fig. 1.3 G), usually by a lack of hairs, while paracytic stomata are common on both leaf surfaces. Arm-palisade cells are frequent in the mesophyll. Hydropoten are frequent on floating or submerged leaves (Mayr 1915). Mechanical tissues are little developed and the ground tissue is extensively aerenchymatous, the air spaces being traversed by well-developed transverse diaphragms which often consist of highly specialized cells. Even the roots are often made conspic-uously septate by diaphragms. Secretory canals, with latex-like contents (latici-fers) are common in the leaf and stem but tanniniferous deposits are rare. Crystals are frequent in the leaf mesophyll. The vascular bundle of the petiole and peduncle include a well-developed protoxylem lacuna, with a characteris-tic U-shaped arrangement of metaxylem in TS (Fig. 1.6 c). Vessels are re-stricted to the roots as single central strands.

FAMILY DESCRIPTION

VEGETATIVE MORPHOLOGY

Axis

Habit dependent on development of axes of four main types, but all based on a common sympodial architectural plan. The following illustrate examples: (i) Vegetative axis a short erect corm, with a rosette of emergent or submerged leaves (e.g. *Limnophyton*, *Alisma* (Fig. 1.1 A, B) larger spp of *Echinodorus*). (ii) Vegetative axis a horizontal rhizome (e.g. *Sagittaria lancifolia*, cf. Fig. 1.3 A, B and Lieu, 1979c). (iii) Axillary stolons. (iv) Vegetative axis elaborated by a stoloniferous extension of the terminal meristem, the axis (pseudostolon) a proliferated inflorescence indicated both by facultative development and intermediate states. The form of the plant largely determined by the environ-ment (e.g. *Baldellia*, *Damasonium*, *Luronium*, *Ranalisma*, and smaller spp of *Echinodorus*, cf. Charlton 1968; Charlton and Ahmed, 1973b).

Regenerative **branching** basically sympodial by ± equal division (bifurca-tion) of the vegetative meristem, one product (morphologically terminal) being the inflorescence meristem, the other product (morphologically axillary) the vegetative renewal shoot (Lieu, 1979b,c). Phyllotaxis of the parent meristem continued in the vegetative renewal shoot, its first leaf being a scale leaf or occasionally a foliage leaf (e.g. *Ranalisma humile*, *Echinodorus* subgenus *Echi-nodorus*). Vegetative meristems usually present in the axils of the remaining vegetative leaves but usually inhibited, the meristem in the penultimate leaf axil often proliferating the sympodium (e.g. Fig. 1.3 c–Eb); branches regularly developing as 'stolons' in *Sagittaria* spp according to Charlton (1973) and

Lieu (1979b). Pseudostolons, e.g. in *Echinodorus tenellus*, developmentally equivalent to inflorescences but extending horizontally, the flowers partly or wholly replaced by vegetative meristems (Charlton 1968; Charlton and Ahmed 1973b).

Leaf

Typically with a long petiole (Fig. 1.3 C, F), sheathing at the base. Blade broadly ovate-lanceolate (e.g. *Alisma*, Fig. 1.1 D), cordate (e.g. *Caldesia*), or hastate (e.g. *Limnophyton*, *Sagittaria sagittifolia*). Submerged blades commonly simpler in form, those of *Sagittaria sagittifolia* strap-shaped when growing in flowing water (Arber 1920). Apical pore absent or at most obscurely developed in older leaves (e.g. of *Echinodorus* spp) according to Stant (1964, cf. Meyer 1932a). Vernation of broader blades involute, the two halves of the blade rolled separately and adaxially. Venation including prominent major veins convergent at the leaf apex, with 1–2 orders of minor veins forming a closed reticulum (Fig. 1.1 D).

VEGETATIVE ANATOMY

Squamules

Present in leaf axils as a palisade-like series of scales (over 100 per node in larger spp, e.g. Fig. 1.3 G); scales densely cytoplasmic.

Leaf blade

Studied extensively by Meyer (1935a–e) with a detailed comparison of emergent and strap-shaped leaves (Meyer 1935d). Usually dorsiventral, but ± isolateral in submerged leaves, e.g. of *Sagittaria sagittifolia* (Fig. 1.5 E), *Wiesneria* (Fig. 1.8 D). **Hairs** absent from most taxa, but leaves conspicuously pubescent in *Limnophyton angolense* (Carter 1960) and *Sagittaria lancifolia* var. *pubescens*; hairs on abaxial epidermis of *Echinodorus macrophyllus* (Fig. 1.5 K, L) described by Stant (1964) as simple, unicellular or branched multicellular trichomes, and by Meyer (1932a) as *Büschelhaare* and also recorded for *E. grandiflorus* var. *floribundus*. Unicellular hairs recorded on the petiole of *Lophotocarpus guayanensis* by Meyer (1934). *Sagittaria japonica* with raised groups of cells in costal regions, resembling hair bases.

Hydropoten (Mayr 1915) observed in the epidermis of spp of *Alisma*, *Caldesia*, *Damasonium*, *Echinodorus*, *Lophotocarpus*, *Luronium*, *Sagittaria* as irregular aggregates of flattened epidermal cells with specialized contents.

Cuticle thin, often smooth but frequently recorded by Meyer as granular; appearing irregular in thin sections. **Epidermis** often similar on both surfaces (Fig. 1.5 A, B, I, J), either colourless, or chlorophyllous in submerged leaves, always large-celled and thin-walled, costal cells often elongated. Cells in surface view with straight (Fig. 1.6 I, J), curved or broadly sinuous (Fig. 1.5 A, B, I, J) anticlinal walls, the outline usually polygonal or isodiametric; spp with both sinuous and non-sinuous walls occurring in the same genus (e.g. *Alisma plantago-aquatica* cf. *A. parviflora*; *Sagittaria japonica* cf. *S. lancifolia*), or cells on the opposite surfaces of a single leaf contrasted in this respect, as in *Echinodorus* spp. Epidermal cells in submerged leaves rectangular in surface view, elongated parallel to long axis of leaf (e.g. *Baldellia, Wiesneria*; Fig. 1.8 A, B).

Stomata usually present on both surfaces, mostly abaxial in large emergent leaves (e.g. spp of *Echinodorus, Sagittaria*), but mostly adaxial in horizontal, and especially in floating leaves (e.g. *Ranalisma, Luronium*); often absent from submerged leaves (Fig. 1.8 A, B). Stomata typically paracytic, with a pair of narrow lateral subsidiary cells, but sometimes with additional indistinct subsidiary cells, as in *Baldellia* and then often essentially tetracytic. Stomata sunken in *Burnatia enneandra* (Meyer 1932b). Orientation of stomatal pores not v. constant, but frequently parallel to the long axis of the lamina or major (second-order) lateral veins of larger leaves. Guard cells in TS usually asymmetrical, the inner cutinized ledge narrower than the outer ledge, but symmetrical guard cells with equal ledges observed, e.g. in *Ranalisma* and *Sagittaria lancifolia*.

Mesophyll in emergent and floating leaves dorsiventral (Figs. 1.5 D and 1.6 F, H), with 1–3 adaxial palisade layers, the cells sometimes interconnected by lateral projecting lobes or appearing H- or U-shaped in TS (Armpalisadenzellen of Meyer, 1932a–1935e). In floating leaves palisade interrupted by large substomatal chambers (e.g. *Luronium*). Abaxial mesophyll cells isodiametric but lobed and enclosing well-developed intercellular spaces. Mesophyll in submerged leaves ± isolateral, aerenchymatous in its most specialized condition (Fig. 1.8 D, E) and resembling the ground tissue of midrib and petiole of emergent leaves. Air-canals directly contiguous with the epidermis, in the absence of a hypodermal layer, in *Wiesneria* (Fig. 1.8 D). **Veins** sometimes with a distinct parenchyma sheath, but the mechanical tissue in smaller leaves represented, at most, by thin-walled fibres (Fig. 1.6 G); the fibres thick-walled and more numerous in emergent leaves (Fig. 1.5 F). Veins usually independent of surface layers, but bundle sheath extensions continuous with epidermis in *Echinodorus macrophyllus*. Xylem represented by a wide protoxylem lacuna in major veins, the phloem including several wide sieve-tubes; metaxylem tracheids often about 4–5 mm long.

Laticifers common, often associated with the veins or situated in hypodermal layers; absent from *Luronium*. **Tannin** uncommon. **Crystals** not uncommon.

Midrib (and main longitudinal veins) prominent below in larger leaves (Fig. 1.5 C), angular or rounded in TS with collenchymatous ground tissue cells at angles. Ground tissue aerenchymatous, like that of the petiole. Vascular bundles consisting of a main arc, accompanied in larger leaves by minor abaxial vbs and an adaxial arc of inversely orientated vbs.

Petiole

Rounded, angular (Fig. 1.5 G) or channelled in TS, with little mechanical tissue, the ground tissue largely occupied by air-lacunae (Fig. 1.6 A). **Hydropoten** common in floating or submerged leaves. **Cuticle** thin, smooth, or minutely papillose, as in *Sagittaria japonica*. **Epidermis** thin-walled. **Stomata** usually present at the distal ends of emergent leaves. Hypodermal layers compact, chlorenchymatous distally and somewhat resembling the palisade of the lamina (Fig. 1.6 E); often collenchymatous next to the peripheral vbs in petioles of larger-leaved spp (e.g. *Alisma, Echinodorus, Sagittaria*; Fig. 1.5 G, H). **Air-lacunae** delimited by longitudinal uniseriate plates of parenchyma and traversed by uniseriate diaphragms of lobed or stellate cells enclosing narrow or pore-like intercellular spaces. Cells of the most specialized diaphragms e.g. in *Sagittaria* spp (Fig. 1.6 B), and *Echinodorus macrophyllus* (Meyer 1932a) thick-walled and with pronounced horizontal peg-like extensions coincident with those of adjacent cells. Points of contact between the ends of the pegs thin-walled and traversed by cytoplasmic extensions. Diaphragms including frequent laticifers (Fig. 1.6 E) and transverse vascular commissures.

Vascular system (Fig. 1.6 A) including a prominent arc of major vbs suspended within the central aerenchyma, together with a series of alternately large and small, minor, peripheral vbs, those of the adaxial series inversely orientated. Larger petioles, as in *Sagittaria* spp, sometimes with additional abaxial arcs below the main arc. Vascular system reduced in the petiole of smaller leaves (e.g. *Baldellia, Damasonium, Ranalisma*), the major arc including only 3–5 vbs, the peripheral system sometimes even absent (e.g. in *Ranalisma*, Fig. 1.6 K; *Wiesneria* Fig. 1.8 E, F). Larger vbs (e.g. Fig. 1.6 C) either without or at most with an indistinctly differentiated parenchyma sheath, fibrous sheathing tissue developed to a limited extent in larger petioles. Protoxylem lacuna well developed, up to 200 μm wide in larger leaves (e.g. *Limnophyton*), partly enclosed on the lower side by an arc of metaxylem tracheids, wider tracheids (e.g. in *Limnophyton*) resembling lacunae in TS because of their thin secondary walls. Phloem appearing in TS as a shallow U-shaped band. Sieve-tubes and companion cells conspicuous, especially in spp with quite thick-walled conjunctive tissue (e.g. *Echinodorus*). Peripheral vbs (Fig. 1.6 D) often including well-developed fibres next to the xylem and more especially to the phloem; vascular tissue reduced, but a protoxylem lacuna usually present, together with one or more wider metaxylem elements in large vbs. **Laticifers** common in all taxa except *Luronium*, associated with the larger central vbs, as well as with the peripheral chlorenchyma, and also in the ground tissue within junctions of parenchymatous plates at corners of air-lacunae.

Leaf sheath

Similar anatomically to distal part of leaf axis but with well-developed sheathing wings; colourless towards leaf insertion.

AXIS

Peduncle

Either rounded (Fig. 1.7 G) or angular, often triangular (Fig. 1.7 E, F) in TS, the surface layers resembling those of the petiolar ground tissue. **Ground tissue** either with a central cavity (*Alisma*, Fig. 1.7 F, older peduncle in *Damasonium*) or aerenchymatous, resembling that of the petiole, in most taxa (Fig. 1.7 E, G). **Vascular system** sometimes consisting of bundles scattered throughout the ground tissue, with the smallest at the periphery, and the largest at the centre (e.g. *Limnophyton, Sagittaria*; Fig. 1.7 E) there being no distinction between the cortex and central cylinder. Alternatively, in *Alisma* (Fig. 1.7 F), *Baldellia, Damasonium* (Fig. 1.7 G), with a single cylinder of large and small vbs contiguous to a narrow mechanical cylinder of thin-walled fibres 1–3 cells wide, forming a boundary between the cortex and central cylinder, but without an endodermis. Peduncle in *Wiesneria* including a single central vb and a few (four) peripheral vbs (Fig. 1.8 C).

Individual vbs indistinctly sheathed by compact parenchyma cells; a mechanical sheath either being absent or represented by a few fibres next to phloem of the outer vbs (e.g. *Limnophyton*). Protoxylem lacuna well-developed, up to 200 μm wide in *Limnophyton*. Lacuna usually partly enclosed by a band of metaxylem tracheids, U-shaped in TS. Tracheids in larger peduncles wide, with indistinct wall thickenings and in TS resembling further xylem lacunae. Laticifers with a similar distribution to that in the petiole.

Rhizome

Elongated, as in *Sagittaria* spp, or a short erect rootstock. Rhizome in *Burnatia enneandra* made up of a series of spherical segments (apparently annual) separated by narrow constrictions (Charlton 1976), the surface clothed with the fibrous remains of cortical bundles and leaf traces, and protected by a periderm-like layer developed from a multiple endodermis. **Epidermis** in older axes of other taxa often replaced by a suberised periderm up to six cells wide, the cells sometimes radially seriated. **Ground tissue** usually uniform, starch-containing, but cortical tissue lacunose in *Limnophyton, Ranalisma, Wiesneria*, with transverse diaphragms or even made up almost entirely of lobed, diaphragm-type cells (*Limnophyton*).

Endodermis indistinctly differentiated as a suberised layer, either with Casparian strips (e.g. *Ranalisma*), or often without them (e.g. *Limnophyton*). Endodermis in *Burnatia* described by Charlton (1976) as multiple, including enlarged idioblasts; interrupted by specialized air-space tissue at departure of each leaf trace. **Vascular system** consisting of an irregular plexus on the inner side of the endodermis accompanied by irregularly scattered central vbs, the vascular tissue consisting of irregular, often indistinctly amphivasal strands. Xylem including short, irregular, spirally sculptured tracheids, with a single or double helix.

ROOT

Commonly branched (Laessle 1953), rarely with distal tubers, as in populations of *Echinodorus radicans*. **Epidermis** persistent, root-hairs not recorded. **Exodermis** 1–few cells wide with slightly thickened and usually suberised walls (Fig. 1.7 B). **Cortex** with only the innermost and outermost layers compact, the broad middle cortex lacunose, usually with radial air-lacunae separated by uniseriate plates of long cells. Air-canals usually transversely septate (septa absent from *Alisma, Echinodorus cordifolius*), the septa being separated by 1–3 series of axially elongated cells and developed as uniseriate diaphragms of regularly arranged cells. The most specialized diaphragms, e.g. those of *Limnophyton, Sagittaria* (Fig. 1.7 C) represented by somewhat thick-walled flat cells with regular tangential lobes and conspicuously pitted on the tangential walls. Inner cortical cells regularly arranged, sometimes thickened (e.g. *Sagittaria*). **Endodermis** in *Sagittaria* (Fig. 1.7 D) usually remaining thin-walled although having conspicuous Casparian strips (*Sagittaria*) but endodermis becoming somewhat thick-walled in other taxa. **Stele** narrow (Figs. 1.7 A and 1.8 G). **Pericycle** always thin-walled, not very distinct and never including conducting elements. Conducting tissues including a central wide metaxylem vessel (Fig. 1.7 D) surrounded by 4–12 peripheral protoxylem poles, sometimes with narrow radially-extended metaxylem, alternating with phloem strands each with a single conspicuous sieve-tube and associated file of companion cells. Number of xylem poles determined largely by root diameter (e.g. four in narrow roots of *Luronium*). Peripheral xylem elements obscure in the narrow stele of *Wiesneria* (Fig. 1.8 G), apparently replaced by thick-walled cells next to central vessel. **Laticifers** usually absent, but observed in the exodermis of *Limnophyton* and *Wiesneria*.

SECRETORY, STORAGE, AND CONDUCTING ELEMENTS

Secretory canals ('laticifers'). Considered to be a diagnostic feature of the family by most authors (e.g. Buchenau 1903; Meyer 1932–1935*f*; Stant 1964), present in aerial parts of all taxa examined, except *Luronium natans* according to Stant and *Burnatia* according to Meyer (1932*c*); least common in roots and possibly infrequent in the rhizome. Laticifers forming a reticulum and each provided with a central cavity, perhaps arising schizogenously or alternatively produced by the breakdown of the end walls of a central cell, as suggested by Stant (1964). Cavity of the laticifers becoming lined by an epithelial layer of 4–8 series of narrow cells, these cells the probable source of the milky latex released from the laticifers when cut. *Puncta pellucida* and *lineae pellucidae*, common in *Echinodorus* and used as a diagnostic character for certain species (Buchenau 1903*a*; Fassett 1955), appearing to represent regions where secretory canals make contact with the epidermis. Their structure and distribution discussed further by Meyer (1932*a*).

[In the absence of a development study, as Metcalfe (1967) points out, there is continuing uncertainty about whether these elements should be described as laticifers in the true sense. It should be noted that Leblois (1887) described the similar elements in Limnocharitaceae (q.v.) as having a schizogenous origin.]

Crystals. Recorded for the lamina of *Limnophyton*, petiole and lamina of *Sagittaria* spp, petiole of *Baldellia, Damasonium* and *Wiesneria* and root of *Echinodorus macrophyllus* by Stant (1964) but widely distributed in the leaf according to Meyer (1932*a*, 1935*f*). Crystals mainly rhombohedral or rod-shaped styloids, occasionally aggregated and forming druse-like structures in submerged leaves of *Sagittaria sagittifolia*.

Tannin. Not or little developed, but recorded, e.g. for the leaves of *Luronium* and *Wiesneria*, as well as the root of *Echinodorus*.

Starch. Common, especially in the rhizome, but also the petiole, commonly in association with larger vbs. Starch grains usually somewhat eccentric and sometimes dimorphic but according to Stant (1964) the range of variation too great to be of systematic value within the family.

Xylem. Vessels restricted to roots in all species examined, e.g. in *Sagittaria lancifolia* mean length of vessel elements 1386 μm (range 654–2109 μm), mean width 113 μm (86–132 μm) but usually narrower. Perforation plates simple to scalariform, usually on transverse or somewhat oblique end-walls. Lateral pitting alternate to opposite, or scalariform. Tracheids in other parts of the plant usually with spiral or annular thickenings. Protoxylem typically disrupted and replaced by a wider protoxylem lacuna.

Phloem. Sieve-tubes with simple sieve-plates present in all parts.

REPRODUCTIVE MORPHOLOGY

Inflorescence

Probably always terminal but evicted precociously (Lieu, 1979*b,c*). Basic ground plan uniform, but two types recognized by Charlton (1973); one restricted to *Ranalisma*, involving immediate sympodial development below a terminal flower; but axis otherwise monopodial, with sympodial flower complexes as the ultimate units. Two contrasted types of orientation occur:

(i) **Inflorescence erect,** not usually proliferating by vegetative meristems (e.g. *Alisma*; Fig. 1.1 B), with a long basal internode and one or more successive pseudowhorls, each of three, often fused, bracts (Fig. 1.1 E, F) and separated by further long internodes. Pseudowhorls subtending either further branches, to produce a paniculate inflorescence, or solitary flowers to produce a spicate inflorescence (e.g. in some *Sagittaria* spp, Fig. 1.4 A, B). Flowers normally prophyllate, the prophylls often either subtending further branches (e.g. Fig. 1.1 F, H), or usually flowers and so initiating helicoid cincinni of up to eight flowers (Fig. 1.1 G). Flower prophylls absent from *Burnatia*, *Sagittaria*. Branching within the inflorescence and especially the cincinni by precocious bifurcation, resembling that initiating the vegetative meristem.

(ii) **Inflorescence horizontal,** ± stoloniferous, with one or more vegetative meristems in the axils of bracts or prophylls or terminating cincinni, the vegetative meristems developing rooted leafy rosettes thereby proliferating the plant vegetatively. Inflorescence axis in extreme submerged forms becoming a pseudostolon, the flowers being replaced by vegetative buds, with one member of each whorl of bracts (scale leaves) subtending a vegetative meristem, e.g. in winter form of *Luronium natans*, in *Echinodorus magdalensis*, *E. tenellus* (Charlton 1968, 1973).

[There is a continuum between these contrasted inflorescence types represented by *Echinodorus* spp, said to have 'dimorphic' inflorescences, according to whether the inflorescence is erect or decumbent.]

Inflorescence of *Ranalisma* (Charlton and Ahmed 1973*a*) sympodial, with a succession of solitary terminal flowers each producing a pair of subopposite scales (bracts), the lower of them subtending the renewal shoot. Vegetative buds developing in the axils of prophylls on each renewal shoot; flower suppressed in submerged forms.

[Charlton and Ahmed (1973*a, b*) emphasize the uniqueness of the *Rana-*

lisma inflorescence within the family, but it is related to the usual plan in having the terminal flowering units (flowers terminating inflorescences in many Alismataceae) and in the number of bracts in a pseudowhorl reduced from three to two.

For further details of inflorescence construction in the family and especially a discussion of symmetry, see Charlton (1973).]

Flower (Figs. 1.2 and 1.4)

Bisexual, or unisexual by abortion, the non-functional organs usually represented by staminodes or carpellodes. Basic organization (e.g. Fig. 1.2 c) trimerous, as shown in developmental studies by Sattler and Singh (1978), Singh and Sattler (1972, 1973, 1977) with two alternate trimerous **perianth** whorls differentiated into sepals and deciduous petals (Figs. 1.2 A, B and 1.4 C, D, F). [The whorls may possibly be pseudowhorls, since an order of succession in each whorl may be detected.] Petals rarely reduced or absent (*Burnatia*). **Stamens** (Figs. 1.2 A, B and 1.4 D, E) with well-developed filaments, tetrasporangiate, extrorse, opening by longitudinal slits; filaments hairy in *Sagittaria lancifolia* (Fig. 1.4 E). [The stamens are typically initiated in antepetalous pairs on the same three primordia which also produce the petals thereby forming a petal–stamen complex (CA primordia of Sattler and Singh 1978). There is no developmental evidence to support the earlier contention of antesepalous stamen pairs (e.g. Salisbury 1926). If further stamens are subsequently developed, they are interpolated between the three original pairs. There may be three, six, or numerous stamens (Fig. 1.4 C, D, I) according to the taxon, but they are always centripetal in order of development (Fig. 3, p. 50).] **Carpels** continuing the trimerous arrangement e.g. in *Alisma* as groups of three on three 'primary gynoecial primordia' (Singh and Sattler 1972). Additional carpels interpolated to form a single whorl at maturity, or still more carpels interpolated progressively, making them numerous (e.g. in *Echinodorus,* Sattler and Singh 1978) or v. numerous (e.g. in the female flower of *Sagittaria*, Fig. 1.4 F, H; Singh and Sattler 1973, 1977). Carpels (Figs. 1.2 A, B and 1.4 F, G) either free from each other or slightly united at the base when mature, these variations depending on the size and shape of the floral receptacle as indicated in the detailed discussion by Kaul (1976a). Individual carpels exhibiting varying degrees of closure in different taxa. Style short, often lateral; ovule usually solitary (except in *Damasonium* and sometimes *Burnatia*) and basal.

[Sattler and Singh (1978) conclude that there is no evidence for the spiral generation of flower parts in the Alismatales.]

REPRODUCTIVE ANATOMY

Flower

According to Singh (1966a), the floral vasculature shows a number of common features. The outer perianth segments are supplied from a single trace (*Alisma*) to as many as 12 traces (*Sagittaria*), their number often being distinctive and somewhat diagnostic for the genera. The number of traces to the inner perianth segments is fewer (three in *Baldellia* and *Luronium*, one in *Alisma, Limnophyton,* and *Sagittaria*). The number of traces diverging from the receptacular vascular bundle is increased by branching in the receptacular cortex or in the base of the appendage so that the perianth segment always includes numerous vbs, e.g. in *Alisma* the single trace ultimately gives rise to a larger number. *Alisma* is distinctive because the marginal vbs of the inner perianth segments arise as branches derived ultimately from the traces to the outer perianth segments.

Each stamen receives a single, independently derived vb, described by Singh as concentric. There is thus no evidence of a common initial vb supplying several stamens which would support the interpretation of Eichler (1875) and others who claim that the stamens are secondarily proliferated ('dedoublement').

Despite the variation in gynoecial morphology (in numbers and degree of basal fusion), each carpel receives a single vb which bifurcates to give a dorsal trace, extending into the style, and a ventral trace which supplies the ovule directly. The dorsal trace normally remains unbranched, but in *Alisma* it bifurcates shortly above its origin, the two branches proceeding independently to the style on each side of a median secretory canal. Kaul (1976*a*) shows a wider range of carpel vasculature than this simple account indicates, the most elaborate expression being shown by *Damasonium polyspermum,* with several ovules.

Singh makes no mention of secretory canals in *Luronium* (cf. vegetative parts), but otherwise in the taxa he studied they form an irregular to regular system, e.g. in *Alisma,* commonly alternating with the vbs in the perianth segments. Secretory canals are recorded in the carpel walls in *Alisma, Baldellia, Limnophyton,* and *Sagittaria* and by Charlton and Ahmed (1973*a*) in *Ranalisma.*

Tracheary elements within the floral vbs are few and commonly represented by a xylem lacuna. In the taxa with unisexual flowers Singh reports similar vasculature in the two sexes, with no vestige of vascular traces to missing organs. The vascular system of the receptacle is completely continuous into the ultimate appendages in all types of flower.

In the larger flowers of *Lophotocarpus calycinus* and *Sagittaria latifolia,* studied by Kaul (1967*a*), the same principles of vasculature can be recognized even though the appendages are either bigger (perianth segments) or more numerous (hundreds of carpels) with consequent greater numerical complexity of the receptacular system. In contrast, the flower of *Alisma triviale,* studied developmentally by Singh and Sattler (1972), is so small that they were able to determine the sequence of initiation of strands to different kinds of appendage and the way in which they become interconnected. In their study of development in *Sagittaria cuneata,* Singh and Sattler (1977) provide no comparable information about vascular development.

Taxonomic and Morphological Notes

Interfamilial relationships

A close relationship between the Alismataceae, Limnocharitaceae, and Butomaceae is suggested by evidence from vegetative morphology, vegetative anatomy, floral morphology, and development, and supports the segregation of these families as an order Alismales (Takhtajan 1966). There is a strong architectural resemblance, notably in their branching by bifurcation of a vegetative meristem, a process which may be interpreted as modified sympodial growth, with precocious development of a vegetative lateral meristem which evicts the morphologically terminal inflorescence meristem (Wilder 1975). This in turn reflects a feature common to other helobian families (Wilder 1975), but here it is associated with very similar organizational plans so that part-by-part morphological comparisons of plant form between members of the different families is easily made. The basic pattern of inflorescence construction in the Alismataceae (Charlton 1973) can be connected to that in the Limnocharitaceae, notably in the strong tendency for pseudostolon development (Wilder 1975). *Ranalisma* in particular is closely comparable with *Hydrocleys* (Charlton and Ahmed 1973b). However, its status as a member of the Alismataceae is not in dispute, on the basis of vegetative anatomy (Stant 1964) and floral morphology (Charlton and Ahmed 1973b).

Anatomically and morphologically the Alismataceae resemble the Limnocharitaceae most strongly and the Butomaceae least closely (see Limnocharitaceae, p. 96). However, the anatomical and morphological continuum among the families is virtually complete, they are most sharply segregated by reproductive features.

Intrafamilial relationships

Stant (1964) made a quantitative estimate of relationships among the species of Alismataceae which she examined, taking anatomical characters at random and assuming their equal weight as measures of relationships. This provided numerical assessment of similarities (Sneath and Sokal 1962) and resulted in a ranking according to similarity coefficients. However, the result largely separated taxa according to the degree of elaboration of their leaves, i.e. taxa with more elaborate, emergent leaves being separated from taxa with simple or submerged leaves. The similarity coefficient therefore reflects the ecological status of a taxon as much as its systematic position. This suggests that the anatomical features of the family are strongly correlated with the environment, a fact tacitly assumed for all hydrophytes. The addition of reproductive features, not so strongly correlated with the plant's environment would strengthen the numerical approach and make it less subjective.

Systematic leaf anatomy

The most extensive studies on vegetative anatomy of Alismataceae in relation to systematics are those by F. J. Meyer presented in a series of papers (1932a–1935f) continued by F. Mayr (1943). The studies were restricted to the leaf, but included some comparisons of different forms of leaves in a given taxon (e.g. Meyer 1935a). Among the characters considered to have diagnostic value are the sculpturing of the cuticle (whether granular or smooth), the shape of the epidermal cells in surface view, the distribution and sometimes the shape of the hydropoten, the presence or absence of hairs, the structure and degree of differentiation of the mesophyll (although much influenced by the leaf environment cf. emergent, floating and submersed

leaves), the number and arrangement of the vascular bundles, the distribution
and degree of branching of the laticifers, the frequency and kinds of crystals.
It seems, however, that only small samples, or even only a single leaf was
studied and the method of presentation makes it difficult to draw reliable
conclusions. In general, however, it seems that taxa can be distinguished
mainly at the specific rather than at the generic level and the range of variation
is considerable. No future work on the anatomy of this family can afford
to neglect these detailed observations.

In the detailed study by Björkqvist (1967, 1968) of the leaf of *Alisma*,
which included an examination of all the nine recognized species, together
with transplant experiments to study environmental modification, it was concluded that anatomical characteristics are of restricted taxonomic importance.

Phylogeny of the Alismataceae

The family occupies an important place in the interpretation of the evolutionary hierarchy of the flowering plants. This is because the family has
been thought to represent (i) a primitive group of monocotyledons because
of its apocarpy and presumed spiral arrangement of floral parts, (ii) a group
derived from the Ranalean dicotyledons, either through the Nymphaeales
(Takhtajan 1966) or in particular from the family Ranunculaceae (Hutchinson
1959). Hutchinson saw *Ranalisma* as a genus sharing features with both
the Ranunculaceae and Alismataceae.

Maheshwari (1962) pointed out the numerous differences between Alismataceae and Ranunculaceae. Recent extensive morphological and anatomical
study renders these classical interpretations still more untenable.

(i) Vegetative organization in the Alismataceae is precise and complex
(e.g. Charlton 1968, 1973; Charlton and Ahmed 1973*b*; Wilder 1975). *Ranalisma* is typical of the Alismatidae in this respect, showing a link between
this family and *Hydrocleys* (Charlton and Ahmed 1973*b*). No similar organizational pattern is known for the Ranunculaceae.

(ii) Analysis of floral development in which early and ephemeral stages
of appendage initiation have been scrutinized carefully (see summary in Sattler
and Singh 1978) show that the flowers in all species examined are basically
trimerous, with no evidence for spiral initiation of anthers or carpels. The
presence of three 'CA primordia' (i.e. primary primordia which each serve
for the initiation of a petal and opposed pair of stamens) is a strongly developed
characteristic and a basic feature in the trimerous organization (Fig. 3. p.
50). Floral development lends support to the idea that elaboration in numbers
of stamens and carpels is a secondary development and not a primary or
primitive feature in the family.

(iii) In their vegetative anatomy the two families are dissimilar (Metcalfe
1963; Meyer 1932*a*; Stant 1964), even when taking only the hydrophytic
Ranunculaceae for comparison; the surveys are by no means exhaustive but
adequate for the conclusions presented (e.g. Table 1.1).

The widely scattered small genera of Alismataceae could indicate a family
of relicts, but with speciation still continuing, notably in *Echinodorus* and
Sagittaria. Certainly there are highly specialized taxa, thus the distinctive

TABLE 1.1

Major anatomical differences in vegetative anatomy between Ranunculaceae and Alismataceae

Character	Ranunculaceae	Alismataceae
Squamules	absent	present
Hairs	frequent	rare
Stomata	anomocytic	paracytic
Aerenchyma	not elaborate	highly elaborate
'Laticifers'	absent	present
Protoxylem lacunae	not common	always developed
Vessels	in all parts	in roots only
	(highly specialized)	(v. unspecialized)

rhizome morphology and anatomy of *Burnatia* are interpreted as adaptations to alternate wet and dry seasons (Charlton 1976).

Stant concludes that the two families resemble each other only in root structure, and then in features which can be interpreted as secondary adaptations to an aquatic environment. Meyer (1932a) notes that both families possess hydathodes (apical pores) a feature shared by many aquatic families, but Stant (1964) does not confirm this observation. *Ranalisma* is also shown to be anatomically typical of the Alismataceae. It is therefore very unlikely that the existing Ranunculaceae could have given rise to the Alismataceae, especially as the tracheary elements are more highly specialized in the former than the latter. If there has to be a common ancestor it is too remote to be of significance in understanding angiosperm diversification.

The only reasonable conclusion is that the Alismataceae are a derived and highly specialized group which provide no useful evidence for the phylogeny of the monocotyledons, or for the evolutionary relationships between dicotyledons and monocotyledons.

MATERIAL EXAMINED†

The anatomical information is based entirely on the account by Stant (1964) in which the following material was examined:

Alisma parviflora Rafin.; stem, leaf.
Alisma plantago-aquatica L.; all parts.
Baldellia ranunculoides Parl. Wicken Fen; C. R. Metcalfe II.IX.52; leaf, stem.
Damasonium stellatum Thuill.; leaf, stem.

* Material marked with an asterisk was from specimens cultivated (in part) at Kew, for sources of other specimens see Stant (1964).
† Material of '*Sagittaria variabilis* Engelm.' cultivated at Kew, described by Stant (1964) has subsequently been shown to be that of an aroid, probably *Peltandra virginica* L., and has not been considered in this account. This re-identification accounts for the features anomalous to the Alismataceae reported by Stant for this sp.

Echinodorus cordifolius Griseb. South of Rushmore, Virginia; J. J. Baldwin s.n.; all parts.
*Echinodorus macrophyllus Micheli; leaf, stem.
Limnophyton obtusifolium Miq. Agra, India; B. L. Gupta 24.XI.57; Bharatput; B. M. Johri XII.56; all parts.
*Luronium natans Rafin.; all parts.
Ranalisma humile (Kuntze) Hutch. Nigeria; Milne-Redhead s.n.; leaf, stem.
*Sagittaria sp cultivar 'Japonica'; all parts.
*Sagittaria lancifolia L.; leaf, stem.
*Sagittaria sagittifolia L.; all parts.
Wiesneria schweinfurthii Hook. Kambia, Sierra Leone (no collector); all parts.

Species Reported on in the Literature

Arber (1918*b*): *Sagittaria montevidensis, S. sagittifolia*: leaf.
Arber (1922*b*): *Sagittaria sagittifolia*: leaf.
Björkqvist (1967): *Alisma*: generalized account of leaf anatomy based on study of nine spp.
Bloedel and Hirsch (1979): *Sagittaria sagittifolia*: leaf development.
Buchenau (1857): *Alisma plantago, Sagittaria sagittifolia*: floral development.
Charlton (1973): comparative survey of inflorescence morphology.
Charlton (1976): *Burnatia enneandra*: rhizome.
Charlton and Ahmed (1973*a, b*): *Ranalisma humile*: floral anatomy, vegetative morphology.
Costantin (1885*c*): *Sagittaria* sp: leaf morphology.
Costantin (1886*b*): *Alisma natans, A. plantago, Sagittaria sagittifolia*: leaf.
Dibbern (1903): *Sagittaria sagittifolia*: inflorescence axis.
Duval-Jouve (1873*b*): *Alisma plantago, Sagittaria lancifolia, S. sagittifolia*: diaphragms.
François (1908): *Alisma plantago, Sagittaria sagittifolia*: seedling morphology and anatomy.
Gérard (1881): *Damasonium stellatum*: seedling.
Govindarajalu (1967): *Sagittaria guayanensis* ssp *lappula*: vegetative anatomy.
Guttenberg and Jakuszeit (1957): *Alisma plantago*: root development.
Kaul (1967*a*): *Lophotocarpus calycinus, Sagittaria latifolia*: floral anatomy and development.
Kaul (1976*a*): comparative survey of carpel morphology.
Le Blanc (1912): *Sagittaria sagittifolia*: aerenchyma.
Lieu (1979*b*): *Alisma gramineum, A. triviale, Sagittaria cuneata, S. latifolia, S. 'microphylla,' S. 'sinensis,' S. 'subulata'*: shoot morphology.
Lieu (1979*c*): *Sagittaria lancifolia*: shoot morphology.
Lee and Hsin-Ying (1958): *Sagittaria sinensis*: root.
Leins and Stadler (1973): *Echinodorus macrophyllus, E. intermedius, Alisma plantago-aquatica, Sagittaria lancifolia*: androecial development.
Mayr (1915): *Caldesia parnassifolia, Damasonium alisma, Echinodorus humilis, E. radicans, E. subulatus, E.* sp, *Elisma natans, Lophotocarpus guyanensis* var. *typicus, L. seubertianus, Sagittaria chilensis, S. graminea* var. *chapmani, S. pugioniformis, S. sagittifolia*: hydropoten.

Mayr (1943): *Caldesia parnassifolia* f. *natans* & f. *terrestris*: leaf.
Meyer (1932*a*): *Echinodorus grandiflorus* var. *floribundus, E. humilis, E. macrophyllus, E. ranunculoides, E. tenellus*: leaf.
Meyer (1932*b*): *Burnatia enneandra* (as *Rautanenia schinzii*): leaf.
Meyer (1932*c*): *Burnatia enneandra*: leaf.
Meyer (1934): *Limnophyton obtusifolium, Lophotocarpus* (= *Sagittaria*) *guayanensis* var. *lappula*, var. *madagascariensis*, var. *typicus, L. seubertianus*: leaf.
Meyer (1935*a*): *Alisma plantago* var. *michaletii* f. *latifolium* & f. *stenophyllum, Elisma natans, Sagittaria chilensis, S. eatonii, S. engelmanniana, S. graminea, S. lancifolia* var. *major, S. montevidensis, S. platyphylla, S. sagittifolia, S. subulata*: leaf.
Meyer (1935*b*): *Alisma plantago* var. *michaletii* f. *latifolium, Sagittaria graminea* var. *chapmani*: hydropoten.
Meyer (1935*c*): *Alisma plantago-aquatica* var. *arcuatum* & f. *latifolium*, f. *natans*, f. *stenophylla, Damasonium alisma* var. *compactum* & f. *natans*, f. *spathulatum*, f. *terrestre, D. minus*: leaf.
Meyer (1934*d*): *Wiesneria schweinfurthii, W. triandra*: leaf.
Meyer (1935*e*): *Alisma plantago-aquatica* var. *arcuatum*, f. *angustissimum, Damasonium alisma* f. *graminifolium, Elisma* (= *Luronium*) *natans, Sagittaria eatonii, S. graminea* var. *chapmani, S. natans, S. platyphylla, S. subulata*: leaf.
Paliwal and Lavania (1978): *Sagittaria guayanensis*: epidermis and stomata.
Raunkiaer (1895–9): *Alisma plantago, Luronium natans* (as *Elisma natans*), *Sagittaria sagittifolia*: leaf
Salisbury (1926): *Alisma plantago, Echinodorus ranunculoides, Sagittaria sagittifolia*: floral morphology.
Sattler and Singh (1978): *Echinodorus amazonicus*: floral development.
Schenck (1886*b*): *Alisma natans, A. plantago*: leaf.
Severin (1932): *Sagittaria latifolia*: root.
Singh (1966*a*): *Alisma lanceolatum, A. plantago-aquatica, A. reniforme, A. subcordatum, A. triviale, Baldellia ranunculoides, Limnophyton obtusifolium, Luronium natans, Sagittaria graminea, S. guayanensis, S. lancifolia, S. latifolia, S. sagittifolia*: floral anatomy.
Singh and Sattler (1972): *Alisma triviale*: floral development.
Singh and Sattler (1973): *Sagittaria latifolia*: floral development.
Singh and Sattler (1977*a*): *Sagittaria cuneata*: inflorescence and floral development.
Stant (1964): see Material Examined, p. 69.
Tarnavschi and Nedelcu (1973): *Sagittaria sagittifolia*: leaf.
van Tieghem and Douliot (1888): *Alisma natans, Damasonium stellatum, Sagittaria lancifolia, S. sagittifolia*: lateral root initiation.

SIGNIFICANT LITERATURE—ALISMATACEAE

Arber (1918*b*, 1919, 1920, 1921, 1922*b*, 1925*a, b*); Björkqvist (1967, 1968); Bloedel and Hirsch (1979); Buchenau (1857, 1882, 1889, 1903*a*); Carter (1960); Charlton (1968, 1973, 1974, 1976, 1979*a, b*); Charlton and Ahmed (1973*a, b*); Costantin (1885*c*, 1886*b*); Dibbern (1903); Duval-Jouve (1873*b*); Eber (1934); Eichler (1875); Fassett (1955); François (1908); Gérard (1881); Gibson (1905); Glück (1905); Goebel (1896); Govindarajalu (1967); von Guttenberg and Jakuszeit (1957); den Hartog (1957); Hutchinson (1959); Jadin (1888); Kaul (1967*a*, 1976*a*, 1978, 1979); Klinge (1880); Kristen (1969); Kroemer (1903); Laessle (1953); Le Blanc (1912); Lee and Hsin-Ying (1958); Leins and Stadler (1973); Lieu (1979*b, c*); Lyr and Streitberg (1955); Maheshwari (1962); Mayr (1915, 1943); Messeri (1925); Metcalfe (1963); Meyer (1932*a, b, c, d,* 1934, 1935*a, b, c, d, e, f*); Paliwal and Lavania (1978); Raunkiaer (1895–9); Salisbury (1926); Sattler and Singh (1978); Schenck (1886*b*); Severin (1932); Singh (1966*a*); Singh and Sattler (1972, 1973, 1977a); Stant (1964); Tarnavschi and Nedelcu (1973); van Tieghem and Douliot (1888); Wilder (1975); Wooten (1971).

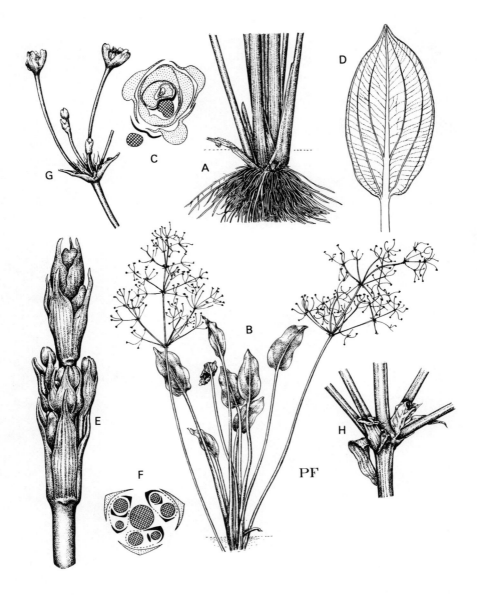

FIG. 1.1. ALISMATACEAE. *Alisma triviale* Pursh. Habit. (Lake Champlain New York. P. B. Tomlinson *s.n.*)

A. Rootstock ($\times \frac{1}{2}$); dotted line indicates level of section in C.

B. Above-ground parts, leafy rosette with two inflorescences ($\times \frac{1}{6}$).

C. Plan of leaf and peduncle arrangement at level of dotted line in A; leaves–stippled; prophylls—solid black; peduncles—hatched.

D. Leaf blade and major veins ($\times \frac{1}{2}$).

E. Young inflorescence ($\times 3$) showing tiers of first-order bracts each with an axillant branch complex.

F. Diagram of branch arrangement at node in E to illustrate initiation of complexes, same shading as in C; prophylls may appear bilobed.

G. Distal part of inflorescence ($\times \frac{3}{2}$), with terminal flower and reduced branch complex of ultimate whorl of bracts.

H. Older branch complex ($\times 3$) with bracts and base of branch complex.

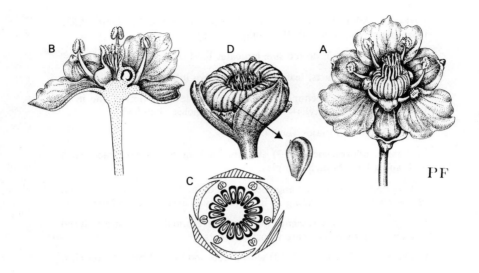

F IG . 1.2. A LISMATACEAE. *Alisma triviale* Pursh. Floral morphology.

A. Flower from above ($\times$ 4).

B. Flower in LS ($\times$ 4).

C. Floral diagram.

D. Fruiting head ($\times$ 4) with single achene inset.

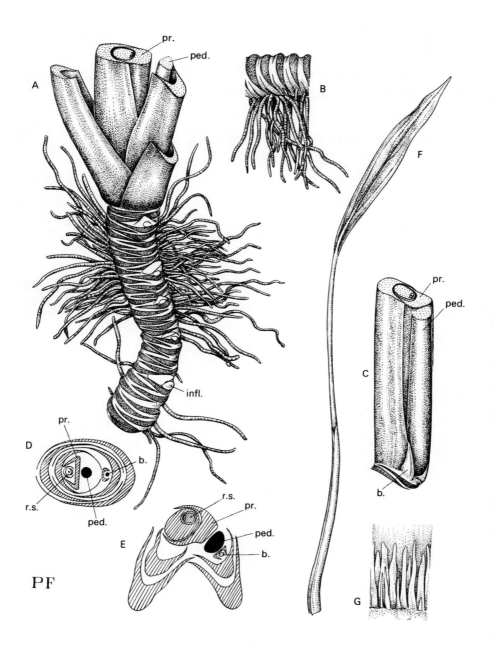

PF

FIG. 1.3. ALISMATACEAE. *Sagittaria lancifolia* L. (Fairchild Tropical Garden, Miami, Florida P. B. Tomlinson *s.n.*) Habit.

A. Rhizome from above (× ½), with erect leaves cut off to show scars of previous terminal inflorescences (infl.).

B. Rhizome from side (× ½), with leaf scars showing regular distichy; roots restricted to lower surface.

C. Part of leaf rosette (× ½), older leaves removed, distal part of remaining leaves cut off to show prophyll and undeveloped bud in axil of penultimate leaf.

D. Diagram of leaf arrangement at level of branching.

E. Actual leaf arrangement as observed at base of leafy rosette in specimen C.

F. Entire mature foliage leaf (× ⅛).

G. Palisade of squamules at node (× 3), subtending leaf removed.

Lettering in A, C–E = b.—vegetative bud; r.s.—renewal shoot, ped.—peduncle; pr.—prophyll.

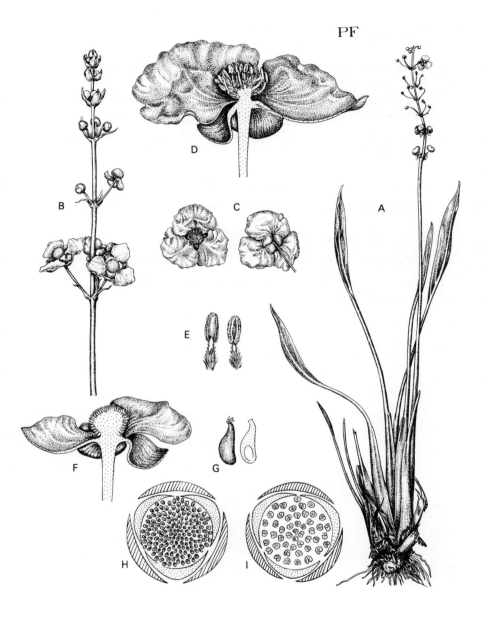

PF

A
B
C
D
E
F
G
H
I

80

FIG. 1.4. ALISMATACEAE. *Sagittaria lancifolia* L. Habit and flower morphology.

A. Habit ($\times$ ½).

B. Distal part of inflorescence ($\times$ ¼), with lower flowers (female) open.

C. Male flower ($\times$ ½), from above (left) and below (right).

D. Male flower ($\times$ ³⁄₂) in LS.

E. Stamens ($\times$ 3) from front and back.

F. Female flower ($\times$ ³⁄₂) in LS.

G. Carpels ($\times$ 12), entire and in LS.

H. Floral diagram female flower.

I. Floral diagram male flower.

Key to lettering used in Figs. 1.5–1.8.

ch.	chlorenchyma.
co.	cortex.
col.	collenchyma.
d.c.	diaphragm cell.
e.c.	epidermal cell.
end.	endodermis.
ex.	exodermis.
f.	fibre.
g.t.	parenchymatous ground tissue.
h.	hair.
h.p.	hollow pith.
i.a.s.	intercellular air-space.
l.	lacuna.
la.	laticifer (= secretory canal).
mx.	metaxylem.
n.	nucleus.
pal.	palisade.
p.b.s.	parenchyma bundle sheath.
ph.	phloem.
p.l.	piliferous layer.
px.	protoxylem.
px.c.	protoxylem canal.
s.	stoma
s.t.	sieve-tube.
v.	vessel.
v.b.	vascular bundle.

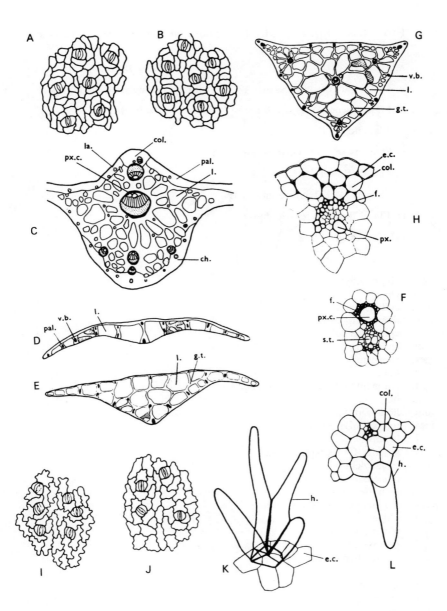

FIG. 1.5. ALISMATACEAE. *Sagittaria* (A–H) and *Echinodorus* (I–L), leaf anatomy. (After Stant 1964.)

A–B. *S. lancifolia,* epidermis in surface view (× 80).
  A. Abaxial. B. Adaxial.

C. *S. japonica,* midrib in TS (× 25).

D–H. *S. sagittifolia.*
  D. TS floating leaf (× 10). E. TS submerged leaf (× 10). F. TS lateral vb floating leaf (× 170). G. TS petiole (× 7). H. TS vb from periphery of petiole (× 160).

I–L. *Echinodorus macrophyllus.* I, J. Epidermis in surface view (× 80).
  I. Abaxial. J. Adaxial. K. Hair from abaxial epidermis over main vein (× 160). L. Abaxial hair from midrib in TS (× 160).

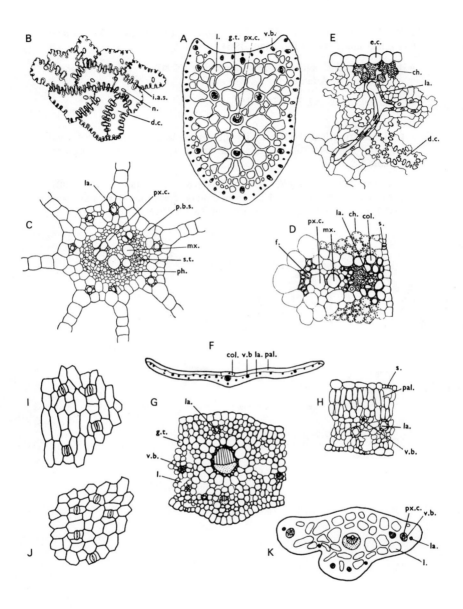

FIG. 1.6. ALISMATACEAE. *Sagittaria* (A–E) petiole anatomy; *Ranalisma humile* (F–K), leaf anatomy. (After Stant 1964.)

A. *S. lancifolia*, TS upper petiole (× 11).

B. *S. lancifolia*, detail of diaphragm cells in surface view (× 160).

C. *S. lancifolia*, TS central vb (× 80).

D. *S. lancifolia*, TS peripheral vb and chlorenchyma (× 160).

E. *S. sagittifolia*, TS peripheral ground tissue of petiole, with superficial chlorenchyma and branching secretory canal (× 115).

F. TS lamina (× 11) to show distribution of palisade tissue (pal.), secretory canal (la.), vbs (v.b.) and collenchyma (col.).

G. TS lamina midrib (× 80).

H. TS lamina towards margin (× 80).

I. Abaxial epidermis, surface view (× 80).

J. Adaxial epidermis, surface view (× 80).

K. TS petiole (× 40).

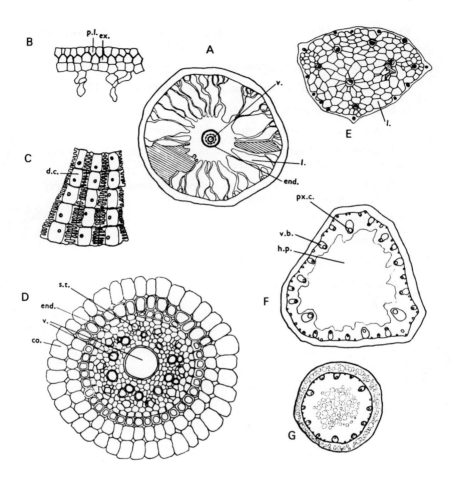

FIG. 1.7. ALISMATACEAE. Root and peduncle anatomy. (After Stant 1964.)

A–D. *Sagittaria lancifolia*, TS root.

A. TS root (× 25), transverse diaphragms hatched. B. TS outer layers (× 80). C. TS cortex (× 80) diaphragm cells in surface view.

D. TS stele and inner cortex (× 160).

E–G. TS peduncle.

E. *Sagittaria sagittifolia* (× 6), diaphragms hatched. F. *Alisma plantago-aquatica* (× 10). G. *Damasonium stellatum* (× 10).

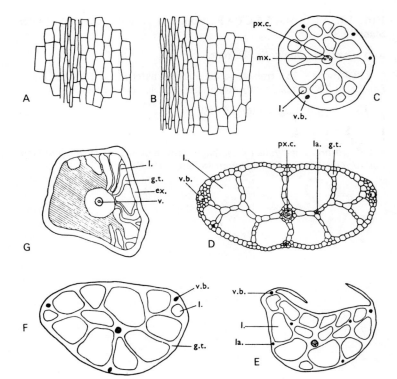

FIG. 1.8. ALISMATACEAE. *Wiesneria schweinfurthii,* anatomy. (After Stant 1964.)

A. Abaxial epidermis, surface view (× 70).

B. Adaxial epidermis, surface view (× 70).

C. TS peduncle (× 20).

D. TS distal part of leaf (× 40).

E. TS leaf base (× 20), with sheathing margins.

F. TS leaf at middle level (× 10).

G. TS root (× 40) diaphragm hatched.

Family 2
LIMNOCHARITACEAE (Takhtajan 1954)
(Figs. 2.1–2.4)

SUMMARY

THIS family comprises 3 (–5) genera and about 12 species widely cultivated and naturalized in the tropics and subtropics. It consists of emergent (*Limnocharis* Humb. and Bonpl.—New World (Fig. 2.1 A), *Tenagocharis* Hochst. (*Butomopsis* Kunth)—Old World) or facultatively free-floating (*Hydrocleys* L. C. Rich., including *Ostenia* Buchenau—New World) aquatic herbs with petiolate and ± broadly-bladed leaves with an apical pore (Fig. 2.2 B). The status of *Elattosis* Gagnep. (Tonkin) is not fully understood.* The vegetative axis is sympodial (e.g. Fig. 2.1 H), without axillary buds except for those producing distal renewal shoots, each unit of the sympodium ending in a complex cymose inflorescence (e.g. Fig. 2.1 C, H). This retains a further capacity for vegetative spread, e.g. by stoloniferous extension in *Hydrocleys* (Fig. 2.1 I).

The flowers are basically trimerous, with a biseriate perianth clearly differentiated into calyx and petaloid corolla (Fig. 2.1 D, F). The stamens are three, six–numerous, and commonly develop centrifugally, the outer (last-formed) stamens sometimes being represented by staminodes. The carpels are three, four, five, six, or more (up to 20 in *Limnocharis*), ± free but closely aggregated with a ± sessile capitate stigma. The numerous ovules are anatropous with laminar placentation (Fig. 2.1 F, i.e. scattered on the inner surface of the carpel). The pollen grains have four—numerous pores (Argue 1973). The follicular fruits open by ventral splits to scatter the numerous ribbed or warted seeds. Endosperm is absent, and the embryo curved; germination is epigeal, and the radicle developed (Kaul 1978).

Anatomically the family is distinguished (Stant 1967) by well-developed secretory canals ('laticifers') (cf. Butomaceae), aerenchyma and squamules. The stomata are paracytic. Hairs are absent. The root cortex is differentiated into long- and short-cells. The vascular tissue is reduced but there are vessels with simple to scalariform perforation plates in the roots, and the protoxylem is usually represented by a well-developed lacuna. Hydropoten are recorded for *Hydrocleys*.

FAMILY DESCRIPTION

VEGETATIVE MORPHOLOGY

Habit

Only known in detail for *Hydrocleys nymphoides* (Charlton and Ahmed 1973*b*) and *Limnocharis flava* (Wilder 1974*a*), as follows:

1. *Limnocharis flava* (Fig. 2.1 A–C) Vegetative axis erect with congested internodes, sympodial with one–several (usually two or three) foliage leaves

---

* *Elattosis* is possibly a depauperate form of *Tenagocharis latifolia* (S. Hooper, pers. comm.)

(not subtending axillary buds) on each unit; branching by ± equal bifurcation of the vegetative apex to produce (i) the vegetative renewal shoot, (ii) an inflorescence. [This branching is interpreted by Wilder (1974a) as arising via a precocious axillary bud because of the association of the vegetative meristem with the ultimate foliage leaf, a position comparable to the more obvious axillary bud of the renewal shoot in *Hydrocleys*.]

Inflorescence meristem (without a basal prophyll) essentially dividing as a cincinnus via similar precocious bifurcations but with the ultimate division resulting in a vegetative meristem. Further (supernumerary) vegetative meristems sometimes produced in the axils of the lower bracts. Change of symmetry not regular. Vegetative proliferation initiated by the erect inflorescence falling into a horizontal position, the vegetative meristems then developing into new rooted sympodia.

2. *Hydrocleys nymphoides* (Fig. 2.1 G–I). Plants rooted at the nodes in wet soil but distally extended and free-floating, the leaves either erect or lying on the surface of the water. Axis creeping, often submerged, sympodial, each unit consisting of an aggregate of foliage leaves and ending in a terminal stoloniferous inflorescence capable of vegetative proliferation (Fig. 2.1 H, I). Each sympodial unit arising in the axil of the ultimate foliage leaf of the previous unit (axillary buds absent from other nodes), the renewal shoot including a foliage leaf as a prophyll and continuing the vegetative growth of the axis. Inflorescence horizontal and extended by its first internode before developing a cymose complex in association with the terminal flowers, further complexes separated by extended internodes being formed subsequently. Each cymose cluster including a vegetative meristem in the axil of the most distal bract, this proliferative shoot having either a scale-leaf or a foliage leaf as the prophyll. [The system is elaborated beyond a simple cincinnus because there are two bracts (one empty) on the peduncle of the first flower, as well as the bract (subtending the proliferative meristem) at the base of the extended internode (Fig. 2.1 G,H). Successive complexes in the resulting pseudomonopodial system show an alternation of symmetry.]

Leaf

Phyllotaxis spirodistichous. Leaf differentiated into an open sheathing base, a long petiole and ovate or cordate blade (Fig. 2.1 A); ligule absent. **Vernation** involute, the two halves of the blade rolled separately above the well-developed midrib. **Petiole** triquetrous (e.g. *Limnocharis*; Fig. 2.3 A) or round (e.g. *Hydrocleys*; Fig. 2.3 B) with conspicuous transverse diaphragms. **Venation** (Stant 1967) convergent, reticulated, the major veins connected by a second and third order of lateral veins, but without free vein-endings (Fig. 2.2 A), the major veins united apically below the pore (Fig. 2.2 B).

VEGETATIVE ANATOMY

Leaf

Leaf blade. Hairs absent. **Hydropoten** frequent on abaxial epidermis of *Hydrocleys* (Mayr 1915), consisting of 1–3 cells with dark contents encircled by cells with non-sinuous walls. **Epidermis** thin-walled, the outer surface somewhat thickened and cutinized, without well-differentiated costal regions, sometimes including chloroplasts. Cells polygonal, ± isodiametric in surface view, in *Hydrocleys* with deeply sinuous anticlinal walls especially on adaxial surface (Fig. 2.2 c). **Stomata** paracytic, occurring on both surfaces, but most frequent on adaxial surface in *Hydrocleys*; absent from indistinctly differentiated costal regions. Stomatal pore orientated at an acute angle to the long axis of the leaf, ± parallel to the first-order lateral veins; each stoma with a pair of narrow lateral subsidiary cells, guard cells slightly sunken in *Limnocharis*, asymmetric in TS, the outer ledge most prominent, the inner ledge narrow or absent (as from *Hydrocleys* Fig. 2.2 D). **Mesophyll** markedly dorsiventral (Fig. 2.2 E), with 1 (–2) adaxial palisade layers, the palisade cells sometimes with peg-like outwardly directed extensions; spongy mesophyll, with fewer chloroplasts, v. lacunose and forming an irregular uniseriate network of parenchyma enclosing larger or smaller air-cavities, the cavities traversed by mostly uniseriate transverse diaphragms of lobed or stellate cells enclosing regular and quite large intercellular spaces (Fig. 2.2 G), diaphragms sometimes including transverse veins or secretory canals. Abaxial hypodermis compact and forming a ± continuous layer in thicker parts of lamina. **Midrib** (Fig. 2.2 F) prominent below, with well-developed aerenchyma, including an adaxial series of inverted vbs; abaxial series of vbs present in midrib and main veins. Large vbs of midrib and main veins each with a well-developed **protoxylem lacuna** and a crescentic arc of metaxylem tracheids conforming in TS outline to the broad arc of phloem (Fig. 2.2 H), the metaxylem less well developed in *Hydrocleys* than in *Limnocharis*. Bundle sheath represented by parenchyma together with thin-walled fibres above xylem and especially below phloem (Fig. 2.2 H). Smaller bundles and smaller veins of blade with few or no sheathing fibres and reduced conducting tissue, the xylem represented by narrow tracheids with scalariform or spiral wall thickenings. **Secretory canals** abundant, forming a continuous network, commonly associated with vbs or towards either surface in mesophyll. **Crystals** and **tannin** absent.

Apical pore (Sauvageau 1893; Stant 1967) formed abaxially by dissolution of mesophyll cells, the cuticle and epidermis persisting above an extensive series of tracheary elements, representing the confluent, but partly disorganized, ends of the major veins.

Petiole (Fig. 2.3). Angular (Fig. 2.3 A) or circular (Fig. 2.3 B) in TS. **Stomata** present on all surfaces in *Limnocharis* but absent from *Hydrocleys*. Peripheral 1–4 layers chlorenchymatous in *Limnocharis* but colourless in *Hydrocleys* (Fig. 2.3 C), the cells somewhat collenchymatous at the angles in *Limnocharis*. Central **ground tissue** consisting of a network of uniseriate parenchymatous or chlorenchymatous plates enclosing large air lacunae (Fig. 2.3 D). Lacunae traversed by frequent diaphragms of ± stellate cells, resembling those of the lamina but wider. **Vascular system** including a main V-shaped arc of large central vbs (*c.* seven in *Limnocharis*; Fig. 2.3 A; three in *Hydrocleys*; Fig. 2.3 B) together with numerous vbs embedded in the peripheral ground tissue. In *Limnocharis*, in addition, an abaxial and an adaxial vb, two or three small bundles in the lateral angles and numerous rudimentary vbs scattered in the ground tissue. Larger vbs with a conspicuous protoxylem lacuna and an arc of metaxylem tracheids, the latter sometimes absent from *Hydrocleys* (Fig. 2.3 E). Smaller vbs with reduced vascular tissue, but usually still retaining a protoxylem lacuna (Fig. 2.3 C). Bundle sheath fibres scarcely developed or absent. **Secretory canals** abundant, associated with central vbs and also present in the peripheral ground tissue (Fig. 2.3 B, C, E).

Leaf sheath. Similar to petiole but with lateral sheathing wings.

AXIS

Inflorescence axis (Fig. 2.4 A–D). The following description refers to the elongated first internode of the inflorescence in *Limnocharis* or to the stoloniferous equivalent of the inflorescence complex in *Hydrocleys*.

1. *Limnocharis* (Fig. 2.4 A). Angular (triangular) in TS. **Epidermis** and **stomata** as in petiole. Outer layers of ground tissue chlorenchymatous (collenchymatous in the angles), the bulk of the ground tissue consisting of a network of uniseriate parenchyma plates enclosing air-cavities but connected to a solid vasculated core of rounded parenchyma cells. Uniseriate diaphragms traversing the lacunae at intervals, frequently including **secretory canals** or transverse vascular commissures. **Vascular system** consisting of six principal vbs in the central parenchymatous core, five smaller vbs towards the periphery and numerous small vbs in the cortical chlorenchyma, with scattered small vbs elsewhere.

2. *Hydrocleys* (Fig. 2.4 B–D). (Stant 1967, distinguishes between the 'peduncle' and 'stolon,' which are the same morphologically, and comments on their anatomical similarity.) Circular in TS. Peripheral zone of 5–6 layers of compact parenchyma or collenchyma (Fig. 2.4 C) transitional to lacunose aerenchyma of central ground tissue with its numerous transverse diaphragms, the intercellular spaces between the diaphragm cells somewhat wider than in *Limnocharis*. **Vascular system** including a series of *c.* six central vbs and numerous small peripheral vbs embedded in the surface layers (Fig. 2.4 B).

Vascular bundles in both genera without a fibrous sheath but usually including a wide protoxylem lacuna (e.g. Fig. 2.4 D), or protoxylem tracheids in smaller vbs. Metaxylem tracheids developed extensively only in *Limnocharis*; phloem ± crescentic. **Secretory canals** abundant, those in *Hydrocleys* mainly associated with vbs; in *Limnocharis* consisting of a medullary circle of 7–10 canals associated with the central vbs, with others scattered in the cortical and peripheral ground tissue.

Rhizome (in *Limnocharis*). **Epidermis** persistent but supplemented by several suberised superficial cell layers. **Cortex** wide, lacunose, consisting of a parenchymatous reticulum without well differentiated diaphragms, the cells often somewhat stellate. Independent cortical vascular system differentiated. **Endodermis** with Casparian strips demarcating stele clearly. **Stele** including a well-developed outer vascular system, immediately within the endodermis, of anastomosing strands continuous with the root traces. Central xylem strands irregular, anastomosing, the metaxylem tracheids numerous, short and irregular with spiral-scalariform wall thickenings. Tannin and crystals absent.

ROOT (Fig. 2.4 E–G)

Branched. **Epidermis** thin-walled and partly persistent, at least in *Hydrocleys*, with regularly developed trichoblasts sometimes becoming root-hairs. **Exodermis** 2–3-layered, walls slightly thickened. **Cortex** fairly wide with a tendency to develop a lacunose region by collapse of radial files of cells in middle cortex (Fig. 2.4 F). Cortical cells differentiated into long-cells and short-cells, the latter tending to form indistinct horizontal plates (Fig. 2.4 G). **Endodermis** thin-walled, but with conspicuous Casparian strips, the cells coincident radially with those of the innermost cortical layer; pericycle little differentiated, uniseriate (Fig. 2.4 E). **Stele** very regular with a wide central metaxylem vessel and up to ten (*Limnocharis*) or four (*Hydrocleys*) protoxylem tracheids alternating with phloem strands next to the pericycle, each strand consisting of a single sieve-tube and associated companion cells (Fig. 2.4 E). **Secretory canals** absent.

SECRETORY, STORAGE, AND CONDUCTING ELEMENTS

Secretory canals ('laticifers'): abundant in leaf and stem as elongated cavities surrounded by a sheath; seen in TS as 4–6 slightly thick-walled cells, the canals continuous as a reticulum. Canals arising by division of a mother cell, first into four, and subsequently into more, daughter cells the latter separating to form an enlarged intercellular cavity (Leblois 1887).

Crystals: not observed.

Tannin: scarcely developed.

Starch: common in the rhizome of *Limnocharis* as small centric rounded grains, somewhat larger grains present in ground tissue of the petiole.

Tracheary elements: vessels present only in the root, elements either with transverse to slightly oblique perforation plates with few scalariform thickening bars, or perforations simple. Vessel elements of the order of 800 μm long and 35–90 μm wide. Tracheids in other parts of the plant thin-walled, with spiral or scalariformly-reticulate wall thickenings.

REPRODUCTIVE MORPHOLOGY

Individual flowers in *Hydrocleys* and *Limnocharis* originating by a $\pm$ equal division of the reproductive meristem producing a precocious bud in the axil of the bract (prophyll) of each cincinnus unit.

Flowers in *Hydrocleys nymphoides* borne in a cymose unit (cincinni, or 'bostrycoid complexes' of Charlton and Ahmed 1973*b*) between each pair of extended internodes (Fig. 2.1 H), with the spiral reversed in each successive complex. Entire axis ending in a vegetative meristem after the formation of several units. *Limnocharis flava* producing a single cincinnus at the end of the extended peduncle (Fig. 2.1 C); first two bracts subopposite, often bilobed and conspicuous; from 2–20 flowers produced, the axis then terminating in a vegetative meristem (Wilder 1974a; cf. also Micheli 1881; Ronte 1891; Wagner 1918). Symmetry relations discussed in detail by Wilder (1974a). Supernumerary vegetative buds sometimes present in the axils of the first two bracts.

Development of the flowers of *Limnocharis* and *Hydrocleys* studied by Ronte (1891), Saunders (1929), Kaul (1967b, 1968a), but most precisely by Sattler and Singh (1973, 1977), thereby demonstrating a trimerous plan of construction with centrifugal initiation of the stamens (see Fig. 3, p. 50).

[In *Hydrocleys* sepals and petals originate in two trimerous whorls but the members of each whorl arise in sequence, the petals completing their differentiation relatively late. The first whorl of six stamens arises as three pairs opposite each petal primordium in the form of an incipient CA primordium (p. 47) on the triangular floral apex, the three CA primordia arising in rapid succession but in the reverse order to that of the sepal primordia. Subsequent development of the androecium is centrifugal, the second three pairs of stamens are alternate with and situated outside the first. This is repeated in subsequent whorls, the outermost appendages differentiating as flattened, awl-shaped staminodes. Centrifugal development is promoted by the rim of tissue developed in association with the three CA primordia. The central six (sometimes five) carpels are formed on the flanks of the floral apex on the inner side of the first whorl of stamen primordia, and they form a single whorl, or possibly two incipient whorls, as described by Buchenau (1903b) and Eber (1934). Each carpel is initially horseshoe-shaped and remains open at maturity, but with the margins appressed. It eventually differentiates a short style and a stigmatic crest. Ovules arise on the radial walls of each carpel, each ovule becoming anatropous.

Sattler and Singh (1977), in extending the observations of Ronte (1891) and Kaul (1967b) on the large flowers of *Limnocharis flava*, found continuing evidence of basic trimery, but no CA primordia (Fig. 3, p. 50). The androecium originates from three antesepalous 'primary androecial primordia' each of them being opposite a sepal and giving rise to three stamens. The primary primordia fuse to form an 'androecial ring' (=Ringwall of Leins and Stadler 1973) on which additional stamen primordia are formed in the antesepalous region. The androecial ring becomes extended basally, producing in centrifugal order further whorls of stamens and finally staminodes. The trimerous arrangement of the gynoecium is represented by the first carpel primordia, which originate in three antesepalous groups of three, with additional carpels (up to a total of 20) interpolated later between the groups. The gynoecium matures as a ring of carpels with little trace of the original trimerous condition. The individual carpels develop as in *Hydrocleys*.]

REPRODUCTIVE ANATOMY

Studies by Kaul (1967b, 1968a) on the floral anatomy of *Hydrocleys* and *Limnocharis* show their structural similarity. There are six central pedicel vbs which form the main vascular supply. Three of these supply the sepals via a ring of vascular tissue at the base of the receptacle and the remaining three supply the petals, stamens, and carpels via another ring of vascular tissue at a somewhat higher level. Kaul emphasizes that the androecial vascular system is branched and that it arises from the upper ring. Individual branches supply each stamen and staminode, this dendritic vasculature being unknown in other monocotyledons except the Alismataceae. The placental supply to each carpel is derived from a basal branch, or series of branches, of the single ventral carpel bundle; there is a single (*Hydrocleys*) or are three (*Limnocharis*) dorsal bundles. Procambium development is said to be acropetal throughout (cf. Sattler and Singh 1973, 1977).

TAXONOMIC NOTES

In the restricted sense, the Limnocharitaceae constitute a natural assemblage, closely allied to, but clearly demarcated from the Alismataceae, on the one hand, and the Butomaceae (*Butomus*) on the other. Information about *Tenagocharis* is limited, but Sauvageau's (1893) brief notes indicate it is similar to *Hydrocleys* and *Limnocharis*. Differences between *Butomus* and Limnocharitaceae (s.s.) are listed in Table 2.1. Pichon (1946), in recognizing the discreteness of *Butomus* and creating the narrowly defined family Butomaceae to contain it, still retained all other taxa in the Alismataceae, but with two subfamilies Limnocharitées and Alismatées. It seems most logical to adopt the narrower family definitions of Takhtajan (1966).

Stant (1967) suggests that the methods of numerical taxonomy cannot be applied satisfactorily within a small group. Using characters of vegetative

TABLE 2.1

Comparison of Butomaceae (Butomus) and Limnocharitaceae (see Eckardt 1964)

Character	Butomaceae (e.g. Pichon 1946)	Limnocharitaceae (e.g. Takhtajan 1966)
Leaves	linear, bladeless	bladed, petiolate
Apical pore	absent	present
Ground tissue of leaf	with spirally-thickened cells	without spirally-thickened cells
Secretory canals	absent	present
Root cortex	without short cells	with short cells
Inflorescence	never stoloniferous, rarely proliferative	± stoloniferous, proliferative
Perianth	perianth uniform	calyx and corolla differentiated
Stamens	always 9	6–many
Stamen development	centripetal	often centrifugal
Staminodes	absent	sometimes present
Pollen	monosulcate	porate
Chromosomes	$n = 13$, small	$n = 7, 8, 10$, large
Embryo sac	*Polygonum*-type	*Allium*-type
Embryo	straight	curved

anatomy the range of correlation percentages is not large and the Butomaceae, in the wide sense, resemble Alismataceae as much as the individual genera resemble each other. This undoubtedly reflects the number of common features determined as a response to the common habitat of these plants. These include aerenchymatous construction, structure and distribution of vascular bundles and distribution of tracheary elements. On the other hand, distinctive features such as sympodial construction and method of branching, similar cymose inflorescence construction and the basic floral plan suggest that the three families, Alismataceae, Butomaceae and Limnocharitaceae, have a monophyletic ancestry. The similar placentation of Butomaceae and Limnocharitaceae suggests a close affinity, but the pollen is dissimilar (Argue 1973; Erdtman 1952).

MATERIAL EXAMINED

Hydrocleys nymphoides Buchenau, Hort. Tropical Dept. R.B.G., Kew; all vegetative parts.

Limnocharis flava Buchenau, Hort. Tropical Dept. R.B.G., Kew; all vegetative parts.

SPECIES REPORTED ON IN THE LITERATURE

Carbonell and Arrillaga (1959): leaf.

Charlton and Ahmed (1973*b*): *Hydrocleys nymphoides:* developmental morphology.

Costantin (1885*a*, 1886*b*): *Limnocharis flava* (as *L. humboldtii*): leaf.

Ernst (1872): *Hydrocleis* nymphoides:* leaf.

Kaul (1967*b*): *Limnocharis flava:* floral morphology and anatomy.

Kaul (1968*a*): *Hydrocleis nymphoides:* floral morphology and anatomy.

Kaul (1976*b*): *Hydrocleis nymphoides:* leaf.

Kaul (1978): *Limnocharis flava:* seedling morphology.

Leblois (1887): *Limnocharis flava* (as *Hydrocleys humboldtii*): development of secretory canals.

Leins and Stadler (1973): *Hydrocleis nymphoides:* androecial development.

Mayr (1915): *Hydrocleis commersoni:* hydropoten.

Ronte (1891): *Hydrocleis nymphoides, Limnocharis flava:* floral morphology.

Sattler and Singh (1973): *Hydrocleis nymphoides:* floral ontogeny.

Sattler and Singh (1977): *Limnocharis flava:* floral ontogeny.

Sauvageau (1893): *Hydrocleis nymphoides, H. martii, H. parviflora, Limnocharis flava, Tenagocharis latifolia:* leaf.

Şerbănescu-Jitariu (1973*a*): *Limnocharis flava:* gynoecium, seedling morphology.

Stant (1967): *Hydrocleys nymphoides, Limnocharis flava:* all vegetative parts.

van Tieghem (1870–1): *Limnocharis flava* (as *Hydrocleis humboldtii*): root.

van Tieghem and Douliot (1888): *Hydrocleis nymphoides:* origin of lateral roots.

Troll (1932): *Limnocharis flava:* floral morphology.

Wilder (1974*a*): *Limnocharis flava:* developmental morphology.

* The original spelling used by the author is retained.

Significant Literature—Limnocharitaceae

Argue (1973); Buchenau (1903*b*); Carbonell and Arrillaga (1959); Charlton and Ahmed (1973*b*); Costantin (1885*a*, 1886*b*); Eber (1934); Eckardt (1964); Erdtman (1952); Ernst (1872); Hutchinson (1959); Kaul (1967*b*, 1968*a*, 1978); Leblois (1887); Leins and Stadler (1973); Mayr (1915); Micheli (1881); Pichon (1946); Richard (1815); Ronte (1891); Salisbury (1926); Sattler and Singh (1973, 1977); Saunders (1929); Sauvageau (1893); Şerbănescu-Jitariu (1973*a*); Singh (1966*c*); Stant (1967); Takhtajan (1966); van Tieghem (1870–1); van Tieghem and Douliot (1888); Troll (1932); Wagner (1918); Wilder (1974*a*).

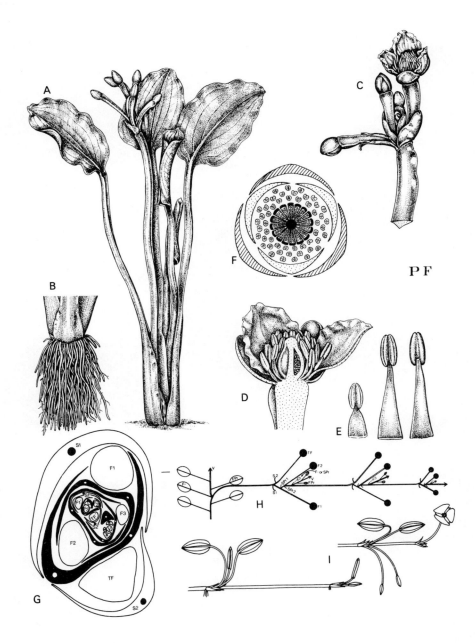

FIG. 2.1. **LIMNOCHARITACEAE**. Habit and floral morphology.

A–F. *Limnocharis flava* (L.) Buchen. (Supplied by G. J. Wilder, from commercial sources.)

A. Habit ($\times$ ¼). Aboveground parts. B. Underground rootstock ($\times$ ½). C. Cymose inflorescence, with bracts ($\times$ ½). D. Flower LS ($\times$ ⅔). E. Stamens ($\times$ 4), the smallest from the inner whorl. F. Floral diagram.

G–I. *Hydrocleys nymphoides* (H.B. ex Willd.) Buchen. (After Charlton and Ahmed 1973.)

G. Composite camera lucida drawing of transverse sections through an inflorescence to show relative arrangement of parts. H. Diagram showing architectural features of the shoot system. I. Illustration of flowering (right) and vegetative (left) stoloniferous shoots.

Explanation of labelling: TF—terminal flower; F_1, F_2, F_3—lateral flowers in order of development; S_1, S_2—first and second bracts of each branch complex; V—vegetative meristem; F+—foliage leaf subtending renewal shoot; FPr—foliage leaf as prophyll of renewal shoot.

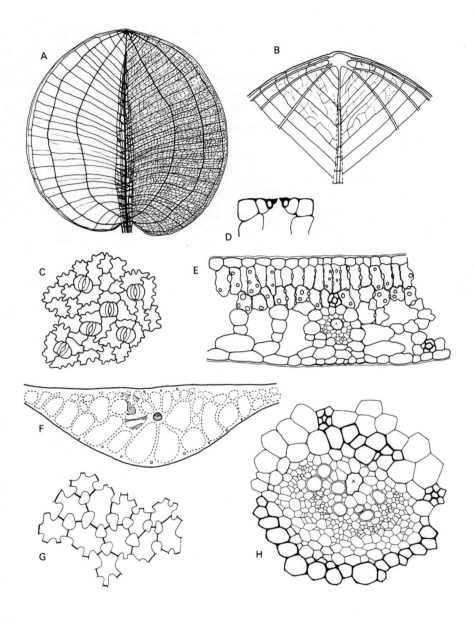

FIG. 2.2. LIMNOCHARITACEAE. *Hydrocleys nymphoides.* Leaf anatomy. (From Stant 1967.)

A. Whole lamina (× ⅔), left half with main veins, right with detailed reticulation.

B. Details of apex (× 2), with veins converged into apical pore.

C. Adaxial epidermis in surface view (× 110).

D. Stoma from adaxial epidermis in TS (× 260).

E. Leaf blade towards margin in TS (× 120).

F. TS midrib (× 9), diaphragms (cut obliquely) cross-hatched.

G. Diaphragm cells from midrib in surface view (× 160).

H. Central vb of midrib in TS (× 160).

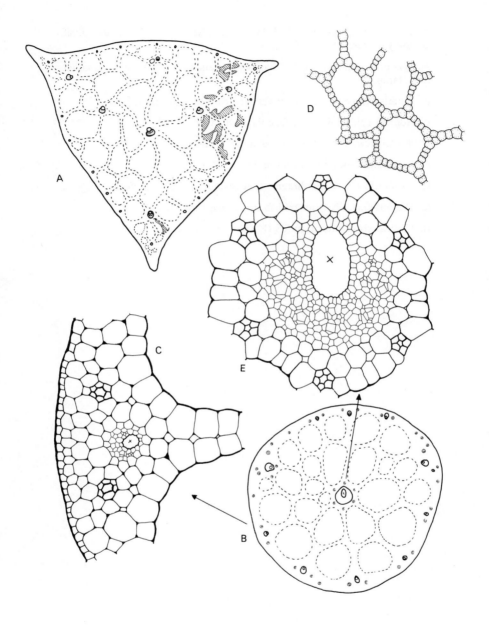

FIG. 2.3. LIMNOCHARITACEAE. Petiole anatomy, all TS. (From Stant 1967.)

A. *Limnocharis flava*, petiole (× 8), transverse diaphragms (cut obliquely) cross-hatched.

B–E. *Hydrocleys nymphoides.*
  B. Petiole (× 20). C. Peripheral layers, including small peripheral vb and associated lacunae (× 120). D. Details of lacunose ground tissue (× 135). E. Central vb (× 165).

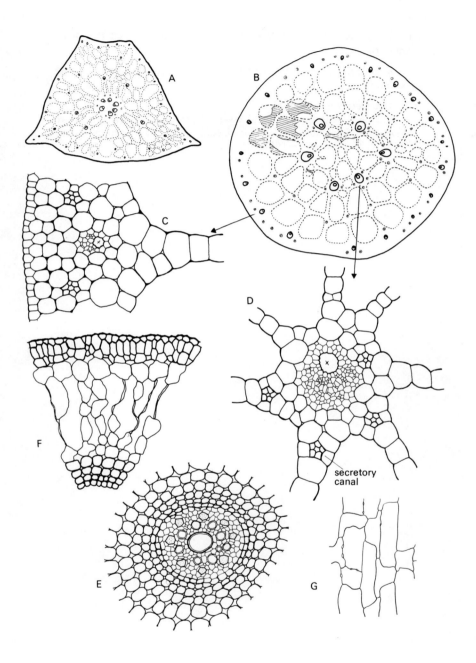

secretory
canal

FIG. 2.4. LIMNOCHARITACEAE. Peduncle (A–D) and root (E-G) anatomy, all TS except G. (From Stant 1967, except G.)

A. *Limnocharis flava* (× 4).

B–D. *Hydrocleys nymphoides.*
   B. Peduncle (× 12), diaphragms (cut obliquely) cross-hatched. C. Peripheral tissues including a small vb and associated laticifers (× 115). D. Central vb (× 65).

E–G. *Limnocharis flava,* root.

E. Stele and inner cortex, sieve-tubes stippled (× 160). F. Cortex and superficial layers (× 120). G. Cortex in LS to show long- and short-cells (× 130).

Family 3
BUTOMACEAE
(in the sense of Pichon 1946 not L. C. Richard 1815)
(Figs. 3.1–3.5; Plate 9)

SUMMARY

IN a restricted sense, this is a monotypic family consisting only of *Butomus umbellatus* L., a rather uniform species with a distribution in temperate Eurasia but extensively naturalized in eastern North America. The plants grow in marshes and on lake shores. The axis is often submerged and only the leaves and flowers are emergent (Fig. 3.2 A). The axis (Fig. 3.1 A) is essentially a creeping, monopodial rhizome (never stoloniferous) bearing erect distichous linear, triquetrous foliage leaves. The inflorescences arise from the rhizome at regular intervals and vegetative branching is initiated from the numerous axillary buds. The umbel-like inflorescences (Fig. 3.2 B) have three major bracts, each subtending a cymose complex of long-pedicellate flowers. The protandrous flowers are essentially trimerous and multicyclic (Fig. 3.2 C–F), with six coloured tepals in two whorls, nine stamens (the outer six forming a single series) and six essentially open carpels with numerous anatropous ovules. The placentation is laminar (Fig. 3.2 D–E). The styles are short, with a ventral stigmatic region. The fruit is a follicle releasing numerous ribbed seeds. The embryo is straight.

Anatomical features include abundant axillary squamules (Arber 1925*b*), absence of hairs, no secretory canals or laticifers, and stomata with a pair of lateral subsidiary cells (Fig. 3.4 C). The leaf and peduncle have well-developed air-lacunae, and protoxylem lacunae are also well developed. Vessels with simple perforation plates occur in the roots. Crystals are common in the transverse diaphragms in the leaf. Tannin is infrequent.

FAMILY DESCRIPTION
VEGETATIVE MORPHOLOGY (Fig. 3.1)

Rhizome axis horizontal, dorsiventral, with subdistichous foliage leaves (Fig. 3.1 A). Rhizome apex situated towards the upper side of the axis in a position enabling it to grow through the leaf bases (Weber 1950), the leaves becoming progressively more displaced (Fig. 3.1 B, C). **Proliferative rhizome branching** normally to produce ± axillary vegetative buds (non-precocious buds of Wilder, 1974*a*), enclosed by the forward edge of each leaf sheath. Buds with prophylls and forming two lateral series along the rhizome, the plane of distichy in the bud changing through 90° after the initiation of the adaxial prophyll. Axillary buds usually remaining dormant, but capable of developing as short-stalked bulbils, the latter serving for vegetative dispersal when detached. Axillary buds occasionally extending as branch rhizomes. **Regenerative rhizome branching** precociously, by bifurcation (dichotomy) of the vegetative apex in an oblique plane (Lieu 1979*c*), one product continuing as the vegetative shoot, and the other as an inflorescence, the latter developing as if subtended by the leaf formed most immediately before it (Fig. 3.1 B). Inflorescences produced on alternate sides of the axis usually separated from one another by an odd number (5, 7, 9) of foliage leaves. Appendages on opposite sides of the axis exhibiting a regular mirror-image symmetry (Wilder 1974a).

Leaves with an open sheathing base, narrowly winged below the distal part, triquetrous (Fig. 3.3 A–C) and equated by some workers (e.g. Stant 1967) with the petiole of bladed leaves in related groups (e.g. *Limnocharis*) as a phyllodic structure. **Squamules** numerous in leaf axils.

VEGETATIVE ANATOMY (Figs. 3.3–3.4)

Leaf

Hairs absent. **Epidermis** differentiated into narrow costal bands 2–4 cells wide. Costal bands overlying the peripheral vascular strands and consisting of long, narrow cells. Intercostal bands consisting of shorter to almost isodiametric cells (Fig. 3.4 A, B). Epidermal cells all rectangular with non-sinuous walls. **Stomata** present on all faces but most abundant abaxially, in 1–2 files in the intercostal bands. Each stoma with a pair of lateral subsidiary cells (Fig. 3.4 C), the latter initiated by a single, often oblique, division (Gupta *et al.* 1975). Guard cells asymmetric with only the outer ledge well developed. **Hypodermis** distinct only towards the leaf base. Outer mesophyll otherwise differentiated as 1–4 layers of chlorenchyma, the cells of the outermost layer being palisade-like (Fig. 3.3 G). Groups of collenchymatous cells replacing the chlorenchyma outside peripheral vbs. **Central tissue** including a network of uniseriate parenchymatous plates enclosing large **air-lacunae** (Fig. 3.3 D). Parenchymatous cells of the plates thin-walled and frequently containing chloroplasts; tannin cells also occasional. Air-lacunae traversed at intervals by transverse, or slightly oblique, uniseriate diaphragms of densely cytoplasmic cells enclosing narrow intercellular spaces. Cells of the diaphragm containing clusters of small crystals (Fig. 3.3 F). Wider diaphragms including transverse veins.

Vascular system essentially made up of a central series of large vbs and a peripheral system of smaller vbs, the latter embedded in the peripheral chlorenchyma (Fig. 3.3 A–C). Five discrete series of vbs recognizable, at least towards the base of the leaf: (i) a V-shaped arc of nine major vbs, (ii) a secondary arc of three abaxial vbs, (iii) a solitary vb near the abaxial angle of the leaf, (iv) numerous peripheral vbs forming a hypodermal series, (v) minute or rudimentary vbs or fibrous strands including narrow fibre-tracheids with spiral thickenings, scattered in the central ground tissue at junctions between the parenchyma plates (Fig. 3.3 D, E). Larger vbs (i–iii) normally orientated. Larger hypodermal vbs (iv) with protoxylem inwardly directed, i.e. adaxial bundles inversely orientated. Smaller hypodermal vbs with vascular tissue reduced (Fig. 3.3 G) or represented by fibres. Outer sheaths of larger bundles parenchymatous, but discontinuous laterally, interrupted by fibres forming caps above and below vb. Fibrous caps relatively well developed in peripheral (iv) vbs (Fig. 3.4 E), often continuous with the epidermis. Large

central vbs each with a wide **protoxylem lacuna,** the metaxylem including
an irregular series of tracheids partly enclosing the phloem (Fig. 3.4 D).
Peripheral vbs including a conspicuous metaxylem element or a protoxylem
lacuna and reduced phloem (Fig. 3.4 E).

Axis (peduncle) (Fig. 3.5 A–G)

Circular in TS (Fig. 3.5 A). **Epidermis** including occasional stomata, the
epidermal cells sometimes with chloroplasts. **Cortex** (Fig. 3.5 B): outer part
narrow, including a peripheral chlorenchymatous layer 5–6 cells wide, the
cells rounded with irregular intercellular spaces; inner part consisting of
thicker-walled, more compact colourless parenchyma. Cortex bounded on
its inner face by a lignified **fibrous sheath** 3–6 cells wide (Fig. 3.5 B). **Central
ground tissue** lacunose (Fig. 3.5 G), with occasional transverse diaphragms.
Vascular system consisting of scattered vbs, the innermost largest but interspersed with narrow vbs; outer vbs alternately large and small and in contact
with the sclerenchymatous cylinder (Fig. 3.5 B, C). Each vb with a wide
protoxylem lacuna, sometimes including the remains of disorganized tracheids. Metaxylem forming a wide, U-shaped arc partly enclosing the phloem,
the extent of the enclosure varying with the size of the vb (Fig. 3.5 D–F).
Bundle sheath fibrous but incomplete, 1–2-layered but up to four cells wide
around peripheral vbs and continuous with sclerenchyma sheath.

Axis (rhizome).

Epidermis persistent. **Cortex** narrow, outer part consisting of loosely-
packed parenchyma with numerous irregular air-spaces, the latter being especially well developed in the middle cortex. Cortex delimited internally by a
layer of compact thick-walled and lignosuberized cells. **Endodermis** absent.
Stele indistinctly differentiated, the vascular system being represented by
vbs of varying sizes and regularity. Recently entered leaf traces collateral
and with well developed sheaths of lignified fibres next to the phloem. Root
traces inserted towards the periphery of the stele, but not forming a well
developed vascular plexus. Smaller central vbs with reduced vascular tissue;
larger vbs irregular. **Tannins** in some cells of the ground tissue. **Crystals**
present in the outer cortical layers. **Starch** abundant.

Root (Fig. 3.5 H–J).

Branched, arising from lower surface of rhizome; said to be contractile
(Weber, 1950). **Epidermis** lost from older roots; **exodermis** uniseriate, outer
wall slightly thickened (Fig. 3.5 I). **Cortex** not differentiated into long- and
short-cells; outer cortex to within 2–3 layers of exodermis with wide radial
air-lacunae (Fig. 3.5 H) arising by separation and collapse of radial cell files,
the lacunae separated by uniseriate plates of incompletely collapsed cells.
Inner cortex of 6–7 regular layers of concentrically arranged radially-flattened
cells; transverse diaphragms absent. **Endodermis** (Fig. 3.5 J) uniformly but
only slightly thick-walled. **Pericycle** indistinctly differentiated from the stelar
ground tissues, but mainly uniseriate and somewhat thick-walled opposite
the phloem. **Xylem** including a single wide central vessel (rarely two) surrounded by eight narrow protoxylem files alternating with **phloem** strands,

each of the latter including a wide sieve-tube (Fig. 3.5 J). Conjunctive tissue somewhat thick-walled and including fibres and long narrow lignified parenchyma cells.

SECRETORY, STORAGE, AND CONDUCTING ELEMENTS

Secretory canals: absent.

Crystals. Common as aggregates in transverse diaphragms of leaf (Fig. 3.3 F) and also recorded in the rhizome by Stant (1967), as irregularly polygonal clusters or as solitary styloids in the innermost cortical layers.

Tannin infrequent, but most common in ground tissue of rhizome.

Starch common in rhizome, as rounded or triangular centric grains up to 14 μm in diameter.

Tracheary elements: vessels restricted to the metaxylem of the root; vessel elements of the order of 700 μm long and 100 μm diameter, with simple transverse perforation plates. Tracheary elements elsewhere represented by thin-walled tracheids with annular, spiral or scalariform-reticulate pitting, those of the rhizome very irregular.

REPRODUCTIVE MORPHOLOGY AND ANATOMY

Inflorescence and floral morphology (Fig. 3.2; Plate 9)

Inflorescence initiated by oblique bifurcation of the vegetative apex (essentially an equal dichotomy), the inflorescence axis not developing a prophyll but apparently occupying the axil of the preceding leaf (Fig. 3.1 B), which itself subtends no bud (Wilder 1974a; Charlton and Ahmed 1973b). Inflorescences on opposite sides of the rhizome with mirror-image symmetry. Peduncle (Fig. 3.2 A) essentially the elongated first internode of the inflorescence axis, after producing the first appendages its apex becoming a flower. First-order bracts (primary bracts of Wilder) three, produced in rapid sequence, each subtending a complex cymose system of flower-bearing buds and associated bracteoles (prophylls). The flower complex subtended by each bract exhibiting precisely the same symmetry and sequence of flower initiation. Flower buds produced directly as derivatives of the parent meristem, sometimes with an associated bract, but simulating a dichotomy. Trifurcations (producing triplets) in early stages of branching interpreted by Wilder as two successive bifurcations following in rapid sequence. Overall sequence of branching clearly cymose, with an initial regular series of triplets but ultimately of monochasia showing a one bract–one branch relationship (for details of actual symmetrical patterns see Wilder 1974a). [Since flowers tend to mature synchronously the initial sequence of development is obscured and an umbel is formed.]

Vegetative **bulbils** reported by Lohammar (1954) as replacing flowers in the umbel.

Flowers (Fig. 3.2 C–G) bisexual, studied developmentally by Singh and Sattler (1974); essentially trimerous (except for the outer whorl of six stamens) and v. slightly epigynous according to Singh and Sattler (1974), the floral parts in each series appearing in rapid sequence (Plate 9). Flowers protandrous; the stamens dehiscing in the order outer six, inner three. Nectar from septal nectaries (Böhmker 1917) secreted between adjacent carpels (Fig. 3.2 F). Carpel receptivity later indicated by slight spreading of papillate stigmas. **Perianth** petaloid, in two distinct whorls (Fig. 3.2 C–E). **Stamens** nine (Plate 9), the outer series of six originating in pairs opposite the 'petals' (inner tepal whorl), but without clear 'CA primordia', the inner series of three originating opposite the 'sepals' (outer tepal whorl). Filaments flattened, with basifixed dithecous anthers dehiscing laterally. **Carpels** (Fig. 3.2 F) six, free, originating in two alternate whorls of three each (Plate 9), the lowest whorl antesepalous, each initially horseshoe-shaped, the marginal meristem later appressed but never fused, the carpel eventually narrowed apically to a short style with a stigmatic crest (Fig. 3.2 G). Ovules numerous, on the inner radial wall of the carpel (Fig. 3.2 D, E), each anatropous and bitegmic. Fruits (Fig. 3.2 H) developing as follicles each derived from a carpel dehiscing by widening of the ventral suture. Seeds numerous, ribbed (Fig. 3.2 I). Endosperm absent, embryo straight, the cotyledon described as 'terminal'.

Floral anatomy

Procambial differentiation in the floral buds acropetal, according to Singh and Sattler (1974), each appendage shortly after inception receiving a procambial strand as a branch from an existing strand, the existing vbs branching and fusing at progressively higher levels.

Pedicel supplied by many traces, these becoming organized first into a ring of about 12 vbs and then aggregated in the receptacle to form a continuous vascular cylinder, supplying (i) three traces to each of the six perianth segments, the tepal traces dividing to produce 7–9 vbs in the tepal itself; (ii) a single trace to each stamen; (iii) a dorsal and ventral trace to each carpel. Dorsal vbs continuing unbranched into the style; ventral vbs V-shaped and on radii alternate with the dorsal bundles, the arms of the V each split into 5–7 vbs supplying traces to the numerous ovules on the inner wall of adjacent carpels, i.e. adjacent lateral margins of the carpels having a common ventral vb. Ventral vbs also continuous into the style. Glandular, densely staining tissue on outer surface of adjacent carpels representing nectaries. Vbs in the flower retaining xylem without developing xylem lacunae.

[Singh (1966b) concludes that the vascular supply of the carpel of *Butomus* and *Hydrocleys* is very similar to that of certain Hydrocharitaceae (e.g. *Hydrocharis*), but dissimilar to that of Alismataceae.]

TAXONOMIC NOTES

The genus *Butomus* has had a somewhat disputed taxonomic history because of the number of distinctive features it shows. Richard (1815) established the family Butomaceae to include *Butomus, Limnocharis,* and *Hydrocleys* (= *Hydrocleis*). Pichon (1946) removed all genera except *Butomus* from the family, and included them in the tribe Limnocharitinae of the Alismataceae. Takhtajan (1966) subsequently established this tribe as a separate family Limnocharitaceae. Thus the Butomaceae in the restricted sense of Pichon

(1946) and Takhtajan (1966) is designed to include only the genus *Butomus,* which is itself monotypic. Hutchinson (1959) retains the larger view of the family and includes five genera, which are here treated (with the exception of *Butomus*) in the family Limnocharitaceae.

The evidence which supports the segregation of the genus as a discrete family is presented elsewhere in comparison with Limnocharitaceae (p. 96). Some of the evidence depends on morphological interpretation. Thus the linear triquetrous leaf of *Butomus,* which seems so dissimilar to the bladed leaves of Limnocharitaceae, can be regarded as a phyllode representing only the petiole of a normal bladed leaf (Arber 1918*b*). The similar anatomy of the organs in the two families supports this interpretation (Stant 1967). It is difficult to decide whether the rhizome of *Butomus* is monopodial or sympodial but most evidence is for the former interpretation (Wilder, 1974*a*, Lieu, 1979*c*). In its proliferative vegetative branching it is monopodial since all buds are essentially axillary (non-precocious buds of Wilder 1974*a*). The same phyllotaxis is continuous along the vegetative axis, but the continuity does not extend into the infloresence (Charlton and Ahmed 1973*b*; Wilder 1974*a*; Lieu 1979*c*). However, the inflorescence lacks a basal prophyll and has a distinctive method of development, essentially involving dichotomy of the vegetative meristem, so that it cannot easily be equated with a lateral shoot. It is clear that this specialized morphology is a further example of the way in which the normal parent–daughter axis relationship is transcended in the Helobiae, a modification shown extensively in the further branching of the inflorescence meristem itself (Wilder 1974*a*). This provides further evidence of the highly specialized morphology of these aquatic monocotyledons.

MATERIAL EXAMINED

Butomus umbellatus L. Cultivated, Aquatic Gardens, R. B. G., Kew; all parts. Shelbourne Bay, Lake Champlain, New York, P. B. Tomlinson *s. n.*; all parts.

SPECIES REPORTED ON IN THE LITERATURE

The following authors have examined *Butomus umbellatus* L., as indicated:

Arber (1925*b*): squamules.
Böhmker (1917): septal nectaries.
Buchenau (1857): floral development.
Charlton and Ahmed (1973*b*): inflorescence morphology.
Duval-Jouve (1873*b*): diaphragms in leaf.
Erdtman (1952): pollen.
François (1908): seedling.
Gupta *et al.* (1975): stomatal development.
Hasman and Inanç (1957): all vegetative parts.
Leinfellner (1973): carpel morphology.
Lieu (1979*c*): shoot morphology.

Lohammar (1954): inflorescence morphology.
Salisbury (1926): floral morphology.
Sauvageau (1893): leaf.
Şerbănescu-Jitariu (1964): carpel development.
Singh (1966b): floral anatomy.
Singh and Sattler (1974): floral development
Stant (1967): all vegetative parts.
van Tieghem and Douliot (1888): lateral root initiation.
Weber (1950): rhizome morphology.
Wilder (1974a): inflorescence development.

SIGNIFICANT LITERATURE—BUTOMACEAE

Arber (1918b, 1925a, b), Böhmker (1917); Buchenau (1857, 1882, 1903b); Charlton and Ahmed (1973b); Costantin (1884); Duval-Jouve (1873b); Erdtman (1952); François (1908); Gupta *et al.* (1975); Hasman and Inanç (1957); Hutchinson (1959); Leinfellner (1973); Lieu (1979c); Lohammar (1954); Micheli (1881); Pichon (1946); Richard (1815); Salisbury (1926); Sauvageau (1893); Şerbănescu-Jitariu (1964); Singh (1966b); Singh and Sattler (1974); Stant (1967); Takhtajan (1966); van Tieghem and Douliot (1888); Weber (1950); Wilder (1974a).

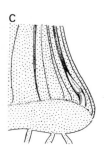

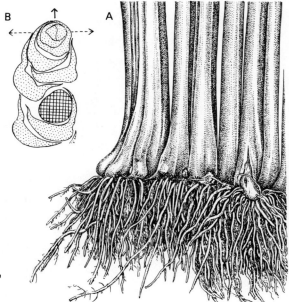

PF

FIG. 3.1. BUTOMACEAE. *Butomus umbellatus* (Shelbourne Bay, Lake Champlain, P. B. Tomlinson *s.n.*)

A. Rhizome and leaf bases from the side ($\times \frac{1}{2}$).

B. Diagram of leaf arrangement in A seen from above ($\times$ 1). Solid arrow indicates direction of growth; dotted arrows indicate initial plane of distichy.

C. LS rhizome apex, with young inflorescence ($\times$ 1).

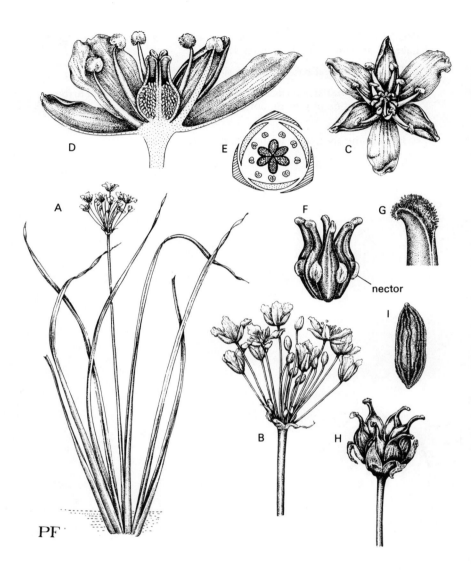

nectar

PF

118

FIG. 3.2. BUTOMACEAE. *Butomus umbellatus* (Same locality as Fig. 3.1.)

A. Aerial parts ($\times \frac{2}{3}$).

B. Inflorescence ($\times \frac{1}{2}$).

C. Flower from above at male phase ($\times \frac{3}{2}$).

D. Flower in LS ($\times 3$) at male phase.

E. Floral diagram.

F. Ovary showing secreted nectar ($\times 4$) at female phase.

G. Details of receptive stigma ($\times 15$).

H. Fruiting head ($\times \frac{3}{2}$).

I. Single seed ($\times 18$).

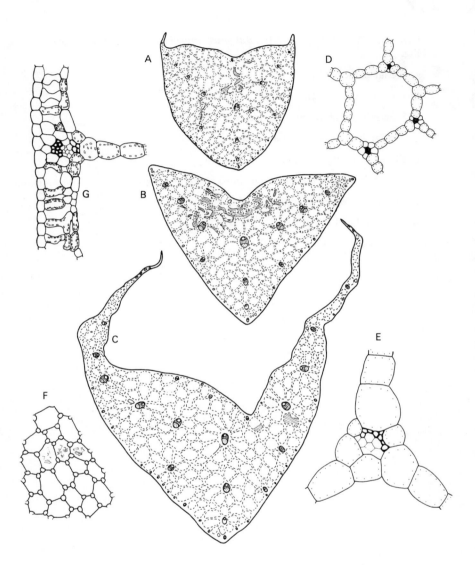

120

FIG. 3.3. BUTOMACEAE. *Butomus umbellatus.* Leaf anatomy.
(From Stant 1967.)

A–C. Leaf in TS at three levels. A. Top ($\times$ 3). B. Middle ($\times$ 2). C. Base
($\times$ 2).

D. TS lacunose ground tissue from B ($\times$ 20).

E. Details of small central vb at junction of cell plates in TS ($\times$ 90).

F. Cells of transverse diaphragm in surface view, crystals shown in a
few cells ($\times$ 90).

G. TS outer layers of leaf with chlorenchyma and a small peripheral vb
($\times$ 50).

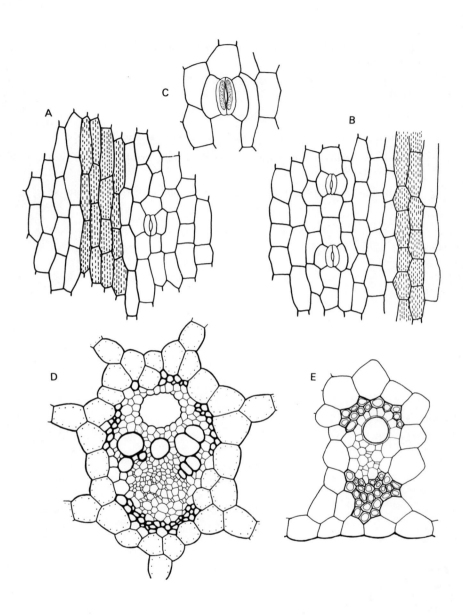

122

FIG. 3.4. BUTOMACEAE. *Butomus umbellatus.* Leaf anatomy. (From Stant 1967.)

A–C. Epidermis in surface view. A. Adaxial epidermis (× 50). B. Abaxial epidermis (× 50). C. Detail of stoma in surface view (× 90).

D. TS large central vb (× 50).

E. TS larger peripheral vb (× 90).

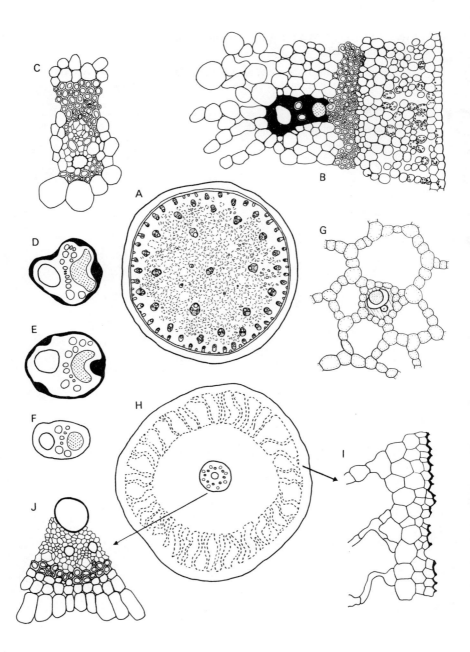

124

FIG. 3.5. BUTOMACEAE. *Butomus umbellatus.* Anatomy of scape (= peduncle) A–G, and root, H–J. (After Stant 1967.)

A. TS peduncle (× 3).

B. TS outer layers (× 40).

C. TS peripheral vb (× 50).

D–F. Diagrammatic TS central vbs (× 20) to show range of variation, F is towards the periphery.

G. TS ground tissue, including a vb in diagrammatic outline (× 20).

H. Root, diagrammatic TS (× 10).

I. TS peripheral layers of root (× 50), the epidermis absent.

J. TS part of root stele (× 50), with central metaxylem vessel, endodermis–pericycle region thick-walled.

Family 4
HYDROCHARITACEAE (A. L. de Jussieu 1789)
(Figs. 4.1–4.15; Plates 1 c, 2 a–b, 3 d, 4 a–b, 5)

SUMMARY

THIS is a diverse, cosmopolitan family of about 15 genera and over 100 species (den Hartog 1957b) which occupies a wide range of habitats in fresh or saltwater. *Enhalus, Halophila,* and *Thalassia* are wholly marine (seagrasses). The plants vary from amphibious types with emergent leaves (*Limnobium*) to those that are obligately (*Stratiotes*) or facultatively (*Hydrocharis*) floating. Other taxa are most commonly submerged but they have either emergent or wholly submerged flowers. The vegetative parts either have the form of a leafy rosette (e.g. *Blyxa* spp, *Ottelia*), or are stoloniferous (*Hydrocharis, Limnobium, Stratiotes, Vallisneria*) or rhizomatous (*Enhalus, Thalassia*). The shoot system is dimorphic in *Thalassia.*

Branching commonly takes place by apical bifurcation which can be interpreted as precocious axillary branching (Wilder 1974a, b). The plants are sometimes filamentous with whorls of short, narrow leaves (e.g. *Elodea* and related genera). The leaves are either spirally or distichously arranged, never ligulate; they fall into at least five main types but with frequent transitions between them (cf. Ancibor 1979). The leaves may be:

(i) Linear, strap-shaped, with a sheathing base; often very long (> 1 m) (e.g. *Enhalus, Thalassia, Vallisneria*).
(ii) Linear-lanceolate, short with a single midvein, insertion narrow (Anachariteae).
(iii) Undifferentiated or with an indistinct petiole, lanceolate, moderately long with several veins, insertion broad (*Blyxa, Ottelia*).
(iv) Distinctly petiolate, with a sheathing ('stipulate') base, and a broad often cordate blade (*Hydrocharis, Limnobium*). *Halophila* is often similar but the leaves are diminutive and without a sheathing base.
(v) Scale-leaves and prophylls (equivalent to bladeless sheaths, e.g. *Thalassia*).

Squamules which vary from two–numerous in each leaf axil are present in all taxa, and the number and sometimes their shape are diagnostic for individual taxa (Gibson 1905; Ancibor 1979).

The plants are mainly monoecious or dioecious, and have actinomorphic, unisexual flowers (bisexual in *Blyxa, Elodea, Ottelia* spp). The inflorescence is either pedunculate or sessile and essentially cymose (Kaul 1970), the flowers being enclosed by two free or, more usually, fused bracts to form a ± tubular 'spathe.' The flowers are solitary (e.g. female flowers in several taxa) or more usually several to numerous (very numerous in male inflorescences of, e.g. *Enhalus, Vallisneria*). The female flowers are sessile or on long pedicels and commonly provided with a long hypanthium. The male flowers are sometimes released and become free-floating when functionally mature (e.g. *Elodea, Enhalus, Nechamandra, Vallisneria*). Others extend to the water surface by long peduncles. Perianth segments are 2–3 or six in number, and when they are six they are either differentiated into sepals and ± showy petals (e.g. *Egeria, Hydrocharis, Limnobium, Stratiotes*) or without showy petals (e.g. *Blyxa, Lagarosiphon*). The stamens may be single (*Maidenia*), paired (some *Halophila* spp) but more commonly there are three–many per flower, usually arranged in several trimerous series. In female flowers they are sometimes represented by staminodes (e.g. *Hydrocharis*). The outer stamens are doubled in *Stratiotes* and species of *Ottelia*. The ovary is inferior, of two–several (–15) carpels which become connate during development to produce a unilocular structure with several dissepiments equal in number to the ± fused walls of adjacent carpels (Troll 1931*b*). In the most specialized ovaries the placentae are essentially parietal (e.g. *Halophila*) or the ovules are uniformly distributed over the ovary chamber (e.g. *Vallisneria*). The styles are usually as many as the carpels, short and commonly divided distally into two distinct, sometimes flattened stigmas. The numerous ovules are essentially distributed over the inner surface of the original carpels, they are usually anatropous and bitegmic. The fruit is fleshy and the numerous seeds are released by decay or irregular splitting of the pericarp. The seeds are without endosperm. The embryo is usually straight but sometimes with an enlarged hypocotyl (e.g. *Hydrocharis* and marine genera).

Anatomical characters include paracytic stomata (on emergent or floating leaves), well-developed aerenchyma with diaphragms of stellate cells which are often provided with specialized thick walls. Hairs are largely represented by unicellular prickle cells at the leaf margin (less commonly on the leaf surface), the prickles sometimes mounted on prominent emergences. Tannin cells are sometimes common and specifically located within the transverse diaphragms. The roots are commonly made septate by transverse diaphragms. The root-hairs of certain taxa (but especially *Hydrocharis* and *Stratiotes*) are characterized by enlarged, internally protuberant bases with endocytic organization.

Taxonomic and Morphological Notes for the Family

The family is so diverse in its vegetative morphology and anatomy that it is scarcely possible to provide a concise summary of its characters. The diversity is reflected by the large number of small (usually monotypic) tribes into which the family is divided, the genera themselves usually being small or even monospecific (e.g. *Hydrilla, Nechamandra, Maidenia, Stratiotes*). There is little agreement in the arrangement of these tribes. A summary of two contrasted systems, presented respectively by Ascherson and Gürke (1889) and J. E. Dandy in Hutchinson (1959) illustrates this well (Table 4.1). The former system emphasizes morphology, the latter is a more biological classification since it is based, in the first instance, on the method of pollination.

In order to accommodate this diversity, the genera (arranged as in Dandy's system) are all described individually, except for those in the tribe Anachariteae which form a natural alliance because of their distinctive vegetative morphology and uniform anatomy. A transition to (or from) other forms with basal rosettes of leaves is seen in the section *Caulescentes* of *Blyxa*. *Maidenia* has not been examined here and is not described.

TABLE 4.1

Hydrocharitaceae–contrasted arrangement in two classifications

A. After Ascherson and Gürke (1889)	B. After Hutchinson (1959)
Carpels usually 3:	Pollination at or above surface of water:
I. HALOPHILOIDEAE (petals absent)	SUBFAMILY 1. VALLISNERIOIDEAE
*Halophila**	TRIBE 1. LIMNOBIEAE
II. VALLISNERIOIDEAE (petals present)	*Hydrocharis,*
	Limnobium (inc. *Hydromystria*)
1. HYDRILLEAE	TRIBE 2. STRATIOTEAE
Hydrilla, Elodea	*Stratiotes*
2. VALLISNERIEAE	TRIBE 3. ENHALEAE
Lagarosiphon, Vallisneria	*Enhalus**
3. BLYXEAE	TRIBE 4. OTTELIEAE
Blyxa	*Ottelia* (inc. *Boottia*)
Carpels 6–15	TRIBE 5. ANACHARITEAE
	Egeria, Elodea, Hydrilla,
III. THALASSIOIDEAE (leaves distichous)	*Lagarosiphon, Nechamandra*
Enhalus, Thalassia**	TRIBE 6. VALLISNERIEAE
IV. STRATIOIDEAE (leaves polystichous)	*Vallisneria*
	TRIBE 7. BLYXEAE
1. STRATIOTEAE	*Blyxa*
Stratiotes	
2. OTTELIEAE	
Boottia, Ottelia	Pollination under water:
3. HYDROCHARITEAE	
Hydromystria, Limnobium, Hydrocharis	SUBFAMILY 2. THALASSIOIDEAE
	One genus *Thalassia**
	SUBFAMILY 3. HALOPHILOIDEAE
	One genus *Halophila**

* Marine.

Key to genera of Hydrocharitaceae indicating major morphological and anatomical diagnostic features (partly after Cook *et al.* 1974).

I. Plants marine, *rhizomatous* (i.e. horizontal axes wholly buried in substrate, relatively thick, fleshy and with short internodes); leaves strap-shaped; roots unbranched, septate; stomata absent   1.

II. Plants marine or of fresh water, *stoloniferous* (i.e. horizontal axes superficial, slender, usually represented by a single internode separating foliage leaf clusters); leaves commonly petiolate; stomata present or absent   2.

III. Plants of fresh water, with leafy *rosettes* (i.e. axes congested, not obviously branched); stomata present or absent   4.

IV. Plants of fresh water with extended, branched *filamentous* axes (i.e. leaves dispersed, internodes evident); stomata absent   5.

1A. Rhizomes dimorphic, with extended horizontal, scale-bearing *long-shoots* and erect, foliage-bearing *short-shoots*; flowers wholly submerged; leaf veins in a single series   *Thalassia* König

1B. Rhizomes monomorphic, bearing foliage leaves; flowers rising to water

surface at functional maturity; leaf veins in two series, the adaxial
series inverted *Enhalus* L. C. Rich.
2A. Plants rooted, not free-floating 3
2B. Plants floating 4A
3A. Marine; roots unbranched; flowers submerged, perianth uniseriate, three–lobed or absent; stamens three; leaves thin, blade of two cell layers, with obvious transverse veins and either petiolate with a broad blade or undifferentiated with a broad base; stomata absent
Halophila Du Petit-Thouars
3B. Fresh water; roots branched; flowers emergent, perianth biseriate, usually with showy petals; leaves petiolate, floating or emergent, with a broad cordate or ovate blade with many cell layers, transverse veins not obvious; stomata present.
3B′. Petals scarcely exceeding sepals, or absent; roots dimorphic
Limnobium L. C. Rich.
3B″. Petals exceeding sepals; roots monomorphic *Hydrocharis* L.
3C. Fresh water; roots unbranched; flowers functioning at water surface; male flowers very numerous and free-floating, perianth biseriate or uniseriate, petals vestigial or absent; stamens 1–2 (–3); leaves strap-shaped with a sheathing base; stomata absent *Vallisneria* L.
4A. Plants floating; flowers with showy petals; leaves rigid, lanceolate with spinous-serrate margins; rosettes overwintering as submerged, calcified buds; stomata present *Stratiotes* L.
4B. Plants usually submerged; leaves not rigid; spathe neither inflated nor winged; stigmas three, entire; leaves linear-lanceolate, blade not involute
Blyxa Du Petit-Thouars ex L. C. Rich. (sect. *Blyxa*)
4C. Plants usually submerged; leaves not rigid; spathe inflated, winged or spiny; styles 3–15 each bifid, narrow; leaves lanceolate or broadly ovate, often petiolate, blade usually involute *Ottelia* Pers.
5A. Leaves with a broad insertion, spirally or distichously arranged 6.
5B. Leaves with a narrow insertion, spirally arranged or in whorls 7.
6A. Leaves spirally arranged; midrib prominent; male flowers not released at maturity *Blyxa* (sect. *Caulescentes* Koidz.)
6B. Leaves distichously arranged; midrib not prominent; male flowers free-floating at maturity *Nechamandra* Planch.
7A. Leaves (at least distally) spirally arranged, not in whorls 8.
7B. Leaves (at least distally) opposite or sub-opposite, in whorls or pseudo-whorls 9.
8A. Male spathes of two united bracts, enclosing numerous male flowers; perianth of female flowers biseriate *Lagarosiphon* Harvey
8B. Male spathes of two distinct bracts, enclosing numerous male flowers each represented by a single stamen; perianth of female flowers uniseriate *Maidenia* Rendle
9A. Petals three times as long and broad as sepals; flowers with nectaries
Egeria Planch.
9B. Petals scarcely exceeding sepals, or absent; flowers without nectaries
10.
10A. Male spathes containing numerous flowers *Lagarosiphon* Harvey

10B. Male spathes each containing one flower; stamens always three; plants
      always with unisexual flowers   *Hydrilla* L. C. Rich.
10C. Male spathes smooth, each containing one flower; stamens usually six
      or nine; plants sometimes with hermaphrodite flowers
                                                        Elodea Michx.

Floral biology

The diversity of inflorescences and floral morphology in the family (Kaul
1968*b*, 1969, 1970) largely depends on the many types of floral mechanisms
that are to be seen (see especially Cook 1981; Ernst-Schwarzenbach 1945,
1953, 1956; Troll 1931*a*, and the classical studies on *Vallisneria*, p. 155). Information has been extensively summarized by den Hartog (1957*b*). The different
floral mechanisms include cleistogamy, autogamy, and a series of devices
which promote outcrossing. Entomophily, anemophily, and hydrophily all
occur and pollination may be aerial, at the air–water interphase (but without
the pollen being wetted), or wholly submerged. In a few taxa pollen is trans-
ported to the stigma directly on the water surface ('semi-aquatic surface
pollination' of den Hartog, in some *Elodea* spp). *Enhalus* is unique as a
seagrass because pollination occurs at the water surface, the pollen remaining
dry (p. 143).

The effects of this diversity on pollen structure has scarcely been examined
but is of major interest.

Branching

Shoot morphology in Hydrocharitaceae is frequently complex and has produced considerable controversy. Studies by Wilder (1974*b, c*) on *Limnobium*
and *Vallisneria* indicate the complexity involved. Wilder concludes that the
erect vegetative axes in these species are sympodial, whereas there is uncertainty about whether stolons have monopodial or sympodial growth. Comparative morphology has allowed these taxa to be accommodated in a larger
scheme in which a basic architectural pattern is perceived for the Alismatidae
as a whole, with reduction in number of parts a frequent process (Wilder
1975).

This interpretation has been challenged to a certain extent by Brunaud
(1976, 1977—see also Bugnon and Joffrin 1962, 1963) who interprets branching as entirely monopodial in the Hydrocharitaceae, on the basis of studies
on *Elodea*, *Hydrocharis*, *Stratiotes*, and *Vallisneria*. The basis for the dispute
is the fact that branching of shoot apices is usually by an equal bifurcation,
a phenomenon observed widely in other Alismatidae. Where there is distichous
phyllotaxis, morphologically there is always a leaf in a position subtending
both products of bifurcation so the problem is to establish which of the
products is the original apical meristem and which is the new lateral meristem.
Apart from the comparative evidence, the existence of a 'shell-zone' at the
base of one meristem was used to identify a putative lateral shoot in *Limnobium* by Wilder (1974*b*). Such a shell-zone does not occur in other taxa
which have been examined. The evidence produced by Brunaud is less substantial since he uses interpretative diagrams. It is possible that the problem is

not resolvable by normally accepted procedures, as is the conclusion of Cutter (1964).

The point of major *biological* interest is that precocious branching or bifurcation is a normal feature of many Hydrocharitaceae and plays a major role in their precise organization. *Thalassia* is a very specialized example in which the lateral shoot is leaf-opposed, originating on the shoot apex in such a position (Tomlinson and Bailey 1972). It seems better to accept this observation, than to dismiss it as superficial (as does Brunaud, 1976, without offering more precise data) because it cannot easily be fitted into our normal concept of axillary branching.

Elodea (and presumably related genera) certainly seems a special example. The lateral vegetative meristem is axillary. However, the floral primordium is precocious because it can be recognized above the youngest leaf primordia (Brunaud 1976). Because of the extended conical shape of the apical meristem of *Elodea* (Lance-Nougarède and Loiseau 1960; Stant 1952) it is not produced by a bifurcation process and the shoot is reasonably interpreted as a monopodium. The flower's axillary position further supports this.

Inflorescence morphology

This has been reviewed comparatively by Kaul (1970), especially in relation to the diverse pollination mechanisms in the family. Kaul notes the frequent marked dimorphism of inflorescences in dioecious species. He concludes that the primitive condition is represented by *Limnobium*, in which there is a terminal flower with a pair of lateral cincinni, each cincinnus subtended by one of the members of the basal pair of bracts. In this system, each flower is associated with a bracteole. An essentially similar construction occurs in *Stratiotes*. Progressive specialization has involved condensation so that flowers become aggregated on the inflorescence axis in various ways (e.g. *Enhalus*) or by aggregation of flower pedicels, as in the male inflorescence of *Nechamandra* and especially *Vallisneria*. In *Boottia* there appear to be more than two cincinni below the terminal flower in vigorous specimens. The one-flowered inflorescence is considered to represent a derived and reduced condition, developed independently in several taxa (e.g. *Ottelia, Blyxa, Halophila, Hydrilla, Elodea* spp) with indications of phylogenetically earlier complexity in the female inflorescences of *Hydrocharis* and *Limnobium* which may have vestiges of the basal cincinni. *Thalassia* seems unusual in that there may be several flowers, each with its own bract pair (spathe) in the axil of a single foliage leaf (Tomlinson 1969a). More information about inflorescence development would be useful for comparative purposes.

Floral anatomy

The most extensive work is by Singh (1966d), and Kaul (1968b) who appear to have worked independently of each other. Between them they studied representatives of all genera except *Maidenia*. Tomlinson's (1969a) study of *Thalassia* was produced without knowledge of this work. Singh's and Kaul's accounts differ in some features, reflecting real differences in the samples they studied, but also differences in interpretation. Variability in numbers of floral parts is recorded by numerous observers.

Despite the great diversity of size and form in the flowers in the family, certain principles of construction can be recognized. There is a marked tendency for the vascular system of the reproductive parts to anastomose at the level of insertion of the paired bracts, and in female flowers at the base of the ovary ('receptacular plexus' of Kaul), and at the top of the hypanthium ('stylar plexus' of Kaul). Bracts are usually supplied by a single median trace derived directly from the central vascular plexus, together with lateral bundles derived from peripheral regions of the axis. This is to be expected in view of the homology of leaves with bracts.

The various analyses which have been made suggest a reduction series. Tepals may be supplied with several traces in the less specialized genera with aerial pollination, but a single or even no trace in genera with small flowers. Kaul interprets the morphology of the ovary as illustrating a reduction series, beginning with genera with large multicarpellate flowers, like *Boottia*, *Ottelia*, *Hydrocharis*, and *Limnobium* and ending with genera considered to have three carpels, like *Blyxa*, *Vallisneria*, and *Elodea*, in which placentation appears to be parietal. The ultimate reduction is seen in *Halophila* and *Hydrilla*. Both Singh and Kaul recognize dorsal and ventral bundles, the reduction series involving fusion of ventral traces belonging to adjacent carpels and, in the most specialized flowers, loss of the presumed dorsal bundle. This reduction also is paralleled by reduction in the number of ovules per carpel. *Thalassia* seems to represent a specialized example because there appear to be fewer ovules than carpels (Tomlinson 1969a).

Classically, the gynoecium of Hydrocharitaceae has been interpreted as syncarpous, a concept supported by Singh from the evidence of floral anatomy and to some extent by Kaul (1968b) in his concept of fused ventral bundles. Troll (1931b), on the other hand, on the basis of comparative and developmental morphology considers the ovary to be 'pseudo-coenocarpous', i.e. strictly apocarpous but with the carpels secondarily aggregated and enveloped by receptacular tissue. This concept is discussed in detail by Kaul.

However, studies of floral development in Hydrocharitaceae which would provide useful evidence in these conflicting interpretations, are very limited (e.g. Caspary 1858; Rohrbach 1871; Montesantos 1912; Troll 1931b; Kaul 1969). It is clear that the morphology of the hydrocharitaceous flower is still very incompletely understood.

GENERIC DESCRIPTIONS

HYDROCHARIS L.
(Fig. 4.1 A–F)

A genus of three or more species, usually free-floating in fresh water but sometimes rooted in mud, with rosettes of long petiolate cordate spongy leaves with sheathing bases, and a pair of scales subtending each rosette, successive rosettes separated from one another by long stoloniferous internodes. Leaves floating or emergent.

VEGETATIVE MORPHOLOGY

Plants stoloniferous, with pairs of scales at intervals, the pairs separated from each other by long internodes, a single unbranched root being associated with each scale pair. Foliage leaves on erect shoots developed in association with scales.

VEGETATIVE ANATOMY

Leaf blade

Rolled in bud with involute vernation; up to 15 principal veins connected by transverse irregular and often anastomosing veins, but never with free vein-endings. Apical pore present. Squamules two per leaf axil. Leaf dorsiventral. **Hairs** absent. **Epidermis** of thin-walled cells, irregularly polygonal in surface view, with straight anticlinal walls on the abaxial surface (Fig. 4.1 F); adaxial anticlinal walls sinuous in surface view (Fig. 4.1 A), or markedly invaginated (e.g. *H. morsus-ranae*). Abaxial somewhat larger than adaxial epidermal cells. **Stomata** present, most common on the adaxial surface (e.g. *H. asiatica*), absent from the abaxial surface (e.g. in *H. morsus-ranae*). Stomata usually paracytic, the guard cells asymmetrical with only the outer ledge developed. **Mesophyll** including three layers of loose adaxial palisade cells (Fig. 4.1 B–D), interrupted by well developed substomatal chambers; abaxial mesophyll lacunose, honeycomb-like in paradermal view (Fig. 4.1 E) with transverse diaphragms of stellate cells, the diaphragms commonly with large tannin cells. Mechanical tissue absent except for **fibres** next to the major veins.

Midvein with a well developed outer sheath of slightly thick-walled or even fibre-like cells. Protoxylem lacuna well developed, metaxylem tracheids narrow and forming a layer next to the phloem. Lateral veins inversely orientated, situated in the palisade. Midrib region somewhat thickened and including minor inverted vbs.

Petiole

Circular in TS, with a slightly thickened epidermis of radially extended cells. **Ground tissue** lacunose with a network of supporting cells, including transverse diaphragms. **Vascular system** including a major arc of (3–) 5 central vbs together with a circle of peripheral minor vbs. Each vb with a parenchymatous sheath, a protoxylem lacuna and wider metaxylem tracheids; phloem with rather indistinct sieve-tubes. Peripheral vbs with reduced vascular tissue.

Stem (stolon)

Circular in TS. **Epidermis** shallow with outer wall slightly thickened. **Cortex** including 1 (–2) series of wide air-canals and a series of narrow vbs each with a single thin-walled metaxylem element. **Stele** surrounded either by an endodermis or by an endodermoid layer with no Casparian strips. Vascular tissue including two wide protoxylem lacunae along one axis, together with associated phloem representing one pair of vbs. A second pair at right angles represented by phloem alone or at the most with the addition of a thin-walled metaxylem element, the phloem *in toto* essentially forming a cylinder with wide sieve-tubes.

Stem (*rhizome-axis supporting leafy rosette*)

Epidermis of narrow uniform cells. **Cortex** compact or somewhat lacunose. Central vbs reduced and aggregated in the stem centre.

Root

Root-hairs commonly developed from initials formed by asymmetric division, each with a much enlarged base penetrating the exodermis, the cytoplasm resembling that described for *Stratiotes* and much investigated (e.g. by Cutter and Feldman 1970*a*, *b*; Cutter and Hung 1972; Hesse 1904; Wilson 1936; Ziegenspeck 1927). **Exodermis** uniseriate, the outer cortical layer immediately within also uniseriate, middle cortex with a single series of radial air-canals segmented by transverse diaphragms of ± stellate cells. **Endodermis** inconspicuous, small-celled, thin-walled. **Stele** narrow; pericycle uniseriate, distinct. Vascular tissue typically triarch, usually with three vessels alternating with three phloem strands, but additional sieve-tubes present towards the centre, according to Ancibor (1979). Vessels recorded in the roots of *H. morsus-ranae* by Ancibor (1979), the elements having long, v. oblique scalariform perforation plates with many (50+) thickening bars.

REPRODUCTIVE MORPHOLOGY

Plants monoecious. Inflorescence sessile (female) or stalked (male) with a pair of bracts (male), or a tubular structure (female) forming a conspicuous spathe enclosing a single female or 1–4 male flowers. Flowers either pedicellate (male), or with a long hypanthium (female), nectariferous. Sepals three, short and somewhat petaloid. Petals three, larger than the sepals, broad, white, and conspicuous. Stamens 9–12, the anthers bilocular, each often with a sterile appendage. Ovary inferior with six flat bifid styles, the single loculus usually having six prominent ovuliferous dissepiments. Staminodia and pistillodia developed. Fruit a globose, fleshy capsule, splitting irregularly, containing numerous elliptic spiny seeds.

REPRODUCTIVE ANATOMY

Female flower (Singh 1966*d*). Pedicel vascular system anastomosing at the base of the ovary; six prominent bundles, identified as dorsal carpel bundles, recognizable above level of anastomosing. Numerous ventral bundles supplying ovules. Anastomosing evident again at the top of the ovary, between peripheral vbs and remains of carpel vbs; the vascular complex giving rise to vascular supply to perianth, nectaries and styles. [Singh (1966*d*) interprets the three nectaries, which lie opposite the perianth segments of the inner whorl, as representing the expanded bases of styles. He also describes the carpels as constituting two whorls, an interpretation supported by the developmental observations of Rohrbach (1871).]

MATERIAL EXAMINED

This account is based mainly on that by Ancibor (1979) without examination of further material.

SPECIES REPORTED ON IN THE LITERATURE
(refers to *Hydrocharis morsus-ranae* unless otherwise stated)

Ancibor (1979): *H. asiatica, H. chevalieri, H. dubia, H. morsus-ranae*: all parts.
Arber (1922*b*): leaf.
Blass (1890): phloem.
Brunaud (1977): branching.
Bugnon and Joffrin (1963): branching and shoot morphology.
Chatin (1856): general morphology and anatomy.
Chauveaud (1897*a*): root.
Cutter and Feldman (1970*a, b*): trichoblasts (root-hair initials).
Cutter and Hung (1972): root-hair initiation.
Dibbern (1903): inflorescence morphology.
Gibson (1905): squamules.
Janczewski (1874): root development.
Kaul (1968*b*): floral anatomy.
Kaul (1970): inflorescence morphology.
Lorenz (1903): winter-bud morphology.
Matsubara (1931): winter-bud development.
Raunkiaer (1895–9): all parts.
Richard (1812): floral morphology.
Richards and Blakemore (1975): turion development.
Rohrbach (1871): all parts; developmental morphology.
Saunders (1929): floral anatomy.
Sauvageau (1887): root.
Schilling (1894): mucilage secretion.
Shinobu (1954): *H. asiatica*: stomata.
Singh (1966*d*): *H. dubia*: floral anatomy.
Solereder (1913): *H. asiatica, H. morsus-ranae*: leaf.
Solereder (1914): *H. parnassifolia, H. parvula*: leaf.
van Tieghem (1870–1): root.
van Tieghem and Douliot (1888): origin of lateral roots.
Wilder (1974*a*): developmental morphology.
Wilson (1936): root-hairs.
Ziegenspeck (1927): root-hair development.

LIMNOBIUM L.C. Rich.
(Figs. 4.2 and 4.3)

A genus of three species of free-floating herbs, or less commonly rooted in mud, with leafy rosettes extended by stolons (Figs. 4.2 A, B and 4.3 A), each rosette associated with a pair of scales. Leaves petiolate, with a broad, cordate, sometimes spongy blade (Fig. 4.2 C) and a basal stipular sheath. Blade with 7–11 main veins connected by frequent transverse veins; minor longitudinal veins inconspicuous. Vernation convolute.

VEGETATIVE MORPHOLOGY

(After Richard 1812; Rohrbach 1871; Montesantos 1912; but mostly after Wilder 1974b). Upright axes with rosettes of foliage leaves arranged spirodistichously (Fig. 4.2 A, B). Every alternate leaf associated with a complex primordium originating by bifurcation of vegetative apex, and interpretable as a precocious development of an axillary bud in a sympodial system. Each complex primordium in turn producing immediately three meristems ('components'). One of them ('vegetative component') becoming a stolon, a second ('fertile component') essentially becoming an inflorescence, the third primordium ('sterile component') normally remaining undeveloped. Inflorescence either wholly staminate or carpellate, or sometimes remaining 'neutral' by producing vegetative buds only. Vegetative buds also occurring at times on flower-bearing inflorescences.

Stolon initiated as the first internode (hypopodium) of the axis to be developed by the 'vegetative component' (Fig. 4.3 A). Two scale leaves produced on the stolon serving to protect it during extension, the stolon bifurcating again to produce a vegetative rosette and a second 'complex primordium'. Roots numerous in association with each vegetative rosette, dimorphic (major and minor), branched, producing an extensive fibrous system. Squamules numerous, in groups of two at each node (e.g. Fig. 4.3 D).

[For a full discussion of the precise symmetrical relationships and developmental details, see Wilder (1974b). It should be noted that each type of vegetative axis always produces another dissimilar to itself. Vegetative rosettes produce only stolons, stolons produce only vegetative axes (a comparable situation occurs in *Thalassia*).]

VEGETATIVE ANATOMY

Leaf blade

Dorsiventral. Hairs absent. **Epidermal** cells slightly thick-walled, isodiametric, and irregularly polygonal in surface view. Adaxial epidermal cells in surface view with distinctly sinuous walls, abaxial cells ± straight sided. Stomata present on both surfaces, but up to six times more frequent adaxially; those of the abaxial surface largest. Stomata paracytic, the guard cells asymmetric. **Mesophyll** with 2–3 adaxial palisade layers and large substomatal chambers; abaxial mesophyll markedly lacunose, with a honeycomb arrangement in surface view, traversed by transverse diaphragms of stellate cells and frequent tannin cells.

Mechanical tissue represented by **fibres** next to midrib. Major veins situated in longitudinal girders; **midrib** including an adaxial series of inverted vbs.

Stem (stolon)

Outline in TS circular, with a small-celled **epidermis** and peripheral **collenchyma**. Ground tissue lacunose, consisting of a network of parenchyma with transverse diaphragms. **Vascular system** including a central series of larger vbs and a peripheral series of smaller vbs.

Stem (rhizome)

Closely resembling those of other genera with erect rosette-producing stems.

Root

Root-hairs abundant, large, with endocytic basal structures recalling those in the root-hairs of *Stratiotes* (Delay 1941). **Exodermis** suberized in old roots. Middle cortex with a single series of septate radial air-canals, traversed by transverse septa consisting of stellate cells. **Stele** surrounded by a thin-walled endodermis. Xylem represented either by a central group of elements or by a more peripheral series. Phloem strands on radii alternating with protoxylem poles; sieve-tubes mainly solitary and next to (but not within) the pericycle.

Vessels in root said by Ancibor (1979) to have oblique scalariform perforation plates with 20–30 thickening bars.

REPRODUCTIVE MORPHOLOGY

Plants monoecious. Inflorescence with two (sometimes only one in female) basal free bracts, initially enclosing developing flowers (Fig. 4.3 B, C), the axis ending in a flower and producing usually two (sometimes one, rarely three) monochasia on each side of the terminal flower, with several (up to 12) flowers in each staminate monochasium (Fig. 4.3 B–E). Basal flowers frequently aborting in male inflorescences. Female inflorescences with few (1–4) flowers per monochasium. Flowers extending above water surface at maturity.

Male flowers (Fig. 4.3 F–G) with an extended receptacle; six perianth segments in two discrete whorls (Fig. 4.3 F). Stamens (6–) 9 (–12) in three discrete whorls. Female flowers (Fig. 4.3 H–K) with six perianth segments in two congested whorls (Fig. 4.3 H, I), six or more staminodes; ovary inferior with several (6–9) locules enclosing numerous ovules in laminar placentae (Fig. 4.3 J, K). Each carpel with a short style bearing two long hairy stigmas clustered in the flower centre (Fig. 4.3 H, I). **Fruit** (Fig. 4.2 D, E) a fleshy capsule, bending downward and maturing beneath the mud. Seed (Fig. 4.2 F) with a spiny testa developing a mucilaginous coating.

REPRODUCTIVE ANATOMY

(After Kaul 1968*b*). **Female flower** with ring of vascular tissue giving rise to (i) staminodial and perianth traces, (ii) dorsal and fused ventral carpel traces. Placental bundles derived from reticulum connecting dorsal and ventral traces together with some bundles directly from the vascular plexus. Dorsal and fused ventral traces supplying styles without forming a stylar plexus.

Male flower with a receptacular plexus supplying sepal traces and a single trace to each stamen. Petal traces derived ± directly from pedicel bundles.

MATERIAL EXAMINED

This account is based on the description by Ancibor (1979) without special study of further material.

SPECIES REPORTED ON IN THE LITERATURE

Ancibor (1979): *Limnobium spongia* (as *L. bosci*), *L. stoloniferum**: all parts.
    * *L. laevigatum* according to C. D. K. Cook (personal communication).
Brunaud (1977): *L. stoloniferum*: branching.
Dammer (1888): *L. stoloniferum*: general morphology.
Delay (1941): *L. bogotense*: root-hairs.
Drawert (1938): *L. bogotense* (as *Trianea bogotense*): hydropoten.
Kaul (1968b): *L. spongia*: floral anatomy.
Kaul (1970): *L. spongia*: inflorescence morphology.
Kaul (1976b): *L. spongia*: leaf.
Montesantos (1912): *L. spongia* (as *L. bosci*): general morphology and anatomy; development.
Perner and Losado-Villasante (1956): *L. bogotensis* (as *Trianea bogotensis*): root-hairs.
Richard (1812): *L. spongia* (as *L. bosci*): morphology.
Schilling (1894): *L. bogotensis* (as *Trianea bogotensis*): mucilage secretion.
Solereder (1913): *L. stoloniferum* (as *Trianea bogotensis*): leaf.
Wilder (1974b): *L. spongia*: developmental morphology.

STRATIOTES L.
(Fig. 4.9 A)

A single existing species (*S. aloides* L.), but known as a frequent fossil (Chandler 1923; Palamarev 1979), with floating leafy rosettes, propagated by stoloniferous offsets. Stolons ultimately forming terminal winter buds and sinking to the bottom of the pond via accumulated lime; overwintering and rising to the pond surface the following spring. Leaves linear, up to 40 cm long, usually with seven main longitudinal veins and more numerous minor veins. Marginal teeth prominent, often calcified. Squamules two or more per leaf axil, pointed and tanniniferous.

VEGETATIVE MORPHOLOGY

Leafy rosette usually spirodistichous, producing lateral axillary shoot complexes at regular intervals, the lateral meristem arising by bifurcation of the shoot apex but considered by Brunaud (1976) to be a precocious axillary bud, rosette consequently a monopodium. Lateral complex usually trifurcating; central meristem becoming a developed stolon; lateral meristems either becoming inflorescences or additional stolons, the latter commonly being suppressed. Each axis usually having a basal pair of scales subtending further lateral meristems; distally scales and foliage leaves developing in stoloniferous shoots, bracts in reproductive shoots.

VEGETATIVE ANATOMY

Leaf blade

Thickened basally and ± heart-shaped in TS with a prominent angular midrib. **Hairs** represented by numerous marginal prickles, each mounted on a prominent group of marginal cells (Fig. 4.9 A). Cuticle thin, striate. **Epidermis** colourless, of ± isodiametric polygonal or somewhat longitudinally extended cells, the anticlinal walls straight in surface view. **Stomata** present distally on certain leaves, equally abundant on both surfaces, usually paracytic, but somewhat variable. Guard cells asymmetric in TS, only the outer ledge developed. **Mesophyll** almost uniformly chlorenchymatous; isolateral, with a palisade-like layer adjacent to each surface. Air-canals in the centre of the mesophyll forming 1–2 series distally, but multiseriate proximally; alternating irregularly with major veins. Transverse diaphragms present, commonly including **tannin cells,** especially towards the leaf base.

Vascular system arranged in three series of bundles: (i) major veins embedded in wide parenchymatous girders extending to both leaf surfaces, the vein equidistant from each surface, (ii) and (iii) a series of minor veins, one abaxial (ii), the other adaxial (iii), within shallow girders mostly continuous with one surface. Adaxial series of vbs with inverted orientation. Larger vbs each enclosed by an outer parenchymatous sheath, and an inner sheath represented by a series of thin-walled fibres, usually most fully developed below the phloem. Protoxylem lacuna narrow, metaxylem represented by two wide lateral and several narrower series of tracheids. Larger metaxylem elements thin-walled, with annular or scalariform wall thickenings.

Midrib becoming collenchymatous in the abaxial angle at the leaf base. Mesophyll in the same region more uniformly lacunose. Protoxylem lacuna wide, metaxylem inconspicuous or absent.

Stem (stolon)

Epidermis thick-walled, small-celled. Outer cortex collenchymatous, but ground tissue somewhat lacunose towards the stem centre. **Vascular system** including one large central and c. eight peripheral vbs of varying size with rather irregularly arranged protoxylem, sometimes forming a lacuna. Large central vb without a distinct endodermis, apparently multipolar with several protoxylem lacunae.

Peduncle

With lacunose ground tissue, including a peripheral ring of vbs and several scattered central vbs (comparable to the multipolar vb of the stolon).

Stem (rootstock)

Outline irregular, surface layers becoming suberized. **Cortex** wide, lacunose, with narrow air-canals traversed by irregular transverse septa, each septum including a large tannin cell. **Endodermis** indistinct. **Stele** broad; central vbs irregularly anastomosing and including short tracheids.

Root

Unbranched. **Root-hairs** numerous, wide, often long, their bases being deeply sunken within the exodermis and exhibiting a markedly endocytic elaboration. Each hair base surrounded by a layer of small endodermoid cells, with large nuclei and obscure Casparian strips on the radial walls (Guttenberg 1968). **Exodermis** becoming suberized. **Cortex** compact; collenchymatous at the periphery, middle cortex with short radial air-canals, transverse diaphragms infrequent. **Endodermis** thin-walled. **Stele** polyarch with a peripheral series of phloem strands and an inner series of wide metaxylem elements with thin walls.

Secretory and conducting elements

Crystals present in mesophyll of leaf as rod-like structures, often aggregated. **Tannin** frequent, notably in diaphragms of leaf and stem. **Phloem** with wide sieve-tubes, sometimes interrupting the pericycle in the root. **Xylem** vessels recorded by Ancibor (1979) as having elements up to 500 μm wide, with v. oblique scalariform perforation plates.

REPRODUCTIVE MORPHOLOGY

Plants dioecious. Inflorescences each with a long stalk supporting a pair of overlapping bracts with a spiny keel. Flowers numerous in each male inflorescence, associated with numerous bracteoles; female flower solitary. Male flower with three green sepals, three large broad white petals; stamens up to 13, with short filaments, surrounded by numerous staminodes (up to a total of 36 stamens and staminodes, according to Richard 1812). Female flower with perianth like that of the male; with numerous staminodes. Ovary inferior with numerous ovules on marginal dissepiments. Styles six, each with a pair of linear stigmas. Fruit globose, seeds few.

REPRODUCTIVE ANATOMY

(After Kaul 1968b.) **Female flower** with each carpel including a dorsal and two unfused ventral bundles; placental bundles derived from dorsal and ventral bundles but those of adjacent carpels unfused. Stylar bundles derived from distal continuation of ventral and some placental bundles. Perianth and staminodia supplied directly from receptacular plexus. **Male flower** having a single trace to each stamen, but three traces to each nectary staminodium.

MATERIAL EXAMINED

This account is mainly based on the description of *Stratiotes aloides* L. by Ancibor (1979).

Species Reported on in the Literature
(all refer to *Stratiotes aloides*)

Ancibor (1979): vegetative anatomy.
Arber (1914): root development.
Beckerowa (1934): root-hairs.
Blass (1890): phloem.
Brunaud (1976, 1977): shoot morphology and branching.
Costantin (1886*b*): leaf.
Gibson (1905): squamules.
Janczewski (1874): root development.
Kaul (1968*b*): floral anatomy.
Kaul (1970): inflorescence morphology.
Kny (1878): root-hairs.
Raunkiaer (1895–9): stem and leaf.
Richard (1812): floral morphology.
Rohrbach (1871): vegetative morphology and development, reproductive morphology and development.
Salisbury (1926): floral morphology.
Schencke (1893): all parts.
Solereder (1913): leaf.
Streitberg (1954): leaf morphology.
Theorin (1905): trichomes.
van Tieghem and Douliot (1888): origin of lateral roots.
Veres (1908): general morphology.
Wilson (1936): root-hairs.

ENHALUS L.C. Rich.
(Fig. 4.4 and Plate 4 A, B)

A genus consisting of one dioecious species *E. acoroides* (L. f.) Rich. ex Steud., a seagrass occurring in tropical waters of the Indian and Pacific Oceans, from Madagascar to Australasia (den Hartog 1957*b*, 1970*a*); often extensive in shallow water and exposed at low tides. Flowering often in monthly cycles, and apparently as a response to the tidal fluctuation (Svedelius 1904; Troll 1931*b*). Pollination at the water surface. Submerged male flowers when released floating to the surface and there encountering the floating female flowers, much as in *Vallisneria*.

Vegetative Morphology

Vegetative axes **monomorphic** and monopodially branched, creeping horizontally in the substrate. Rhizomes dorsiventral with distichously arranged foliage leaves, the plane of distichy vertical, the lateral buds being present only in the axils of each of the upper leaves, and either developing infrequently and irregularly as proliferative vegetative axes, or possibly as inflorescences. Details of branching not known. Older parts clothed with persistent stiff fibrous remains of leaf sheaths, representing the massive sclerenchymatous sheaths of vbs. Roots arising from the lower surface, thick, spongy and un-branched.

VEGETATIVE ANATOMY

Leaf blade (Fig. 4.4)

Leaves eligulate, strap-shaped, the blade up to 1.5 cm wide and over 1 m long, with an open sheathing base, the blade alone usually deciduous or broken. **Squamules** abundant, 10–12 per leaf axil, each elongated, and up to six cells thick, broad and with frequent tannin cells. Leaf apex rounded, without an apical pore. Leaf margin thickened and flanged adaxially towards the base (Fig. 4.4 A). **Hairs** represented by unicellular marginal prickles; scale-like outgrowths of the lamina surface reported in some samples by Tomlinson (1980). Cuticle thin. **Epidermis** somewhat thick-walled, especially the outer wall, densely chlorophyllous; cells isodiametric or slightly elongated in surface view, sometimes tanniniferous; fine calcium oxalate crystals common. **Stomata** absent.

Mesophyll lacunose, with 1–2 longitudinal air-lacunae between each adjacent pair of veins (Fig. 4.4 A, B). Lacunae delimited above and below by 1 (–2) hypodermal cell layers and laterally either by uniseriate partitions up to six cells high or by multiseriate girders including the veins. Mesophyll cells large, elongated, and with only diffuse chloroplasts. Lacunae separated laterally by uniseriate partitions of narrow lobed cells enclosing small intercellular spaces, the partitions sometimes including transverse veins. Tannin cells frequent, mainly hypodermal.

Veins up to 30 in number, including a large abaxial, but not protruding, midvein, a large marginal or submarginal vein, and smaller lateral veins forming a separate adaxial and abaxial system. Adaxial veins with inversely orientated vascular tissue (Fig. 4.4 C). Midvein (Fig. 4.4 B) with an indistinct parenchyma sheath, sheathing fibres well developed below. Protoxylem lacuna in the midvein well developed. Phloem with rather irregularly developed sieve-tubes. Lateral veins accompanied by well developed fibres next to phloem but vascular tissue reduced, the protoxylem lacuna usually being absent. Tannin cells occasional in vascular bundles.

Leaf sheath

Colourless, with sheathing wings, biseriate at the margin. Hypodermal mesophyll up to six cells deep. Air-lacunae well developed. **Veins** forming a single abaxial layer, each with a massive abaxial fibrous bundle sheath, especially that vein becoming the sub-marginal vein of the blade at its distal end. Adaxial vein system of the blade not continuous basally into the sheath, but fusing with the abaxial system via transverse commissures at the mouth of the sheath.

Rhizome (Plate 4 A, B)

Epidermis persistent, but outer cortical layers lignosuberized with age to form an indistinct protective layer of dead cells. **Cortex** with wide intercellular spaces, but not lacunose and without transverse diaphragms. Cortex apparently either lacking an independent cortical vascular system but traversed by leaf traces extending into the stele, or at most with obscure vbs in the outer cortex, representing basal continuations of the lateral veins of the leaf (Cunnington 1912).

Stele narrow, delimited by a thin-walled endodermis. Pericycle indistinct, interrupted by numerous narrow vbs. Central vbs irregular, the xylem with narrow, spirally or reticulately thickened tracheids. Tannin cells abundant; starch often common.

Peduncle

With markedly lacunose ground tissue; peripheral cortex including a single series of narrow vbs; central vascular system represented by two large vbs.

Root

Unbranched, up to 2 mm in diameter, arising close to the rhizome apex in the pericycle region. Epidermis with root-hairs forming a short, dense indumentum. Exodermis of 1–3 layers of compact cells. Middle cortex becoming lacunose by enlargement of radial intercellular spaces and collapse of radial cell files. Transverse diaphragms developed, usually biseriate but rather irregular, the cells with characteristic lobes at the corners, filling the intercellular spaces. Stele narrow, delimited by the thin-walled endodermis, pericycle uninterrupted. Vascular system usually hexarch with each xylem pole represented by spirally thickened and slightly lignified tracheids and alternating with the phloem; each phloem strand consisting of a single, rather indistinct sieve-tube. Centre of the stele undifferentiated. Vessel elements with v. oblique scalariform perforation plates with 7–10 thickening bars recorded by Ancibor (1979).

REPRODUCTIVE MORPHOLOGY

Female flower solitary on a long peduncle and enclosed by the two overlapping, scarcely fused bracts, floating on the surface at maturity while still attached. Outer surface of hypanthium covered with numerous scaly protuberances. Sepals three, recurved, petals three, erect; staminodes present. Styles six and forked just above the base. Ovary compressed, with several anatropous ovules attached basally on the six incomplete dissepiments. Flowers drawn beneath the water surface after pollination, the fruit developing submerged. Male flowers in shortly pedunculate inflorescences remaining ± enclosed by the leaf sheaths; numerous and initially enclosed by the two fused bracts; each flower thinly pedicellate, with six tepals in two series and three stamens. Pollen grains large, few, trinucleate. Male flowers becoming detached and eventually released by the splitting of the bracts, then floating with the tepals recurved. For details of pollination see Svedelius (1904), Troll (1931b), and den Hartog (1957b, 1970). Fruit a fleshy capsule with a warty, spinous surface; capsule dehiscing irregularly to release the few angular seeds. Embryo with a massive hypocotyl, the radicle said to be undeveloped (Cunnington 1912).

REPRODUCTIVE ANATOMY

(After Singh 1966*d*; Kaul 1968*b*; see also Kaushik 1940.)

Male flowers each with a single vb derived from a central mass of vascular tissue formed above the departure of the bract traces. Each floral appendage receiving a single, unbranched trace. Tannin cells common. **Female flower** with two rings of vbs at the base of the ovary (i.e. above the departure of the bract traces), a central ring of six and a peripheral ring of nine vbs. Central ring supplying carpels, each carpel with five traces, the ventral pair supplying the ovule traces. The dorsal bundles extend into the stigmatic lobes. The outer ring of vbs supplies the perianth traces at the top of the ovary.

Fruit anatomy not described, except for brief notes by Cunnington (1912).

TAXONOMIC NOTES

Pollen morphology (Erdtman 1952) supports the concept of *Enhalus* being closely related to *Vallisneria* but not to *Thalassia.* However, the differences in pollination mechanisms have to be taken into account.

MATERIAL EXAMINED

The account is based on the same material examined by Ancibor (1979); together with the following:

Enhalus acoroides (L. f.) Rich. ex Steud. Tarauma, Papua New Guinea; P. B. Tomlinson 31.x.74N; all vegetative parts.

SPECIES REPORTED ON IN THE LITERATURE
(all refer to *Enhalus acoroides*)

Ancibor (1979): all parts.
Cunnington (1912): all parts.
den Hartog (1970*a*): general morphology.
Kaul (1968*b*): floral anatomy.
Kaul (1970): inflorescence morphology.
Kaushik (1940): anatomy of female flower.
Magnus (1871): leaf, stem.
Miki (1934*a*): leaf, stem, floral morphology.
Sauvageau (1890*d*): leaf.
Singh (1966*d*): floral anatomy.
Solereder (1913) (as *E. koenigii*): leaf.
Svedelius (1904): floral morphology and biology.
Tomlinson (1974*a*): rhizome morphology.
Tomlinson (1980): leaf.
Troll (1931*a*): general morphology, leaf development; floral development and biology.
Troll (1931*b*): floral morphology and development.

[The material described by Chatin (1856) was probably of *Posidonia* and not *Enhalus* (Cunnington 1912).]

OTTELIA Pers. (including *Boottia* Wall.)
(Fig. 4.1 G–M; Plates 2 A, B and 3 D)

A genus of about 40 phenotypically plastic species, widely distributed in warmer climates as plants of fresh, often running water but occupying somewhat diverse habitats, e.g. fixed to the substrate and either wholly submerged or with floating leaves. Usually consisting of a leafy rosette with a short erect axis, never stoloniferous. Leaves varying from long-petiolate with a broad cordate blade to undifferentiated and frequently ribbon-like; vernation of bladed leaves involute. Roots unbranched.

VEGETATIVE MORPHOLOGY

No detailed descriptions available.

VEGETATIVE ANATOMY

Leaf blade (Fig. 4.1 G–I)

Venation parallel with few transverse commissures (*O. ulvifolia*), or reticulate but with 7–15 main veins (e.g. *O. alismoides*). Leaf margin smooth or with **prickle-hairs** supported by large emergences; similar structures sometimes also present on leaf surface (e.g. *O. muricata*). Cuticle thin. **Epidermis** uniformly of polygonal cells, the two surfaces identical in submerged leaves (Fig. 4.1 H, I) but strikingly contrasted in floating leaves, anticlinal walls in surface view straight or somewhat sinuous (e.g. *O. cylindrica, O. yunnanensis*); anticlinal walls with conspicuous invaginations in e.g. *O. ovalifolia*. **Stomata** absent from submerged leaves, present adaxially on floating leaves; paracytic, asymmetric in TS with only the outer ledge developed. **Mesophyll** of floating leaves with large substomatal cavities between 2–3 palisade layers, but otherwise mesophyll lacunose with transverse but perforated diaphragms (Kaul 1976*b*). Mechanical tissue represented only by sclerotic tissue supporting the veins. Mesophyll of submerged leaves with a series of wide lacunae separated by a honeycomb lattice of uniseriate girders (Fig. 4.1 G–I); hypodermis uniseriate towards midrib, absent from margin. **Veins** arranged in a single series in the mesophyll, but in several series in the midrib; often inverted. Leaf anatomy of narrow leaves of *O. muricata* and *O. scabra* more like that of a petiole.

Petiole (Fig. 4.1 J, K)

Outline triangular or ± oval; epidermis small-celled. Ground tissue lacunose, with transverse diaphragms, vbs in a central arc of 3–7, together with a series of peripheral, adaxially inverted vbs. Diaphragm cells with thick walls, e.g. in *O. alismoides* (Fig. 4.1 K).

Peduncle (Plate 2A, B)

Oval in outline, with a lacunose ground tissue (Plate 2A) and a peripheral series of vbs, those towards the centre the largest (Plate 2B). Fruiting peduncle with more extensive peripheral collenchyma, and the ground tissue lacunae often arranged in a pattern.

Root (Fig. 4.1 L, M)

Epidermis ephemeral and replaced by a persistent **exodermis.** Cortex with a central lacunose region traversed by transverse septa consisting of cruciform cells (Fig. 4.1 L, M); inner cortical layer thick-walled. **Stele** with a central xylem lacuna, the peripheral xylem strands said by Ancibor (1979) to include vessels with 15–30 thickening bars in the scalariform perforation plates.

REPRODUCTIVE MORPHOLOGY

Plants monoecious or dioecious, flowers unisexual or bisexual (rarely cleistogamous, Ernst-Schwarzenbach 1956). Inflorescence with a stout stalk, spathe typically ribbed or winged and enclosing a single flower. Sepals three, green; petals three, longer than the sepals, white or coloured. Stamens 6–15; staminodia frequent. Ovary inferior, oblong, narrowed at the top, unilocular with six dissepiments. Styles 6–15, bifid, stylodia present in male flowers. Fruit oblong, fleshy, splitting, or rotting to release numerous minute seeds.

REPRODUCTIVE ANATOMY

(After Singh 1966*d*; Kaul 1968*b*.) In *O. alismoides* (with perfect flowers), anastomosing vascular system at base of ovary, above level of vascular supply to spathes. Six carpel dorsal and six carpel ventral vbs identified, some of these supplying the ovule traces via 'placental strands'. Further anastomosing at the top of the ovary producing the vascular system to the perianth, stamen, staminodes, and styles. Each perianth segment receiving three traces, each stamen and style receiving a single trace. Staminodes having either a short trace, or none.

In *Boottia cordata* (dioecious) only **male flowers** examined by Singh (1966), with 12 stamens arranged in four whorls, and six stylodia arranged in two whorls. Vascular tissues derived from a ± complete vascular cylinder at the base of the flower. Perianth segments each receiving three vbs, stamens and stylodia each receiving a single vb. **Female flower** (according to Kaul 1968*b*) with a dorsal and fused ventral bundle supplying each carpel, the placental supply arising partly from these but partly from the receptacular plexus directly. Stylar plexus well-developed.

MATERIAL EXAMINED

This account is based mainly on that by Ancibor (1979) with examination of additional material:

Ottelia alismoides (L.) Pers.; Botanic Gardens, Lae, Papua New Guinea; P. B. Tomlinson 28. x. 74.

SPECIES REPORTED ON IN THE LITERATURE

Ancibor (1979): *Ottelia alismoides, O. cylindrica, O. fisherii, O. muricata, O. ovalifolia, O. polygonifolia, O. scabra, O. somalensis, O. ulvifolia, O. yunnanensis*: mainly leaf.

Chatin (1856): *O. alismoides*: vegetative anatomy.

Datta and Biswas (1976): *O. alismoides*: brief notes on anatomy of root, scape, leaf.

Ernst-Schwarzenbach (1956): *O. alismoides, O. ovalifolia*: floral morphology, anther anatomy.

Kaul (1968b): *O. alismoides, O. cordata* (as *Boottia cordata*): floral anatomy.

Kaul (1969): *O. alismoides, O. cordata* (as *Boottia cordata*) and their synthetic hybrid: floral anatomy.

Kaul (1970): *O. alismoides, O. cordata* (as *Boottia cordata*): inflorescence morphology.

Kaul (1976b): *O. cordata* (as *Boottia cordata*): leaf.

Majumdar (1938): stem.

Montesantos (1912): *O. alismoides*: morphology, root, floral development.

Richard (1912): *O. alismoides*: floral morphology.

Singh (1966d): *O. alismoides, O. cordata* (as *Boottia cordata*): floral anatomy.

Solereder (1913): *O. alismoides, O. baumii, O. japonica. O. ulvaefolia*: leaf.

    And under *Boottia, B. aschersoniana, B. kunehensis, schinziana, B.*: leaf.

Swamy (1963): *O. alismoides*: embryo development.

TRIBE ANACHARITEAE
(Figs. 4.5 and 4.6)

Including the five genera *Egeria* Planch. (two spp), *Elodea* Michx. (17 spp), *Hydrilla* L. C. Rich. (one sp), *Lagarosiphon* Harvey (17 spp), *Nechamandra* Planch.* (one sp). Plants widely distributed in temperate and tropical fresh waters, often introduced and becoming weedy in eutrophic conditions. Submerged and either free-floating or rooted on the bottom in shallow water; pollination aerial (e.g. *Elodea, Lagarosiphon*) or at the surface of the water (e.g. *Hydrilla*). Sometimes overwintering by specialized buds (hibernaculae).

VEGETATIVE MORPHOLOGY

Stems (Fig. 4.5 A, C) extended with narrow (to 3 mm diameter) axillary branches at regular but distant intervals (5–15 nodes); plants sometimes described as stoloniferous (*Elodea*) without further details. Stem segmented in *Egeria* (Jacobs 1946), the segments each with a branch and breaking immediately above a double leaf whorl, facilitating vegetative dispersal. Leaves alternate, sub-opposite, or whorled, when spirally arranged often appearing in pseudowhorls of 2–5 or more by differential elongation of internodes (Fig. 4.5 A). Leaves linear to broadly ovate with narrow insertions, undifferentiated and each usually with a single central vein (e.g. Fig. 4.5 E). Leaf in *Nechaman-*

---

* *Nechamandra* is more appropriately considered with *Vallisneria* (C. D. K. Cook, personal communication).

dra widened at the base with an encircling insertion, and with 1–3 lateral veins on each side of central vein. **Squamules** two (e.g. Fig 4.5 D), laterally at each node, the thickness and length of the squamule said to be somewhat specifically diagnostic (Ancibor 1979). Roots fine, unbranched, widely spaced, usually arising at branching nodes; often remaining unextended.

Shoot apex studied extensively because of its unusual elongated shape and accessibility (e.g. Brunaud 1976; Dale 1957; Stant 1952).

VEGETATIVE ANATOMY

Leaf blade

From narrowly linear to broadly ovate. **Hairs** represented by unicellular apically-directed prickle-hairs at the margin (Fig. 4.6 M); similar hairs, either solitary or in aggregates, on enlarged adaxial protuberances of the midrib in *Hydrilla* (Figs. 4.5 E and 4.6 G). **Hydropoten** recorded by Ancibor (1979), e.g. in *Egeria densa*, but not by Drawert (1938). Cuticle thin. Lamina mostly biseriate and represented by the two chlorenchymatous epidermal layers (except in the thickened midrib), with narrow intercellular spaces between the layers (Fig. 4.6 A–D). Leaf in *Nechamandra* multiseriate with wide longitudinal, transversely septate air-canals. **Epidermal cells** rectangular in surface view, usually somewhat elongated (Fig. 4.6 E, F). Outer epidermal wall slightly thickened, especially in *Lagarosiphon major*; adaxial epidermal cells always larger than abaxial (Fig. 4.6 C, E, F). **Stomata** absent. Mechanical tissue either absent or represented either by somewhat thickened narrow marginal cells (e.g. *Nechamandra, Hydrilla*), by the cells of the midrib (e.g. *Egeria*), or by occasional fibres (e.g. *Lagarosiphon*). **Midrib** thickened, usually narrow with 1–2 mesophyll layers on each side of the midvein, but midrib region occasionally broad, lacunose, and occupying up to half width of blade (e.g. *Lagarosiphon crispus*). Midvein including a narrow protoxylem lacuna, narrow phloem elements and sometimes a few abaxial fibres (e.g. *Elodea*).

Stem (Fig. 4.6 H–J, N–O)

Hairs and **stomata** absent. Cuticle thin. **Epidermis** undifferentiated, the outer walls slightly thickened. **Cortex** wide (Fig. 4.6 H, N), 1–2 outermost layers somewhat collenchymatous, middle cortex including several series of intercellular air-lacunae extending the whole length of the internode without interruption; mechanical tissue absent. Chlorenchymatous throughout, the chloroplasts in the stele smaller than those in the cortex. **Nodal regions** usually distinguished by transverse diaphragms of short, sometimes thick-walled and often starch-filled cells together with short tracheary elements in the stele. Cortical vascular tissue either absent (*Egeria, Hydrilla*, Fig. 4.6H) or represented by a single cylinder of about six reduced vbs in the outer cortex (e.g. *Elodea* (Fig. 4.6 N, O), *Nechamandra, Lagarosiphon*), each vb represented by 3–4 narrow elongated cells or perhaps by a functional sieve-tube surrounded by a sheath of parenchyma. Cortical vascular tissues continuous with that of the stele at the node (Dale 1957). **Stele** narrow, delimited by a thin-walled endodermis, the Casparian strip sometimes obscure (e.g.

Elodea); vascular tissue represented by a conspicuous central protoxylem lacuna, the latter sometimes including tracheids with annular or spiral-reticulate wall thickenings, together with occasional peripheral metaxylem elements (e.g. *Hydrilla*, Fig. 4.6 J). Phloem surrounding the central xylem lacuna and represented by numerous conspicuous sieve-tubes, the outermost next to the endodermis. Features of sieve-tubes described by Blass (1890) and Currier and Shih (1968).

Buried portion of stem in *Lagarosiphon major* including thick-walled cortical cells and relatively well developed metaxylem, according to Ancibor (1979).

Root (Fig. 4.6 K, L)

Epidermis with or without root-hairs. **Root-hairs** studied developmentally and experimentally, e.g. by Leavitt (1904), Wilson (1936), and especially Cormack (1937). Basal portion of the root-hairs in *Elodea canadensis* and *Lagarosiphon major* with 'endocytic' wall elaboration according to Ancibor (1979; see also Kroemer 1903). **Exodermis** uniseriate, undifferentiated, or becoming suberized in older roots. Outer **cortex** collenchymatous in *E. canadensis* according to Ancibor (1979). Middle cortex made somewhat lacunose by irregular enlargement of the intercellular spaces but diaphragms absent. **Stele** (Fig. 4.6 L) narrow, delimited by a thin-walled endodermis. **Pericycle** uniseriate, distinct. **Xylem** represented by a central protoxylem cavity and sometimes also by incompletely differentiated cells alternating with sieve-tubes (Fig. 4.6 L).

Phloem consisting of three or more equally spaced strands, each with a single sieve-tube next to the pericycle.

Secretory and storage elements

Crystals infrequent. **Tannin** common in the epidermis of the leaves, stem and nodes as well as in the cortex of the stem and root. **Starch** mostly in cortical ground tissue, but especially in nodal diaphragms of the stem.

REPRODUCTIVE MORPHOLOGY

Plants usually dioecious, or monoecious. Flowers lateral at infrequent but regular intervals usually associated with one leaf of a pseudowhorl (Fig. 4.5 c), mostly unisexual, occasionally bisexual, either solitary and enclosed by a tubular spathe, or in male flowers of *Egeria* 2–4 aggregated within a common spathe. In *Lagarosiphon* and *Nechamandra* with up to 50 male flowers or 1–3 female flowers. Mouth of the spathe oblique, often bidentate. Male flowers sessile, usually becoming detached and floating to the surface. Female flowers with a v. long hypanthium capable of raising the receptive organs to the water surface when functionally mature. Perianth usually in two separate imbricate trimerous whorls (petals two in *Nechamandra*), the outer commonly sepaloid, the inner usually petaloid, but both whorls petaloid, e.g. in *Hydrilla* (Fig. 4.5 B). Stamens three, two fertile, one sterile (*Nechamandra*), three (*Hydrilla*, Fig. 4.5 H–J), six, three fertile, three sterile (*Lagarosiphon*), six or nine (e.g. *Elodea, Egeria*), commonly in whorls but separated on an extended receptacle, sometimes represented by staminodes (e.g. *Lagaro-*

siphon). Pollen in tetrads in *Elodea*. Female flowers sometimes with stami-nodes (e.g. *Elodea, Lagarosiphon*). Stigmas or stylodia rarely recorded in male flowers of *Elodea* (e.g. Wylie, 1904; Singh 1966*d*). Ovary inferior, nar-rowly cylindrical with few ovules usually on three parietal placentas. Stigmas three, either short, papillose at the base of the perianth cup (e.g. *Hydrilla,* Fig. 4.5 F, or expanded and bifid (e.g. *Elodea, Lagarosiphon*). Fruit a narrow, thin-walled capsule. Seeds few, oblong elliptic to fusiform.

REPRODUCTIVE ANATOMY

The following notes, from Singh (1966*d*), Kaul (1968*b*), indicate something of the range of vascular anatomy in this group, which can be interpreted as a progressive reduction series.

Nechamandra alternifolia (dioecious)

According to Singh (1966*d*), structure and vascular supply of peduncle similar in male and female flower, with a single trace to each bract derived from an aggregate of vascular tissue below the numerous male flowers or solitary female flower. Male flowers each with a single pedicel trace; perianth segments (four in two whorls) without a vascular supply; two stamens each with a single trace. Female flowers with two dorsal and two ventral vbs, the latter entirely supplying the ovules. Dorsal bundles distally supplying the three styles; the three perianth segments also without vascular tissue.

[On the basis of this evidence, Singh interprets the ovary as being composed of two carpels, but this may be a peculiarity of his collection because Kaul (1968*b*) describes three carpels, each with a single bundle interpreted as the fused ventral bundles of adjacent carpels, the dorsal bundles being com-pletely absent.]

Elodea canadensis (dioecious)

According to Singh (1966*d*), the male inflorescence including four or more flowers enclosed in a common spathe. Spathe with two vbs. Each male flower with a single trace, producing three separate vbs in the pedicel. Perianth segments of both whorls each supplied by a single vb. Stamen traces arising in three successive whorls, each stamen receiving a single vb. Stylodia without vascular tissue. According to Kaul (1968*b*), female flower without dorsal bundles to the three carpels; ovular supply represented by a single bundle opposite each placental ridge; petals and staminodia supplied from stylar plexus.

Hydrilla verticillata (dioecious cf. Fig. 4.5 F–J)

According to Singh (1966*d*), structure and vascular supply of peduncle and spathe similar in male and female flower, with a pair of vbs entering the tubular spathe. Male flowers with a single trace to each outer perianth member, but inner perianth whorl remaining unvasculated. Stamens each supplied by a single trace. Female flower with vascular system reduced to three vbs. In the ovary these bundles situated opposite and supplying each placenta, distally the bundle supplying a single trace to the outer perianth whorl and the styles. Inner perianth whorl not vasculated. Staminodia reported by Kaul (1968b), who also records traces to inner perianth whorl in both sexes.

FLORAL BIOLOGY

The floral morphology of taxa in the group is best understood with reference to the method of pollination; the studies of Ernst-Schwarzenbach (1945) are particularly relevant (see summary by den Hartog 1957b).

Pollination may be aerophilous, taking place above the water surface, e.g. *Egeria* (= *Elodea*) *densa* said to be pollinated by flies (den Hartog 1957b), or at the water surface but without the pollen itself coming into contact with water. In *Hydrilla* the male flowers float to the water surface and anthesis is the result of sudden erection of the stamens as the perianth parts are abruptly turned inside out. In *Lagarosiphon* male and female flowers come directly into contact (cf. *Vallisneria, Enhalus*).

In *Elodea* pollination is said to be hydrophilous since the pollen floats to the stigma on the water surface. The pollen is released explosively in *E. callitrichoides* and *E. occidentalis*. According to Wylie (1904), the pollen in *Elodea canadensis* is not entirely wetted because the spiny exine retains enough air to allow the grains to float. If the air is eliminated (as by dilute alcohol) the grains sink in water.

TAXONOMIC NOTES

Although the anatomy of this group is very uniform Ancibor suggests a number of features that have specific or even generic diagnostic value, e.g. squamule shape, epidermal cell size, distribution and shape of prickle hairs, fibre distribution in the leaf, cortical vascular system of the stem. However, little is known about variation of these characters in a given taxon since existing observations are based on few samples.

Nechamandra is probably incorrectly placed in this tribe; its closest affinity seems to be with *Vallisneria* (C. D. K. Cook, personal communication). In view of the abundance of many species in this group as water weeds an extended study of their vegetative development would be useful.

MATERIAL EXAMINED

This account is based mainly on that by Ancibor (1979), using some of the same material.

SPECIES REPORTED ON IN THE LITERATURE

Literature on this group of plants is very extensive; the following citations are to the more significant papers. No attempt has been made to survey the literature in which these plants have been used as experimental subjects.

Ancibor (1979): *Egeria densa, E. najas, Elodea callitrichoides, E. canadensis, E. chilensis, E. granatensis, E. nuttallii, E. potamogeton, Hydrilla verticillata, Lagarosiphon chinensis, L. cordofanus, L. crispus, L. hydrilloides, L. ilicifolium, L. madagascariensis, L. major, L. muscoides, L. schweinfurthii, L. steudneri, Nechamandra alternifolia*: all parts of most spp (C. D. K. Cook has suggested (personal communication) that *Elodea callitrichoides* perhaps = *E. ernstae, Lagarosiphon chinesis* is wrongly named, and *L. madagascariensis* is probably *L. crispus* or *L. densus.*)

Blass (1890): *Elodea canadensis*: phloem.

Brunaud (1976, 1977): *Elodea canadensis*: branching.

Boysen-Jensen (1959): *Egeria densa* (as *Helodea densa*); leaf development.

Caspary (1858, 1861): *Hydrilla verticillata*: all parts, and stem development.

Cordes (1959): *Elodea* sp: lipid-containing cells.

Dale (1957): *Elodea canadensis*: shoot apex, developmental anatomy of the stem.

Drawert (1937, 1938): *Egeria densa* (as *Elodea*), *Elodea canadensis, Hydrilla verticillata*: hydropoten.

Géneau de Lamarlière (1906): *Elodea canadensis*: cuticle.

Grier (1920): *Elodea* spp: propagation.

Herrig (1914): *Egeria densa* (as *Elodea densa*), *Elodea canadensis*: shoot apex.

Holm (1885): *Egeria densa* (as *Elodea densa*): leaf, stem.

Hulbary (1944): *Egeria densa* (as *Elodea densa*): stem, air-space tissue.

Jacobs (1946): *Egeria densa* (as *Anacharis densa*): shoot morphology.

Kaul (1968 b): *Egeria densa* (as *Elodea densa*), *Elodea canadensis, Hydrilla verticillata, Nechamandra alternifolia*: floral anatomy.

Kaul (1970): *Egeria densa* (as *Elodea densa*), *Elodea canadensis, E. nuttallii, Hydrilla verticillata, Nechamandra alternifolia*: inflorescence morphology.

Lance-Nougarède and Loiseau (1960): *Elodea canadensis*: shoot apex.

Matzke (1948, 1949): *Egeria densa* (as *Anacharis densa*): cell shape in apical meristems.

Matzke and Duffy (1956): *Egeria densa* (as *Anacharis densa*): cell shape in apical meristems.

Obermeyer (1964): *Lagarosiphon* spp: general morphology.

Pendland (1979): *Hydrilla verticillata*: leaf ultrastructure.

Pfeiffer (1919): *Egeria densa* (as *Elodea densa*), *Elodea canadensis, E. najas*: all parts.

Planchon (1849): *Nechamandra roxburghii*: morphology.

Raunkiaer (1895–9): *Elodea canadensis*: leaf, stem anatomy, general morphology.

Richard (1812): *Elodea callitrichoides* (as *Anacharis callitrichoides*), *Elodea guayanensis, Hydrilla verticillata* (as *H. ovalifolia*): general illustrated account.

Riede (1920): *Egeria densa* (as *Elodea densa*); *Elodea crispa*: leaf, stem.

Rougier (1972): *Elodea canadensis*: mucilage secretion.

Schenck (1886b): *Elodea canadensis*: leaf, stem, root; *Hydrilla verticillata*: leaf, stem.

Schilling (1894): *Elodea canadensis*: mucilage secretion.
Singh (1966*d*): *Elodea canadensis, Hydrilla verticillata, Nechamandra alternifolia*: floral anatomy.
Solereder (1913): *Egeria densa* (as *Elodea densa*); *Elodea callitrichoides, E. canadensis, E. chilensis, E. granatensis, E. guianensis, E. orinocensis, E. planchoni*, plus three spp of uncertain identity, *Hydrilla verticillata* (three forms), *Lagarosiphon muscoides, L. cordofanus, L. densus, L. madagascariensis, L. schweinfurthii*: leaf.
Stant (1952, 1954): *Egeria densa* (as *Elodea densa*): shoot apex.
Takada (1952): *Egeria densa* (as *Helodea densa*): tannin cells.
Tarnavschi and Nedelcu (1973): *Elodea canadensis*: leaf.
van Tieghem (1870–1) *Elodea canadensis*: root.
van Tieghem and Douliot (1888): *Elodea canadensis*: root development.
Wilson (1936): *Elodea canadensis*: root-hairs.
Wylie (1940): *Elodea canadensis*: floral morphology and biology; floral development.
Yoshida (1958): *Elodea* spp: leaf idioblasts.

MAIDENIA Rendle

One species (*M. rubra*) from N. W. Australia. This seems to belong to the Anachariteae on the basis of its vegetative morphology but is distinguished by its male flowers which are each considered to be reduced to a single stamen and several of them aggregated between two free bracts. The perianth of female flowers is uniseriate.

No material of this plant has been available for anatomical study.

VALLISNERIA L.
(Figs. 4.7–4.9)

A genus of some ten species; submerged plants of fresh water in tropical and temperate latitudes, the erect vegetative rosettes of strap-shaped leaves proliferating extensively by stolons, sometimes interpreted as modified inflorescences (Wilder 1974*b*, 1975). Erect axes extended in *V. caulescens*, from Queensland. Squamules several in each leaf axil, pointed and with marginal tannin cells.

VEGETATIVE MORPHOLOGY

Shoot system

Plants consisting of upright vegetative rosettes of linear strap-shaped leaves, each rosette extended by one (sometimes more) stolons (Fig. 4.7 A), roots restricted to the base of the upright vegetative axis; terminal stolons in winter forming resting shoots (turions). Branching of stolon regular, with successive groups of three scale leaves, the groups separated from one another by long internodes. Two of the scale leaves empty, the third subtending the meristem producing the upright vegetative axis. Upright vegetative axis becoming branched and so producing further stolons, the latter either solitary or developed as a complex, the basal part serving as potential inflorescence primordia. Axes always branching by apical bifurcation. [Interpretation of shoot morphology is somewhat disputed, but the developmental studies of Wilder (1974c) lead him to the conclusion that while the upright axes have sympodial growth, certain stolons may be entirely monopodial. The evidence is presented and discussed in detail by Wilder and his interpretations are, in part, based on comparisons with related taxa (e.g. Wilder 1975).]

VEGETATIVE ANATOMY

Leaf (Fig. 4.9)

Eligulate, narrowly strap-shaped, up to 1 m long, with a sheathing base. Apex rounded, without an apical pore. **Hairs** in all spp examined except *V. asiatica* represented by marginal prickle-hairs (e.g. Fig. 4.8 F), the prickle-hairs in *V. denseserrulata* terminating prominent multicellular emergences. **Squamules** several (up to 14) in each leaf axil, scale-like, developing in sequence on each side of a primary squamule (Wilder 1974c). Cuticle thin. **Epidermis** thin-walled; cells elongated, rectangular in surface view (Fig. 4.9 H), those of the two surfaces sometimes differing in size. **Stomata** absent. **Mesophyll** (Figs. 4.9 B, D) lacunose, with two series of canals towards the midrib, but one towards margin with up to eight lacunae between each pair of major longitudinal veins. **Hypodermis** including v. long tannin cells. Transverse diaphragms of stellate cells, often including a large isodiametric tannin cell (Figs. 4.9 F, G). Mechanical tissue represented by abaxial **fibres** next to the midvein and larger lateral veins; adaxial fibres infrequent. **Veins** 7–11, prominent midrib not developed but midvein largest (Fig. 4.9 E); transverse veins conspicuous. Veins in parenchymatous buttresses each provided with a well developed protoxylem lacuna and abaxial phloem, the vascular tissues being indistinctly sheathed by parenchyma; fibres well-developed above and below larger veins (Fig. 4.9 E). Occasional minute longitudinal veins with narrow cells alternating with major veins.

Stem (erect vegetative axis)

Epidermis small-celled. **Cortex** lacunose, with narrow air-canals and transverse diaphragms; outer cortex slightly thick-walled. Endodermis not differentiated. **Central cylinder** with irregular, congested vbs.

Stolon (Fig. 4.9 I)

Epidermis thin-walled, outer cortical layer somewhat collenchymatous. Ground tissue lacunose with numerous wide air-canals separated by a network of parenchymatous cells; transverse diaphragms developed. **Vascular system** represented by a ± single series of collateral vbs of varying size. Protoxylem lacuna well developed in larger vbs, metaxylem represented by a few reticulately thickened thin-walled tracheids. Phloem well developed and with wide sieve-tubes.

Root

Unbranched. **Epidermis** with root-hairs (Leavitt 1904). Outer **cortex** of thin-walled compact cells; middle cortex somewhat lacunose with radially extended canals traversed by uniseriate diaphragms of characteristically flat cells with extended radial lobes contacting those of adjacent cells. **Endodermis** thin-walled, inconspicuous. **Stele** narrow; central protoxylem lacuna surrounded by a single layer of thin-walled cells and four phloem strands, each strand represented by a single sieve-tube.

Conducting, secretory, and storage elements

Tannin not uncommon, especially in the epidermis and transverse diaphragms of leaf. **Starch** common in the vegetative axis, the grains large, numerous and ellipsoidal. **Tracheary elements** (cf. Scherer 1904) described by Ancibor (1979) as including vessels with oblique scalariform perforations.

REPRODUCTIVE MORPHOLOGY

Plants dioecious, inflorescences axillary, associated with leaves in a vegetative rosette. **Male inflorescences** solitary or several at a node (Fig. 4.7 D), short-stalked with numerous minute flowers enclosed in a flask-shaped spathe, the flowers abscissing at maturity and released by irregular rupture of the spathe to float to the water surface (Fig. 4.7 B, C). Sepals three, petals 0–1 (–2), minute; stamens 1–2 (–3) with bilocular anthers, the missing stamens sometimes represented by minute staminodes. **Female inflorescence** solitary at a node or proliferating (Fig. 4.8 A) with a v. long peduncle terminating in a tubular spathe, the mouth with two round lobes, enclosing a single flower (Fig. 4.8 B, C). Sepals three; petals three, minute; staminodes three, alternating with the petals (Fig. 4.8 E). Ovary inferior, linear, styles three, each with two papillose stigmatic lobes. Ovules v. numerous (Fig. 4.8 D). Peduncle usually spirally contracted after pollination (Fig. 4.8 A). Fruit cylindrical with numerous elliptical seeds.

REPRODUCTIVE ANATOMY

(After Singh 1966*d*; Kaul 1968*b*). Spathe receiving three vbs in male but six in female inflorescence. Each male flower receiving a single trace, the trace bifurcating to supply each of the stamens. Perianth and staminode without vascular tissue. Female flower with three vbs at the base of the ovary, each bundle dividing into two, interpreted as a dorsal and ventral carpellary bundle. Dorsal bundles vestigial, ventral bundles first supplying ovular traces, and later and distally a trace to each outer tepal and two traces to each style.

FLORAL BIOLOGY

[The floral mechanism of *Vallisneria* is a classic example of floral specialization but it has been the source of some controversy. In *V. spiralis,* from the early studies of Chatin (1855–6) and Kerner von Marilaun (1895), it was recognized that the pollen from the divergent anthers of the wind-blown free-floating male flowers is brought into contact with the erect stigmas of the attached female flowers. In *V. americana*, Wylie (1917) was the first to show that the mechanism is different since the female flowers lie along the surface of the water, while the stamens are held erect in the free-floating male flowers. The female flowers bob below the water surface by wave action, but they remain dry because the sepals close around an air bubble. Neighbouring male flowers are tipped into the meniscus formed by the female flower as it sinks and are included in the air-bubble. The anthers thus strike the stigmas (cf. *Enhalus*). Svedelius (1932) pointed out that Wylie did not recognize he was dealing with a different species when he disputed earlier accounts.]

MATERIAL EXAMINED

This is based mainly on that of Ancibor (1979), without examination of additional material.

SPECIES REPORTED ON IN THE LITERATURE

Ancibor (1979): *Vallisneria aethiopica, V. asiatica, V. denseserrulata, V. gigantea, V. gracilis, V. neotropicalis, V. spiralis*: all parts.
Blass (1890): *V. spiralis*: phloem.
Brunaud (1976): *V. spiralis*: branching.
Bugnon and Joffrin (1962): *V. spiralis*: shoot morphology and branching.
Chatin (1855, 1856): *V. spiralis*: all vegetative parts.
Drawert (1937, 1938): *V. spiralis*: hydropoten.
Duval-Jouve (1873*b*): *V. spiralis*: diaphragms in leaf.
Falkenberg (1876): *V. spiralis*: stem.
Gibson (1905): *V. spiralis*: squamules.
Kaul (1968*b*): *V. americana*: floral anatomy.
Laessle (1953): *V. spiralis*: root morphology.
Miki (1934*b*): *V. asiatica*: leaf, stem.
Müller (1875): *V. spiralis*: development.
Richard (1812): *V. spiralis*: floral morphology.
Rohrbach (1871): *V. spiralis*: morphology.
Schenck (1886*b*): *V. spiralis*: all vegetative parts.
Schilling (1894): *V. spiralis*: mucilage secretion.
Singh (1966*d*): *V. americana, V. spiralis*: floral anatomy.
Solereder (1913): *V. alternifolia, V. spiralis*: leaf.
van Tieghem (1870–1): *V. spiralis*: root.
van Tieghem and Douliot (1888): *V. spiralis*: origin of lateral roots.
Wilder (1974*c*): *V. americana*: vegetative morphology.

BLYXA Du Petit-Thouars ex L.C. Rich.

A genus of nine species, submerged aquatics, with leafy rosettes arising either (sect. *Blyxa*) from congested, rhizomatous or stoloniferous axes or (sect. *Caulescentes*) along an extended stem (den Hartog 1957*b*). Morphological details of vegetative branching not known. Squamules two or more per leaf axil, acute but varying in length and width in different spp.

VEGETATIVE MORPHOLOGY

No existing detailed accounts.

VEGETATIVE ANATOMY

Leaf blade

Linear or lanceolate, undifferentiated, with 3–7 major longitudinal veins. **Hairs** frequent as uniseriate marginal prickle hairs, the size and shape somewhat diagnostic for different spp, but hairs absent from *B. radicans.* **Cuticle** thin. **Epidermis** in surface view with isodiametric or somewhat elongated cells and ± rectangular, only the outer wall slightly thickened; cells of each surface ± same size. **Stomata** absent. **Mesophyll** well developed, thickest in midrib region and tapered marginally, lacunose with two series of longitudinal air-canals in the broad midrib region, but reduced to one series marginally. Canals separated by uniseriate longitudinal partitions of thin- or thick-walled cells and with frequent transverse diaphragms, sometimes including a transverse vein. Mechanical tissue represented by strands or irregular groups of marginal **fibres,** together with occasional thick-walled fibres associated with the phloem of the main veins. **Veins** situated adaxially in somewhat enlarged partitions. Vascular tissues with well developed protoxylem lacunae, sometimes including thin-walled metaxylem tracheids.

Stem

Epidermis with large cells. **Cortex** including peripheral collenchyma. Outer cortex including a series of narrow vbs with reduced vascular tissues. Middle cortex lacunose with several series of longitudinal air-canals, traversed by frequent transverse septa of stellate cells. **Stele** delimited by a thin-walled endodermis; including irregular vbs.

Root

Unbranched in all spp examined. Root-hairs present on young roots, the base with characteristic endocytic structures. **Exodermis** two-layered. Middle **cortex** usually lacunose, with radial longitudinal partitions developed between alternate collapsed radial files of cells. Transverse partitions also present, usually uniseriate and composed of regularly cruciate cells. **Stele** delimited by a thin-walled endodermis. Vascular tissues in narrower roots including a central protoxylem lacuna surrounded by 4–5 phloem strands within the pericycle, each strand represented by a single sieve-tube. Larger roots with a more typical polyarch structure, the metaxylem tracheary elements moderately thick-walled and forming a central series.

Secretory and conducting elements

Crystals few. **Tannin** common in ground tissues and epidermal cells. **Tracheary elements** including vessels, according to Ancibor (1979), the elements long, narrow with v. oblique perforation plates with several (5–10) thickening bars.

REPRODUCTIVE MORPHOLOGY

Flowers unisexual or bisexual, plants rarely hermaphroditic or monoecious but usually dioecious. Inflorescences sessile or stalked, with a tubular spathe enclosing 1 (–2) flowers, or spathes of dioecious plants with several male flowers. Male flowers pedicellate. Female flowers with a long beaked hypanthium. Sepals three, green, narrow; petals three, white, linear, sometimes papillose. Stamens 3–6 or nine. Ovary inferior, linear, with three parietal placentas. Stigmas three, petaloid. Staminodia sometimes present in female flowers; pistillodia sometimes present in male flowers. Fruit linear with several smooth, spiny (e.g. *B. echinosperma*) or winged seeds.

REPRODUCTIVE ANATOMY

(After Singh 1966*d*). *Blyxa octandra.* (with unisexual flowers (dioecious)). Each male flower receiving three vbs from the central mass of vascular tissue developed above the departure of the leaf traces. Each tepal receiving three traces, each stamen and pistillode a single trace. Female flower with six peripheral vbs below the ovary, these representing three dorsal carpel vbs and three secondary marginal vbs; remaining central vascular tissue continuous apically as three ventral strands supplying the ovules. Peripheral bundles continuous in the ovary wall to the top of the flower supplying, with the remains of the ventral traces, the vascular system of the tepals and styles.

B. echinosperma (with hermaphrodite flowers)

This is described by Singh as being similar to *B. octandra* except for minor differences related to the perfect flowers and the different numbers of parts.

B. alternifolia (after Kaul 1968*b*)

Carpel vascular supply described as an outer and inner bundle on the same radius, opposite each placenta; outer bundles supplying sepals and stamens; inner bundles (interpreted as fused ventral bundles of two adjacent carpels) supplying ovules. Vascular supply to inner perianth parts and styles form weak stylar plexus. In *B. aubertii* an additional bundle alternating with radial pairs opposite placenta [Kaul concludes that there is no dorsal bundle in the two species examined].

FLORAL BIOLOGY

[The study by Cook (1981) of *B. octandra* reveals a distinctive and unusual pollination mechanism, probably characteristic of all dioecious species. In the male flower the pollen is deposited on the adaxial surfaces of the petals in bud and presented to pollinating insects when the petals spread. The female flowers resemble the males only superficially since the three styles are petal-like, the true petals apparently being vestigial. The flowers are short-lived, functioning for a brief period in the morning, transfer of pollen being facilitated by exudation of fluid drops on the petals or stigmas. Pollen transfer is apparently effected by flies.]

MATERIAL EXAMINED

This account is based mainly on the description by Ancibor (1979) without examination of additional material.

SPECIES REPORTED ON IN THE LITERATURE

Ancibor (1979): *Blyxa angustipetala, B. aubertii, B. coreana, B. echinosperma* (= *B. aubertii*), *B. japonica, B. novaguineense, B. octandra, B. radicans, B. senegalensis*: all parts.
Cook (1981): *B. octandra*: floral morphology and biology.
Kaul (1968*b*): *B. aubertii, B. japonica* (as *B. alternifolia*): floral anatomy.
Lakshmanan (1961): *B. octandra* (possibly *B. aubertii*, according to Cook 1981): floral morphology and embryology.
Montesantos (1912): *Blyxa* sp: leaf, root, notes on morphology.
Richard (1812): *B. aubertii, B. roxburghii*: floral morphology.
Singh (1966*d*): *B. echinosperma, B. octandra*: floral anatomy.
Solereder (1913): *B. griffithii, B. octandra, B. radicans*, plus two unidentified spp: leaf.

THALASSIA König
(Figs. 4.10–4.12; Plates 1C and 5)

A genus having two species of seagrasses in tropical waters, *T. testudinum* König in the Caribbean, *T. hemprichii* (Ehrenb.) Aschers. widely distributed from the West Indian Ocean (East Africa) to the western Pacific (den Hartog 1970*a*) and forming extensive marine meadows below low water. Both species very similar in morphology and anatomy and described together in this account.

VEGETATIVE MORPHOLOGY

Axes strictly **dimorphic** (Fig. 4.10 A). Horizontal rhizomatous axes (long-shoots) buried in the substrate, scale-bearing and with extended internodes, giving rise at regular intervals (usually nine, 11, or 13 internodes) to erect shoots (short-shoots) with foliage leaves separated by short internodes. Older short-shoots partially enclosed by persistent remains of leaf bases. Leaves distichous, plane of distichy transverse on the rhizome. Erect shoots (Fig. 4.10 A–D) arising by precocious branching (Fig. 4.10 B, C) of the rhizome apex (Tomlinson and Bailey 1972), the lateral short-shoot meristem leaf-opposed and occupying a distal position on the internode at maturity (Fig. 4.10 D). Vegetative branching of short-shoots irregular, infrequent but always giving rise to new rhizomes; rhizomes never developed directly from a previously existing rhizome. Scale leaves present on the rhizome, with transitions

to foliage leaves at the bases of short-shoots (Fig. 4.10 E). **Squamules** two
per node, each acute. Flowers axillary on short-shoots (Fig. 4.10 G). Roots
mainly restricted to long-shoots (e.g. Fig. 4.10 A), towards lower surface
and originating close to the apical meristem, always unbranched and conspicuously septate, becoming numerous on older short-shoots (e.g. Fig. 4.10 D).

VEGETATIVE ANATOMY

Foliage leaf (Fig. 4.10 E)

Eligulate, with a strap-shaped blade, up to 70 cm long and 1.5 cm wide,
and an open sheathing base (Fig. 4.10 F); the apex rounded, toothed, and
without an apical pore. **Hairs** absent except for marginal multicellular (rarely
unicellular) teeth, those of *T. hemprichii* (Fig. 4.11 D, E) larger than those
of *T. testudinum* (Fig. 4. 11 F, G). **Cuticle** thin, but multiporous (Gessner
1968). **Epidermis** densely chlorenchymatous, often including calcium oxalate
crystals, the plasmalemma markedly convoluted (Jagels 1973, see Plate 1
C). Epidermal cells in surface view ± isodiametric, walls thickened, especially
outer walls. **Mesophyll** developed from a single initial layer (Fig. 4.12), at
maturity including longitudinal air-canals, with 3–4 canals alternating with
each pair of veins (Fig. 4.11 A). Canals separated laterally by uniseriate
partitions up to five cells high. **Hypodermis** one-layered above and below
the lacunae. Mesophyll cells large, elongated and without densely aggregated
chloroplasts. Canals segmented by transverse, usually uniseriate, diaphragms
of small lobed cells, the diaphragms sometimes including transverse veins.
Non-vascular **fibres** well developed above and below but independent of the
lateral veins as narrow unlignified hypodermal strands (Fig. 4.11 B), together
with a more massive marginal strand not continuous proximally into the
sheathing base.

Veins 9–12, equidistant from both surfaces, each situated within a multiseriate girder; midvein large but not protruding, with a few sheathing fibres
immediately above and below (Fig. 4.11 C); protoxylem lacuna conspicuous
but narrow; phloem rather irregular with inconspicuously differentiated sieve-
tubes. Lateral veins smaller, sheathing fibres absent, vascular tissue reduced,
protoxylem lacuna usually absent. Transverse veins conspicuous, retaining
a continuous series of short tracheids running into the longitudinal veins.
Tannin frequent in the mesophyll and veins.

Leaf sheath

Colourless, with well developed marginal wings including lateral veins,
these uniting distally at the level of the ligule. Veins with massive fibrous
sheaths, hypodermal fibrous strands absent. Mesophyll more massive than
in the blade. Detachment of leaf blade said by Tomlinson (1972) to be facilitated by abrupt anatomical transition between blade and sheath.

Leaf development (Fig. 4.12)

Described by Tomlinson (1972) with a regular pattern of segmentation
of the original single mesophyll layer, a pattern common to a number of
linear-leaved aquatic monocotyledons (e.g. *Enhalus, Vallisneria, Zostera*). Air-

lacunae initiated as intercellular spaces between the primordial files of mesophyll cells (detailed description in legend to Fig. 4.12).

Stem (Plate 5)

Anatomy of the rhizome and short-shoot v. similar. **Epidermis** undifferentiated, persistent. **Cortex** wide, outer layers somewhat collenchymatous, middle
cortex extensively lacunose and with irregular transverse diaphragms, the
cortical parenchyma frequently including abundant starch. Outer cortex including continuations of the fibrous bundles of the leaf sheaths, only the
median leaf traces extending into the stele. **Stele** narrow, delimited by a
thin-walled endodermis. Vascular tissue including irregular vascular strands,
the xylem comprising only short tracheids with annular or spiral wall thickenings; tracheids well developed at the nodes.

Root

Rhizosphere apparently represented by a region of local concentration of
nitrogen-fixing microorganisms, the structure of the root being related to
this possible source of nutrients (Patriquin 1973; Tomlinson 1969*a*). **Epidermis**
with numerous root-hairs arising from short-cells (trichoblasts), the base of
the root-hair somewhat sunken in the exodermis. **Exodermis** biseriate, compact, with short pitted cells below the root-hairs; cells longer elsewhere. Middle **cortex** becoming lacunose, with radially collapsed plates of cells delimiting
the longitudinal lacunae. **Transverse diaphragms** conspicuous and represented
by 1–3 seriate plates of lobed cells at regular intervals, affording a transport
pathway across the root cortex. **Endodermis** thin-walled. **Stele** narrow; pericycle uniseriate. Vascular tissue reduced. Phloem usually including 5–7 sieve-
tubes, with associated companion cells, next to the pericycle, the sieve-tubes
alternating with an equal number of subsequently differentiated tracheary
initials. [The tracheary elements may remain undifferentiated distally, or become differentiated proximally as elements with spiral or reticulate wall thickenings.]

REPRODUCTIVE MORPHOLOGY

Plants dioecious. Flowers 1–4 in the axils of leaves on the short-shoots.
Female flowers usually solitary; each with a separate spathe; never in multiples
or enclosed by a common sheathing organ. Each unit then interpretable as
one-flowered inflorescence. Peduncle long, supporting a pair of ± completely
fused bracts (= spathe) opening on one side. **Male flower** with three tepals
and 8–13 sessile stamens, the latter not inserted in obvious trimerous whorls.
Pollen globose but adhering in chains. Female flower with three tepals inserted
on a narrow neck above the ovary. Ovary inferior with a central cavity
and several irregular dissepiments but fewer ovules, the dissepiments continued distally into the short styles, each style ending in a pair of stigmas.
Ovary coarsely warted. **Fruit** (Fig. 4.10 G-I) globose with several large angular
seeds (Fig. 4.10 J) released by irregular splits. **Embryo** with a massive hypocotyl fused with the cotyledon, the seedling axis inserted laterally (Fig.
4.10 K).

REPRODUCTIVE ANATOMY

(After Kaul 1968b; Tomlinson 1969a). Anatomy of peduncle and spathe similar in the two kinds of flower. **Male flower** with three vbs in pedicel, united at the insertion of the appendages, but supplying distally a single vb to each tepal and to each stamen, without evidence for the stamens being in whorls. **Female flower** with three vbs becoming aggregated at the base of the ovary but distally producing three tepal traces and a variable number of ovary traces. Tepal traces remaining distinct in the wall of the ovary and entering tepals at top of hypanthium. Ovary traces split at top of hypanthium supplying a single trace to each stigma. Ovules supplied by strands from base of ovule traces.

[In comparison with other genera described by Singh (1966d), *Thalassia* seems unusual in that the tepal traces are evident at the base of the ovary and distally are not connected with other floral traces, whereas in other Hydrocharitaceae tepal traces are often only identifiable at the top of the ovary.]

MATERIAL EXAMINED

Thalassia hemprichii (Ehrenb.) Aschers. in Peterm.; Tarauma Bay, Port Moresby, Papua New Guinea; P. B. Tomlinson 31.x.74; all vegetative parts.

T. testudinum Banks ex König; Biscayne Bay, Miami, Florida; P. B. Tomlinson *s.n.*; several collections.

SPECIES REPORTED ON IN THE LITERATURE

Kaul (1968b): *Thalassia hemprichii*: floral anatomy.
Miki (1934a): *T. hemprichii*: leaf, stem.
Pascasio and Santos (1930): *T. hemprichii*: general morphology.
Sauvageau (1890d): *T. hemprichii*, *T. testudinum*: leaf.
Solereder (1913): *T. hemprichii*, *T. testudinum*: leaf.
Tomlinson (1969a,b): *T. testudinum*: root anatomy and development; floral morphology and anatomy.
Tomlinson (1972): *T. testudinum*: leaf anatomy and development.
Tomlinson (1974a): *T. testudinum*: rhizome morphology.
Tomlinson (1980): *T. hemprichii*, *T. testudinum*: leaf.
Tomlinson and Bailey (1972): *T. testudinum*: branching.
Tomlinson and Vargo (1966): *T. testudinum*: general morphology.

HALOPHILA Du Petit-Thouars
(Figs. 4.13–4.15)

A genus consisting of about nine species of diminutive marine herbs (leaves usually < 3 cm long) widely distributed in tropical waters; stoloniferous, much-branched and often forming extensive patches in finer-grained substrates.

VEGETATIVE MORPHOLOGY

Stolons with sub-opposite pairs of scales associated with a root, separated from the next consecutive scale pair by a long internode (Figs. 4.13 and 4.14). First (lower) scale of each pair empty, second (upper) scale subtending a lateral bud, the latter producing either (i) an initially erect indeterminate shoot beginning with a pair of petiolate foliage leaves, as in *H. ovalis* (Balfour 1879; e.g. Fig. 4.14 A), or (ii) an erect determinate shoot with a basal pair of scales and either (sect. Americanae, e.g. *H. engelmanni*) a cluster of 4–5 sessile foliage leaves (Fig. 4.13 E, F) or (e.g. *H. stipulacea*) a pinnate series of up to 20 sessile leaves (Fig. 4.13 A–C). The system in (i) sometimes proliferated by direct extension of the lateral shoot. The system in (ii) proliferated by the development of a lateral meristem from the axil of one of the basal scales of the determinate shoot (Fig. 4.13 B). Symmetry of the system changing from one node to the next, each pair of scales being set at a shallow angle to the next (Balfour 1879). Branching apparently precocious and taking place by bifurcation of the apical meristem, but not documented in detail.

Foliage leaves either long-petiolate (e.g. *H. ovalis*; Fig. 4.14 A) or sessile (e.g. *H. engelmannis*; Fig. 4.13 I), asymmetric in *H. stipulacea* with a basal overwrapping lobe (Fig. 4.13 D). Veins conspicuous, including a thickened midrib and usually a pinnate series of lateral veins united by a marginal commissure, the number and inclination of the lateral veins being somewhat diagnostic (cf. Fig. 4.15 A, I). Lateral veins rarely anastomosing with each other (e.g. *H. balfourii*), usually absent from *H. beccarii*. Lamina sometimes conspicuously bullate (Fig. 4.14 A).

VEGETATIVE ANATOMY

Leaf blade

Midrib and sometimes the margin thickened (Fig. 4.15 B–D), the blade otherwise mainly biseriate except in the region of the veins. **Hairs** represented by marginal, often raised, unicellular prickle hairs as in *H. balfourii, H. decipiens, H. engelmannii, H. spinulosa* (Fig. 4.15 H); similar but longer and more slender hairs conspicuous on both surfaces in *H. baillonis, H. decipiens, H. spinulosa* (Fig. 4.15 F, G). **Cuticle** thin. **Epidermis:** cells in surface view ± isodiametric, those of the two surfaces ± equal in size; walls straight (Fig. 4.15 F) but sometimes sinuous in surface view as in *H. johnsonii, H. ovalis* (Fig. 4.15 E), *H. ovata, H. spinulosa*; cells elongated above and below veins. Outer walls slightly thickened, especially above and below the veins. Outer wall with a locally specialized annulus* showing affinity for silver ions in *H. decipiens, H. ovalis, H. ovata* but not *H. spinulosa* (Birch 1974; Solereder 1913). **Stomata** absent. **Mesophyll** developed only in midrib region and around the veins; air-lacunae present only in the midrib region (Fig. 4.15 B). **Median vein** surrounded by a parenchymatous sheath; including a narrow protoxylem lacuna or a single series of tracheids; phloem cells v.

---

* Birch (1974) notes that this annulus is absent from the epidermal cells of the leaves of species of *Cymodocea, Enhalus, Halodule, Syringodium, Thalassia,* and *Zostera* which he studied.

narrow and inconspicuously differentiated. Lateral veins largely composed of narrow elongated, undifferentiated cells. Mechanical tissue represented by slightly thickened epidermal cells above and below the midrib. Occasional **fibres** developed below the midvein.

Scale leaves

Colourless, biseriate with three median veins.

Petiole

Where developed, with epidermal walls slightly thickened. Ground tissue somewhat lacunose and including three veins, the median equal to the midrib vein, the two laterals being equal to the marginal veins.

Stem

Outer walls of the **epidermis** thickened. **Cortex:** outer layers sometimes collenchymatous. Middle cortex lacunose, with a single (e.g. *H. decipiens, H. ovalis*) or several *(H. engelmanni)* series of air-canals. Transverse diaphragms absent. Cortical vascular system represented by several (up to 12) strands of narrow cells in the outer cortex. **Endodermis** thin-walled. **Stele** narrow, usually with a central xylem lacuna surrounded by a series (as few as four) of sieve-tubes next to the endodermis.

Root

Unbranched. **Root-hairs** usually abundant, often thick-walled. **Exodermis** uniseriate, compact. Outer **cortex** recorded as collenchymatous by Ancibor (1979). Middle cortex sometimes lacunose as in *H. ovalis*, with one or more concentric series of air-canals represented by enlarged intercellular spaces but otherwise cortex without lacunae. **Endodermis** thin-walled. **Stele** narrow, sometimes including a central protoxylem lacuna, but xylem often undifferentiated. Phloem represented by a few (up to four) sieve-tubes next to the endodermis.

Secretory and storage elements

Tannin and **crystals** little developed; **starch grains** common in the ground tissue of the stolon and sometimes of the petiole.

REPRODUCTIVE MORPHOLOGY
(Fig. 4.14 C–I).

Plants dioecious or monoecious. Flowers unisexual, almost always solitary and associated with scales on the stolon, enclosed by two overlapping, sometimes keeled bracts (= 'spathe'). **Male flower** (Fig. 4.14 F–I) pedicellate with three imbricate tepals enclosing three stamens; anthers sessile. Pollen in long chains. **Female flower** (Fig. 4.14 C–E) ± sessile, ovary inferior with vestigial tepals, style extended into 2–6 (usually three) filiform stigmas. Ovary with three parietal placentae and several ovules. **Fruit** (Fig. 4.14 J) a fleshy capsule with persistent styles and several minute globose seeds.

Reproductive Anatomy

Refers to *H. ovalis*, described by Singh (1966*d*) and Kaul (1968*b*). Peduncular anatomy similar in male and female flowers, with the central xylem lacuna replaced by lignified cells at the base of the spathe. Bracts each supplied by a single trace. In the **male flower** each tepal and stamen receiving a single trace. In the **female flower** four traces produced in two successive pairs and interpreted by Singh as dorsal and ventral bundles of two carpels. Ventral traces supplying ovules, dorsal traces extending through the ovary wall, one splitting into two to produce three stylar traces. Vestigial perianth without vascular tissue. [Kaul recognizes six traces in the ovary wall, interpreted as the dorsal and fused ventral traces of adjacent margins of three carpels, the fused ventrals supplying the ovules. Again Kaul's and Singh's contrasted observations seem to reflect differences between different collections, cf. *Nechamandra*.]

Taxonomic Notes

The species of *Halophila* seem reasonably discrete (den Hartog 1970*a*), but there is continuing nomenclatural confusion so that the names used in the literature may have been misapplied. New species continue to be described (Eisman and McMillan 1980). It seems that there are sufficient microscopic features of vegetative anatomy to allow a clear separation of taxa, although such characters have been little used. An extended study based on a wide sampling would be rewarding.

Material Examined (all vegetative parts)

Halophila baillonis Aschers. After Ancibor (1979).

H. balfurii Solered. After Ancibor (1979).

H. beccarii Aschers. After Ancibor (1979).

H. decipiens Ostenf. var. *pubescens* den Hartog. Townsville, Queensland; W. R. Birch *s.n.*

H. engelmanni Aschers. Florida Bay, in drift.; P. B. Tomlinson *s.n.*

H. johnsonii Eisman & McMillan. Biscayne Bay, Florida; P. B. Tomlinson *s.n.*

H. ovalis (R. Br.) Hook. var. *bullosa* Setchell and forma *major* (= *H. australis* Doty & Stone). Viti Levu, Fiji; P. B. Tomlinson *s.n.*

H. ovata (R. Br.) Hook. f. After Ancibor (1979).

H. spinulosa (R. Br.) Aschers. Townsville, Queensland; W. R. Birch *s.n.*

H. stipulacea Aschers. After Ancibor (1979).

Species Reported on in the Literature

Ancibor (1979): *Halophila baillonis, H. balfurii, H. beccarii, H. decipiens, H. engelmanni, H. ovalis, H. ovata, H. spinulosa, H. stipulacea*: all parts.

Balfour (1879): *H. ovalis, H. stipulacea*: general morphology and anatomy.

Birch (1974): *H. decipiens, H. ovata, H. ovalis, H. spinulosa*: leaf ultrastructure.

Gibson (1905): *H. ovata*: squamules.

Holm (1885): *H. baillonis*: all parts.
Miki (1934*a*): *H. ovalis*: leaf, stem.
Ostenfeld (1916): *H. ovalis, H. spinulosa*: leaf.
Pettitt and Jermy (1975): *H. decipiens*: pollen structure.
Sauvageau (1890*d*): *H. ovalis, H. spinulosa*: leaf.
Singh (1966*d*): *H. ovalis*: floral anatomy.
Solereder (1913): *H. aschersonii, H. beccarii, H. ovalis, H. ovalis* f. *major, H. stipulacea*: leaf.
Swamy and Lakshmanan (1962): *H. ovata*: embryo development.
Tomlinson (1974*a*): *Halophila* spp: rhizome morphology.
Tomlinson (1980): *Halophila* spp: leaf.

SIGNIFICANT LITERATURE—HYDROCHARITACEAE

Ancibor (1979); Arber (1914, 1918*b*, 1921, 1922*b*, 1923); Areschoug (1878, pp. 161–3): Arzt (1937); Ascherson and Gürke (1889); Balfour (1879); Beckerowa (1934); Benedict and Scott (1976); Bennett (1914); Bentham and Hooker (1883); Biebl (1951); Blass (1890); Boysen-Jensen (1959); Brunaud (1976, 1977); Bugnon and Joffrin (1962, 1963); Caspary (1857, 1858, 1861); Castell (1935); Chandler (1923); Chatin (1855, 1856, 1862); Chauveaud (1897*a, b*); Cheadle (1942, 1943*a, b*, 1944); Cheadle and Uhl (1948); Cook (1981); Cook *et al.* (1974); Cordes (1959); Cormack (1937); Costantin (1885*a, b*, 1886*a, b*); Cronquist (1968); Cunnington (1912); Currier and Shih (1968); Cutter (1964); Cutter and Feldman (1970*a, b*); Cutter and Hung (1972); Dale (1951, 1957); Dammer (1888); Datta and Biswas (1976); Daumann (1970); Delay (1941); Diannelidis (1950); Dibbern (1903); Doohan and Newcomb (1976); Drawert (1937, 1938); Duval-Jouve (1873*b*); Eber (1934); Ernst-Schwarzenbach (1945, 1953, 1956); Falkenberg (1876); Forti (1927); Géneau de Lamarlière (1906); Gibson (1905); Grier (1920); Gunning and Pate (1969); Haccius (1952); den Hartog (1970*a*); Harvey (1842); Hegnauer (1963); Herrig (1914); Hesse (1904); Holm (1885); Hulbary (1944); Hutchinson (1959); Irmisch (1858*a*); Islam (1950); Jacobs (1946); Jagels (1973); de Janczewski (1874); Jensen (1959); Kadej (1966); Kattein (1897); Kaul (1968*b*, 1969, 1970, 1976*b*); Kaushik (1940); Kerner von Marilaun (1895); Kny (1878); Kroemer (1903); Kudryashov (1964*a, b*); Laessle (1953); Lakshmanan (1961); Lance-Nougarède and Loiseau (1960); Leavitt (1904); Lorenz (1903); Lyr and Streitberg (1955); Magnus (1871); Majumdar (1938); Marie-Victorin (1931); Maroti (1950); Matsubara (1931); Matzke (1948, 1949); Matzke and Duffy (1956); Mayr (1915); Meyer (1966); Miki (1934*a, b*); Mitra (1955, 1964); Mitroiu (1969); Montesantos (1912); Müller (1875); Ndongala-Nlendi-Ntunga (1976); Obermeyer (1964); Ostenfeld (1916); Palamarev (1979); Pascasio and Santos (1930); Patriquin (1972, 1973); Pendland (1979); Perner and Losada-Villasante (1956); Pettitt (1980); Pettitt and Jermy (1975); Pfeiffer (1919); Planchon (1849); Rao (1951); Raunkiaer (1895–9); Richard (1812); Richards and Blakemore (1975); Riede (1920); Rohrbach (1871); Rougier (1972); Sachet and Fosberg (1974); Salisbury (1926); Saunders (1929); St. John (1961, 1962, 1963, 1964, 1965*a, b*, 1967); Sauvageau (1887, 1888, 1890*d*); Schade and von Guttenberg (1951); Schenck (1886*a, b*); Schenke (1893);

Scherer (1904); Schilling (1894); Sculthorpe (1967); Shinobu (1954); Singh (1966*d*); Smith and Lew (1970); Solereder (1913, 1914); Stant (1952, 1954); Streitberg (1954); Subramanyam (1962); Svedelius (1904, 1932); Swamy (1963); Swamy and Lakshmanan (1962*a*); Takada (1952); Tarnavschi and Nedelcu (1973); Theorin (1905); van Tieghem (1870–1); van Tieghem and Douliot (1888, pp. 334–520); Tomlinson (1969*a*, 1972, 1974*a*, 1980); Tomlinson and Bailey (1972); Tomlinson and Vargo (1966); Troll (1931*a, b*); Vakhmistrov and Kurkova (1979); Veres (1908); Weinrowsky (1898); Wilder (1974*b, c*, 1975); Wilson (1936); Witmer (1937); Wylie (1904, 1913, 1917); Yoshida (1958); Ziegenspeck (1927).

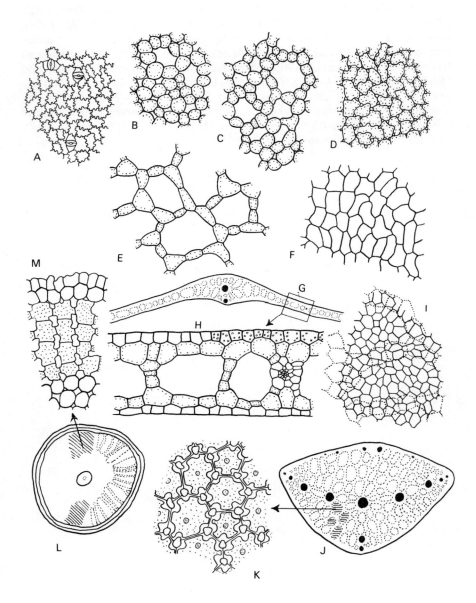

FIG. 4.1. HYDROCHARITACEAE. Leaf and root anatomy.

A–F. *Hydrocharis* cf. *H. asiatica*. Lamina structure in paradermal view (× 60).
A. Adaxial epidermis. B. Adaxial hypodermal layer. C. Uppermost mesophyll layer. D. Second or third mesophyll layer. E. Abaxial hypodermal layer. F. Abaxial epidermis.

G–M. *Ottelia* spp.

G–K. *O. alismoides*. Leaf.
G. TS midrib region of blade (× 8). H. Detail of lamina towards midrib in TS (× 37). I. Epidermis in surface view towards leaf margin, with dotted outline of hypodermal honeycomb of mesophyll cells. J. Petiole (× 8). K. Detail of diaphragm cells in surface view, from TS petiole.

L–M. *Ottelia ovalifolia*. Root.
L. TS (× 30). M. Detail of cells of transverse septa (× 110), in surface view, from TS of root shown hatched in L.

Vbs solid black in G, J. Transverse diaphragms—hatched in J, L.

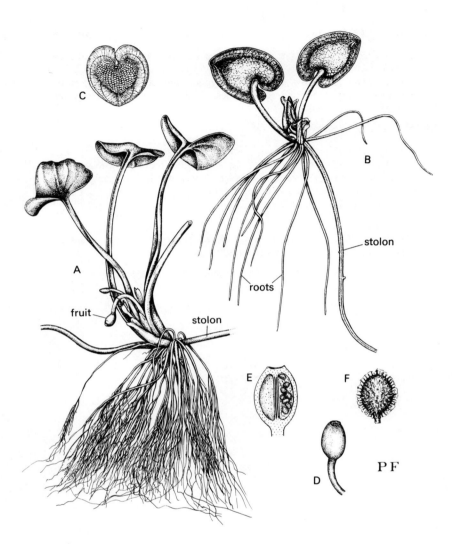

FIG. 4.2. HYDROCHARITACEAE. *Limnobium spongia* (Bosc.) Steud. fruiting specimens. (Lake Miccosukee, Florida, G. J. Wilder 2.vii.72.)

A. Mature leafy shoot (× 1), with developing fruits.

B. Stolon apex with young plantlet (× 3), continuing stolon not yet emerged.

C. Leaf blade (× 1), with the lacunose mesophyll tissue which produces buoyancy.

D. Immature fruit (× 3).

E. Fruit (× 3) in LS.

F. Single seed (× 18), testa spiny with mucilaginous coating.

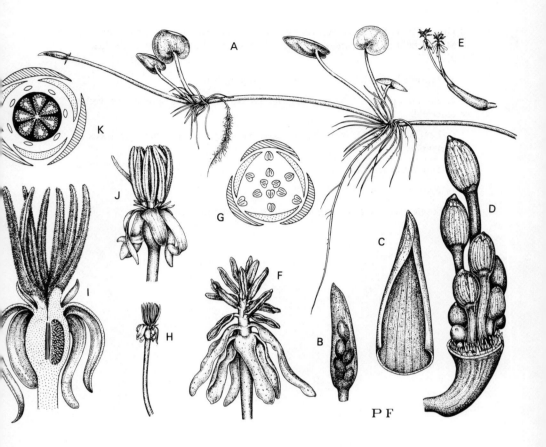

FIG. 4.3. HYDROCHARITACEAE. *Limnobium spongia* (Bosc.) Steud., stolon system and flowers (same locality as Fig. 4.2).

A. Portion of stoloniferous shoot ($\times \frac{1}{2}$) with vegetative rosettes separated by long internodes.

B. Male inflorescence ($\times \frac{3}{2}$) still enclosed by 'spathes'.

C. Detached 'spathe' ($\times$ 3).

D. Male inflorescence ($\times$ 3) with spathes removed to show developing monochasia and basal palisade of squamules.

E. Male inflorescence at anthesis ($\times \frac{1}{2}$) with two exposed male flowers.

F. Single male flower ($\times$ 3).

G. Floral diagram of male flower.

H. Single female flower ($\times \frac{1}{2}$).

I. Female flower ($\times$ 3) in LS.

J. Female flower ($\times \frac{3}{2}$) showing staminodes.

K. Floral diagram of female flower.

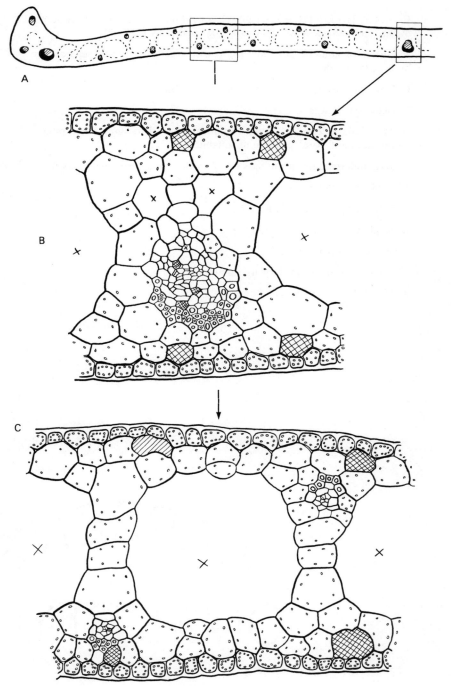

FIG. 4.4. HYDROCHARITACEAE. *Enhalus acoroides*, leaf anatomy. (After Tomlinson 1980.)

A. TS one half of lamina (× 15).

B. Detail of midvein (× 150).

C. Detail of lateral part of lamina (× 150) including abaxial (normally orientated) and adaxial (inversely orientated) minor veins.

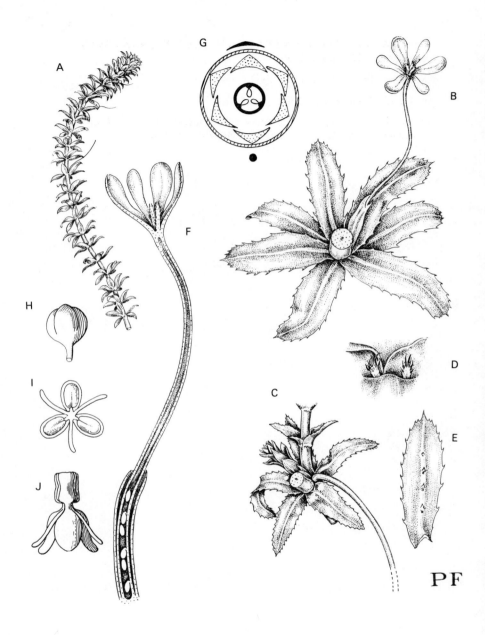

PF

176

FIG. 4.5. HYDROCHARITACEAE. *Hydrilla verticillata* (L.f.) Royle; female plants, roadside canal; Tamiami Trail, Miami, Florida; P. B. Tomlinson 17.X.79.

A–G. Female material.
A. Distal leafy shoot (× ½). B. Detail of one pseudowhorl (× 3) with an axillary flower. C. Pseudowhorl (× ³⁄₂) with two branch axes and associated root. D. Details of leaf insertion with a pair of squamules (× 12). E. Leaf (× 3), from abaxial side to show emergences on midrib. F. Female flower in LS (× 6). G. Floral diagram of female flower.

H–J. Male flower. (After den Hartog 1957b.)
H. Flower bud from side (× 4). I. Flower from above (× 4). J. Flower before anthesis, from side (× 4).

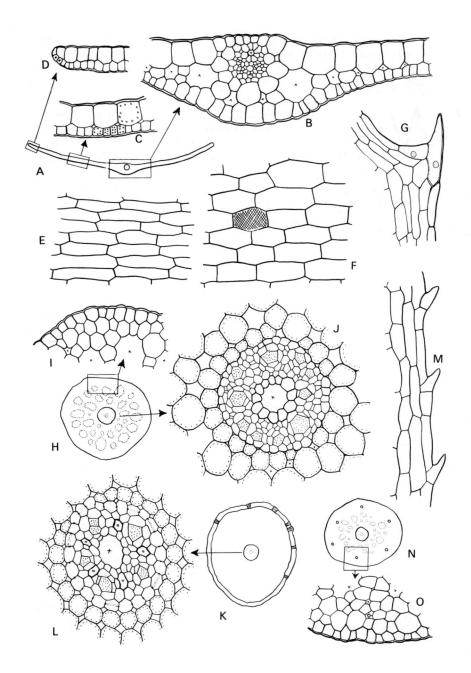

178

FIG. 4.6. HYDROCHARITACEAE—Anachariteae.

A–D. *Lagarosiphon major*; leaf TS. A. Outline (× 20).

B–D. Details (× 130).

  B. Midrib. C. Lamina. D. Margin with abaxial thick-walled cells.

E–J. *Hydrilla verticillata*. E–G. Leaf in surface view (× 130).
  E. Abaxial surface of leaf blade. F. Adaxial surface of leaf blade; tannin cell cross-hatched. G. Marginal prickle-hair.

H–J. Stem in TS.
  H. Outline (× 20). I. Surface layers (× 130) without cortical vbs.
J. Stele (× 250); protoxylem lacuna—x; sieve-tubes—stippled.

K–L. *Lagarosiphon major*. Root in TS.
  K. Outline (× 57). L. Stele (× 250); protoxylem lacuna—x; sieve-tubes stippled; possible vestigial tracheary elements—o.

M–O. *Elodea canadensis*.
  M. Leaf margin in surface view (× 130), with prickle-hairs. N. Outline of stem in TS (× 20). O. Detail of cortex with cortical vb (× 130).

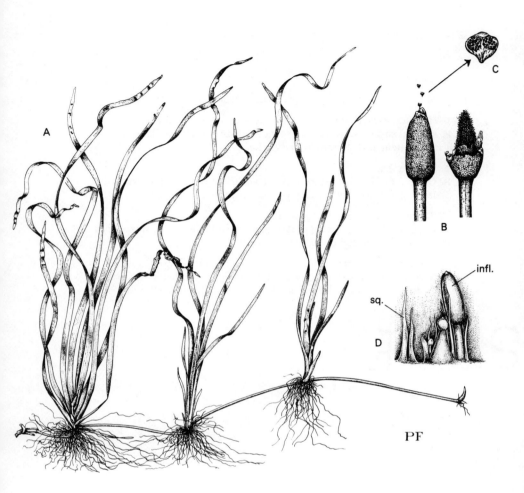

A

C

B

infl.

sq.

D

PF

180

FIG. 4.7. HYDROCHARITACEAE. *Vallisneria americana* Michx.
var. *neotropicalis* Marie-Vict. Male plant, morphology. (North Florida;
P. B. Tomlinson *s.n.*)

A. Habit ($\times$ ¼).

B. Male inflorescences ($\times$ ⅔) to left, early stage, spathes ruptured and
releasing first flowers, to right later stage, many flowers detached.

C. Single male flower ($\times$ 15).

D. Details of node ($\times$ 15) with foliage leaf removed to show squamules
(sq.) and young male inflorescences (infl.) at different developmental
stages.

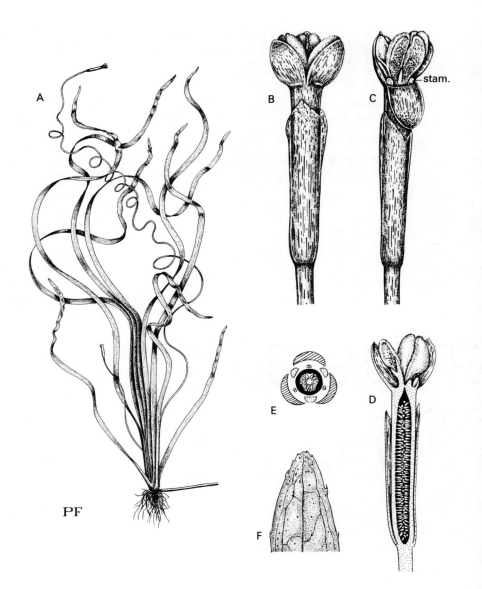

PF

stam.

A B C D E F

182

FIG. 4.8. HYDROCHARITACEAE. *Vallisneria americana* Michx. var. *neotropicalis* Marie-Vict. Female plant, morphology (same population as in Fig. 4.3).

A. Erect shoot ($\times \frac{1}{4}$) with young and older inflorescences.

B. Flower from side ($\times$ 3) enclosed in bilobed 'spathe'.

C. Flower ($\times$ 3) with one sepal pulled back to show stigmas and staminodes (stam.)

D. Flower in LS ($\times$ 3).

E. Floral diagram (petals stippled, alternating with staminodes).

F. Leaf tip ($\times$ 4).

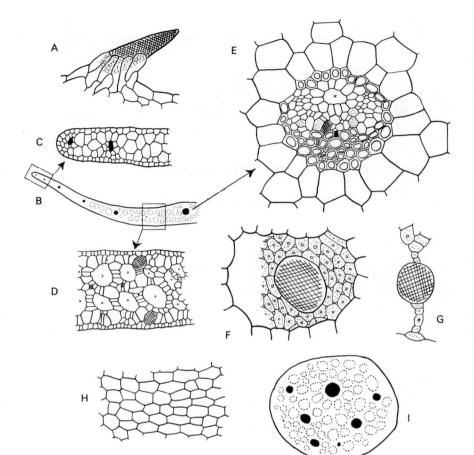

Fɪɢ. 4.9. HYDROCHARITACEAE.

A. *Stratiotes aloides.* Prickle-hair from leaf margin ($\times$ 150).

B–ɪ. *Vallisneria americana* var *neotropicalis.*

B–ʜ. Leaf.
   B. TS half of lamina ($\times$ 9). C. Detail of margin ($\times$ 80). D. Detail of lamina ($\times$ 80). E. Median vb ($\times$ 150). F. Transverse diaphragm ($\times$ 150) in longitudinal air-canal of lamina, with included tannin cell, from TS of lamina. G. Same in transverse view from LS of lamina ($\times$ 150). H. Epidermis in surface view ($\times$ 80).

ɪ. Stolon in TS ($\times$ 11).

   Vbs—solid black in B, C, and ɪ; tannin cells—cross hatched in D, F, and G; sieve-tubes—stippled in E.

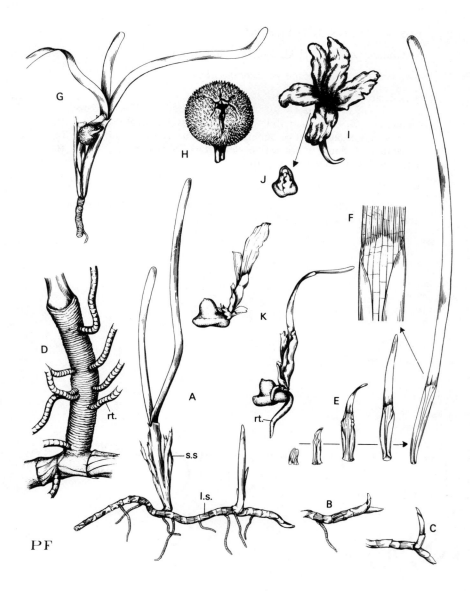

PF

186

FIG. 4.10. HYDROCHARITACEAE. *Thalassia testudinum* Banks ex König (Biscayne Bay, Miami, Florida; P. B. Tomlinson, mixed collections).

A. Habit (× ½) showing dimorphism between rhizomatous long-shoot (l.s.) and erect short-shoot (s.s.).

B, C. Two stages in development of a short-shoot from within rhizome apex (× ½).

D. Details of old, unbranched short-shoot (× 2) to show characteristic internodal insertion, the erect axis has been cleared of old leaf bases to show node scars. Remains of scale-leaves on rhizome, rt.—adventitious root.

E. Series of leaves from young short-shoot (× ½) to show progressive sequence of increasingly elaborate leaves, with right-hand one a typical foliage leaf.

F. Detail of eligulate mouth of leaf (× 2).

G. Short-shoot with axillary fruit (× ½).

H. Mature fruit (× 1).

I. Dehisced fruit, seeds dispersed (× 1).

J. Seed (× 1).

K. Two stages in early development of seed (× 1). rt.—seedling root.

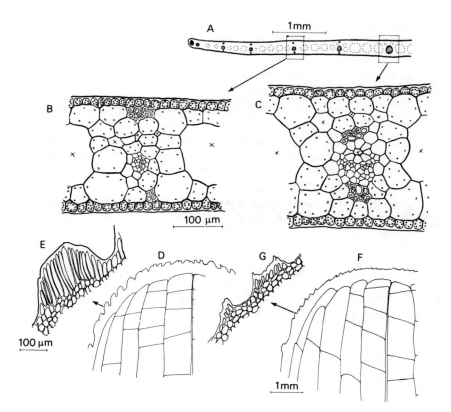

FIG. 4.11. HYDROCHARITACEAE *Thalassia* spp. Leaf blade. (After Tomlinson 1980.)

A–E. *Thalassia hemprichii.*
   A. TS one half of blade (× 15). B. Detail of lateral vein (× 110). C. Detail of midvein. D. Leaf apex in outline, with venation (× 10). E. Detail of marginal tooth.

F–G. *Thalassia testudinum.*
   F. Leaf apex in outline with venation. G. Detail of marginal teeth (× 80). (B and C; D and F; E and G—same magnification.)

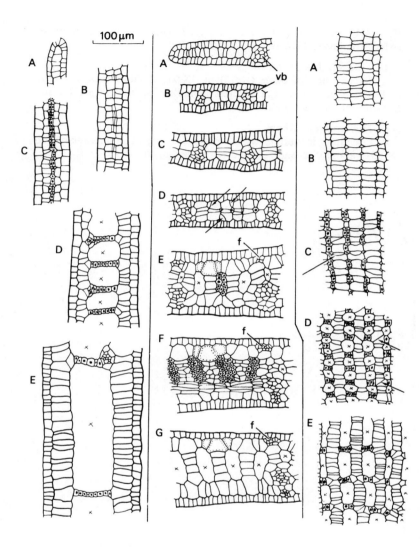

100 µm

FIG. 4.12. HYDROCHARITACEAE. *Thalassia testudinum.* Details of mesophyll histogenesis. (After Tomlinson 1972.) The series represents camera lucida drawings of leaf primordia at different stages of development in three planes mutually at right angles; *left,* in radial longitudinal section; *centre* in transverse section; and *right* in tangential longitudinal (i.e. paradermal) section. Diaphragm cells are shown with solid nuclei, air-canals and intercellular spaces are marked with an X.

Left. A. Uniseriate mesophyll. B. First paradermal divisions. C. Initiation of diaphragms, air-canals, and hypodermis. D. Expansion of air-canals and diaphragms, differentiation of transverse commissures. E. Elongation of blade with wide separation of diaphragms and transverse commissures.

Centre. A. Uniseriate mesophyll with differentiation of midvein (vb). B. Differentiation of lateral veins. C. First division of mesophyll cells to establish girders. D. Divisions (arrows) to establish 'roof' of air canals which are just initiated. E. Enlargement of air canals and establishment of transverse diaphragms, one of which is seen in surface view; fibres (f.) differentiated. F. Similar stage in region of a transverse commissure. G. In region between diaphragms. (In E, F, and G the dotted cells represent the products of a single initial cell.)

Right. A. Uniseriate mesophyll. B. Initiation of air-canals by widening of intercellular spaces. C. Initiation of diaphragms and first divisions (arrow) which extend the girders. D. Enlargement of air-canals and lengthening of girders; oblique divisions (arrows) to further support diaphragms. E. Lengthening of air canals and girders by further transverse divisions.

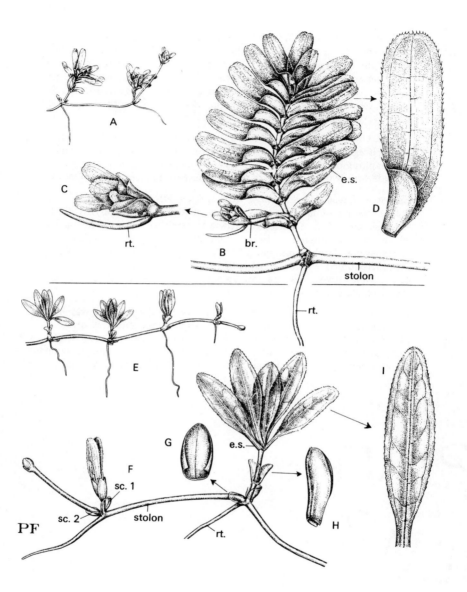

A

C

rt.

B

br.

e.s.

D

stolon

rt.

E

I

G

e.s.

F

sc. 1

sc. 2

stolon

rt.

H

PF

192

FIG. 4.13. HYDROCHARITACEAE. *Halophila,* vegetative mor-
phology, A–D (above) *Halophila spinulosa* (R. Br.) Aschers. (W. R. Birch,
Queensland); E–I (below) *H. engelmanni* Aschers. (Florida Bay, P. B.
Tomlinson *s.n.*)

A. Habit ($\times$ $\frac{1}{2}$).

B. Details ($\times$ $\frac{3}{2}$) of erect shoot (bracts on stolon removed) to show determinate erect shoot (e.s.) and lateral branch (br.) of proliferative shoot in
axil of scale pair on erect shoot, rt.—adventitious root.

C. Shoot apex ($\times$ 3), erect shoot developing from stolon scale axil.

D. Single leaf ($\times$ 3).

E. Habit ($\times$ $\frac{1}{2}$).

F. Details ($\times$ $\frac{3}{2}$) to show same essential morphology as in B, but without
proliferative shoots. (sc.1 and sc.2—successive scales on stolon)

G. Scale from stolon ($\times$ 3).

H. Scale from erect shoot ($\times$ 3).

I. Foliage leaf ($\times$ 3).

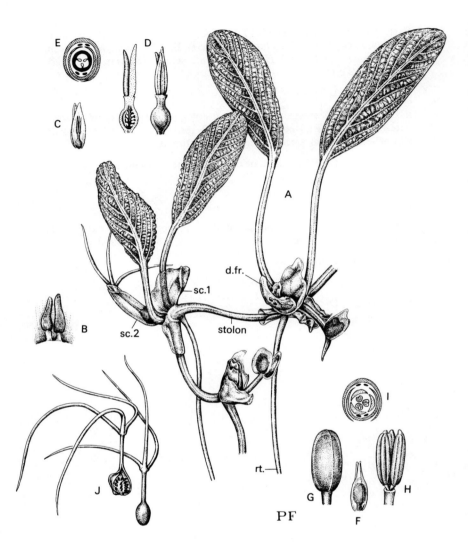

Fɪɢ. 4.14. HYDROCHARITACEAE. *Halophila ovalis* (R. Br.) Hook. ssp *bullosa* Setchell (= *H. minor*). Vegetative and reproductive morphology. (Deumba and Wainiyambia, Viti Levu, Fiji, P. B. Tomlinson *s.n.*)

A. Habit with fruit (× 3), the fruit at the oldest node dehisced (d. fr.); leaves of youngest expanded shoot partly eaten; sc.1 and sc.2 the successive scales on the stolon.

B. Detail of squamules in scale axil (× 20).

c. Female flower enclosed by bracts (× 3).

D. Female flower with bracts removed (× 9), to left in LS.

E. Floral diagram of female flower, squamules—solid black.

F. Young male flower enclosed by bracts (× 3).

G. Mature male flower with enclosing bracts (× 6).

H. Male flower with bracts removed (× 12).

ɪ. Floral diagram of male flower, squamules—solid black.

J. Fruits (× 3), the left-hand one dehiscing.

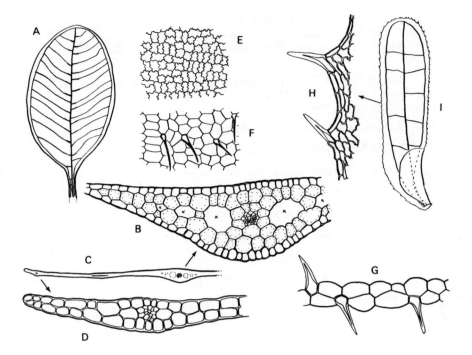

FIG. 4.15. HYDROCHARITACEAE. *Halophila*, leaf anatomy. (After Tomlinson 1980.)

A–E. *H. ovalis.*

A. Lamina in outline to show venation (× 2 ½). B. Detail of midvein (× 100). C. TS one half of lamina (× 15). D. Detail of margin (× 100). E. Epidermis in surface view (× 100).

F–I. *H. stipulacea.*

F. Epidermis in surface view including hairs (× 100). G. TS lamina (× 150). H. Marginal prickle hairs (× 100). I. Outline of lamina to show venation.

Family 5
APONOGETONACEAE (J. C. Agardh 1858)
(Figs. 5.1–5.6; Plate 3c, Plate 10)

SUMMARY

THIS family is represented by the genus *Aponogeton* L.f., consisting of some 40 species of aquatic herbs, restricted to the Old World, and extending from West Africa to Australia. The family is especially well developed in Madagascar. Many of the species are very localized (van Bruggen 1968–73; Sculthorpe 1967). The family has no commercial value, but several species are cultivated. One (*A. distachyos*)* has escaped from cultivation in several countries. The leaves usually have long petioles with open shortly-sheathing bases; the blade is usually well developed and it either floats on the surface or is submerged, both types sometimes occurring in the same species. Less commonly the leaves are strap-shaped or even filiform. The short, erect, fleshy axis is usually cormous, but rarely rhizomatous or stoloniferous. The plants are usually bisexual (rarely dioecious), with spicate inflorescences; the spike is sometimes double. The flower is ebracteate, with 1–3 (usually two) petaloid tepals or a single abaxial, bract-like tepal; it is rarely naked. There are six or more stamens, often in indistinct whorls. There are three or more free carpels, each with a short style and enclosing (1–) 2 or more (up to eight) basal or submarginal anatropous ovules. The follicular fruit has one – several seeds. Endosperm is absent and the embryo is straight.

Anatomical features include a well developed system of air-lacunae in the petiole and inflorescence axis. The anatomy of leaves is sharply differentiated according to situation: (i) The floating leaves have paracytic stomata, largely restricted to the adaxial surface of the lamina; the mesophyll is well developed and includes multiseriate adaxial palisade tissue. (ii) The submerged leaves usually lack stomata, and the mesophyll is reduced, often to only two layers and without differentiation of a palisade tissue. Both types of leaf may occur on one species (e.g. Fig. 5.2 A). The venation of the two types of leaves is apparently also distinct, the floating leaves developing a reticulum of ultimate closed veins which is absent from submerged leaves. The epidermis in both types of leaf consists of irregularly polygonal cells, not arranged in obvious longitudinal files. The epidermis of submerged leaves frequently includes chloroplasts. Hydropoten are present as localized epidermal structures, frequent on both surfaces of the lamina and petiole of submerged leaves, and on the abaxial surface of the lamina and petiole of floating leaves. Fenestrate leaves are produced in *A. fenestralis* and *A. bernieranus* by the breakdown of the intercostal mesophyll during the development of submerged leaves. Crystals are developed abundantly only in the mesophyll of floating leaves. Starch grains are always compound. Articulated laticifers are developed in association with veins of the leaf and in the inflorescence axis. Vessels are apparently absent.

* This species has had at least four spelling variants, *distachyus, distachyon, distachyum,* and *distachyos.* The last is used consistently throughout this account, variation in the literature being disregarded.

198

FAMILY DESCRIPTION
VEGETATIVE MORPHOLOGY

Leaf arrangement

First described in detail by Engler (1889*a*) for *A. distachyos*, thereby correcting earlier accounts by Planchon (1844) and Dutailly (1875). A further, more detailed revision of Engler's interpretation called for by Riede's (1920) later and more extended observations. Leaves in more or less opposite pairs, although superficially appearing four-ranked as illustrated by Dutailly. Each pair including an outer (lower) leaf *(n)* enclosing the inner (upper) leaf (*n* + 1). A lateral inflorescence axis also enclosed by the margin of the inner leaf. Successive leaf pairs arranged as mirror images of each other, with the inflorescence of each leaf pair borne either to the right or left. Angle of divergence between successive pairs somewhat greater than 90°, resulting in the apparent four-ranked arrangement. This arrangement regarded by Engler as a sympodium with congested internodes, each unit of the sympodium consisting of a shoot bearing two foliage leaves, but no specialized prophyll, and ending in a terminal inflorescence. Each unit then said to arise in the axil of the second leaf (*n* + 1) of the previous unit.

[Serguéeff (1907) noted that leaves of the same pair succeed each other almost immediately (within 1–3 days) but interval between successive leaves of different pairs is much longer (4–17 days). This is regarded as evidence supporting Engler's interpretation of the shoot as a sympodium.

Riede's interpretation of the shoot as a monopodium is based on the analysis of serial sections of a number of shoots at different stages of development. Evidence for monopodial branching: (i) there is no prophyll on each unit, as would be expected if each unit was the branch of a sympodium,* (ii) the shoot apex grows continually, with no evidence of the periodic eviction which sympodial growth requires, (iii) repeating cycles in the angular divergence of successive leaves can be demonstrated. Riede also showed that developmentally the inflorescence cannot be terminal since it is associated with the first (lower) leaf (*n*) of each pair, only becoming enclosed by the lateral margin of leaf *n* + 1 at a late stage of development. Riede's conclusion is that *Aponogeton* has a unique leaf arrangement but one that is analogous to that of *Najas* or *Halophila*. Engler, considering shoot construction in *Aponogeton* to be sympodial, drew attention to similarities between it and certain Araceae, but this analogy may lack substance.]

Leaf morphology (Figs. 5.1 A, and 5.2 A)

Leaf usually subdivided into a shortly sheathing but open base, a long petiole, and an ovate-lanceolate lamina, the latter sometimes slightly cordate at the base and emarginate apically. Leaves uniformly strap-shaped, e.g. in *A. vallisnerioides*, *A. rigidifolius*, and sometimes in juvenile forms of other spp; leaves filiform in *A. troupini*. Lamina fenestrate in *A. fenestralis* and *A. bernieranus*. According to Arber (1922*b*), lamina formed by outgrowths

---

* However, one might not expect a modified prophyll if there was continuity of growth from one unit to the next.

from lateral margin of ventral surfaces of 'petiole'. Two halves of lamina rolled in bud with adaxial surfaces innermost. Lamina in *A. ulvaceus* with undulate margins or even helically twisted (Czaja 1930; Riede 1920); bullate in *A. bernieranus.*

Axis

Usually short and erect, but rhizomatous, e.g. in *A. rigidifolius*, or stoloniferous, e.g. in *A. undulatus* (van Bruggen 1970). Corm up to 8 × 5 cm in *A. echinatus*. Development and general morphology described for *A. distachyos* by Dutailly (1875) and Serguéeff (1907). Erect, tuberous rhizome originating as a swelling of the plumular axis, the cotyledon and primary root abscissed early. Tuber described as a hypocotyledonary segment by Riede but this probably inaccurate according to Serguéeff's observations. Adventitious **roots** forming a circular series below the insertion of the plumular leaves and subsequently below each periodic leaf cluster. Early growth of axis oblique, leading to ultimate development of an erect axis with regular constrictions, each constriction representing a seasonal period of dormancy after a period of vigorous growth. Each constriction also marked by a ring of roots and, distally, by the persistent remains of old leaves. Each unit becoming separated from the previous one by a suberised, cork-like layer. Tissues between each cork layer finally collapsing in older rhizomes. [Whether this seasonal development is related to the phyllotactic cycles described by Riede is not known.]

Rhizome frequently branched but in an unspecified manner, described as a 'bifurcation' by Dutailly (see Taxonomic Notes, p. 210), but possibly the result of replacement of inflorescences by vegetative buds or by development of buds in axils of normally empty leaves, as observed by Riede. [Although growth takes place by adding distal increments to an erect axis, a uniform level in the soil is maintained. This suggests that roots may be contractile. Contractile roots, however, have not so far been recorded in *Aponogeton.*]

Regeneration of *A. distachyos* stimulated artificially by Riede by removal of above-ground parts and injury of shoot apex, thereby inducing adventitious buds to develop from meristematic regions arising close to vbs.

Vegetative Anatomy

Leaf apex

Stomata absent. Apex not normally with an open **pore** except in old leaves, as observed by Riede (1920). Longitudinal veins uniting below apex in a group of congested tracheids. Tracheidal characteristics sometimes developing in the epidermal cells themselves. Apical cells of *A. distachyos* stated by Minden (1899) to be smaller than nearby epidermal cells. Apical tissue colourless and including 1–2 spiral tracheids. Apical cells disintegrating or shrivelling in old leaves leading to the development of an open pore after the cuticle becomes torn. No secretion through the pore observed by Minden. Secretion through an apical pore thought by Riede to be unnecessary because replaced by secretion through hydropoten.

Leaf surface

Hairs absent except for **squamules** in the axils of each leaf. Squamules not always in regular pairs, as suggested by Engler, but according to Serguéeff (1907) squamules paired in the axil of each plumular leaf, one being enclosed by each leaf margin. Squamules 2–3 cells thick. Later-formed squamules developing as semicircular outgrowths, dissected distally to form a complete investment around the base of each leaf pair and associated inflorescence. Squamules including tannin cells and oil droplets.

Hydropoten (Mayr 1915) observed in all spp studied; more or less restricted to submerged parts according to Riede, i.e. on both surfaces of the lamina of submerged leaves, on the abaxial surface of floating leaves and on the petioles of both types of leaf. Each hydropote (Fig. 5.4 L) a localized area, usually more or less elliptical in surface view, and consisting of small, irregular epidermal cells. Hydropoten made conspicuous to the naked eye by their relatively dense cytoplasmic contents, associated with yellowish-brown, possibly resinous material. Epidermal hydropoten cells somewhat collapsed and shallow in material examined, but according to Mayr deeper than normal epidermal cells. Hypodermal cells below hydropoten also smaller and shallower than hypodermal cells elsewhere. Hydropoten typically situated below longitudinal veins of first and third order, but not restricted to these positions; those on petioles commonly above veins.

Development of hydropoten studied by Riede in *A. distachyos*. Hydropoten not evident at start of unrolling of the leaf, but appearing in basipetal sequence on the lamina, originating first on the midrib at any one level and extending over the lamina during unrolling. Hydropoten on petiole appearing acropetally. Hydropoten first recognized as small groups or files of plasma-rich cells, not initially differing from surrounding cells but, by rapid division in planes perpendicular and parallel to long axis of the leaf, developing a characteristic appearance, followed by the deposition of impregnating wall substances and modification of the cuticle.

In fresh leaves, hydropoten characteristically associated with an abundant growth of microorganisms.

The anatomy of the lamina of floating and submerged leaves is sufficiently distinct to merit separate description.

Leaf, lamina of floating leaves (Fig. 5.4 A–M)

Examples of spp with floating leaves including *A. angustifolium*, *A. dinteri*, *A. distachyos*, *A. echinatus*, *A. elongatum*, *A. spathaceum*, *A. subconjugatus*. Lamina markedly dorsiventral. **Wax** coat recorded by Serguéeff (1907) for *A. distachyos*. **Epidermis** usually colourless, rarely including small chloroplasts. Adaxial epidermis in surface view (Fig. 5.4 B) consisting of irregularly polygonal cells, not arranged in distinct longitudinal files; the distinctive surface appearance of the epidermis in *Aponogeton* presumably resulting from late and irregular cell division in the epidermal mother cells, the outline of the primoridal protodermal cell sometimes recognizable in mature leaves according to Solereder and Meyer (1933). Cells above larger veins rectangular, narrow, elongated parallel to veins and otherwise more regular than cells elsewhere. Epidermis in TS, shallow, thin-walled, outer wall scarcely thick-ened. Abaxial epidermal cells similar to, but somewhat larger than, those of adaxial surface (Fig. 5.4 c). Small chloroplasts observed in the abaxial epidermis of *A. angustifolium.*

Stomata (Fig. 5.4 B, I–K) abundant on the adaxial surface but not infrequent abaxially although presumably then non-functional. Stomata recorded by Riede on the abaxial surface towards the apex of submerged juvenile leaves (cf. Fig. 5.5 A) and regarded as phylogenetically significant (see p. 210). Guard cells (Fig. 5.4 H) thin-walled, each with a prominent outer ledge, the inner ledge less pronounced. Guard cells orientated with the stomatal pore more or less parallel to the long axis of the leaf; in *A. spathaceum* varying in length from 25–40 μm, a range encompassing that recorded for the whole genus by Solereder and Meyer. Each stoma with a pair of indistinct lateral subsidiary cells (Fig. 5.4 I) somewhat narrower than but otherwise scarcely distinct from the remaining epidermal cells. One (Fig. 5.4 J) or both (Fig. 5.4 K) subsidiary cells commonly segmented by a median anticlinal division, the subsidiary cells then becoming obscure and not recognized as distinct by Serguéeff. [A developmental clarification is needed.] **Hypodermis** not differentiated as distinct colourless layers except for a somewhat compact abaxial layer (Fig. 5.4 E). **Mesophyll** (Fig. 5.4 F) as much as ten-layered; adaxial 2–5 layers constituting a well developed **palisade tissue** of anticlinally extended, more or less cylindrical cells densely packed with large discoid chloroplasts. Palisade cells congested but interrupted by a well developed chamber below each stoma (Fig. 5.4 D). Palisade continuous across midrib. **Spongy tissue** of 2–5 abaxial mesophyll layers usually sharply delimited from the palisade. Cells large, including fewer and often smaller chloroplasts com-pared with palisade layers; cells loosely lobed and enclosing a well developed intercellular space system.

Venation (Fig. 5.4 A) showing three distinct orders of veins: (i) longitudinal first-order veins, usually 5–7 on each side of midrib, uniting distally beneath apex (Fig. 5.1 A); (ii) second-order veins as a series of numerous transverse commissures perpendicular to the longitudinal veins and parallel to each other, closely connecting the longitudinal veins; (iii) third-order veins, as a closed anastomosing reticulum connecting the second-order veins predomi-nantly in a longitudinal direction.

Veins equidistant from each surface, not connected to surface layers by bundle sheath extensions. Each first-order vein surrounded by a colourless sheath of elongated parenchyma cells, the abaxial cells around the phloem forming an indistinct mechanical sheath of slightly thick-walled, somewhat narrow but scarcely sclerotic cells. **Xylem** including a single wide adaxial tracheal element with persistent spiral wall thickenings, separated by xylem parenchyma from a series of narrow metaxylem elements in the middle of the vein. **Phloem** strand wide, with conspicuous wide sieve-tubes separated from the metaxylem elements by conjunctive parenchyma. Successively nar-rower veins of higher orders progressively simpler; ultimate third-order veins (Fig. 5.4 G) indistinctly sheathed by elongated chlorenchyma cells and each including a single narrow tracheal element and a narrow phloem strand. Narrow metaxylem of first-order veins continuous with the xylem of the second-order veins. **Laticifers** associated with bundle sheath parenchyma, described below (p. 206).

Leaf margin usually rounded and not otherwise specially differentiated from the rest of the lamina, but in *A. angustifolium* (Fig. 5.4 M) including a conspicuous **mechanical strand** of thick-walled, somewhat collenchymatous epidermal and hypodermal cells, continuous downwards into the petiole margin. Marginal strand also recorded for this species by Solereder and Meyer (1933) and possibly diagnostic.

Leaf, lamina of submerged leaves (Figs. 5.5 A–C and 5.6 A–F)

Species with submerged leaves including *A. crispum, A. fenestralis, A. monostachyon, A. natans,* and *A. ulvaceus.* Lamina more or less isolateral, except for differences in cell dimensions of upper and lower surface layers; abaxial usually larger than adaxial epidermal and hypodermal cells (cf. Figs. 5.6 B, C, E, F). **Epidermis** usually chlorenchymatous, but with smaller chloroplasts than the mesophyll layers; uniform or rarely with slight differentiation of costal bands of elongated cells above or below the veins; cells in surface view (Fig. 5.6 F) irregularly polygonal; longitudinal files of cells indistinct or absent. Cells thin-walled, outer wall scarcely thickened.

Stomata either absent from, or not infrequent on, the adaxial surface, e.g. in *A. crispum* (Fig. 5.5 A), and on juvenile submerged leaves of several spp with floating leaves, according to Riede. Stomata resembling those of floating leaves but with closed pores and presumably non-functional.

Mesophyll 2–5-layered between the veins, the number of mesophyll layers varying between different spp. Hypodermal layer or layers always large-celled, compact and densely filled with discoid chloroplasts (Fig. 5.6 C, E). Middle mesophyll layer (if present) consisting of flattened but lobed cells forming a well developed intercellular space system (Fig. 5.6 D).

The following variations in the construction of the intercostal mesophyll observed or recorded but little information available about the anatomy of different leaves on a single plant, except in Riede's (1920) work.

A. ulvaceus. Mesophyll two-layered in both adult and juvenile leaves according to Czaja (1930), but a third central layer in the adult leaf suggested by his illustrations, thereby confirming Solereder and Meyer's record of three-layered mesophyll in this sp. Juvenile leaves in this sp slightly dorsiventral according to Czaja with most chloroplasts in adaxial mesophyll layer; adult leaves remaining isolateral. [This may be because juvenile leaves are more sensitive to light. Riede records a 2–3-layered mesophyll in juvenile leaves of this species.]

A. natans. Mesophyll 2–3-layered, locally two-layered and then resembling adult leaves of *A. ulvaceus* in intercostal regions remote from veins (Fig. 5.5 C), but three-layered closer to veins (Fig. 5.5 B) and then more like the following spp. Another leaf of *A. ulvaceus* seen with a 3–4-layered mesophyll and local doubling of central mesophyll layer.

A. monostachyon. Intercostal mesophyll three-layered throughout, with a single compact hypodermal layer beneath each epidermis enclosing a central layer of lobed cells (Fig. 5.6 A–F).

A. crispum. Mesophyll in some leaves five-layered, including two compact, large-celled layers below each epidermis, but cells of each outer hypodermal layer smaller than those of the inner layer, and all enclosing a single central layer of lobed cells. Other leaves of this sp showing a three-layered mesophyll like that of *A. monostachyon.*

A. (Ouvirandra) fenestralis, and to a lesser extent *A. bernieranus,* distinguished by the development of **fenestrate leaves** with rounded perforations in the lamina between the first- and second-order veins. Juvenile leaves of *A. fenestralis* said by Serguéeff (1907) to be imperforate; lamina of succeeding leaves at first perforated irregularly in middle region (the normal condition in *A. bernieranus).* Adult leaves uniformly perforated throughout, the leaf tissue largely consisting of the vascular skeleton. **Perforations** late to develop, according to Serguéeff, in two-layered regions of the mesophyll between veins. Circular-elliptical sites of future perforations distinguishable as brown subepidermal zones caused by suberization of the cell walls and the secretion of brown substances into the intercellular spaces. Central cells of the perforation site becoming separated from one another, beginning first on upper surface of the lamina. Perforations then becoming enlarged passively by expansion of leaf, the margins of the perforations becoming sealed largely by intercellular secretions rather than by a secondary epidermis. **Hydropoten** abundant, according to Riede.

Venation of submerged leaves consisting only of longitudinal (first-order) veins connected to each other by transverse (second-order) veins, and so differing conspicuously from the venation of floating leaves (Fig. 5.4 A) in the almost total absence of third-order veins. A third-order reticulum developed to a slight extent between longitudinal veins immediately adjacent to the midrib in some spp (e.g. *A. crispum*) but never becoming a conspicuous feature of lamina as a whole. Veins situated in the middle layer of mesophyll between the hypodermal layers, the mesophyll often being deepest near to veins and producing a slight ribbing of the leaf surface. Anatomy of the veins as already described for floating leaves.

Leaf axis in both submerged and floating leaves (Fig. 5.5 D–J)

Floating leaves with a thickened **midrib** (Fig. 5.5 E). Adaxial epidermis and palisade of the midrib continuous with those of the lamina; epidermis perforated by stomata. Epidermis otherwise similar to first layers above and below the larger veins of the lamina. Abaxial mesophyll, in both kinds of leaf, colourless, lacunose, and resembling that described below for the petiole. Veins including a wide, conspicuous, median vb and two or more somewhat smaller, lateral bundles. Midrib of floating leaves including a distal continuation of the adaxial and abaxial bundles as described under petiole below.

Petiole of floating leaves long, flexible, and more or less triangular in TS (Fig. 5.5 F). **Epidermis** of thin-walled cells, with occasional stomata on the adaxial surface. Epidermal cells in short, but often v. indistinct, longitudinal files in surface view; cells elongated, more or less rectangular or with oblique end-walls. **Ground parenchyma** chlorenchymatous; cells wide and axially elongated. One–two hypodermal layers compact (Fig. 5.5 G) but remaining mesophyll forming a v. lacunose tissue with longitudinal **air-canals** separated by uniseriate plates of parenchyma (Fig. 5.5 J). Larger **veins** enclosed by compact layers of parenchyma. Longitudinal air-canals, traversed at frequent but irregular intervals by uniseriate transverse diaphragms composed of small, densely cytoplasmic, stellately lobed cells (Fig. 5.5 H); the intercellular spaces between the lobes of the diaphragm cells providing gaseous continuity along the lacunae.

Vascular system of floating leaves (e.g. Fig. 5.5 F) represented by: (i) a single arc of dominant major vascular bundles, equidistant from each surface, with a wide median bundle and successively narrower lateral bundles; (ii) several abaxial and a few (2–3) adaxial subsidiary vascular bundles in the peripheral compact hypodermal tissue (Fig. 5.5 I). Subsidiary bundles in *A. distachyos* sometimes forming an almost complete peripheral series. **Median vein** (Fig. 5.5 K) without well developed mechanical sheathing tissues, although cells next to the phloem and to a certain extent those next to the xylem narrow and somewhat collenchymatous. **Xylem** of largest veins including a wide xylem lacuna and several narrow metaxylem tracheids connecting via transverse veins with the xylem of other longitudinal bundles (Fig. 5.5 K). Xylem of successively narrower lateral veins with progressively more persistent wide protoxylem elements with unaltered spiral wall thickenings in place of the xylem lacuna, the number of narrow metaxylem elements being reduced at the same time. **Peripheral subsidiary veins** (Fig. 5.5 I) including one or more narrow tracheal elements, often inversely or obliquely orientated regardless of the position of the bundle in relation to the major veins. Longitudinal veins connected by irregular, transversely anastomosing veins always crossing the lacunae via a transverse diaphragm. Despite the absence of mechanical tissue throughout the petiole, all walls of the ground parenchyma cells seem to be thickened in petioles of old leaves of *A. distachyos*. Ground parenchyma next to the veins often including large starch grains. **Tannin** sometimes conspicuous in cells of lacunar ground parenchyma (Fig. 5.5 J).

Petiole of submerged leaves usually narrower than that of floating leaves, but otherwise identical anatomically except for a reduced number of vbs (Fig. 5.5 D). Individual vbs smaller than those of floating leaves, the xylem being represented only by a single wide tracheal element. Abaxial and adaxial subsidiary bundles absent from all spp examined except for 1–2 small abaxial bundles in *A. natans*. Subsidiary bundles well developed in *A. fenestralis* according to Serguéeff.

[**Growth responses** of submerged leaves, which involve reorientation of the blade under differing light intensities, have been studied by Riede and Czaja. According to Serguéeff this response is the result of growth in the region of the petiole just below the insertion of the blade. Czaja showed that in *A. ulvaceus* juvenile leaves were more sensitive than adult.]

Leaf sheath largely distinguished from petiole by 2–3-layered lateral wings without vascular tissue, according to Singh (1965).

Corm (Plate 3c)

Including well developed peripheral **etagen periderm** interrupted by numerous root and leaf traces. Periderm in *A. fenestralis* represented only by suberized layers without etagen-divisions, according to Solereder and Meyer (1933). **Cortex** indistinctly delimited from the central cylinder by a narrow plexus of vascular tissues with narrow xylem and phloem cells aggregated into indistinct vbs but not surrounded by an endodermis except for possible slight suberization of innermost cortical layers. The vascular tissues otherwise scattered as discrete collateral strands throughout the cortex and **central cylinder** and forming an irregular vascular reticulum. Vascular strands v. irregular and occasionally amphivasal. Amphiphloic bundles recorded by Solereder and Meyer. Vascular system described and illustrated by Dutailly (1875) as being more or less regular, with two concentric series of vbs around a central plexus, but this regularity neither observed in *A. distachyos* and *A. subconjugatus* nor recorded by other investigators.

Ground tissue of uniform, compact parenchyma, in younger parts densely filled with large compound starch grains. Laticifers noted as absent from the rhizome by Serguéeff, a fact confirmed in material examined here.

Root (Fig. 5.6 G–J)

Roots forming an annular series below the leaf insertions on each seasonal rhizome segment. Roots possibly contractile but this property not yet clearly demonstrated.

Casual observations on *A. subconjugatus* suggest root dimorphism, with wider roots possibly contractile. **Root-hairs** developed from short epidermal cells (trichoblasts); branched in *A. fenestralis* according to Serguéeff, but only in roots in contact with a solid body, e.g. its own rhizome. **Exodermis** of 2 (–3) compact layers of small, relatively short cells with minutely plicate longitudinal walls. **Cortex** v. wide in proportion to narrow central stele (Fig. 5.6 G). Cortical cells uniformly wide and radially elongated, rounded to form a well developed intercellular space system. Cortex in roots of *A. angustifolium* including wide radial **air-lacunae** (Fig. 5.6 I) developed by collapse of cells in two middle cortical layers. Similar air-lacunae regarded as a general feature of old roots by Serguéeff. Inner cortex small-celled in all roots, with regularly concentric and radiating cell layers. **Stele** narrow, delimited from the cortex by a narrow, thin-walled **endodermis** with conspicuous Casparian thickenings on the radial walls. Endodermal cells with U-shaped wall thickenings recorded in *A. distachyos* by van Tieghem (1870–1). **Pericycle** uniseriate, usually distinct and never interrupted by vascular tissues. **Phloem** restricted to narrow strands, usually next to pericycle, alternating with protoxylem strands if present; each phloem strand usually consisting of one sieve-tube and associated companion-cells; sieve-tubes v. inconspicuous in narrow roots of *A. angustifolium*. **Xylem** in large roots, as represented by *A. elongatum* (Fig. 5.6 H), including one (rarely more as illustrated by van Tieghem) central wide tracheal element, with 2–5 irregular radiating series of narrow protoxylem elements extending to the pericycle. Xylem in smaller roots, as represented by *A. angustifolium* (Fig. 5.6 J), including only one wide central element and without narrow protoxylem elements. Xylem represented only by a central cavity in roots of *A. crispum* and *natans* according to Singh (1965c).

SECRETORY, STORAGE, AND CONDUCTING ELEMENTS

Laticifers: first observed and described in the leaf of *A. distachyos* and *A. fenestralis* by Serguéeff (1907) and subsequently noted by other authors (e.g. Riede) in other spp; possibly universally present in leaf and inflorescence but absent from rhizome and root. Laticifers articulated, consisting of long cells with opaque, granular contents and large nuclei. End walls of adjacent laticifers overlapping extensively but never breaking down. Contents in *A. fenestralis* giving reactions interpreted by Serguéeff as indicating the presence of rubber-like substances. Contents of laticifers in *A. distachyos* readily soluble in alcohol except for a resinous residue. Laticifers consequently not obvious in TS and readily overlooked, or described as tannin-filled cells, as by Singh. Laticifers 3–10 around each larger vb of the petiole and longitudinal veins of the lamina, forming an indistinct, incomplete sheath, continuous along transverse veins of lamina in *A. fenestralis*.

Crystals: abundant only in the mesophyll of floating leaves; most abundant in the palisade parenchyma as fine or coarse aggregations of rhombohedral crystals, presumably of calcium oxalate.

Tannin: frequent but never abundant; in isolated cells, e.g. in partition cells of the lacunose petiole (Fig. 5.5 J); common next to vascular tissues. Elongated tannin cells in the root described as tannin-laticifers (*laticifères à tannin*) by Serguéeff, but not to be confused with true laticifers of leaves.

Anthocyanin: not uncommon in abaxial tissues of the lamina of submerged leaves, as indicated by Serguéeff for *A. fenestralis*.

Starch: large, compound grains abundant in the ground parenchyma of rhizomes, and also common in cells around the larger vbs of the leaf.

Xylem: vessels not observed; elements throughout represented by narrow tracheids with annular or spiral wall thickenings. Large xylem elements in major vbs of the leaf apparently undergoing early disorganization and each then represented in the mature organ by a wide lacuna (Fig. 5.5 K). Central metaxylem elements in the root stele v. long, with spiral wall thickenings and tapering imperforate end walls. Lateral wall thickenings in these elements developed only after the breakdown of the nucleus (Serguéeff 1907).

REPRODUCTIVE MORPHOLOGY

Inflorescences solitary, erect, associated with every alternate leaf in the manner described above (see leaf arrangement, p. 199). **Peduncle** long, naked, ending in a simple (Fig. 5.2 A) or compound (Fig. 5.3 A) flower-bearing axis (spike) enclosed by a deciduous (persistent in *A. loriae*) spathe-like **bract** (Fig. 5.3 B). Flower-bearing axis either single or double, but sometimes further divided, the flowers either more or less uniformly distributed around an axis with radial symmetry (Fig. 5.2 B) or on the inner side of a flattened axis with dorsiventral symmetry (Fig. 5.3 A). Flowers either congested or more diffuse. In *A. ranunculiflorus* inflorescence floating and supported by sterile lower flowers, each represented by a single tepal (Guillarmod and Marais 1972).

Each flower accompanied by (1–) 2 (–3) petaloid organs, of variously interpreted morphology, here referred to as tepals for descriptive purposes. cf. Plate 10.

Diagnostic characters for the species and for subdividing the genus include the arrangement of the flowers, and the symmetry and number of axes in the plant, as indicated in the following artificial key:

1. Spike (flower-bearing axis) radially symmetrical   2
2. Axis single (e.g. *A. natans*; Fig. 5.2 A, B)
2A. Axis double (sometimes more than two) (e.g. *A. fenestralis*)
1A. Spike dorsiventrally symmetrical   3
3. Axis single (e.g. *A. gracilis*)
3A. Axis double (sometimes more than two) (e.g. *A. distachyos*; Fig. 5.3 A)

Flowers usually perfect, rarely female by reduction and apomictic in *A. junceus*; plants dioecious in *A. troupini* and *A. dioecus*. Each flower with usually two lateral tepals (Figs. 5.1 B–E and 5.2 B–D) (sometimes three in distal flowers); tepal solitary (e.g. Fig. 5.3 C–E) (but sometimes two), enlarged and bract-like in *A. distachyos*; flowers rarely naked (e.g. female flowers of *A. dioecus* and *A. troupini*). Stamens six (e.g. Fig. 5.1 B, C) or more in indistinct whorls, number of whorls more numerous in *A. distachyos* (Fig. 5.3 C, D; up to five), the number of stamens apparently further increased by proliferation or bifurcation to a maximum of 23 recorded in this sp. Carpels usually three (Fig. 5.2 B, C) but often more in *A. distachyos* (usually four, e.g. Fig. 5.3 C, D; cf. Fig. 5.1 C) and in female flowers of other spp. Carpels free, each with a short style and stigma, the carpel margins always free when young but wholly or partly fused at maturity. Ovules two– several, inserted either basally when few but usually in two series, one series on each carpel margin (Figs. 5.1 B, E; 5.2 C, D; and 5.3 C, E). Ovules bitegmic, anatropous, erect or pendulous. Integument single in *A. distachyos*, *A. quadrangularis*. Fruit follicular, one– several seeded. Seeds very variable, often with a spongy testa permitting them to float. Embryo straight with a large cylindrical or flattened cotyledon, the plumule more or less lateral. Seedling at an early stage developing a corm-like axis by the swelling of either the hypocotyl or the plumular axis.

REPRODUCTIVE ANATOMY

Inflorescence axis

Cylindrical, irregularly fluted. Ground parenchyma lacunose, like that of the petiole except for 3–4 compact hypodermal layers. **Vascular system** including an irregular peripheral series of small vbs more or less included within the compact hypodermal layers, each vb resembling a peripheral subsidiary bundle of the petiole. Vascular supply to sheathing bract solely derived from this peripheral series, according to Singh (1965c). Large central vbs usually four, included within the lacunose ground tissue.

Flower

(Data taken mainly from the description by Singh (1965c) and Singh and Sattler (1977b).

A. natans. [This species and *A. undulatus* (also *A. crispum*) are examples of species with radially symmetrical inflorescence axes.] Central vascular tissues above the spathe aggregated into a complete vascular cylinder, continuous above with ten central vbs and a ring of peripheral vbs. Vascular supply to each flower formed by the fusion of a few traces from both the peripheral and central systems and forming a mass of vascular tissue. Total number of vbs in the axis decreasing distally.

Vascular system of the flower consisting of: (i) two unbranched traces, one for each tepal; (ii) approximately six unbranched traces, one for each stamen; (iii) vascular aggregate forming three traces by 'bridge formation', one trace going to each carpel. Each carpel trace forming a dorsal bundle extending unbranched to the base of the stigma, and a ventral bundle initially normally orientated but becoming inversely orientated distally. Each ventral bundle dividing to supply the ovular traces.

A. distachyos. Vascular supply of twin axes identical, and derived from the vascular supply to the peduncle by equal division. Tepal irrigated by many traces derived from the vascular plexus below the flower. Some of these traces often forming a second, inversely orientated series, thereby making the tepal anatomically bifacial. Numerous stamens each supplied by a single trace. Dorsal carpel traces derived from remaining tissue of the vascular cylinder, independently of ventral carpel traces. Ventral traces appearing on same radius as dorsal traces, but at a higher level; 2–3 additional bundles forming a total of 4–6 lateral bundles between the dorsal and ventral sides. [The two ventral traces to each carpel are thus independent of each other from the start; they supply the ovules. The dorsal and lateral bundles extend into the style, the former reaching as far as the stigma, and the latter ending blindly below.

These differences in tepal and carpel anatomy provide the basis for the recognition of two groups of species within *Aponogeton*, according to Singh. Whether any other species resemble *A. distachyos* is not known; otherwise these observations further emphasize the distinctiveness of *A. distachyos*.

In some discussions of carpel morphology (e.g. by Eber 1934; Saunders 1929), great emphasis is given to the 'lateral' carpellary bundles of *A. distachyos*. However, these should be contrasted with the 3-trace system of other spp (e.g. *A. natans*).]

Taxonomic and Morphological Notes

Species with either submerged or floating leaves

Aponogeton includes spp with either wholly submerged leaves or wholly floating leaves. There are corresponding sharp differences in leaf anatomy which may be useful taxonomically if a species has only one leaf type. Variation in the number of mesophyll layers in submerged leaves is another potential taxonomic character still awaiting investigation. However, reports on whether a particular species has either floating or submerged leaves are sometimes conflicting. Solereder and Meyer (1933), examining herbarium material, recorded that the anatomy of the leaves of *A. angustifolium*, *A. gracilis*, and *A. heudelotii* indicates that they are floating leaves, contrary to the account by Krause and Engler (1906) who record no floating leaves in these spp. The explanation may be simply that a species can have submerged juvenile and floating adult leaves, although this possibility seems not to have been investigated. Riede, on the other hand, records stomata in the juvenile leaves both in a number of species with wholly submerged leaves (*A. fenestralis, A. natans, A. ulvaceus*) as well as in species with floating leaves. Stomata may even occur in adult submerged leaves (e.g. in *A. crispum*, Fig. 5.5 A). Riede concludes from this similarity that species with wholly submerged leaves are phylogenetically derived from those that develop adult floating leaves. These problems can only be solved by a much more extensive knowledge of the range of variation in leaf anatomy within a single species.

Czaja (1930), Glück (1924), Riede (1920), and Serguéeff (1907) have discussed the phototropic and other responses of leaves in *Aponogeton*. Functional differences between juvenile and adult leaves have been observed by Czaja.

Morphological interpretations of the inflorescence—branching

Whether the inflorescence axis is double or single seems simply to be a matter of whether the apex bifurcates or not. Riede observed stages in this forking process, with further splitting of the peduncle when more than two axes are produced. It seems reasonable to regard this process as a dichotomy or even trichotomy . . . etc., since such bifurcation has been recorded in other helobial families (Wilder 1974*c*). Wilder believes that branching of the vegetative axis in *Aponogeton* takes place in this way in the light of the description given by Serguéeff (1907), but my own reading of the literature shows that only Dutailly (1875)—cited by Serguéeff—specifically mentioned this possibility, but without actual documentation. This important aspect of development in *Aponogeton* needs detailed examination. According to Riede the 'spathe' itself is a double structure made up of two separate 'prophylls' which fuse at an early stage. This affords an analogy with certain Hydrocharitaceae. Their development apparently precedes the division of the axis itself.

Petaloid bracts

Numerous early authors (see Serguéeff) interpreted the 1–2 petaloid structures of most *Aponogeton* species as bracts subtending naked flowers. This interpretation is most obviously acceptable for *A. distachyos* in which there is usually only one conspicuous structure per flower. Uhl (1947) on the other hand, considered the flower to be a condensed branch system. However, it must be remembered that paired bracts or bracteoles are otherwise unknown in monocotyledons. Also there is evidence of a rudimentary third structure in some species normally with two appendages (Riede), and even in *A. distachyos* there may be two structures. This makes more acceptable the alternative (and simpler) explanation, originally proposed by Eichler (1875), that these organs represent the remains of a perianth. Eichler suggests that they represent a reduction from a six-membered perianth, but there seems to be no reason why the organ should not have been three-membered, a condition which is known to occur in the flowers in distal parts of the inflorescence axes of some spp (e.g. *A. angustifolium* according to Krause and Engler). This interpretation of the structures as tepals has been accepted by several later authors (e.g. Krause and Engler, Riede, Singh) but Serguéeff accepted the earlier view.

Flower (Plate 10)

The presence of six stamens and three carpels in many flowers in *Aponogeton* so strongly recalls the basic floral plan for the monocotyledons that it is only reasonable to regard the anomolous perianth as reduced from a three-lobed structure. This raises problems of description. Bearing the usual trimerous construction in mind, the high and variable number of stamens in the flower of *A. distachyos* together with the variation in the number of its carpels, is noteworthy. Serguéeff refers to this proliferation as the result of 'multiplication' of three original stamens, in a manner which has been recorded in *Butomus*. The six stamens normal for *Aponogeton* are themselves regarded as derived by bifurcation of an original 'primitive' three. Much more evidence than is presented by Serguéeff is needed before this interpretation can be accepted.

The basic floral morphology of *Aponogeton* has been greatly clarified by a developmental study by Singh and Sattler (1977*b*). They confirmed that the flower is basically trimerous in its organization although they found only a rudiment of the third (adaxial) tepal. However, there is no evidence for the development of secondary (CA) primordia (Plate 10) as in some Alismataceae.

Problems which arise in interpreting the morphology of the reproductive parts in Aponogetonaceae would be minimized if a wider range of species were examined. Too often interpretations are based largely on a study of the commonly cultivated *A. distachyos*, simply because it is most readily available, although it seems to show anomalies of construction which suggest that it is not typical for the family as a whole. Developmental studies on a wider range of species are much to be desired.

Affinities of Aponogetonaceae

The helobian affinity of the Aponogetonaceae is not in doubt, on the general evidence of their embryology. However, the position of the family within the Helobiae is not at all clear on the basis of present evidence. The suggestion of Engler that the Araceae represent one line of affinity (on the basis of inflorescence construction, leaf fenestration, presence of laticifers, . . . etc.) is much weakened if shoot construction is interpreted as monopodial (Riede) and not regarded as sympodial and aroid-like (Engler). Interestingly, however, Singh and Sattler (1977*b*) in comparing the floral development of *Acorus* (Araceae) with that of *Aponogeton*, point out that there are striking resemblances between the two genera. On the other hand, the strong resemblance between *Aponogeton* and *Scheuchzeria*, suggested by Uhl on the basis of carpel vasculature, is further supported by the more detailed investigations of floral anatomy by Singh (1965*c*), continued in the developmental studies by Singh and Sattler (1977*b*). However, Yamashita (1976*b*) notes that the first root on the embryo is above the suspensor in *Scheuchzeria*, but lateral in *Aponogeton* (p. 23). There is little real evidence to support Hutchinson's (1959) derivation of the Zosteraceae from the Aponogetonaceae since this is largely based on a superficial comparison of *Zostera* and *Phyllospadix* with *A. distachyos* alone (not the whole family of the Aponogetonaceae). *Aponogeton distachyos* and *A. quadrangularis* differ from most Helobiae in having only a single integument (Afzelius 1920).

The most direct evidence points to a relationship between the Aponogetonaceae and the Scheuchzeriaceae–Juncaginaceae–Lilaeaceae group of families.
In floral construction, carpel morphology, and the presence of laticifers there
are, however, similarities with the Alismataceae and Butomaceae, whereas
the distinctive phyllotaxis and branching pattern of Aponogetonaceae recall
those of some Hydrocharitaceae. Placing the Aponogetonaceae between these
two groups of families probably best reflects our understanding of them.

MATERIAL EXAMINED

Aponogeton angustifolium Ait. Cape Province, South Africa; Cheadle CA
922; all parts.
A. crispum Thunb. Cultivated at Kew; leaf.
A. distachyos L.f. Cape Province, South Africa; Cheadle CA 797–8. Natural-
ized plants, Auckland, New Zealand; P. B. Tomlinson, 17.x.69; all parts.
Cultivated at Kew; leaf, inflorescence axis.
A. elongatum Muell. Brisbane, Australia; Cheadle CA 455; all parts.
A. monostachyon L. f. Cultivated at Kew; leaf.
A. natans Engl. and Krause. Cultivated at Kew; leaf.
A. spathaceum E. Mey. Cultivated at Kew; leaf, inflorescence.
A. subconjugatus Schum. and Thonn. Ghana; J. B. Hall *s.n.*; all parts.

SPECIES REPORTED ON IN THE LITERATURE

Afzelius (1920): *Aponogeton distachyos, A. guillotii, A. quadrangularis, A.
violaceus, A. ulvaceus*: seed development.
Arber (1921): *A. spathaceum* var. *junceum*: leaf.
Arber (1922*b*): *A. distachyum**: leaf.
Chrysler (1907): *A. fenestralis*: rhizome.
Costantin (1886*b*): *A. fenestralis* (as *Ouvirandra fenestralis*): leaf morphology.
Czaja (1930): *A. ulvaceus*: leaf.
Drawert (1938): *A. distachyus*: hydropoten.
Dutailly (1875): *A. distachyum*: all parts.
Duval-Jouve (1873*b*): *A. distachyon*: lacunar diaphragms in stem.
Glück (1924): *A. distachyum*: leaf morphology.
Kaul (1976*b*): *A. distachyos*: leaf.
Lagerheim (1913): *A. fenestralis, A. guillotii*: leaf morphology.
Mayr (1915): *A. distachyus, A. ulvaceus*: hydropoten.
Minden (1899): *A. distachyus*: leaf apex.
Paliwal (1976): *A. natans*: leaf epidermis.
Riede (1920): *A. dinteri, A. distachyus, A. fenestralis, A. natans, A. ulvaceus*:
all parts.
Serguéeff (1907): *A. distachyus, A. fenestralis*: all parts.
Singh (1965*c*): *A. crispum, A. natans*: all vegetative parts. *A. distachyon*:
no vegetative anatomy but floral parts.
Singh and Sattler (1977*b*): *A. natans, A. undulatus*: floral development.

* The spelling variants given by individual authors are quoted.

Solereder and Meyer (1933): *A. ulvaceus*: fresh material of all parts; together with herbarium material of *A. angustifolius, A. dinteri, A. gracilis, A. heudelotii, A. leptostachys* var. *abyssinicus, A. natans*.

van Tieghem (1870–1): *A. distachyum*: root.

van Tieghem and Douliot (1888): *A. distachyum*: lateral root initiation.

Yamashita (1976*b*): *A. crispus, A. junceus* ssp. *natalensis, A. madagascariensis*: embryo development.

SIGNIFICANT LITERATURE—APONOGETONACEAE

Afzelius (1920); Arber (1921, 1922*b*); van Bruggen (1968, 1969, 1970, 1973); Chrysler (1907); Costantin (1886*b*); Czaja (1930); Drawert (1938); Dutailly (1875); Duval-Jouve (1873*b*); Eber (1934); Eichler (1875); Engler (1887, 1889*a*); Gibson (1905); Glück (1924); Guillarmod and Marais (1972); Hutchinson (1959); Kaul (1976*b*); Krause and Engler (1906); Lagerheim (1913); Mayr (1915); von Minden (1899); Obermeyer (1966*a*); Paliwal (1976); Planchon (1844); Riede (1920); Sane (1939); Saunders (1929); Sculthorpe (1967); Serguéeff (1907); Shah (1972); Singh (1965*c*); Singh and Sattler (1977*b*); Solereder and Meyer (1933); van Tieghem (1870–1); van Tieghem and Douliot (1888); Uhl (1947); Verdoorn (1922); Yamashita (1976*b*).

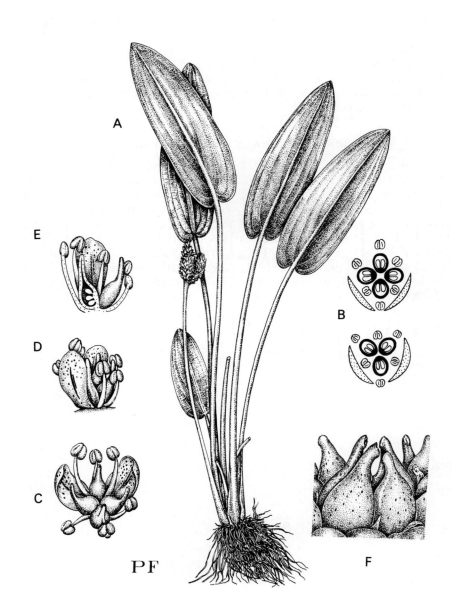

A

E

D

C

B

F

PF

FIG. 5.1. APONOGETONACEAE. *Aponogeton subconjugatus* Schum. and Thonn. (Drawn from fluid-preserved material; Ghana; J. B. Hall *s.n.*)

A. Habit ($\times \frac{1}{3}$) (in nature leaves are floating).

B. Floral diagrams of three-carpellate (lower) and four-carpellate (upper) flowers.

C–E. Flowers ($\times$ 6).
  C. From above (four-carpellate). D. From the side. E. In LS.

F. Detail of fruiting spike with immature fruits ($\times$ 6).

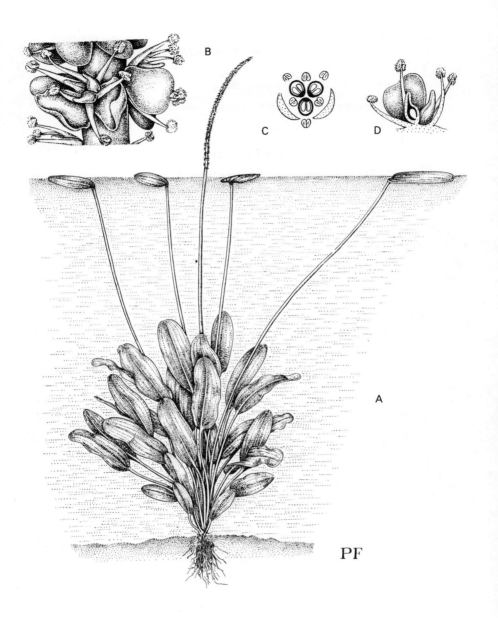

B

C

D

A

PF

216

F IG. 5.2. APONOGETONACEAE. *Aponogeton natans* Engl. and Krause. (Cultivated, Fairchild Tropical Garden, Miami.)

A. Habit of plant (× ¼) with submerged and floating leaves and emergent spike.

B. Detail of flower spike (× 8).

C. Floral diagram, three-carpellate flower.

D. Flower in LS (× 8).

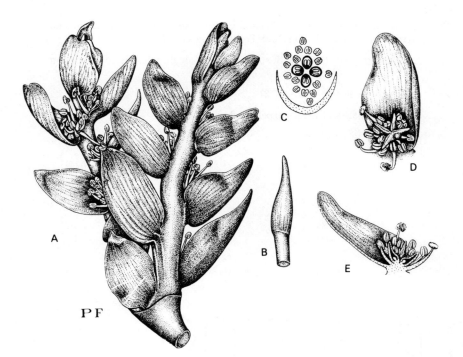

218

FIG. 5.3. APONOGETONACEAE. *Aponogeton distachyos* L.f., details of flowers. (Drawn from fluid-preserved material; naturalized, Auckland, New Zealand; P. B. Tomlinson 17.x.69.)

A. Bifurcate inflorescence with scar of basal bract (× 2).

B. Unexpanded inflorescence enclosed by enveloping bract (× 1).

C. Floral diagram of four-carpellate flower.

D. Flower from above (× 3).

E. Flower in LS (× 3).

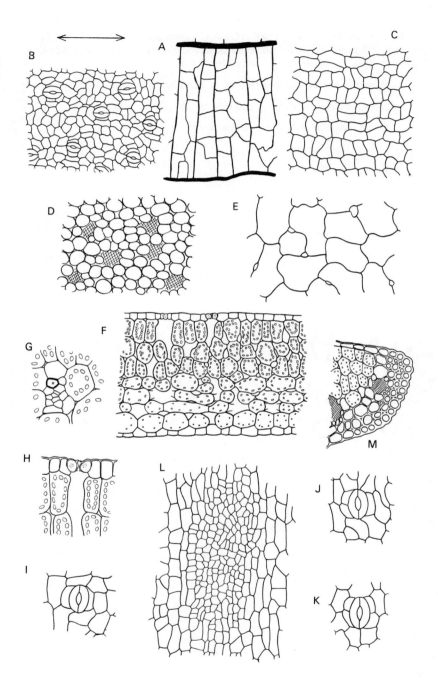

FIG. 5.4. APONOGETONACEAE. Floating leaves. Double-headed arrow indicates longitudinal axis of leaf in A–E.

A–G. *A. spathaceum.*

A. Lamina surface view (× 15) showing arrangement of second- and third-order veins between a part of two first-order (longitudinal) veins. B. Adaxial epidermis, surface view (× 150). C. Abaxial epidermis, surface view (× 150). D. Outermost palisade layer, surface view (× 150). Substomatal chambers cross-hatched. E. Abaxial hypodermal layer, surface view (× 150). F. TS lamina (× 150). G. TS third-order veinlet from lamina (× 270).

H–L. *A. distachyos.*

H. TS stoma from adaxial epidermis (× 310). I–K. Stomata, surface view (× 270) from adaxial epidermis. I. With two undivided lateral subsidiary cells. J. With one and K. with both subsidiary cells divided. L. Hydropote from abaxial epidermis, surface view (× 150).

M. *A. angustifolium.* TS margin of lamina (× 110) to show marginal mechanical tissue. Marginal and submarginal vein cross-hatched.

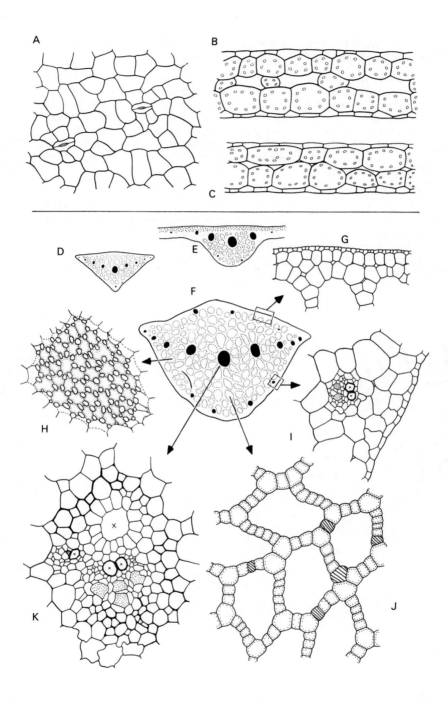

F IG . 5.5. APONOGETONACEAE. Submerged leaves (A–C) and leaf axis (D–K).

A. *A. crispum* adaxial epidermis, surface view (× 150) including non-functional stomata.

B–D. *A. natans* B, C, TS lamina (× 150).
   B. Close to vein, mesophyll three-layered. C. Remote from vein; mesophyll two-layered. D. TS petiole (× 9).

E–K. *A. spathaceum.*
   E. TS midrib region of lamina (× 9); palisade tissue stippled. F. TS petiole (× 9). G. Enlargement of an adaxial part of F (× 60). H. Transverse septum of a single air-lacuna in surface view (× 110). I. TS abaxial subsidiary bundle, enlargement of part of F (× 150). J. Lacunar ground tissue of petiole in TS (× 60); tannin cells hatched. K. TS median vascular bundle of petiole (× 150) including transverse vein to left; sieve-tubes stippled.

In D–F vbs—solid black; air-lacunae—dotted outline.

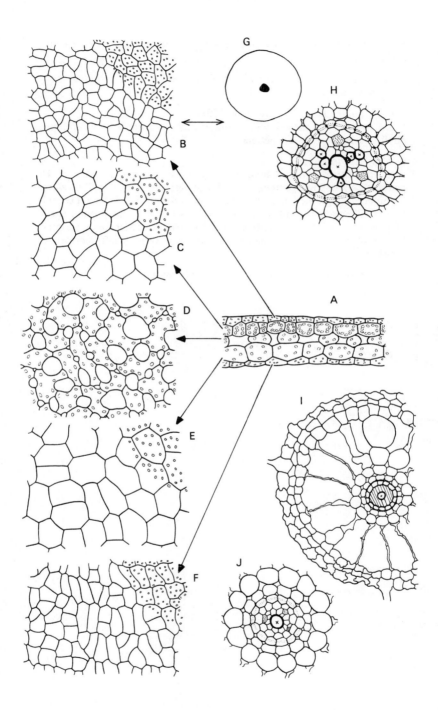

FIG. 5.6. APONOGETONACEAE. Submerged leaf (A–F) and root (G–J).

A–F. *A. monostachyon* (all × 150).

A. TS lamina. B–F. Five successive layers of lamina, from above downwards, in surface view. B. Adaxial epidermis. C. Adaxial hypodermis. D. Central mesophyll layer. E. Abaxial hypodermis. F. Abaxial epidermis. Chloroplasts shown in only a few cells in B–C and E–F to indicate relative sizes. Double-headed arrow indicates longitudinal axis of leaf in B–F.

G–H. *A. elongatum*. TS root.

G. Outline (× 13); stele solid black. H. Details of stele (× 270), xylem triarch.

I–J. *A. angustifolium*. TS root.

I. Root (× 150) without details of stele. J. Stele (× 270). Sieve-tubes stippled in H and J.

SCHEUCHZERIACEAE (Rudolfi 1830)
(Figs. 6.1–6.4, Plate 3B and Plate 11)
(Based on an account, with anatomical illustrations, by Dr M. Y. Stant.)

SUMMARY

THIS monotypic family is represented by *Scheuchzeria palustris* L. (Fig. 6.1 A) from bogs in the colder parts of the northern hemisphere. It is a low, rush-like perennial (10–40 cm tall) with an irregularly sympodial rhizome, each unit of which eventually terminates in an erect flowering shoot. The leaves are ± distichously arranged (Fig. 6.2 A), ligulate (Fig. 6.1 C), with sheathing bases. The blade is obtusely triangular and somewhat centric in section and has a large adaxial apical pore (Fig. 6.1 D) supplied by tracheids. The flowers, subtended by bracts (but without pedicel prophylls), have six tepals, six stamens, and three (occasionally more) large bi-ovulate carpels. The fruit is a follicle, dehiscing ventrally to release the usually solitary seed.

Anatomically the family is distinguished by the absence of squamules. Instead there are abundant hairs restricted to the leaf insertion (Fig. 6.1 E). The anatomy is reduced in relation to the wet habitat, with well developed air-lacunae in the leaf mesophyll, root cortex and throughout the ground tissue of the axis. Crystals occur in hypodermal files next to the vascular bundles of the leaf. Vessels are restricted to the roots.

FAMILY DESCRIPTION

VEGETATIVE MORPHOLOGY

Rhizomatous axis horizontal, clothed by a dense felt of nodal hairs and persistent fibrous remains of older leaves (Fig. 6.1 A); adventitious roots inserted just below the node. Axillary buds infrequent but producing renewal shoots at irregular intervals in an apparently little organized sequence, but usually situated towards the base of the erect shoot. Leaf linear, the blade rounded but flattened adaxially (Fig. 6.2 A, G) with an open sheathing base extending into the short ligule (Fig. 6.1 C). **Apical pore** conspicuous adaxially (Fig. 6.1 D), well supplied by tracheids; the pore representing an interruption of the epidermis and hypodermis lined by small thin-walled cells with a presumed but not demonstrated secretory function (Buchenau 1872; Minden 1899; Montfort 1918; Raunkiaer 1895–9). **Hairs** abundant in leaf axil (Fig. 6.1 E), each usually with a narrow multicellular base of slightly thick-walled, pitted cells extending into a long, uniseriate, distal, thin-walled filament; hairs absent from rest of leaf.

VEGETATIVE ANATOMY

Leaf lamina

Epidermis (Fig. 6.3 A–D) of small rectangular, longitudinally extended cells, each narrow with four straight walls in surface view. Epidermal cells uniformly shallow in TS, outer wall thickened and with a distinct cuticle, but differentiated into narrow costal bands above vbs and wider intercostal bands of somewhat wider cells. **Stomata** restricted to intercostal regions, orientated longitudinally but not in regular files. Each stoma with four subsidiary cells: two lateral and parallel to guard cells, two terminal and either square or somewhat anisodiametric and extended transversely, the degree of differentiation varying in different samples and even within a single leaf. Guard cells not or scarcely sunken, but somewhat deeper than adjacent epidermal cells (Fig. 6.2 B) with distinct inner and outer ledges above slight wall thickenings.

Hypodermis consisting of a single continuous layer of ± colourless, compact, and tangentially-extended cells, interrupted by substomatal chambers, but sometimes replaced at the acute adaxial corner of the leaf by a marginal group of **fibres** (Fig. 6.2 G, H). Costal hypodermal cells containing crystals (Fig. 6.2 E). **Mesophyll** lacunose but including a peripheral 5–6-layered chlorenchymatous region abruptly delimited from a central lacunose region. Outer cells anisodiametric or somewhat palisade-like adaxially, cell walls sometimes lobed, the lobing becoming coarser and the cells more irregular towards the interior of the leaf. Central ground tissue lacunose (Fig. 6.2 G), consisting of a uniseriate parenchymatous network enclosing large air-spaces. Mesophyll cells with few chloroplasts except towards the periphery of the leaf.

Transverse diaphragms (Fig. 6.2 F) occasional, unspecialized and without vbs, consisting of a single layer of slightly lobed, chlorophyll-containing cells with small intercellular spaces.

Vascular bundles in two series (Fig. 6.2 G): (i) a main arc of three large vbs with xylem orientated towards the adaxial face; (ii) smaller peripheral vbs alternately large and small, all with xylem facing towards leaf centre, i.e. with adaxial vbs inverted in relation to others. Large vbs of series (i) with protoxylem extended and disintegrated to form a canal with two diverging arms of metaxylem tracheids forming a wide-angled ∧. One large tracheid sometimes present on each side of the protoxylem canal (Fig. 6.2 c) or 2–6 elements in one or both metaxylem arms (Fig. 6.2 D). **Phloem** consisting of a group of narrow cells enclosed by the xylem ∧. **Bundle sheath** consisting of fibres, those at the phloem pole more abundant, narrower and thicker-walled than those elsewhere and sometimes encroaching into the abaxial face of the phloem itself. Each vb also with an outer parenchymatous sheath of narrow elongated cells not sharply circumscribed from the ground tissue. Smaller vbs of series (ii) with reduced vascular tissue and an extensive fibrous sheath continuous externally with the hypodermis and internally with a 1–2-layered sheath of elongated colourless parenchyma cells. Larger peripheral vbs similar in structure to vbs of series (i) but with xylem consisting entirely of scalariform tracheids, smallest peripheral vbs without vascular tissue and forming purely fibrous strands. According to Peisl (1957), a marginal sclerenchymatous strand, without vascular tissue, sometimes present along one or both edges (cf. Fig. 6.2 H). **Crystals** present as large prismatic, cuboid or hexagonal bundles or druses in hypodermal cell files immediately external to peripheral vbs (Fig. 6.2 E). **Tannin** occasional in the parenchyma of the ground tissue and in the diaphragms, appearing as dark contents in the cells. Tannin cells sometimes differing slightly in size from adjacent cells and referred to as **idioblasts** by Solereder and Meyer (1933).

Leaf sheath

Colourless, open but encircling the stem completely, the margins scarious and biseriate, distally extended into a pronounced ligule (Fig. 6.1 C). **Epidermis** without stomata, the outer cell wall thickened especially on the abaxial surface, each file of epidermal cells with a prominent cuticular ridge. **Ground tissue** lacunose except for 2–3 layers next to the abaxial epidermis; lacunae separated by longitudinal uniseriate partitions and irregular transverse uniseriate diaphragms. **Vascular system** with a conspicuous median and two large lateral vbs embedded in the ground tissue towards the adaxial surface, the vascular tissue resembling that in type (i) vbs of blade. Abaxial series of smaller vbs, alternating with narrow fibrous strands as in type (ii) vbs of blade. Parenchymatous bundle sheath cells with granular contents. **Tannin cells** occasional.

AXIS

Flowering stem (Fig. 6.3 E–J)

Tapering distally and apparently ending directly in a flower (Fig. 6.1 B). **Epidermis** of lignified, slightly thick-walled cells, cuticle ridged. **Stomata** present, guard cells slightly sunken, subsidiary cells slightly protruding. **Cortex** including three peripheral layers of closely packed angular parenchyma cells, interrupted below stomata; cells of inner compact layers more rounded and merging into parenchymatous network of the lacunose system. **Pith** similarly lacunose, but with smaller air-cavities. Vascular cylinder (stele), separating cortex from pith, consisting of a fibrous sheath or of a cylinder of adjacent vbs coalescing to form a **sclerenchymatous cylinder** (Fig. 6.3 J), outer limit of cylinder indicated by a single layer of slightly lignified endodermoid parenchyma cells but without Casparian strips.

Cortical vascular system consisting of an inner ring of four or more large and an outer ring of 14 or more small, strands, the smallest strands having reduced vascular tissue, or even being entirely fibrous. Larger vbs (possibly representing direct leaf traces) similar to those in leaf, and provided with a protoxylem canal, ∧-shaped metaxylem and phloem, the abaxial fibres tending to penetrate into the phloem. **Stelar vascular system** consisting of an outer circle of vbs connected by sclerenchyma to form a continuous cylinder. An inner ring of several independent vbs also present towards the periphery of the pith (Fig. 6.3 J). Vbs (Fig. 6.3 G, H) each including a protoxylem canal, ∧-shaped metaxylem and a phloem group; a well developed fibrous sheath present next to the phloem, but with several irregularities related to the size and extent of bundle anastomoses (e.g. Fig. 6.3 E, H). **Crystals** absent.

Rhizome (Fig. 6.4 A–D, Plate 3B)

Up to 8 mm in diameter. **Epidermis** of tabular cells with lignified, U-shaped thickenings on the inner tangential and radial walls in older parts. **Stomata** absent. **Cortex** with an outer zone of lignified angular fibres, up to eight cells deep, with 2–4 layers of compact parenchyma within (Fig. 6.4 D). Remaining cortex consisting of a uniseriate parenchymatous network surrounding wide air-lacunae, the lacunae traversed by irregular uniseriate diaphragms with narrow intercellular spaces. Parenchyma frequently filled with finely granular **starch.** Isolated **tannin cells** conspicuous within the network. Presumed leaf traces within cortex few, sometimes reduced to peripheral fibrous strands, but otherwise similar to those of leaf.

Inner cortex abruptly delimited from the **stele** by a single parenchyma layer and by an **endodermis,** the cells of the endodermis eventually becoming thickened on their tangential and radial walls. Outer stelar vbs, each with a ± continuous fibrous sheath, coalescing to form an irregular vascular cylinder (Fig. 6.4 A). Larger vbs at periphery of pith separated from the vascular cylinder. Metaxylem of each vb V-shaped, the arms diverging towards the endodermis (Fig. 6.4 B) or sometimes almost completely encircling the crescentic phloem to form amphivasal vbs (Fig. 6.4 C). Protoxylem of larger vbs forming an irregular lacuna. Metaxylem consisting of wide tracheids with indistinct, tapering, end walls and scalariform thickening on lateral walls. **Crystals** absent. **Tannin** present occasionally in cells with dark contents.

ROOT (Fig. 6.4 E–H)

Adventitious roots arising immediately below the nodes, branched, up to 1 mm in diameter. First-order roots exhibiting the following structure. **Epidermis** thin-walled, lost from older roots. **Exodermis** consisting of 5–6 layers of cells, the two outer layers suberised and lignified, and the inner layers consisting of thick-walled, slightly lignified fibres (Fig. 6.4 H). **Cortex** consisting of radial, ± uniseriate files of parenchyma cells, the outermost cells being widest; external cells of files mostly contracted or completely collapsed to form radial strands separated by large air-cavities (Fig. 6.4 G). Inner cortex of 3–4 layers of rounded cells arranged in regular concentric circles. **Endodermis** with inner and radial cell walls becoming thickened and conspicuously U-shaped in TS (Fig. 6.4 F). **Pericycle** an indistinct layer of small cells radially alternating with endodermal cells. **Vascular system** consisting of a ring of 6–8 xylem strands alternating with phloem groups (Fig. 6.4 E–G). Pith fibrous. Xylem strands each consisting of a radial series of 1–5 tracheary elements, the innermost widest. Phloem strands each consisting of a conspicuous exarch sieve-tube adjoining the pericycle and surrounded on the three inner faces by a small group of undifferentiated elements.

SECRETORY, STORAGE, AND CONDUCTING ELEMENTS

Crystals. Present in hypodermal cell files external to peripheral vbs in leaf.

Tannin. Occasional in leaf, leaf sheath, and rhizome.

Starch observed in rhizome.

Xylem. Represented in leaf and stem by long tracheids with tapering end walls, the lateral wall pitting finely scalariform. Elements in the rhizome widest (up to 40 μm diameter).

[Cheadle (1942, see also Cheadle and Uhl 1948) includes Scheuchzeriaceae among his list of families with **vessels** in the root only, on the basis of long metaxylem elements with long scalariform plates on distinct but very long end walls. Observation supports this interpretation, since the thickening bars are narrow and show the irregularity associated with dissolved pit membranes.]

REPRODUCTIVE MORPHOLOGY AND ANATOMY

Morphology (cf. Plate 11)

Inflorescence a simple raceme, terminating the distal extremity of each axis unit, but apparently always ending in a terminal flower (Fig. 6.1 B). Lateral flowers subtended by progressively more reduced bracts, the floral axis itself being without a basal prophyll (bracteole). **Flowers** each with six tepals in apparently alternating whorls, six stamens and usually three ± free carpels (Fig. 6.1 F–H). Variation in carpel number from 3–6 (4–5, up to 13) recorded by Şerbănescu-Jitariu (1966). Each carpel with two ± basal but ventrally attached ovules, usually only one of them developing. Stigmatic surfaces represented by distal dorsal papillose hairs.

Anatomy

(i) Flowering axis, continuous basally with rhizome, described above.

(ii) Flower, according to Uhl (1947), with a common trace to each stamen–tepal pair, dividing tangentially at the level of stamen attachment to produce a tepal trace and stamen trace. The dorsal carpel traces above this level diverging in a spiral sequence, each continuing about three-quarters the length of the carpel, spreading out and finally dividing to send a branch to each of the two short stylar regions. The two ventral vbs each providing an ovular trace and sometimes bundles to the lateral wall of the carpel. [On the basis of vascular anatomy, Uhl (1947) interprets the floral parts as being spirally arranged (not in whorls) and considers it inappropriate to describe the structure as a simple flower (see below).]

TAXONOMIC NOTES

The genus has had separate family status since 1830 and has been so maintained by recent authors (e.g. Hutchinson 1959; Takhtajan 1966), and variously associated with the Alismataceae, Petrosaviaceae, Hydrocharitaceae, and especially with the Juncaginaceae, in which it is sometimes included. Buchenau (1903c) placed the family Juncaginaceae (*Scheuchzeria, Triglochin, Maundia, Tetroncium, Lilaea*) immediately preceding the Alismataceae and Butomaceae. Hutchinson regarded *Scheuchzeria* as a reduced type of the Alismataceae tending towards the Liliaceae and possibly representing the point of departure of a stock of more terrestrial monocotyledons from primitive aquatic plants. It would be just as logical to accept this sequence of events in reverse.

Relationship with Juncaginaceae and Lilaeaceae

The Scheuchzeriaceae differ from the Juncaginaceae and Lilaeaceae in a large number of important features which are listed in Table 6.1. These include a number of anatomical and morphological features which are certainly of equivalent value to those that distinguish several other helobial families. *Scheuchzeria* is unique in the Helobiae in lacking squamules; they are probably replaced functionally by the axillary hairs.

On the other hand there are certain structural resemblances, including leaf morphology (an open sheath), branched roots, a terminal flower on the inflorescence (reported in examples from all three families), the absence of hairs on the leaf blade, limited development of tannin and similar distribution and structure of vessels. They all have a comparable carpel organization, somewhat similar floral anatomy and organization of vascular bundles in the leaves. The bundle sheath shows a tendency to develop a sheathing layer with U-shaped wall thickenings, a feature most pronounced in species of *Triglochin*.

A feature that strongly supports the taxonomically close affinities of these three families is the common occurrence of the distinctive cyanogenic glucoside triglochinin (e.g. Ruijgrok 1974) not known to occur in exactly the

TABLE 6.1

Comparison of three families with putative relationships

	Scheuchzeriaceae	Juncaginaceae	Lilaeaceae
Cormous axis	−	+	+
Leaf apical pore	+	−	−
Ligule	+	+ or −	+
Squamules	−	+	+
Marginal leaf fibres	(+)	−	−
Tetracytic stomata	+	−	−
Hypodermal crystals	+	−	−
Laticifers	−	−	+
Root, central metaxylem vessel	−	+ or −	+
Flowers*	bisexual	bisexual or unisexual	bisexual or unisexual
Floral bracts	+†	−	+†
Flower parts	trimerous	dimerous or trimerous	monomerous
Perianth	+‡	+	−
Carpels/flower	3 (to 6 or more)	3–6	1
Ovules/carpel	2	1	1
Fruiting carpels	follicular	achenous	achenous

* The nature of the flower is open to differing interpretations, but the basic structural differences are clear.

† In *Scheuchzeria* these are leaf-like, in *Lilaea* tepal-like, their development is correspondingly different.

‡ The tepal in *Scheuchzeria* is comparable in its development to the 'bract' of *Lilaea*. It is possible that these organs are homologous.

same form in other flowering plants, although a methyl ester of triglochinin occurs in *Thalictrum aquilegifolium*, Ranunculaceae.

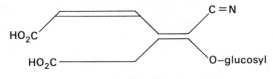

Triglochinin

Most structural resemblances between the families otherwise largely reflect their aerenchymatous construction and are related to their aquatic habitat.

The consensus of evidence therefore supports the juxtaposition of these three taxa as being closely related and very possibly monophyletic, but the structural divergence between them warrants the maintenance of them as discrete families.

Relationships with other Helobiae

According to Erdtman (1952), pollen grains similar to those of *Scheuchzeria* occur in the Potamogetonaceae, but not in the Alismataceae. Reference to Erdtman (1943) suggests some resemblance between the diads of *Butomus* and *Scheuchzeria*. An estimation of resemblances between *Scheuchzeria* and members of the families Alismataceae and Butomaceae (s.1.) was made according to the principles and methods adopted by Sneath and Sokal (1973) and outlined in accounts of these two families (Stant 1964, 1967). The characters selected for punched card descriptions of the Alismataceae were used for this comparison.

The mean of the correlation coefficients of *Scheuchzeria* with genera belonging to the Butomaceae (s.1.) is about 10 per cent higher than the mean calculated for members of the Alismataceae. This suggests that *Scheuchzeria* is more closely related to the Butomaceae (s.1.) than to the Alismataceae. However, on examining the correlation figures for individuals of the Butomaceae it can be seen that the higher mean largely results from a 79 per cent affinity between *Scheuchzeria* and *Butomus*. If *Butomus* is excluded from the data (i.e. the comparison made with the family Limnocharitaceae), *Scheuchzeria* is observed to exhibit an approximately equal affinity with both families. This special resemblance between *Scheuchzeria* and *Butomus* almost undoubtedly reflects their common linear leaf morphology and does not imply any special taxonomic affinity.

Within the Alismataceae, *Scheuchzeria* has most similarity to *Luronium*, based on this method.

Interpretation of floral morphology

Uhl (1947) interprets the flower of *Scheuchzeria* as a reduced inflorescence composed of six staminate flowers (each reduced to a single stamen) with individual subtending bracts (= tepals) and either one or more central carpellate flowers. The evidence for this is the presumed spiral arrangement of the appendages indicated by the sequence of origin of vascular traces to different floral appendages. *Scheuchzeria* was said to be more primitive than *Triglochin* because of the spirally arranged carpels which have free ventral edges only fused laterally with adjacent carpels (cf. also Şerbănescu-Jitariu 1966).

Floral development (a note added by U. Posluszny on the basis of Plate 11)

An analysis of floral development in *Scheuchzeria* does not support Uhl's interpretation. Each floral unit originates as a floral apex in the axil of a leaf-like bract (Plate 11, Fig. A). Floral organs are initiated in an acropetal sequence. First to form is the outer trimerous whorl of tepals followed by the inner whorl of tepals and shortly afterwards, opposite each tepal, the stamen primordia are initiated (Plate 11, Figs. B and C). Each trimerous whorl of tepal and stamen primordia arises almost simultaneously. The final organs to form on the floral apex are the three carpels, again initiated almost simultaneously (Plate 11, Figs. D and E). There is no indication of a spiral arrangement in the formation of the carpels. These observations compare very favourably with similar observations on floral development in *Triglochin striata* by Lieu (1979a).

MATERIAL EXAMINED

Scheuchzeria palustris L. from Scotland[x] (C. R. Metcalfe), Sweden[x] (R. Florin) and Tunis, USA (V. I. Cheadle); all parts; all material at Kew[x] represented in Kew slide collection. *Scheuchzeria palustris* L. Arkill Hills, Ontario, Canada; U. Posluszny 5.VII.78; all parts.

SPECIES REPORTED ON IN THE LITERATURE

(All refer to *Scheuchzeria palustris*)
Buchenau (1872): leaf (apical pore).
Cheadle (1942): tracheary elements.
Eber (1934): floral morphology.
Metsävainio (1931): root.
Minden (1899): leaf (apical pore).
Montfort (1918): leaf (apical pore).
Peisl (1957): leaf.
Raunkiaer (1895–9): leaf, inflorescence axis.
Ravn (1894–5): fruit.
Şerbănescu-Jitariu (1966): floral and fruit anatomy.
Solereder and Meyer (1933): leaf, rhizome, inflorescence axis, root.
Uhl (1947): floral anatomy and morphology.
Yamashita (1976b): note on embryo development.

SIGNIFICANT LITERATURE—SCHEUCHZERIACEAE

Arber (1918a, 1918b); Buchenau (1872, 1903c); Cheadle (1942); Cheadle and Uhl (1948); Eames (1961); Eber (1934); Erdtman (1943, 1952); Hofmeister (1861); Hutchinson (1959); Metsävainio (1931); von Minden (1899); Montfort (1918); Peisl (1957); Raunkiaer (1895–9); Ravn (1894–5); Ruijgrok (1974); Şerbănescu-Jitariu (1966); Sneath and Sokal (1973); Solereder and Meyer (1933); Stant (1964, 1967); Stenar (1935); Uhl (1947); Yamashita (1976b).

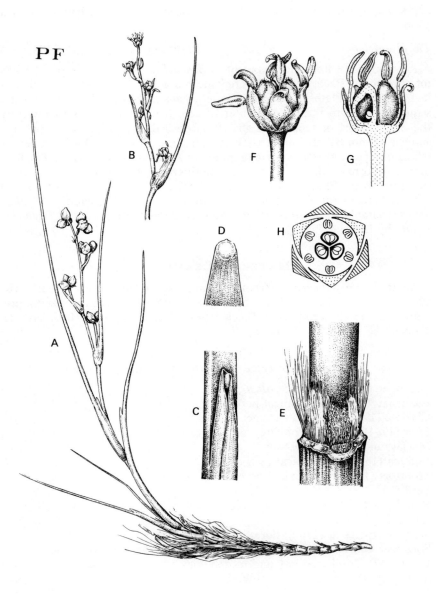

PF

FIG. 6.1. SCHEUCHZERIACEAE. *Scheuchzeria palustris* L. Habit and floral morphology. (Arkill Hills, Ontario; Posluszny 5.vi and 5.vii.78.)

A. Habit of fruiting plant (× ⅓).

B. Inflorescence at anthesis, note terminal flower (× ½).

C. Detail of leaf ligule (× 3).

D. Detail of leaf apical pore (× 9).

E. Node with leaf removed (× 4) to show multicellular filamentous hairs in leaf axil and surrounding axillary bud.

F. Flower at anthesis from side (× 3).

G. Flower in LS (× 3).

H. Floral diagram.

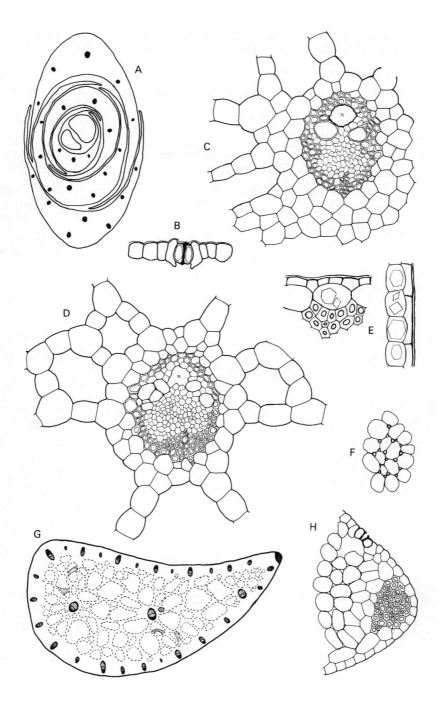

FIG. 6.2. SCHEUCHZERIACEAE. *Scheuchzeria palustris.* Leaf.

A. TS shoot, leaf arrangement ($\times$ 3).

B. TS epidermis, stoma ($\times$ 140).

C. TS lateral vascular bundle from main arc adjacent to marginal fibre bundle ($\times$ 95).

D. TS central vascular bundle and ground tissue ($\times$ 95).

E. Crystals in hypodermis ($\times$ 420), to left TS epidermis, hypodermis, and fibres of vascular bundle. To right LS epidermis and hypodermal crystal cells.

F. Diaphragm cells in surface view ($\times$ 90).

G. TS whole leaf, lacunose ground tissue, vascular bundles ($\times$ 10).

H. TS margin with fibrous bundle ($\times$ 95).

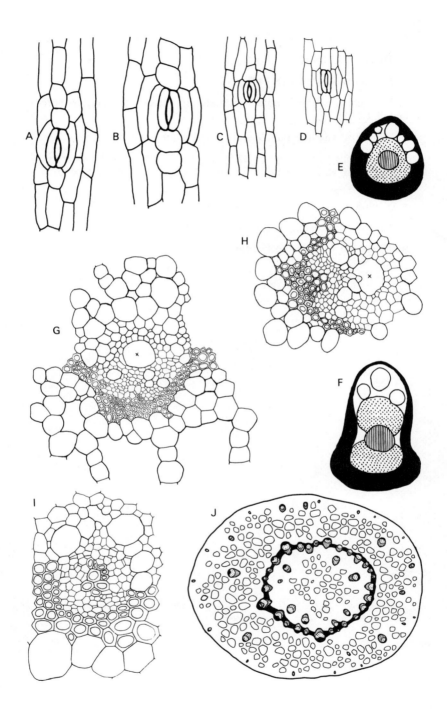

238

FIG. 6.3. SCHEUCHZERIACEAE. *Scheuchzeria palustris.* Leaf (A–D), inflorescence axis (E–J).

A–D. Surface view, leaf epidermis with stomata showing variations in arrangement of subsidiary cells. A–B ($\times$ 140), C–D ($\times$ 75).

E–F. Tissue distribution in representative stem vascular bundles ($\times$ 95).

G–I. TS vascular bundles of inflorescence axis to show variation in tissue arrangement, G ($\times$ 95), H ($\times$ 75), I ($\times$ 140).

J. TS stem ($\times$ 10).

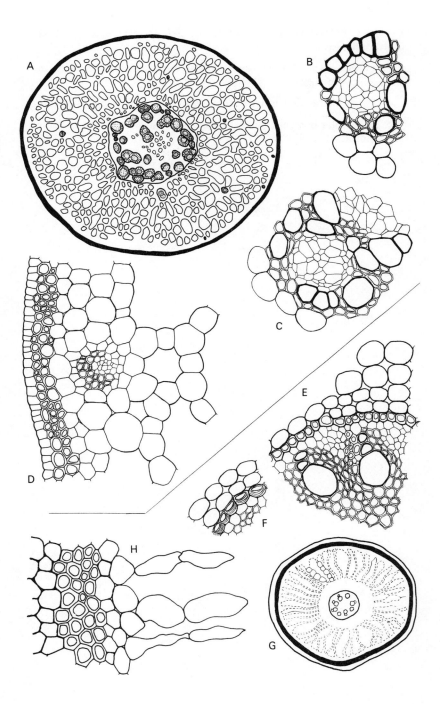

240

FIG. 6.4. SCHEUCHZERIACEAE. *Scheuchzeria palustris.* Rhizome (A–D), root (E–H).

A. TS rhizome, vascular bundles, and lacunose ground tissue (× 10).

B–C. TS vascular bundles, with ± amphivasal arrangement (× 140).

D. TS outer layers, subepidermal band of sclerenchyma, lacunose cortex with small pripheral vascular bundle (× 75).

E. TS root, inner cortex, and outer stele; endodermis and two groups of protoxylem and phloem (× 140).

F. TS endodermis, pericycle, and inner cortical layers from more mature root than E (× 135).

G. TS root (× 20).

H. TS outer layers, exodermis, and outer cortex (× 145).

Family 7
JUNCAGINACEAE (L. C. Richard 1808)
(Figs. 7.1–7.4)

SUMMARY

THIS family of about 20 species in the broad sense consists of two genera, *Tetroncium* Willd. (dioecious, with dimerous flowers, temperate South America) and *Triglochin* L. (hermaphrodite, with trimerous flowers, cosmopolitan but well represented in Australia). *Triglochin* in a broad sense is commonly segregated into smaller sections or genera. These are *Triglochin* in a restricted sense, with three sterile and three fertile, or ± six fertile carpels partly adnate to the floral axis, *Cycnogeton* Endl. with 3–6 free carpels, none imperfect, and *Maundia* F. Muell. with 3–4 fertile carpels, united after anthesis. The plants are often characteristic of wetter sites or salt marshes, e.g. *Triglochin maritimum*, or they may even grow in flowing water, where they have strap-shaped submerged leaves. More diminutive Australian species, e.g. *Triglochin calcipta*, grow in drier sites and they may even be annuals. Some species, e.g. *Triglochin striata*, are widely distributed. The plants are typically tufted with linear leaves arising from an erect, sympodially-branched, cormous axis. The axis is sometimes proliferated by stoloniferous offsets (*Triglochin palustris*, *T. striata*). The flowers are in spikes which terminate one out of the two axes produced by bifurcation of the apical meristem. The spikes are ebracteate and the flowers anemophilous, without a nectary (Daumann 1970), protogynous, green, and each of them has two alternate trimerous (dimerous) perianth whorls. The tepals are opposite the stamens (or staminodes). Sometimes there are fewer than six stamens, by abortion. The pollen is described as binucleate by Gardner (1976). There are 2–6 (–8) carpels, which are ± free or united around a central receptacular column after anthesis. They are either all fertile or three of them are sterile. Each carpel usually has a single basal anatropous ovule, but in *Maundia* the ovule is orthotropous and attached apically. The fruit consists of a series of achenes, separating from the central column or the three sterile carpels. The achenes are sometimes basally horned, e.g. in *T. calcipta*. The seeds have no endosperm, and the embryo is straight.

In their vegetative anatomy the plants are characterized by paracytic or tetracytic stomata; they lack hairs (except for numerous axillary squamules) and the mesophyll is lacunose. The main vascular bundles of the leaf usually have a sclerenchymatous sheath (the outermost sheathing layer commonly with U-shaped wall thickenings), but mechanical tissue is not otherwise developed. Vessels are restricted to the roots. Tannin and crystals are little developed.

FAMILY DESCRIPTION
VEGETATIVE MORPHOLOGY

Habit (Fig. 7.1)

Leaves distichous or spirodistichous, on a short, erect, cormous or somewhat rhizomatous axis (Fig. 7.1 A) (described as bulbous in *T. bulbosum*,

T. milnei by Horn af Rantzen 1961). Apical meristem (in those spp specially investigated) bifurcating at intervals, one product (a presumed *precocious* axillary renewal shoot) continuing the further vegetative development, the other product (the presumed original terminal meristem) developing as a determinate flowering spike (Lieu 1979*a*). Prophyll on the continuing vegetative shoot represented by an adaxial foliage leaf continuing the distichy of the parent axis without interruption. Axes in some spp (e.g. *T. palustris, T. striata*) proliferating via axillary, scale-bearing stolons (Fig. 7.1 B), usually arising from buds produced *non-precociously* in the axils of distal foliage leaves on each renewal shoot (Fig. 7.1 D), the long hypopodium of the stolon carrying the prophyll scale to a position near the middle of the extension shoot (Fig. 7.1 C). Distal scale leaves (up to six) inserted in a plane at right angles to that of the prophyll, the axis eventually turning erect to establish a new rosette of foliage leaves.

Leaf

Foliage leaves of three main types developed: (i) With a linear or sub-terete blade; the mouth of the sheath ligulate (e.g. *Triglochin* sect. *Triglochin*; Fig. 7.1 E). (ii) With a dorsiventrally flat blade, eligulate; plants growing in flowing water, the leaves becoming ± strap-shaped (e.g. *Triglochin* sect. *Cycnogeton*). (iii) With a laterally flattened blade, i.e. equitant or ensiform (*Tetroncium*).

In bulbous spp (e.g. *Triglochin bulbosum*) axis surrounded by persistent fibrous remains of leaf sheaths.

VEGETATIVE ANATOMY

Foliage leaf blade

Apical pore not developed. **Hairs** absent. **Squamules** well developed, numerous in each leaf axil, described by Hill (1900) in *Triglochin maritimum* as having an endodermoid layer facilitating their abscission. **Cuticle** fairly thick, commonly longitudinally ribbed or striate. Lamina either dorsiventral and with the palisade chlorenchyma limited to the adaxial surface (*Cycnogeton, Maundia*; Fig. 7.3 D, E) or in terete-leaved spp ± isolateral, with chlorenchyma uniformly distributed around the whole blade (Fig. 7.3 F). **Epidermis** unequally thick-walled, the outer wall usually much thicker than the inner (Fig. 7.3 H); in surface view differentiated into narrow bands of rather short wide cells, with stomata, and wider bands of narrowly rectangular, elongated cells, without stomata, (Fig. 7.3 A–C). **Stomata** equally common on both surfaces, or most abundant on the adaxial surface of dorsiventral leaves; paracytic (less commonly tetracytic), always with a pair of narrow lateral subsidiary cells and sometimes, but irregularly, short terminal subsidiary cells (Fig. 7.3 A–C). Guard cells (Fig. 7.3 H) not or slightly sunken, usually ± symmetrical in TS with a pair of equal (less commonly unequal) ledges.

Hypodermis rarely developed as a colourless layer (*Triglochin maritimum*), more usually superficial mesophyll layers in terete-leaved spp developed as a palisade-like **chlorenchyma** interrupted by wide substomatal chambers;

chlorenchyma restricted to adaxial surface in dorsiventral leaves. Central **mesophyll** ± colourless, lacunose to varying degrees, e.g. in spp with smaller, terete leaves with small intercellular spaces; in larger-leaved taxa forming a reticulum of uniseriate longitudinal plates delimiting wide air-canals, the canals irregularly septate via uniseriate plates of small, stellately lobed cells. Mechanical tissue independent of the veins not developed.

Vascular system: a large median vb with 2–6 additional lateral vbs forming a major arc either in the centre of the leaf or towards the upper surface (Fig. 7.3 D–G). Small vbs in terete-leaved spp forming a peripheral series close to the epidermis, with the xylem always towards the leaf centre, i.e. the adaxial vbs inverted (Fig. 7.3 F), but in dorsiventral leaves adaxial vbs normally orientated, abaxial vbs then few and rather remote from the abaxial epidermis (Fig. 7.3 D, E). Leaf anatomy in *Tetroncium* only superficially described, but, according to Arber (1924), with a single series of vbs in the lamina, adjacent vbs with ± alternately inverted orientation.

Major vbs enclosed by a parenchymatous sheath of mesophyll cells. Mechanical sheath either poorly developed, e.g. *C. procera*, and represented by narrow thin-walled cells, or more commonly represented by thick-walled cells, e.g. *Triglochin striata*, the outer cell layer sometimes continuous and provided with U-shaped wall thickenings (Fig. 7.3 J). Phloem of v. narrow elements separated from the xylem by a narrow layer of conjunctive parenchyma. Xylem including a number of wide tracheids arranged in a ∧; protoxylem not usually represented by a wide lacuna, except in floating leaves of *Cycnogeton*. Smaller vbs with reduced vascular tissue.

Foliage leaf sheath (Fig. 7.3 G)

Sheath colourless, becoming biseriate laterally. **Epidermis** ± without stomata; cells on the abaxial surface shallow, thick-walled and with a ridged cuticle, cells on the adaxial surface thin-walled. **Mesophyll** lacunose (Fig. 7.3 I), as in the blade, sometimes including starch close to the veins. **Vascular system** usually represented by a single series of vbs continuous with those of the blade, the peripheral (minor) part of the system represented at most by basal extensions of a few adaxial or abaxial vbs (Fig. 7.3 G). Veins as in the blade, but mechanical tissue better developed, e.g. in *Triglochin striata* the outer sheath of cells with U-shaped thickenings conspicuous and supplemented by 2–3 additional sclerenchymatous layers (Fig. 7.3 K).

AXIS

(i) Rhizome

Surface layers becoming lignified and/or suberized; in *Triglochin maritimum* forming a sclerotic layer of thick-walled cells (Hill 1900). Outer cortex in *Cycnogeton procera* including fibrous extensions of the minor leaf veins. **Cortex** wide, undifferentiated and without a separate vascular system, the ground tissue sometimes somewhat lacunose (*Triglochin striata*), but typically starch-filled.

Central cylinder delimited by a ± distinct endodermis, becoming thick-walled with age. Vascular system including a peripheral series of ± congested vbs forming a plexus by the attachment of root traces. Central vbs scattered, mainly collateral but with some tendency to be amphivasal. Vbs most congested in narrower axes, e.g. *Triglochin striata.*

(ii) Stolon (e.g. in *Triglochin striata,* Fig. 7.4 A–D)

Surface layers irregular, the **epidermis** often eroded; 2–3 hypodermal cell layers starch-free, with lignosuberized but only slightly thickened walls forming a conspicuous protective layer (Fig. 7.4 B). **Cortex** lacunose (Fig. 7.4 A), without vbs, cells starch-filled (Fig. 7.4 C). **Endodermis** absent. **Central cylinder** resembling a solenostele with a peripheral mechanical layer 1–2 cells wide. Outer layer of mechanical cells often with unequal wall thickenings (Fig. 7.4 D) like the sheath cells of the leaf vbs. **Phloem** forming an almost continuous cylinder enclosing a discontinuous series of tracheids, with one or more protoxylem poles. Narrow pith occupied by thick-walled sclerenchyma.

(iii) Inflorescence axis

Epidermis similar to that of the leaf blade but with fewer stomata. Outer **cortical layers** somewhat chlorenchymatous but never palisade-like. Cortex narrow, somewhat lacunose. **Mechanical tissue** between the cortex and pith represented by a ring of sclerenchyma, continuous with the smallest outer vbs, or by the partially confluent mechanical sheaths of the outer vbs. Main **vascular system** represented by a single series of collateral vbs with poorly developed mechanical sheaths; metaxylem wide, but conspicuous protoxylem lacunae observed only in *Cycnogeton.* Pith wide, represented by a system of wide air-canals separated from one another by uniseriate plates as in the leaf sheath, the central ground tissue often breaking down and the pith becoming hollow.

ROOT (Fig. 7.4 E–G)

Usually narrow, filiform, branched or unbranched, but in *Cycnogeton* developing somewhat fleshy distal root tubers. **Epidermis** with well developed root-hairs, with enlarged bases; root-hairs sometimes persistent and becoming thick-walled in old roots. **Exodermis** 1–(2–) seriate, the cells angular in TS, initially slightly thick-walled and suberized, but becoming v. thick-walled in old roots, e.g. *Triglochin maritimum, T. palustris.* Exodermal cells below the root-hairs, narrow, shallow and with dense cytoplasmic contents (Fig. 7.4 F). **Cortex** uniform, often lacunose, the extent of the lacunae depending on the degree of collapse shown by the radial files of cells (Fig. 7.4 E). Starch-containing cells sometimes abundant, e.g. *T. palustris,* not collapsing.

Endodermis becoming thick-walled in older roots, with U-shaped wall thickenings, developing first at the phloem poles. **Stele** always narrow (Fig. 7.4 F). **Pericycle** uniseriate but interrupted by phloem. **Vascular tissue** in narrow roots (e.g. of *T. striata,* Fig. 7.4 G) including a wide central xylem vessel, with 4–6 adjacent strands of narrow protoxylem. Phloem strands alternating with the protoxylem, each represented by a single sieve-tube next

to the endodermis and an adjacent companion cell to the inside (Fig. 7.4
G). Xylem in wider roots (e.g. *T. maritimum, Cycnogeton procera*) with several
central metaxylem vessels around a central pith; up to seven protoxylem
poles alternating with the pericyclic sieve-tubes.

Secretory, Storage, and Conducting Elements

Crystals: never abundant.

Tannin: scarcely developed.

Starch: common and often abundant in ground tissue of root, rhizome,
stolon, leaf sheath, and even the leaf blade, as narrow spherical or slightly
angular, simple grains. In the leaf tending to be associated with the veins.

Tracheary elements: Vessels recorded only in roots; elements long with
lengthy, oblique, scalariform perforation plates, not always distinguishable
with certainty from tracheids.

Reproductive Morphology and Anatomy

Inflorescence

Inflorescences spicate (Fig. 7.1 A), terminal on the axis. Produced by equal
bifurcation of the vegetative growing point, the meristem of the future inflorescence always being the larger of the two products (Lieu 1979a). Axis interpretable as a sympodium, with 3–5 foliage leaves between each successive fork.
Terminal flower frequently developed. Flowers in 4–8 orthostiches, each
ebracteate (Fig. 7.1 F), including two pseudowhorls of green tepals, one tepal
usually abaxial, the members of each pseudowhorl originating in sequence
(Hill 1900; Lieu 1979a). Tepals of the inner series originating somewhat
later and higher than those of the lower series. Inner tepal margins extending
in front of the outer stamens although the latter actually formed later (Figs.
7.1 G and 7.2 A, D, E).

Stamens up to six, tetrasporangiate, ± sessile, each opposite a tepal (Figs.
7.1 G–J and 7.2 B, C). Flower sometimes somewhat zygomorphic with only
the lowest (abaxial) stamen functional and larger than remaining staminodes
(e.g. *T. striata*). Carpels up to six (Fig. 7.2 A, D, E) originating in sequence
in two whorls, the inner at a higher level than the outer; lower series of
carpels aborting in *Triglochin* section *Triglochin* (e.g. *T. milnei, T. striata*,
Fig. 7.1 G, H, L). Each fertile carpel originating as a horseshoe-shaped primordium enclosing the ventrally developed ovule, becoming ± enclosed ventrally
and developing stiff stigmatic hairs. Carpels ± free or somewhat fused basally,
often to an extension of the floral axis persisting after fruit dispersal (e.g.
T. maritimum). Ovule solitary, anatropous, attached basally except in *Maundia* (Aston 1973). Achene consisting of the mature carpel wall and a thin
testa and enclosing a straight embryo. Endosperm not persistent. Achenes
in section *Triglochin* separating from the flat persistent remains of the sterile
carpels (Fig. 7.1 K, L).

Floral vasculature (mainly after Uhl 1947; see also Lieu 1979*a*.)

Each flower is supplied by a single trace diverging from the four collateral vbs in the scape. This pedicel trace supplies a single trace to each floral appendage in the order of their insertion, those of the associated tepal plus stamen arising almost simultaneously. Tepal traces remain unbranched; the stamen traces end in the connective. Each carpel also receives a single trace. In the sterile carpels the trace remains unbranched, but in the fertile carpels it divides to give a dorsal bundle, which may extend into the stigmatic region, and a single ventral bundle, which may be weakly developed and serves essentially as an ovular trace or may sometimes be more strongly developed. This applies in *T. maritimum* where the strongly developed ventral bundle forks above the departure of the ovule trace. This is interpreted by Uhl (1947) as the remains of a double ventral bundle, which in species of *Triglochin* shows progressive reduction from this presumed primitive condition. Uhl also described the trace to the staminal unit as having a double origin and regarded this as evidence that the stamen is morphologically a branch.

TAXONOMIC AND MORPHOLOGICAL NOTES

Interpretation of the flower

The interpretation of the flower is disputable because floral bracts are absent, the stamens are antitepalous, the outer stamens are apparently situated lower on the axis than the inner tepals, and because the floral axis is frequently extended in fruit. In the simplest interpretation, each assemblage of tepals, stamens, and carpels is regarded as a bractless flower. Otherwise it is possible to regard the floral axis as a bractless branch, itself bearing unisexual flowers, consisting of either a single bract (= tepal) and its subtended solitary stamen, together with a terminal cluster of unicarpellate flowers which are not subtended by bracts (or a single terminal multicarpellate flower). Similar contrasted interpretations exist for the flower of *Scheuchzeria* and *Lilaea* (cf. Yamashita 1970).

Developmental evidence most strongly supports the first interpretation (Buchenau 1882; Hill 1900; Lieu 1979*a*). The two tepal whorls are initiated in sequence although the sequence is not necessarily consistent, and the stamens are clearly inserted above all of the tepals. The association between the stamens and tepals is secondary and is due to lateral intercalary growth between members of two successive whorls so that each stamen and opposed tepal become inserted on a common extension of the axis and fall as a unit (Figs. 7.1 J and 7.2 C). Marginal growth of the inner tepals results in their extending behind the outer stamens.

Anatomical evidence (e.g. Uhl 1947; Eames 1961) supports the second interpretation chiefly because the traces to the second whorl of tepals arise above the traces to the first whorl of stamens. However, this sequence, together with the discreteness of the vascular traces to tepals and stamens, could be the result of the developmental process described above. *Maundia* is said to lack perianth parts below the lowest whorl of stamens (Markgraf 1936).

Biologically there is an obvious correlation between the arrangement of parts and the bractless condition. Because a bract is lacking, the tepals serve directly as protective structures enclosing the stamens. The asymmetry of each flower, with the enlarged abaxial (lowermost) tepal, certainly supports this functional explanation.

Relationship of Juncaginaceae

The close affinity between Juncaginaceae, *Scheuchzeria*, and *Lilaea* has long been recognized and they have frequently been included as members of a single family. However, differences between them are quite considerable and it is reasonable to treat them as a group of separate, but closely related families. This point of view has been elaborated under Scheuchzeriaceae (p. 231). It should be emphasized that our knowledge of the Juncaginaceae is very limited, because most of the available information relates to the aquatic members of the family, represented in Europe by *Triglochin maritimum* and *T. palustris.* Little is known about either *Tetroncium* or the several Australian species which are not aquatic; the latter are not covered in a recent account of the aquatic plants of Australia (Aston 1973) because they are not marsh plants.

MATERIAL EXAMINED

Cycnogeton procera R. Br. Cogra Swamp and Bell's Creek, New South Wales, R. C. Carolin, 13.ix.69; all parts. New South Wales; H. T. Clifford; aerial parts (in Jodrell slide collection).

Maundia sp. F. Williams 1968; all parts (in Jodrell Laboratory slide collection).

Triglochin maritimum L. Cultivated, Aquatic Garden, Kew; P. B. Tomlinson 22.vii.79.

T. palustris L. Olchin Bay, Ontario, Canada; Anderson 5.vi.78; all parts

T. striata Ruiz and Pavon Montgomery Foundation, Fairchild Tropical Garden, Florida. P. B. Tomlinson 24.iv.72 A; Sea Spray, Victoria, Australia; P. B. Tomlinson *s.n.* Kawakawa Bay, Auckland, New Zealand; P. B. Tomlinson *s.n.*; (all parts).

SPECIES REPORTED ON IN THE LITERATURE*

Andersson (1888): *Triglochin maritimum*: seedling.
Arber (1921): *Tetroncium magellanicum*: leaf.
Arber (1924): *Cycnogeton procera, Triglochin andrewsii, T. bulbosa, T. concinna, T. laxiflora, T. maritimum, T. minutissima, T. palustris, T. stowardii, T. striata*: leaf.
Areschoug (1878): *Triglochin maritimum*: leaf.
Brunkener (1975): *Triglochin maritimum*: pollen development.
Chatin (1856, 1862): *Tetroncium magellanicum*: all parts; *Triglochin maritimum*: all parts; *Triglochin palustris*: leaf.

---

* The spelling variations *Triglochin palustre–palustrus; maritima—maritimum; striata—striatum* found in the literature are ignored.

Chauveaud (1897b): *Triglochin palustris*: root.
Chrysler (1907): *Triglochin maritimum*: stem vasculature.
Cordemoy (1862–3): *Triglochin palustris*: floral development.
Freidenfelt (1904): *Triglochin palustris*: root.
Gérard (1881): *Triglochin palustris*: seedling.
Guillaud (1878): *Triglochin maritimum*: rhizome.
Hill (1900): *Triglochin maritimum*: morphology and anatomy of all parts.
Lieu (1979a): *Triglochin striata*: floral development.
Messeri (1925): *Triglochin laxiflora*: seedling.
Montfort (1918): *Triglochin palustris*: leaf.
Park (1931): *Triglochin maritimum*: seedling.
Raunkiaer (1895–9): *Triglochin maritimum, T. palustris*: stem, leaf morphology and anatomy.
Ravololomaniraka (1972): *Tetroncium magellanicum*: brief note on leaf development.
Şerbănescu-Jitariu (1973b): *Triglochin maritimum, T. palustris*: floral morphology and anatomy, fruit morphology, seedling.
Solereder and Meyer (1933): *Triglochin maritimum*: leaf, rhizome, inflorescence axis, root.
van Tieghem (1897–1): *Triglochin maritimum*: root.
van Tieghem and Douliot (1888): *Triglochin maritimum*: origin of lateral roots.
Uhl (1947): *Triglochin bulbosa, T. maritimum, T. palustris, T. striata*: floral anatomy.
Yamashita (1970): *Triglochin maritimum*: embryology.

SIGNIFICANT LITERATURE—JUNCAGINACEAE

Andersson (1888); Arber (1918b, 1921, 1923, 1924); Areschoug (1878); Aston (1973); Brunkener (1975); Buchenau (1882, 1903c); Buchenau and Hieronymus (1889); Chatin (1856, 1862); Chauveaud (1897b, 1901); Chrysler (1907); de Cordemoy (1862–3); Daumann (1970); Eyjólfsson (1970); Freidenfelt (1904); Gardner (1976); Gérard (1881); Guillaud (1878); Hill (1900); Horn Af Rantzen (1961); Lieu (1979a); Markgraf (1936); Messeri (1925); Montfort (1918); Park (1931); Raunkiaer (1895–9); Ravololomaniraka (1972); Şerbănescu-Jitariu (1973b); Singh (1973); Solereder and Meyer (1933); van Tieghem (1870–1); van Tieghem and Douliot (1888); Uhl (1947); Yamashita (1970).

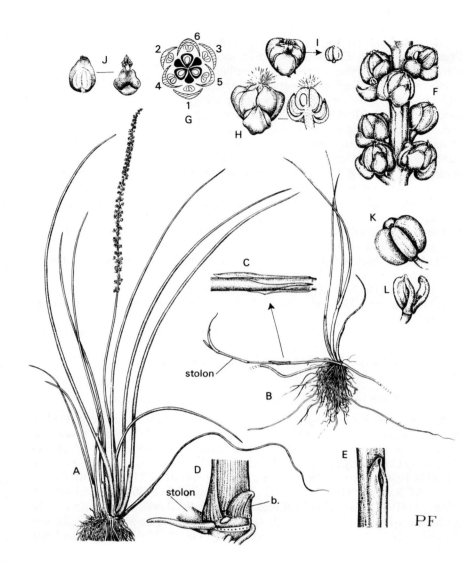

stolon

stolon b.

PF

250

FIG. 7.1. JUNCAGINACEAE. *Triglochin striata* Ruiz and Pavon. Vegetative (A–E) and reproductive (F–L) morphology. (Montgomery Foundation, Fairchild Tropical Garden, Florida; P. B. Tomlinson *s.n.*)

A. Habit ($\times \frac{1}{2}$).

B. Habit of young rosette ($\times \frac{1}{2}$) showing stoloniferous development.

C. Detail of stolon prophyll scale ($\times$ 3) developed at end of hypopodium.

D. Base of young erect shoot ($\times$ 4) with one foliage leaf removed (scar–dotted) to show axillary bud (b.) and open sheathing leaf base.

E. Mouth of leaf sheath ($\times$ 3) with ligule.

F. Part of inflorescence ($\times$ 6), bracts absent.

G. Floral diagram ($\times$ 6), sterile carpels—solid black.

H. Flower ($\times$ 6) in side view (left) and LS (right) at early (female) stage.

I. Flower ($\times$ 6) from above at later stage with stigmatic hairs shrivelled; inset, undehisced anther.

J. Anther and subtending tepal ($\times$ 8), left from outside, undehisced; right, dehisced.

K. Ripe achenes ($\times$ 6).

L. The same after dehiscence of three fertile achenes to leave three sterile carpels ($\times$ 6).

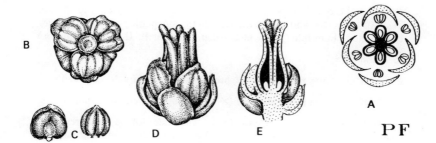

FIG. 7.2. JUNCAGINACEAE. (*Triglochin* (*Cycnogeton*) *procera*
R. Br. Floral morphology. (Via N. W. Uhl, New South Wales, 1954.)

A. Floral diagram.

B. Flower from below (× 6)

C. Stamen and subtending tepal (× 6), from inside (left); from outside
with tepal removed (right).

D. Flower from the side (× 6).

E. Flower in LS (× 6).

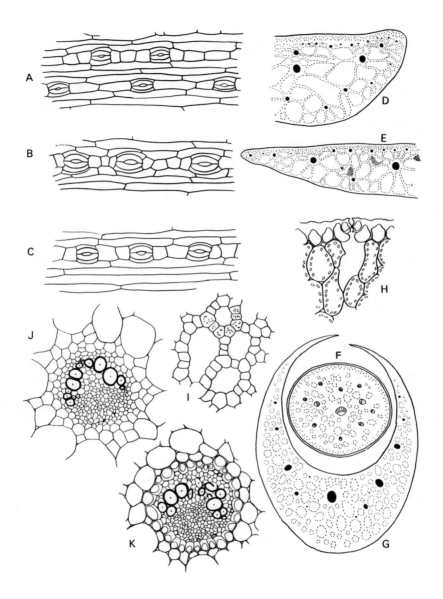

FIG. 7.3. JUNCAGINACEAE. Leaf anatomy.

A–C. Epidermis in surface view (× 130).
   A. *Triglochin palustris*. B. *Cycnogeton procera*. C. *Triglochin maritimum*.

D–H. Leaf blade and sheath in TS, vbs solid black.
   D. *Cycnogeton procera*, one half of blade (× 5). E. *Maundia* sp, one half of blade (× 5). F–G. *Triglochin striata* (× 12). F. blade. G. sheath. H. *Triglochin striata*, stoma in TS (× 270).

I–K. *Triglochin striata*, details of leaf anatomy in TS.
   I. Ground tissue of leaf sheath (× 70), with starch shown in some cells. J. TS median vb leaf blade (× 130). K. TS median vb leaf sheath (× 130).

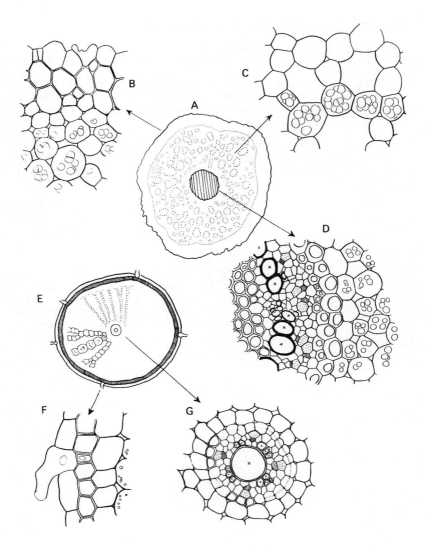

Fɪɢ. 7.4. JUNCAGINACEAE. *Triglochin striata.* Anatomy of stolon
and root.

ᴀ–ᴅ. Stolon in TS

ᴀ. Diagram (× 40). ʙ. Details of surface layers (× 270). ᴄ. Detail of
lacunose cortex (× 270) with included starch grains shown in some cells.
ᴅ. Detail of peripheral tissues of stele (× 270) with starch grains in inner
cortical cells.

ᴇ–9. Root in TS.

ᴇ. Diagram (× 130). ꜰ. Details of surface layers, with base of root
hair and associated specialized exodermal cell (× 270). ɢ. Stele (× 270),
endodermis still largely in primary state.

Family 8
LILAEACEAE* (Dumortier 1829)
(Figs. 8.1–8.3 and Plate 12)
(Based on an account, with anatomical illustrations,
by Dr M. Y. Stant.)

SUMMARY

THIS monotypic family is represented by *Lilaea scilloides* (Poiret) Hauman
(= *L. subulata* H. B. & K.) with a wide latitudinal range in the Americas
from California to Chile and Argentina, in seasonally wet habitats at low
to mid elevations. The plants are sometimes annual, diminutive (to 30 cm),
tufted (Fig. 8.1 A), cormous, with linear, ligulate, distichously arranged leaves
(Fig. 8.1 B) and filiform unbranched roots. The axis is a precise sympodium
and never stoloniferous. The inflorescences are terminal, spicate, bracteate
and appearing leaf-opposed, mostly bisexual, but sometimes unisexual. The
'flower' units are either unisexual or perfect. Each inflorescence has 1–4 basal,
long-styled, naked female flowers, each represented by a single carpel; while
distally and remotely there are clusters of flowers. The lower flowers in each
cluster are female, unicarpellate, and naked, but short-styled. The middle
flowers are hermaphrodite, with a male and female flower subtended by the
same bract, while the most distal flowers are exclusively male and the inflorescence commonly terminates in a male flower. The male flower consists of a
± sessile tetrasporangiate anther. The carpels are uniovulate, and the ovule
is basally attached and anatropous. The fruit is a flattened, ridged, and usually
winged achene (e.g. Fig. 8.1 J, K). The seeds are without endosperm, and
the embryo is torpedo-shaped (Fig. 8.1 L).

Anatomically the plants have a lacunose leaf mesophyll, hypodermal laticifers occur in the leaves and possibly elsewhere in aboveground parts. Squamules are present but hairs absent. The tracheary elements are probably
imperforate. The root stele is diarch, with a single series of central metaxylem
tracheids.

FAMILY DESCRIPTION
VEGETATIVE MORPHOLOGY AND ANATOMY

Habit (Fig. 8.1)

Sympodial growth established early in seedling development (Hieronymus
1882), each growth unit having a terminal inflorescence and a single foliage
leaf (prophyll), the erect cormous axis being produced by superposition of
such units, but with additional buds (presumably from the axil of the prophyll)
producing a branched corm, with several clustered sympodia. Stoloniferous
offsets not developed. **Squamules** present, usually six at each node and a
similar number at base of each inflorescence.

---

* Alternate family name Heterostylaceae Hutchinson, from *Heterostylus gramineus* Hook., a
later synonym for *Lilaea subulata*, but Phalangiaceae would be as logical, from *Phalangium
scilloides* Poiret (1804).

Leaf

Linear, the open tubular sheath with narrow scarious wings extended distally into a short round or bifid ligule (Fig. 8.1 B). Apical pore absent, but leaf apex rounded with numerous small stomata apparently functioning as a hydathode.

(i) **Blade** (Fig. 8.2 A–E). Terete basally, becoming progressively more flattened distally, especially on adaxial side. **Hairs** absent. **Cuticle** either thin and smooth or becoming prominently ridged. **Epidermis** without differentiated costal bands; cells elongated, outer wall becoming thick in some leaves (Fig. 8.2 C). **Stomata** (Fig. 8.2 B) not in discrete files, diffuse; each with two lateral thin-walled subsidiary cells but no distinct terminal subsidiary cells. Guard cells not sunken, each with two equal ledges (Fig. 8.2 C). **Hypodermis** not differentiated, except for inconspicuous **laticifers** (secretory canals) forming a discontinuous series of cells next to epidermis (Fig. 8.2 D). **Mesophyll** rather variable but essentially with outer 1–3 layers of densely chlorophyllous and sometimes palisade-like cells, the number of layers increasing distally; central mesophyll lacunose (Fig. 8.2 A) with wide, irregular longitudinal air-lacunae separated by uniseriate, chlorophyll-bearing partitions. Transverse diaphragms differentiated at infrequent intervals, sometimes including a transverse vein.

Vascular bundles including a central arc of 3–4 major vbs, two of them sub-marginal, suspended in the mesophyll, and a series of alternately larger and smaller subperipheral minor bundles (Fig. 8.2 A); vbs with normal xylem orientation except for occasional adaxial vbs with inverted orientation. Individual large vbs (Fig. 8.2 E) with an indistinct outer colourless parenchymatous sheath but immediately sheathed by a single layer of compact cells, these becoming irregularly thickened (usually on the inner wall) with age. Xylem ± ∧-shaped in TS, with indistinct narrow protoxylem and a group of several wider metaxylem elements; phloem with fairly wide sieve-tubes. Smaller vbs with reduced vascular tissue, the xylem represented by one wide element. **Transverse veins** connecting longitudinal veins at irregular intervals.

(ii) **Sheath** (Fig. 8.2 F, G). Abaxial epidermis with somewhat ridged or papillose cuticle, cells slightly thick-walled and with few stomata in distal part. Adaxial epidermis without stomata, cells flat and thin-walled. Mesophyll lacunose, as in blade, with occasional diaphragms. **Laticifers** as in blade, common next to abaxial surface (Fig. 8.2 G). Vascular system consisting of alternate large and small vbs, becoming arranged distally in the configuration of the blade. Adaxial vascular system absent. Ground tissue like central mesophyll of blade, but colourless basally.

AXIS

Cormous axis

With congested internodes and irregular distribution of vascular tissues. Surface layers irregular and becoming suberized with age. **Cortex** somewhat lacunose with numerous starch grains. Independent cortical vascular system not developed. **Central cylinder** delimited by an indistinct thin-walled **endo-**

dermis, the peripheral vascular tissue immediately within formed by irregularly anastomosing leaf traces. Central vbs collateral but with irregular strands of xylem and phloem.

Inflorescence axis (Fig. 8.3 A–C)

Narrow; **stomata** infrequent or absent; cuticle ridged; the outer epidermal wall scarcely thickened (Fig. 8.3 B). **Vascular system** as a single ring (Fig. 8.3 A) of 6–8 alternately large and small vbs resembling those of the leaf sheath (e.g. Fig. 8.3 C).

ROOT (Fig. 8.3 D–F)

Narrow, unbranched, filiform, up to 1 mm in diameter. **Epidermis** with frequent root-hairs from short specialized cells (trichoblasts). **Exodermis** (Fig. 8.3 F) uniseriate, cells with lignosuberized, slightly thickened walls which appear corrugated in longitudinal view, the exodermal cell files below the root-hairs appearing narrower than those elsewhere. **Cortex** uniform with 4–6 layers of wide cells, the outermost compact; cortex developing radial lacunae (e.g. Fig. 8.3 D) according to Hieronymus (1882) but not always in material examined at Kew. **Endodermis** uniseriate, thin-walled with Casparian strips, sometimes becoming slightly thick-walled opposite the phloem. **Stele** (Fig. 8.3 E) narrow. Vascular tissues diarch (rarely monarch) with a single central wide metaxylem element and two contiguous groups of narrow protoxylem elements alternating with phloem so that there are only 1–3 cell layers between pericycle and central xylem.

SECRETORY, STORAGE, AND CONDUCTING ELEMENTS

Laticifers. Present as regularly spaced elongated subepidermal cells or cell files (occasionally tangential groups) in leaf and inflorescence axis, the cells colourless and with granular contents.
Crystals. Absent.
Tannin. Absent except for occasionally solitary cells with somewhat granular cell contents in mesophyll of leaf.
Starch. Abundant in cortex and pith of cormous axis as rounded or somewhat angular simple or aggregated grains 2.5–8 μm in diameter; sometimes also present in inner cortex of root.
Tracheary elements. Tracheids in leaf and stem with spirally or scalariformly thickened lignified walls. Central metaxylem elements of root similar, but larger cells interpreted as vessels by Cheadle and Tucker (1961).
Phloem. Not examined in detail.

REPRODUCTIVE MORPHOLOGY AND ANATOMY

Morphology (Plate 12 A–F)

'Flower' difficult to define owing to stamens and carpels being solitary or in pairs (Fig. 8.1 C–I); the single sex organ here termed a flower. Basal flowers of inflorescence, **long-styled female**, each consisting of a single naked carpel (without a subtending bract), a long filiform style (to 10 cm), and a short capitate stigma (Fig. 8.1 H, I). Distal, remote flower-bearing part of axis including further naked **short-styled female** flowers each consisting of a single carpel, followed by a series of bisexual units subtended by a scale-like bract (Fig. 8.1 D–F), each unit with the anterior flower **male** and consisting of a single short-stalked tetrasporangiate anther, the posterior flower female and short-styled (Fig. 8.1 G). Distal units wholly male (Fig. 8.1 C), the axis frequently terminated by a large precociously developed male flower, according to Hieronymus (1882) and Agrawal (1952) (see also Plate 12).

Anatomy

Vasculature of distal flower complex studied by Uhl (1947), Agrawal (1952), and Singh (1965d). Each pair of male and female flowers irrigated by a single vascular strand giving rise in rapid succession to (i) a bract trace, (ii) a stamen trace, (iii) an ovary trace; the two former remaining unbranched, the stamen trace ending in the connective. Trace to the ovary branching at the carpel insertion to produce three vbs (interpreted as three dorsal bundles by Uhl), one adaxial vb, and the remaining two abaxial; all three vbs extending to the stigmatic region. [According to Singh, the ovular trace is derived from one of the laterals but Uhl indicates it has an independent origin.]

Trace system to male flower as for bisexual unit but without any vestige of carpellate supply. Vascular supply to naked female flowers (both basal and distal) like that described above, the single trace arising directly from the scape bundle.

Taxonomic and Morphological Notes

There seems to be continuing uncertainty about the number of taxa in this genus and consequently about nomenclature. We have accepted a single broadly defined species, based on *Phalangium scilloides* Poiret (1804) (= *L. subulata* H.B. & K. (1808), = *Heterostylus gramineus* Hooker (1840)). Argentinian forms have been referred to *L. superba* Rojas. Examination of material from a diversity of sources does indeed show a fairly wide structural range, but this is likely to be phenotypic in view of the rather unstable ecological conditions under which the plant is said to grow.

Treated in this broad sense, the species occupies an indeterminate position but is generally considered to be allied to the Scheuchzeriaceae and Juncaginaceae (s.s.). It shares with *Triglochin* the same cyanogenic glucoside triglochinin (Hegnauer and Ruijgrok 1971). Embryological and palynological evidence, summarized by Agrawal (1952), shows that *Lilaea* is closer to *Triglochin* than *Scheuchzeria* since they both have two-celled pollen (not three-celled), one basal anatropous ovule (not two or more) and free nuclear (not helobial) endosperm. Furthermore Yamashita (1970) has shown that *Lilaea* and *Triglochin* share the unusual type of embryo development in which the first root is initiated somewhat laterally, not opposite the suspensor. *Lilaea* and *Triglochin* also have squamules, which are lacking from *Scheuchzeria*. *Lilaea* remains unique in its laticifers, not previously recorded by the numerous investigators who have examined this species.

In stressing the relationship between these three taxa several workers have tried to equate the distinctive floral morphology of *Lilaea* to that of *Triglochin*. Both Uhl (1947) and Yamashita (1970) succeeded by considering both 'flowers' as pseudanthia, the single lateral flower of *Triglochin* then being equated with the whole scape of *Lilaea*, with perianth segments as bracts subtending solitary male flowers of one stamen (*Triglochin*) or a staminate and pistillate flower (perfect 'flowers' of *Lilaea*). These interpretations create as many problems as they resolve. For the present, it is better to retain the separate family status of *Lilaea* until a more extensive comparative study has been made.

Floral development (a note added by U. Posluszny on the basis of Plate 12)

The data obtained through the study of floral development of *Lilaea* support a closer affinity to *Triglochin* than to *Scheuchzeria*. Of particular note are the branching and development of the inflorescence axis (Plate 12, Fig. A, B) and the development of the carpel (Plate 12, Fig. E). Both are very similar to the developmental patterns seen in *Triglochin striata* (Lieu 1979a). Uhl's and Yamashita's interpretation of the 'flower' of *Triglochin* as being equal to the whole scape of *Lilaea* is not supported by floral development. In both, the 'flowers' originate as separate units (apices) on the inflorescence axis (Plate 12, Figs. B, C). Floral organs are initiated on the apices in an acropetal sequence . . . first tepals, followed by stamens and finally carpels (Plate 12, Figs. D, E). Variation in the number of appendages or the complete elimination of either male or female organs is not unusual in aquatic monocotyledons. It is, therefore, possible to link the flower of *Lilaea* with that of *Triglochin*. Even more plausible comparisons can be made between the floral development of *Lilaea* and members of the Potamogetonaceae, especially the genus *Ruppia* (see Posluszny and Sattler 1974b).

MATERIAL EXAMINED

Lilaea scilloides (Poir.) Haum. 8 m. S. of Dixon, California; J. B. Fisher 9.iv.67; Thorne 32264; Gankin 2. Cultivated specimens grown from seed at Kew, University of Liverpool, and Berkeley. Melville 25395, 1962; all parts.

SPECIES REPORTED ON IN THE LITERATURE

Agrawal (1952): *L. subulata*: reproductive morphology.
Campbell (1898): *L. subulata*: floral anatomy and embryology.
Hieronymus (1882): *L. subulata*: all parts.
Singh (1965d): *L. scilloides*: floral anatomy.
Solereder and Meyer (1933): *L. subulata*: leaf.
Uhl (1947): *L. scilloides*: floral anatomy.
Yamashita (1970): *L. subulata*: embryology.

SIGNIFICANT LITERATURE—LILAEACEAE

Agrawal (1952); Arber (1940); Buchenau (1903c); Campbell (1898); Cheadle and Tucker (1961); Hegnauer and Ruijgrok (1971); Hieronymus (1882); Hutchinson (1959); Markgraf (1936); Singh (1965d); Solereder and Meyer (1933); Uhl (1947); Yamashita (1970).

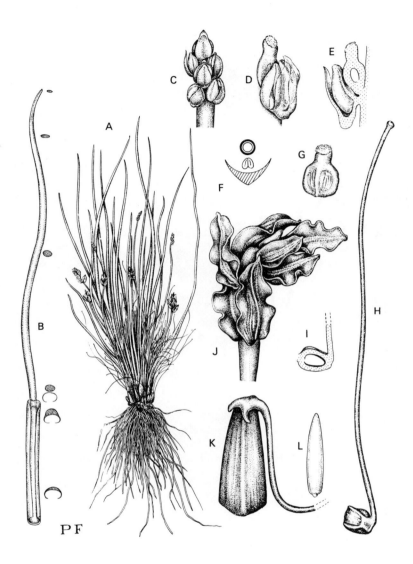

PF

264

FIG. 8.1. LILAEACEAE. *Lilaea scilloides* (Poir.) Haum. (Dixon, California, J. B. Fisher 9.iv.67.)

A. Habit ($\times$ ½).

B. Single leaf ($\times$ 2), TS at successive levels.

C. Distal part of inflorescence with male flowers ($\times$ 5).

D. Bisexual unit from lower part of inflorescence ($\times$ 15), including male flower, short-styled female flower subtended by a bract (cf. F).

E. The same unit in LS ($\times$ 15).

F. Diagram of bisexual unit.

G. Female flower from bisexual unit ($\times$ 15) showing impression of male flower.

H. Long-styled female flower from base of inflorescence ($\times$ 5).

I. LS long-styled female flower ($\times$ 5), only base of style shown.

J. Young winged fruits from distal part of inflorescence ($\times$ 5).

K. Young fruit developing from basal long-styled flower ($\times$ 5).

L. Embryo ($\times$ 8).

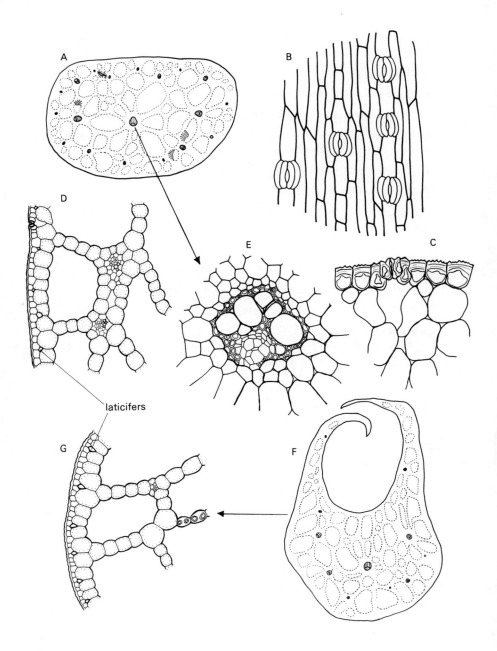

laticifers

FIG. 8.2. LILAEACEAE. *Lilaea scilloides.* Leaf (various sources).

A. Lower leaf blade TS (× 8).

B. Epidermis in surface view (× 70).

C. Stoma TS (× 140) to show thickened epidermal cells, cuticular ridges.

D. Detail of surface layers of lower leaf blade in TS (× 35), with hypodermal laticifers.

E. Vascular bundle from centre of leaf blade (× 70).

F. Leaf base in TS (× 8).

G. Detail of surface layers of leaf base in TS (× 35) including hypodermal laticifers.

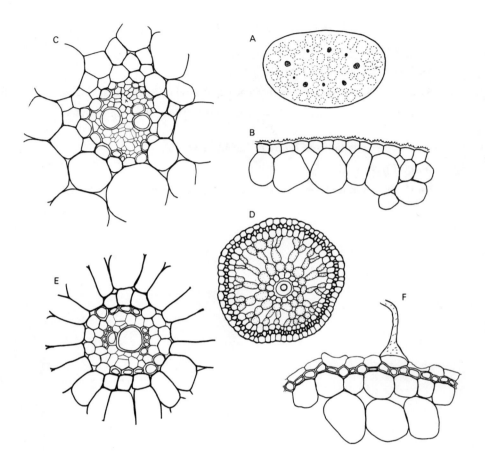

268

FIG. 8.3. LILAEACEAE. *Lilaea scilloides* (various sources).

A–C. Inflorescence axis.

A. TS inflorescence axis (× 8). B. Surface layers (× 110) to show ridged cuticle and hypodermal laticifers. C. TS larger vb (× 110).

D–F. Root.

D. Root TS (× 35), a form with well developed cortical lacunae (details of stele not drawn). E. Root stele TS (× 110). F. Surface layers of root (× 70), including base of root hair and thick-walled exodermis.

Family 9
POTAMOGETONACEAE (Dumortier 1829)
(Figs. 9.1–9.13)
(In a restricted sense and excluding Cymodoceaceae, Posidoniaceae, Zannichelliaceae, and Zosteraceae, but including Ruppiaceae.)

SUMMARY AND KEY TO GENERA

THIS is a cosmopolitan family of mostly submerged, often filamentous pond-weeds, usually occurring in fresh, or sometimes in brackish, water but never truly marine. It includes three genera: *Groenlandia* J. Gay with one species *(G. densa)*, *Ruppia* L. with several closely related species, and *Potamogeton* L. with over 100 species, often widely distributed and poorly circumscribed. *Groenlandia* is obviously closely related to *Potamogeton* and may well be included within it (as *P. densa*). *Ruppia* is more distantly related and there is some justification for regarding it as a separate family, Ruppiaceae. It is here treated as a subfamily Ruppioideae.

The following characters are diagnostic for the family and the key illustrates the major differences between genera.

The habit in the vegetative phase is either monopodial *(Ruppia)* or sympodial *(Groenlandia, Potamogeton)*. It is also sympodial distally in association with flowering. The leaves (except in *Groenlandia*) have a distinct basal tubular sheath, the sheath in *Potamogeton* often being free from the blade as a stipule-like appendage. The blade is linear to broadly ovate, with a distinct thickened midrib in *Potamogeton*. The leaves are wholly submerged in *Ruppia* and *Groenlandia* but the distal leaves are floating in some *Potamogeton* species. Hairs are absent but there are frequent marginal teeth. Squamules are 2–many at each node. Hydropoten have not been recorded (Lyr and Streitberg 1955). The roots are unbranched.

The inflorescence is terminal, spike-like, rising above a subopposite pair of leaves. There are two flowers per spike in *Ruppia* and *Groenlandia*, but in *Potamogeton* there are more (up to 30). Bracts may be present or absent. There are two or four stamens, each stamen with a pair of bilocular thecae. The stamens in *Groenlandia* and *Potamogeton* have a broad tepal adnate to the connective. There are (3–) 4 (–5) carpels in *Groenlandia* and *Potamogeton*, and few to several in *Ruppia*. Each carpel is ovoid with a short stigma and a single, more or less pendulous ovule. The fruit is an achene with *(Ruppia, Potamogeton)* or without *(Groenlandia)* a stony endocarp. The embryo is short in *Ruppia*, curved in *Potamogeton* or with a spirally coiled cotyledon in *Groenlandia*. Germination is epigeal, and the fruit wall splits laterally. The plants are anatomically reduced in relation to their aquatic environment, the reduction being greatest in wholly submerged species, which have well developed air-lacunae and lack pronounced mechanical tissue. The vascular system of the stem is of the same basic type throughout, but it is progressively more reduced from larger to small species in *Potamogeton*. The vascular tissues are reduced, with the xylem largely represented by lacunae. Specialized secretory elements are not developed.

270

Diagnostic key for subfamilies and genera of Potamogetonaceae

1A. Rhizome system well differentiated from the erect system, sympodial; renewal shoots always at the base of erect shoots and developing at (apparently) alternate nodes on a rhizome. Leaves diverse, narrowly linear to broad, heterophylly often marked. Ligular sheath often independent of the blade. Leaf trace system from the stele consisting of at least three separate vbs (one median, two lateral). [Flowers usually more than two per spike, stamens and carpels four per flower; sepaloid tepal adnate to the stamen connective; peduncle short, carpel stalk not elongating after pollination.]   Potamogetonoideae (2)

1B. Rhizome system poorly differentiated from the erect system, monopodial; leaves uniformly linear with an attached sheathing base. Leaf trace system from the stele consisting of a single median bundle. [Flowers consistently two per spike, each with two stamens and few–several (–14) carpels; tepals absent; peduncle elongated; carpel stalk often elongating after pollination.]

Ruppioideae (*Ruppia*, c. 5–10 spp) (Fig. 9.1)

2A. Leaves on erect shoots all in subopposite pairs; without a distinct sheathing base. [Flowers two per spike. Annual. Fruit without a stony endocarp. Cotyledon spirally coiled in the seed. Germination probably not light-dependent.]   *Groenlandia* (1 sp) (Fig. 9.4)

2B. Leaves on erect shoots not all in subopposite pairs; with a distinct sheathing base (which may be free of the blade). [Flowers > 2 per spike. Perennial. Fruit with a stony endocarp. Cotyledon curved. Seed germination probably light-dependent.]

Potamogeton (c. 100 spp) (Figs. 9.6 and 9.7)

POTAMOGETONACEAE–RUPPIOIDEAE

RUPPIA L.
(Figs. 9.1–9.3 and Plate 1 D)

A cosmopolitan genus variously regarded as one species including several distinct subspecies and varieties or a number of species. (Ascherson and Graebner 1907; Gamerro 1968; Mason 1967); but modern studies tending to establish taxa on a cytotaxonomic basis (e.g. Reese 1962, 1963). Plants wholly submerged, usually in brackish coastal waters or equivalent habitats inland; sometimes in hypersaline habitats.

Anatomy of *Ruppia* much reduced in relation to its aquatic habitat. Characters include the following: hairs restricted to uniseriate, few-celled teeth at the margin of the blade, especially at its apex. Well developed air-lacunae in the mesophyll of the leaf and cortex of the stem and root. Stomata absent. Epidermis uniform and constituting the chief photosynthetic part of the leaf. Vascular system in the leaf reduced to a single median vein, directly continuous with the vascular cylinder of the stem, together with obscure lateral veins, one in each leaf margin, extending through one internode in the cortex of the stem without making contact either with the central vb or with the lateral veins of other leaves. Central xylem in the leaf, stem, and root largely represented by a xylem lacuna. Tannin cells common in the epidermis of leaf and in the ground tissues elsewhere.

VEGETATIVE MORPHOLOGY

Plants filamentous, often perennating by means of persistent, buried, but not specially modified, horizontal axes. **Stem** (Fig. 9.1 A) slender, initially horizontal and branching monopodially during an early vegetative period of growth, this part of the axis (rhizome) usually rooted in mud by filamentous, unbranched roots arising singly at the nodes. In subsequent reproductive periods of growth, sympodially branched below the terminal inflorescences (Fig. 9.1 B, C, I), the branches largely erect, restricted to one plane, not rooted, spreading and floating fan-wise in water. **Turions** described for *R. tuberosa*, from Western Australia (Davis and Tomlinson 1974) as distal swellings of shoots, each largely consisting of a single swollen, starch-filled internode. Similar structures illustrated for *R. maritima* by Aston (1973) in Australian material.

Leaves inserted distichously, separated by long internodes, but distal pair immediately below the inflorescence made subopposite by the suppression of the intervening internode (Fig. 9.1 B). Each leaf with a long open tubular sheathing base, opposite margins overwrapping somewhat; mouth of the sheath with a pair of small laterally opposed auricles. Ligule absent. [The leaf base has been interpreted as equivalent to two adnate stipules by analogy with *Potomogeton pectinatus* (Graves 1908; Monoyer 1927) but there is no ontogenetic evidence for this.] Flattened blade up to 150 mm long but rarely exceeding 2 mm in width, with an inconspicuous median vein; apical pore not developed. Apex acute to truncate; outline diagnostically useful. Bicarinate **prophyll**, in adaxial position (Fig. 9.3 I), always commencing each new branch. Subprophyllar internode suppressed thus ensuring the insertion of the prophyll between the main axis and branch. Pairs of shorter leaves (Fig. 9.1 B) with somewhat inflated sheaths enclosing terminal inflorescence, referred to as subfloral leaves (Graves), or folia floralia (Irmisch 1851), but morphologically and anatomically identical with vegetative foliage leaves.

Intravaginal scales (squamules) forming an opposed pair laterally in the axil of each foliage leaf (Fig. 9.3 I); each scale non-vasculated, mostly biseriate throughout, the cells densely cytoplasmic with conspicuous nuclei; distal cells somewhat elongated.

VEGETATIVE ANATOMY

Leaf (Fig. 9.2 A–R)

(i) **Lamina. Uniseriate hairs** (Fig. 9.2 D–G) more or less restricted to the leaf margin, close to the apex of the blade (Fig. 9.2 C), becoming very diffuse basally. Hairs solitary, rarely in twos or threes, each consisting of 1–3 (but up to ten) narrow cells with a few small chloroplasts, distal cell usually the longest.

Stomata absent. **Cuticle** thin, inconspicuous. Epidermis (Fig. 9.2 B) of more or less cubical cells, square or somewhat elongated in surface view and including numerous discoid chloroplasts. Epidermis functioning as the chief photosynthetic tissue. Outer epidermal wall somewhat thicker than the remaining slightly thickened walls (Fig. 9.2 J). Epidermis completely uniform except for frequent **tannin cells**, usually longer and often deeper than normal epidermal cells, often in short series of 2–3 cells. **Mesophyll** (Fig. 9.2 H) including one or more subepidermal cell layers with a few chloroplasts. Cells enlarged, more or less square in TS, but 6–10 times longer than the overlying epidermal cells (Fig. 9.2 A), enclosing a large **air-lacuna** on each side of the midrib. Lacuna extending throughout the length of the blade (Fig. 9.2 K–M) but segmented at irregular, infrequent intervals by uniseriate, non-vasculated **transverse diaphragms** of compact cells with dense cytoplasmic contents. Longitudinal gaseous continuity through air-lacuna maintained by narrow intercellular spaces at the angles between the diaphragm cells. The number of subepidermal mesophyll layers enclosing the lacunae regarded as diagnostically significant by some authors (e.g. uniseriate in *R. cirrhosa* (Fig. 9.2 H), biseriate in *R. spiralis*, 3–5-seriate in *R. obtusa* according to Hagström (1911); 2–3-seriate in *R. truncatifolia* according to Miki (1935a)). A 1–3-seriate subepidermal layer recorded by Sauvageau (1891a), in *R. maritima*, the number of layers decreasing distally. Air-lacunae separated from each other by median buttress, consisting of somewhat narrow elongated mesophyll cells, enclosing the median vein; frequently with additional small irregular lacunae on the adaxial side of the vein (Fig. 9.2 H, K). **Median vascular bundle** (Fig. 9.2 I) enclosed by an indistinct parenchyma sheath with scarcely thickened walls, but without Casparian thickenings although described as an endodermis by Graves (1908) and Sauvageau (1891a); xylem represented by a narrow lacuna; phloem restricted to two small lateral or slightly abaxial strands in the conjunctive parenchyma surrounding the lacuna.

Leaf **margin** rounded, including two small strands of narrow, elongated but otherwise undifferentiated cells in the lower part of the blade. Strands extending distally but unequally, and disappearing distally (cf. Fig. 9.2 H, K–M). Lateral strands extending to within 10 mm of the apex according to Graves, but not uniting with the median vein; scarcely reaching half-way to the apex in material illustrated in Fig. 9.2 H–M. Marginal strands illustrated by all previous authors, but their presence apparently determined by the level at which the blade is examined and possibly also by the vigour of the shoot. Lateral bundles absent from *R. brachypus* according to Sauvageau (1891a).

(ii) **Leaf sheath** (Fig. 9.2 L–R). Narrow marginal wings overlapping considerably, auriculate distally. Dorsal region corresponding anatomically to the blade and including a median vb enclosed by a parenchymatous buttress. Air-lacunae surrounded by one or more subepidermal layers with included lateral bundles (Fig. 9.2 N). **Abaxial epidermis** (Fig. 9.2 P) of elongated chlorenchymatous cells, accompanied by frequent tannin cells. Epidermal cells transitional distally to isodiametric cells of blade. **Adaxial epidermis** (Fig. 9.2 Q) including fewer tannin cells, colourless. Adaxial epidermal cells thin-walled, longer and somewhat narrower than the abaxial cells at same level. Adaxial epidermal cells abruptly transitional distally to the adaxial chlorenchymatous epidermis of the blade. Epidermis of both surfaces, where close to the leaf insertion, colourless and resembling each other. Leaf sheath abruptly biseriate beyond the lateral bundles forming the **marginal wings** (Fig. 9.2 o, R) of the sheath. Cells of the outer layer somewhat wider and larger than those of the inner layer, but cells of the inner layer largest close to the margin. **Vascular bundle** of the sheath continuous into and identical anatomically with that of the lamina.

[For information about relative rates of growth of blade and sheath see Graves (1908).]

(iii) **Prophyll.** Equivalent morphologically to the sheathing portion of the foliage leaf, without its blade, but **biseriate** throughout and wholly without vascular tissue. Prophyll of the branch immediately below the inflorescence short, emarginate, secondarily becoming adherent to the main axis by growth of the main axis itself rather than by growth of the branch to which it belongs (Posluszny and Sattler 1974*b*). [This scarcely justifies regarding it as a spathe, as Graves suggests.]

Stem (Fig. 9.3 A–I)

Vegetative shoots, produced during the post-seedling phase of growth, differing from the later, distal **reproductive shoots** (for details see Graves 1908; Irmisch 1851; Kirchner *et al.* 1908). Vegetative shoots long-persistent under certain conditions. Rhizome and free-floating erect axis morphologically identical and anatomically v. similar.

(i) **Vegetative stem,** branching abundant, monopodial in acropetal order; each foliage leaf subtending a lateral branch originally enclosed by a bicarinate biseriate prophyll; first foliage leaf of branch therefore always on same side of shoot as axillant leaf. Stem (Fig. 9.3 A) more or less circular in TS, including a v. broad cortex surrounding a narrow central stele or axial vb. **Epidermis** uniform, thinly cutinized, slightly thick-walled, often somewhat chlorenchymatous; including tannin cells. Stomata absent. **Cortex** wide, including 1–2 peripheral compact layers of enlarged cells and a broader middle layer with a single irregular series of **air-lacunae.** Lacunae continuous throughout the internode, without transverse diaphragms and separated by irregularly radial uniseriate plates of cortical parenchyma. Nodal cortical parenchyma compact, without pronounced air-lacunae. Cortical vbs (Fig. 9.3 B) represented by a pair of reduced subopposite strands in the outer cortex. Chloroplasts few in the periphery of the cortex, but replaced by starch in the inner compact layers surrounding the central vb. Starch abundant in buried, creeping axes. **Stele** (Fig. 9.3 C), or central vb, delimited from the compact, slightly thick-walled layers of the inner cortex by 1–2 indistinct layers of cells with suberized radial walls but scarcely constituting a distinct endodermis. Vascular tissues of stele including a wide central xylem lacuna surrounded by a more or less continuous but irregular ring of phloem tissue sometimes recognizably divided into four separate strands.

(ii) **Rhizome,** or buried basal axes, differing from aerial stem in being somewhat more robust; parenchyma cells thicker-walled; stele wider in proportion to total diameter, sometimes eccentric; air-lacunae less well developed; chloroplasts largely replaced by starch grains.

(iii) **Course of the vascular bundles in the stem** (after Chrysler 1907; Graves 1908; cf. Singh 1964) (Fig. 9.3 D–I). Median vein of leaf and vascular supply to branch diverging directly from the axial bundle of the stem (Fig. 9.3 D–F). Lateral strand of the leaf becoming a **cortical bundle** of the stem and forming an independent system discontinuous with stele, extending throughout the internode but ending blindly just above next node below (cf. Fig. 9.3 D–G). On the other hand cortical bundles of successive internodes said by Singh to be continuous. Cortical bundles subopposite, usually both directed towards the same side of the stem as the leaf into which they run. Each cortical bundle represented by a few narrow elongated cells without apparent conducting elements (cf. Figs. 9.2 N and 9.3 B), but, according to Graves, tracheal elements present in early stages of development, but later 'resorbed.' Sieve-tubes not observed. [The cortical bundle does not 'disappear' as Solereder and Meyer (1933) suggest.] Distal and proximal extent of the cortical strands apparently variable and perhaps dependent on the vigour of the shoot.

[When Monoyer (1927) later added to the above account he noted greater complexity at the node. Phloem within the stele tends to be aggregated into four separate strands at least near the node (also recorded by Graves 1908; Hagström 1911) and these strands were regarded as four separate vascular bundles or their equivalent. Of these, two in the plane of insertion of the leaves are directly continuous with median bundles of successive leaves (*traces foliaires*). Traced downwards, each divides at the node below to join the remaining two lateral strands (*faisceaux sympodiques*) which continue indefinitely through the axis without having any direct contact with leaves, except via the girdling trace derived from the split median bundle. The root trace at each node unites directly with this anastomosing trace. These strands are individually less distinct than Monoyer's analysis would suggest, since his account depends in part on a theoretical comparison with the vascular system of related plants. Developmentally the strands associated with the leaves differentiate vascular tissue earlier than the lateral strands. It may be significant that Monoyer does not discuss the insertion of the branch trace, although Graves shows this may be larger than the leaf trace in the mature axis (Fig. 9.3 F–I).]

Root (Fig. 9.3 J–L)

Unbranched, arising singly (rarely in pairs), at nodes always from the same side of the stem (i.e. successive root insertions are always mirror images) even in branches of successively higher orders. Root early enclosed by a tubular **coleorrhiza** (largely an outgrowth of the stem epidermis), ruptured by the outgrowth of the root, but persisting as a basal tubular structure. Root stele attached directly to nodal complex in a sublateral position.

Epidermis (Fig. 9.3 J): large-celled, root-hairs often long-persistent; initiated as densely cytoplasmic **trichoblasts** distinct in appearance, but not in size, from adjacent cells in early stages of root development. Trichoblasts maturing as elongated cells with a root-hair developed from distal end. **Cortex** including a two-layered exodermis. Subepidermal layer consisting of narrow cells with longitudinal walls distinctly plicate in surface view (Fig. 9.3 L). Next innermost layer of cells wider, and slightly thick-walled. Middle cortex uniform; parenchyma initially consisting of regular concentric and radiating cells. Cells later separating radially and partially collapsing to produce irregular radial lacunae. Inner cortex compact, small-celled, innermost layer radially coincident with the endodermal cells. **Stele** (Fig. 9.3 J, K) v. narrow, delimited from the cortex by a uniseriate **endodermis** of slightly thick-walled cells, with uniformly lignosuberized walls in mature roots. Rest of stele consisting usually of two layers: (i) pericycle, indistinct, uniseriate, including four (5–6) sieve-tubes next to the endodermis; the companion cells, corresponding to each sieve-tube, cut off to the inside from a common initial; (ii) central xylem lacuna surrounded by a single layer of conjunctive parenchyma.

Secretory, Storage, and Conducting Elements

Tannin. Present as dark brown contents in otherwise more or less unmodified cells in the ground tissue of all parts; especially conspicuous in the epidermis, usually as elongated, often anticlinally extended tannin cells, sometimes in short longitudinal series.

Starch. Mostly restricted to the inner cortex of the stem, but especially abundant and more uniformly distributed throughout the cortex in buried axes (rhizomes).

Phloem in all parts including simple, transverse sieve-plates. Possible occluded phloem observed in stele of stem.

Xylem in the stem and leaf replaced by xylem lacunae, representing either single files of tracheal elements, or several files in stem, differentiated early but ruptured during elongation of the organ. Remnants of mostly annular (rarely spiral or pitted) wall thickenings occasionally observed in the lacunae. Tracheal elements at nodes, with minimal elongation, persistent.

Reproductive Morphology
(Figs. 9.1 B–I)

Inflorescence terminal (Fig. 9.1 B, C) on sympodial parts of the shoot, each at first enclosed by two subopposite foliage leaves (*folia floralia*), one or both leaves subtending renewal shoots with basal scale-like prophylls. [The prophyll of the renewal shoot is not a spathe as suggested by Graves 1908]. **Flowers** two (rarely more), on opposite sides of the inflorescence axis (Fig. 9.1 D), the axis being prolonged v. shortly beyond the uppermost flower (Fig. 9.1 E). Each flower without a subtending bract or perianth [except for a minute scale on the connective of the stamen (Fig. 9.1 F)]. Stamens two; carpels (2–) 4 or more (up to 16). **Stamens** sessile, connective broad with two curved bisporangiate thecae, each partly enveloping the axis. **Carpels** (Fig. 9.1 F) each at first v. short-stalked, with a peltate sessile stigma and a single lateral, or more or less pendulous, bitegmic ovule. Fruit stalk (podogyne) elongating during fruit development (Fig. 9.1 I). **Fruit** (Fig. 9.1 G) an achene (sometimes described as a drupe), not floating, the outer soft layers decaying, the inner sclerotic layers persistent and opening via a small lid at germination. Endosperm absent. **Embryo** with an enlarged hypocotyl, the primary root (radicle) regarded by some authors (e.g. Gamerro 1968) as aborted, the first visible seedling root, at the base of the cotyledon, then interpreted as adventitious. [Developmental and comparative studies by Yamashita (1972) more convincingly demonstrate that this lateral root is the displaced radicle, which is not aborted (cf. Ly Thi Ba *et al.* 1973).]

REPRODUCTIVE ANATOMY

(i) *Peduncle*

According to Graves (1908) anatomy of peduncle identical with that of vegetative stem except for: (a) the absence of cortical bundles; (b) more pronounced air-lacunae; (c) epidermis slightly thicker-walled, including cells with overlapping ends. Elongation of peduncle bringing flowers to the surface of the water said to be effected by a general stem elongation in the absence of a meristem. [The fate of the sieve-tubes under these circumstances is not discussed.]

(ii) *Flower* (after Uhl 1947; Gamerro 1968; Posluszny and Sattler 1974*b*)

Vascular supply to whole flower a single, often v. short, trace branching from the stele. Floral trace producing first two lateral vbs, the stamen traces, one passing to each stamen. [Within the connective this trace is widened and even forks at the level of the two thecae (Posluszny and Sattler 1974*b*). This has been used as evidence that the stamen is actually double (cf. Miki 1937).] Remaining central vascular tissue forming four vbs, one to each carpel, the carpel traces diverging in two successive series in a manner suggesting the carpels to be in two whorls, the pair opposite the stamens being uppermost. Each carpel trace itself dividing once only and producing a short dorsal trace (possibly extending to the base of, but not into, the peltate stigma) and a ventral ovular trace.

[The vascular anatomy of the flower is thus very similar to that of *Potamogeton* (p. 299) except for the weaker dorsal bundle which attains at most only two-thirds of the length of the carpel. This difference may be related to the hydrophilous method of pollination contrasted with the anemophilous or even entomophilous method in *Potamogeton*. The anatomy of the stamen in relation to its method of dehiscence and floating has been described in detail by Gamerro (1968). It includes a functional endothecium and a free cutinized layer inside the theca, which is initially in contact with the endothecium, but, in the floating anther, free and spread out so as to support the recurved theca.]

(iii) *Fruit*

Endocarp of the fruit heavily sclerified, but perforated in the placental region by two ducts of pitted parenchyma, visible externally as two depressions with diagnostically useful outlines.

[For further details of flowers and fruit anatomy in relation to their biology the detailed account of Gamerro (1968) should be consulted.]

POLLINATION

Pollination is unusual in that it takes place on the surface of the water (Gamerro 1968). The peduncle elongates somewhat and the anthers become detached, floating to the surface of the water by virtue of a gas bubble which is attached to them. At the surface the anthers dehisce, releasing the pollen which, though heavier than water (Schwanitz 1967), floats as adherent masses which may be quite conspicuous. The shape and surface texture of the individual pollen grains (Fig. 9.1 H) are related to their ability to float by means of trapped gas bubbles. By wave, wind, or current action the pollen is carried to the stigmas which have been brought to the water surface by continued elongation of the peduncle, so that the female parts of the flowers now lie on the surface of the water. After fertilization the whole inflorescence is drawn beneath the surface of the water (by unequal elongation and spiral contraction) and fruit development is underwater. Elongation of the stalks of individual carpels (e.g. Fig. 9.1 I) occurs to varying degrees in different species during fruit development.

According to McCann (1945, 1978) pollination in *Ruppia* is underwater, but observations made in Florida and in more detail by Gamerro (1968) suggest that the above interpretation is more accurate. It is possible, however, that several methods of pollination occur, depending on the taxon and the locality.]

TAXONOMIC AND MORPHOLOGICAL NOTES

Infrageneric taxa and key to species

Anatomical criteria which have been used by earlier authors in the recognition of taxa within *Ruppia*, whether these are regarded as species or subspecies, have included size of epidermal cells, number of subepidermal mesophyll layers in the leaf and shape of the leaf apex (e.g. Sauvageau 1891*a*; Hagström 1911; Miki 1935*a*). These features, however, do vary on a single plant and may be influenced directly by the environment. A more modern cytotaxonomic survey of *Ruppia*, based on extensive measurements in population samples in northwestern Germany by Reese (1962, 1963), provides comparative data and has established morphological criteria which may be used reliably in diagnosis. A clarification of taxonomic and nomenclatural problems in *Ruppia* over part of its range seems now to have been achieved. Vegetative features which are diagnostically reliable include: leaf width, colour, shape of the leaf apex and average width of epidermal cells. Diagnostic features of reproductive parts include: size of thecae, fruit shape, length of peduncle and fruit stalk and whether the peduncle is spirally wound or not. Mayer (1971) noted that fruit size could be influenced by salinity. Important differences in the biology of reproduction in different taxa are also emphasized (see also Luther 1947). These differences are correlated with chromosome number in such a way that *Ruppia*, in this part of Europe, is segregated by Reese into two species and two varieties as follows:

1. Peduncle not spirally wound, shorter than 5 cm (leaves *c.* 0.5 mm wide, apex pointed, epidermal cells *c.* 12.5–16 μm wide, thecae small, fruit strongly asymmetrical   *R. maritima* L. $(2n = 20)$

    1A. Fruit stalk longer than 5 mm   var. *maritima*
        (= *R. rostellata* Koch; *R. maritima* var. *rostellata* Agardh)

    1B. Fruit stalk shorter than 5 mm   var. *brevirostris* Agardh
        (= *R. brachypus* J. Gay)

2. Peduncle spirally wound, mostly longer than 10 cm (leaves *c.* 1 mm wide, apex rounded, epidermal cells *c.* 16–19 μm wide, thecae large, fruit scarcely asymmetrical)   *R. spiralis* L.* $(2n = 40)$
    (= *R. maritima* of many authors, but not L.)

Similar characters are used by Mason (1967) to distinguish *Ruppia* species in New Zealand, especial emphasis being given to fruit size, shape, and surface texture. The leaf apex is again found to be diagnostically useful. *Ruppia megacarpa* Mason, in this respect, is distinguished from all other species except the Japanese *R. truncatifolia* Miki.

These studies all represent the beginnings of a taxonomic understanding of *Ruppia* on a cosmopolitan basis. They also suggest that, apart from the character of the leaf apex and epidermal cell size, microscopic anatomy is unlikely to be of any value in the infrageneric taxonomy of *Ruppia.*

Affinity of Ruppia with Potamogeton

Many authors have commented on the similarity in leaf structure between *Ruppia* and some *Potamogeton* spp (especially *P. pectinatus*). The branching habit of both *Potamogeton* and *Ruppia* in association with flowering is essentially identical (Irmisch 1851) and the vascular system of the stem of *Ruppia* can be equated with that in the more diminutive spp of *Potamogeton* (see *Course of vascular bundles in stem*, p. 275). Similarly, Uhl (1947) describes the vascular supply to the flower in *Ruppia* as a reduction of that in *Potamogeton*, the chief difference being in the weaker dorsal trace of the carpel. Davis and Tomlinson (1974) have described an Australian form of *Ruppia* with vegetative structures which recall the resting buds (turions) of certain *Potamogeton* spp. There is every indication that *Ruppia* is more closely related to *Potamogeton* than to any other aquatic monocotyledon and in this account it is included as a subfamily within the Potamogetonaceae.

Floral morphology

The interpretation of the 'flower' of *Ruppia* is closely bound up with that of *Potamogeton* because of their considerable superficial likeness. It has been presumed that whatever interpretation applies to one genus should also apply to the other. *Ruppia* differs from *Potamogeton* in having two and and not four stamens, in lacking the conspicuous appendages to the connective which have occasioned so much discussion in morphological literature, and in having a variable number of carpels (from (2–) 4–16). However, in *Ruppia* the connective or common stalk to the anther thecae does develop minute, inconspicuous

---

* According to Gamerro (1968) the correct name for this is *R. cirrhosa* (Petag.) Grande.

appendages (Fig. 9.1 E, F) which have been regarded by some authors (e.g. Graves 1908; Čelakovský 1896) as possible tepal lobes, and by others (e.g. Irmisch 1851; Markgraf 1936) simply as outgrowths of the connective. These interpretations are summarized by Gamerro (1968). Posluszny and Sattler (1974b) show that the appendage is a late outgrowth of the connective and developmentally, positionally and structurally quite unlike the 'tepal' of *Potamogeton*. In general two contrasted views of the *Ruppia* 'flower' have been taken, essentially determined by the interpretation of the morphological nature of this connective organ.

(i) It is a single flower with two stamens and usually four (but often more) carpels. Some authors (e.g. Roze 1894) have said that there are four stamens each with a bilocular anther, the wide separation of each pair of thecae at maturity (and during early development) suggesting this.

(ii) In the view of Uhl (1947) and Miki (1937) the assemblage, usually regarded as a flower, is considered to be an entire first-order inflorescence branch, consisting of two (or four) staminate flowers each in the axil of a vestigial bract ('perianth part'), by which presumably is meant the minute scale on the connective. Each individual carpel is then equivalent to a pistillate flower. There is no developmental evidence for this interpretation (Posluszny and Sattler 1974b).

MATERIAL EXAMINED

Ruppia cirrhosa (Petag.) Grande; Fairchild Tropical Garden, Miami, Florida; P. B. Tomlinson *s.n.*; all parts.

R. rostellata Koch.; Ghana; J. B. Hall *s.n.*; all parts.

R. tuberosa Davis and Tomlinson; Useless Inlet, Shark Bay, Western Australia; J. S. Davis, viii and ix.70; all parts.

SPECIES REPORTED ON IN THE LITERATURE—RUPPIOIDEAE

Chrysler (1907): *Ruppia maritima*: course of vbs in stem.
Davis and Tomlinson (1974): *R. tuberosa*: morphology.
Donà Delle Rose (1946): *R. maritima*: morphology
Gamerro (1968): *R. cirrhosa*: flower and fruit anatomy.
Graves (1908): *R. maritima*: all parts.
Gravis *et al.* (1943): *R. maritima*: seedling.
Hagström (1911): *R. maritima*, *R. obtusa*, *R. spiralis*: stem, leaf.
Hisinger (1887): *R. rostellata*: pathological anatomy.
Irmisch (1851): *R. rostellata*: morphology.
Kirchner *et al.* (1908): *R. maritima*: morphology.
Lüpnitz (1969): *R. maritima*: primary root.
Luther (1947): *R. brachypus*, *R. rostellata*, *R. spiralis*: leaf.
Ly Thi Ba *et al.* (1973): *R. maritima*: embryo development.
Miki (1935a): *R. truncatifolia*: stem, leaf.
Monoyer (1927): *R. maritima*: stem.
Ogden (1974): *R. maritima*: leaf, stem.
Pettitt and Jermy (1975): *R. maritima*: pollen structure.

Philip (1936): *R. rostellata*: stem.
Posluszny and Sattler (1974*b*): *R. maritima* var. *maritima*: floral development.
Pottier (1934): *R. maritima*: developmental anatomy of leaf, stem, and root.
Raunkiaer (1895–9): *R. brachypus, R. rostellata, R. spiralis*: leaf; *R. brachypus*: fruit.
Şerbănescu-Jitariu (1974): *R. maritima*: floral morphology.
Singh (1964): *R. maritima*: nodal anatomy.
Singh (1965*a*): *R. maritima*: floral anatomy.
van Tieghem and Douliot (1888): *R. maritima*: developmental anatomy of root.
Uhl (1947): *R. maritima*: floral anatomy.
Wille (1882): *R. rostellata*: seedling development.
Yamashita (1972): *R. maritima*: embryo development and morphology.

SIGNIFICANT LITERATURE—RUPPIOIDEAE

Arzt (1937); Ascherson and Graebner (1907); Aston (1973); Blass (1890); Chrysler (1907); Čelakovský (1896); Davis and Tomlinson (1974); Donà Delle Rose (1946); Gamerro (1968); Graves (1908); Gravis *et al.* (1943); Hagström (1911); Hisinger (1887); Hutchinson (1959); Irmisch (1851); Kirchner *et al.* (1908); Lüpnitz (1969); Luther (1947); Lyr and Streitberg (1955); Ly Thi Ba *et al.* (1973); Markgraf (1936); Mason (1967); Mayer (1971); McCann (1945, 1978); Miki (1935*a*, 1937); Monoyer (1927); Murbeck (1902); Ogden (1974); Ostenfeld (1915, 1916); Philip (1936); Posluszny and Sattler (1974*b*); Pottier (1934); Raunkiaer (1895–9); Reese (1962, 1963); Roze (1894); Sauvageau (1891*a*); Schwanitz (1967); Şerbănescu-Jitariu (1974); Setchell (1946); Singh (1964, 1965*a*); Solereder and Meyer (1933); Sutton (1919); van Tieghem and Douliot (1888); Uhl (1947); Verhoeven (1979); Wille (1882); Yamashita (1972).

POTAMOGETONACEAE—POTAMOGETONOIDEAE

GROENLANDIA J. Gay
(Figs. 9.4–9.5)

A monotypic genus (*G. densa* Fourr. = *Potamogeton densus* L.) widely distributed in Europe, western Asia, and north Africa.

VEGETATIVE MORPHOLOGY

Habit

Probably annual and not forming specialized overwintering organs. Distal branching below the inflorescence in a single plane, less commonly proximally from occasional axillary buds.

Leaves in subopposite pairs (Fig. 9.4 A–C), successive pairs separated from each other by a distinct internode. Individual leaves lacking a sheath and petiole. Leaf base usually at most slightly auriculate, but the pair of leaves (inflorescence leaves) immediately below the peduncle with pronounced basal lobes (stipular auricles; Fig. 9.4 C). Prophyll at base of branch short, without vascular tissue.

Scales on rhizome sometimes distinctly auriculate. Rhizomes creeping, established early by the seedling axis, with the same sympodial construction as *Potamogeton* spp, i.e. with a solitary scale between each pair of branch-associated scales (Raunkiaer 1895–9).

VEGETATIVE ANATOMY

Foliage leaf

Lamina lanceolate with 5–7 veins at its widest part (Fig. 9.5 C, D), but the number of veins reduced to three at the leaf base, only three veins entering the stem. **Hairs** absent except for marginal unicellular lignified teeth (Fig. 9.5 A). **Squamules** always two at each node (Fig. 9.4 D). **Apical pore** well developed, the pore formed by the breakdown of the epidermis, and continuous with a group of tracheids situated at the end of the median vein. **Epidermis** uniform, except for occasional stomata associated with the apical pore (Fig. 9.5 B), cells polygonal; epidermal as well as mesophyll cells including chloroplasts.

Mesophyll uniseriate between the veins (Fig. 9.5 E), thickened only in the midrib region and around the veins. **Midrib** (Fig. 9.5 C, G) with a well developed lacunar system diminishing distally; vein with a distinct xylem lacuna, often including the remains of tracheal elements; phloem associated with a few abaxial and occasional adaxial thin-walled fibres. Endodermis absent. Lateral veins similar but smaller, without lacunae (Fig. 9.5 D). Marginal **fibrous strand** present (Fig. 9.5 F). Lacunae in the midrib said by Sauvageau (1890c) to be initiated from the uniseriate mesophyll layer by oblique divisions, unlike *Potamogeton*.

Prophyll

Delicate, reduced to two surface layers, including 1–4 fibrous strands of unequal length.

Rhizome scales

Colourless. Margins overwrapping at base; apical pore absent. Reduced to two surface layers except in the region of the midrib; including several fibrous strands with adaxial phloem, but scales sometimes with a mesophyll layer and resembling the base of a foliage leaf.

Stem (Fig. 9.5 H–J)

Hypodermis together with hypodermal and cortical fibrous strands absent (Fig. 9.5 I). Endodermis remaining thin-walled. Stele (Fig. 9.5 J) with four vbs, the two median strands having a common xylem cavity.

Peduncle

Hypodermis uniseriate, vascular system represented by four separate vbs without a common endodermis.

Root (Fig. 9.5 K–M)

Unbranched. Described by Sauvageau (1889c) and Schenck (1886b) as reduced, with a single axile tracheal element. More detailed description by Lüpnitz (1969). **Epidermis** large-celled, early differentiated into short cells with large nuclei (root-hair initials) and longer cells with small nuclei (not developing root-hairs). Exodermis uni- or biseriate (Fig. 9.5 M) cells slightly thick-walled, differentiated into long suberized cells and groups of short non-suberized (transmission) cells, situated 2–4 together immediately below each trichoblast. **Cortex:** outermost cortical layer, immediately within the exodermis (derived from the same subepidermal series of initials as the exodermis), compact, small-celled. Remaining cortical cells uniform, and interspersed by well developed intercellular spaces, the innermost cell layers being in very obvious concentric and radiating series. **Endodermis** thin-walled. **Stele** narrow (Fig. 9.5 K, L). Pericycle uniseriate, including 5–6 sieve-tubes with companion cells cut off on inner side. Xylem represented by 1–2 central thin-walled tracheal elements (Fig. 9.5 L).

Reproductive Morphology

Flowers (Fig. 9.4 E–J) two per spike, usually tetramerous (Fig. 9.4 F, G), but showing frequent trimery (Posluszny and Sattler 1973). Differing from *Potamogeton* in the larger hooded inflated tepals, initially completely enclosing other floral parts (Fig. 9.4 E). Stamen connective with a distal extension (Fig. 9.4 I). Carpel with a pronounced stylar outgrowth, partly covered by the peltate stigma (Fig. 9.4 J). Seed (Fig. 9.4 L) without sclerotic layers, germinating readily. Cotyledon of embryo (Fig. 9.4 M) with 3–4 spiral coils (cf. Zannichelliaceae), the embryo as a whole surrounded as a unit by the testa, not as in *Potamogeton* with the curved embryo having the testa closely applied to its surface throughout.

[In their study of **floral development,** Posluszny and Sattler (1973) note differences which set *Groenlandia* apart from *Potamogeton richardsonii* studied as a representative of *Potamogeton*: bracts absent below flowers, median tepal initiated first; lateral stamens initiated as two separate primordia; little up-growth between tepals and stamens so that they are not obviously conjoined; terminal outgrowth of stamen connective; stylar outgrowth; stamen vascular strand undivided in the stamen connective. However, these authors note that *P. crispus,* as studied by Hegelmaier (1870), was either intermediate in a number of these features, or more closely resembled *Groenlandia.*]

Reproductive Anatomy

(Posluszny and Sattler 1973.) Each flower supplied by a single vb, branching to give a single trace to each tepal, the same traces in turn each supplying the stamen trace. Carpel trace likewise single but dividing to give a branch to the ovule and a dorsal vb continuous into the stigmatic region. Remains of the vascular system visible in the sterile extension of the receptacle, above the insertion of the carpels.

Taxonomic Notes

This genus is evidently very distinct in its leaf morphology and anatomy and especially in the structure of its fruit, and it has been regarded as a link between *Potamogeton* and Zannichelliaceae (*s.s.*). It is described by Hagström (1916) as an isolated section (V. Laterales) within *Eupotamogeton* and is often regarded as a separate genus (*Groenlandia* Gay.). There is nothing strikingly unusual in its anatomy to emphasize its morphological distinctiveness, except for the persistently thin-walled stem endodermis. The stele corresponds to the reduced four-bundle type of *Potamogeton*, although Monoyer (1972) considers it rather distinctive among species with similar stem anatomy in having six distinct phloem strands around the central lacuna. Other authors (e.g. Hagström, Raunkiaer) do not describe this feature.

Posluszny and Sattler (1973) consider that their developmental evidence supports the systematic discreteness of this taxon, and discuss the different kinds of evidence used in interpreting the 'flower' in it and Potamogetonaceae generally, making use of the concept of 'partial homology.'

Material Examined

Groenlandia densa Fourr.; Cultivated Munich Botanic Garden; P. Leins (via Professor R. Sattler); all parts.

Species Reported on in the Literature

This information is listed under *Potamogeton* (p. 303) as *P. densus* in bold type.

POTAMOGETON L.
(Figs. 9.6–9.13)

Introductory Comments

Perhaps no other genus of monocotyledons has been examined anatomically so intensively as this, largely because it has been demonstrated that features of microscopic anatomy are often as useful as morphological ones in identifying sterile specimens and diagnosing species (see Taxonomic Notes, p. 301). The following account does not attempt to add any large body of new information to that which is available (though widely scattered) in the literature. The object here has been to bring together as much as possible of the existing anatomical and morphological information about the genus and to assess it somewhat critically. The detailed examination of several species (see Material Examined) has provided a background to this assessment. Quite obviously anyone who has not dealt extensively with the taxonomy of the genus and who has not considerable field experience with it cannot be relied upon for instant solutions to the enormous number of problems which still exist in classifying Potamogetons. For this reason the present account must be accepted as an anatomical overview of *Potamogeton* rather than a systematic survey, but one which provides some insight into those features which are diagnostically useful.

VEGETATIVE MORPHOLOGY

Habit

Leaves distichous (or spirodistichous). Seedlings with a long hypocotyl having a basal 'foot' (orbis of certain authors, e.g. Lüpnitz 1969) bearing root-hairs. Seedling leaves always alternate (never sub-opposite as in *Groenlandia*), with a narrow blade and tubular sheathing base, the blade and sheath not being free as in adult leaves of many spp; seedling leaves showing a gradual to abrupt transition to the adult foliage. Seedling normally developing a rhizomatous (horizontal) shoot system at an early stage from buds in axils of seedling leaves.

Rhizome (Figs. 9.6 B, C and 9.7 A–E)

Sympodial and very uniform in construction throughout the genus. Each rhizome segment of the sympodium made up of the first two internodes of a renewal shoot, the remainder of the shoot growing erect (Fig. 9.7 A, B). Each branch arising laterally on the previous segment and consisting of: (i) a scale (1) (= prophyll), with a long subprophyllar internode and (ii) a pair of subopposite scales (2 and 3) separated from the prophyll by a second long internode. Prophyll (Fig. 9.7 F) not subtending a bud, but the bud in the axil of scale 2 developing into a regenerative renewal shoot and repeating the construction of the parent axis as the next segment of the sympodium (Fig. 9.7 C–E). Bud in axil of scale 3 either remaining dormant (as a 'reserve bud'), or growing out as a proliferative branch of the rhizome sympodium and repeating its construction. Axis beyond scale 3 developing into an erect shoot bearing a series of leaves transitional from scales to typical adult foliage (Fig. 9.7 A). Length of the transition region dependent on the depth at which the rhizome rooted. Rhizome appearing to consist of alternate single and paired scales, separated by a long internode (Fig. 9.6 B, C). Single scales not associated with branches, but the paired scales associated either with an erect shoot alone or an erect shoot plus rhizome branch. Shoot system elaborated by continued proliferation of the rhizome in this simple way during a growing season.

Erect shoots

With an axillary bud at each node (rarely with multiple buds); buds mostly remaining undeveloped. General habit and appearance of the plant determined by the frequency and distribution of the branches developed from these axillary buds, i.e. the relative proportions of basal and apical branching.

[According to Hagström (1916) differences in habit are diagnostically useful. Branches of erect stems always have a basal scale-like prophyll and they sometimes bear leaves which are transitional between prophylls and the adult foliage.

Erect shoots terminate in a subopposite pair of foliage leaves (inflorescence leaves or '*folia floralia*') and an inflorescence. Renewal shoots arise from the axils of inflorescence leaves. When only one renewal shoot develops, the sympodium remains unbranched (Fig. 9.6 A, F). In some species (e.g. *P. obtusifolius* according to Irmisch 1858*b*), the buds in the axils of the two inflorescence leaves are not alike, one having two, the other only one, basal scale leaves. Additional inflorescences have been described by some authors as arising on short to very short distal branches, their distribution adding to the characteristic habit of the species.

'Land forms' of certain species of *Potamogeton* (e.g. *P. lucens* and *P. natans*) are sometimes produced. This happens for example to submerged forms of *Potamogeton* that chance to grow in a pond that dries up. In this event the plants produce erect shoots with short internodes and the leaves assume a tufted appearance. The plants can then easily be mistaken for terrestrial, rhizomatous monocotyledons.]

Roots

Always restricted to the nodes, either on seedling axes or on rhizome segments, but also common on the lower nodes of erect shoots. Roots **unbranched;** produced singly or more usually in whorls of up to ten.

Leaf

Morphology very diverse, even on a single individual. Certain spp heterophyllous, the most pronounced contrast being between submerged (basal) leaves and floating (distal) leaves. [Anderson (1978) showed that abscisic acid promotes the development of the floating leaf in germinating tubers of *P. nodosus*.]

The following is a brief outline of recorded leaf types:

(i) **Sheath and blade attached.** Leaf blade linear with a tubular sheathing base, as in the cotyledon, seedling leaves or basal leaves (e.g. Fig. 9.6 B) of many spp. This type of leaf persisting in the adult plant in certain groups of spp. with wholly submerged leaves (floating leaves never developed).

(ii) **Sheath and blade detached.** In majority of *Potamogeton* spp, sheath remaining free from the blade as a conspicuous organ variously described as a 'ligule,' 'stipule,' or 'stipular sheath' (see Morphology on p. 300); described here as the ligular sheath (e.g. Fig. 9.6 D).

Leaf blade then either: (a) narrow, linear, and without a distinct petiole, or (b) broad, lanceolate, broadly elliptic to ovate with a distinct thickened midrib, and often also with a distinct petiole (Fig. 9.6 D, F). This last, most elaborate form usually found in the floating leaves of certain spp, the disarticulation between sheath, petiole, and blade then being regarded as mechanically useful. Leaf blade in certain spp (e.g. *P. perfoliatus*) sessile, more or less amplexicaul.

Leaf sheath either open on the ventral side, or forming a closed tube, plicate or not on the ventral side, these differences often providing important diagnostic features for distinguishing sections of the genus. [Whether a sheath is open or closed is best established in young leaves since a sheath which is initially closed may become ruptured and open with age.] Ligular sheath where free from the blade, often providing diagnostically useful features. In certain spp the sheath shredding into its constituent fibrous strands with age, in a diagnostically useful manner, as in *P. perfoliatus* (Fig. 9.6 E), *P. porteri.*

Leaf vernation of three main types (Fig. 2, p. 50), related to leaf shape and useful in distinguishing certain sections, as indicated by Raunkiaer (1895– 9, 1903).

(i) *Adplicate*: leaves linear, narrow, without a broad blade, flat and not rolled in the bud (e.g. *P. obtusifolius*).

(ii) *Convolute*: leaves with a more or less broad blade, the margins rolled in the bud, one half of the blade overwrapping the other (e.g. *P. perfoliatus*, *P. praelongus*).

(iii) *Involute*: leaves with a well developed blade, each half of the blade rolled separately above the midrib in bud (e.g. *P. natans*, *P. polygonifolius*).

Vegetative proliferation and overwintering

Specialized branches variously described as turions, hibernaculae, or winter-buds are developed in many species of *Potamogeton*. They serve as overwintering shoots and/or propagules. They may appear on seedlings soon after germination (e.g. Muenscher 1936), but on adult plants they are formed either with the onset of winter or under other adverse conditions. The physiology of their production has never been investigated closely. They are known mainly in plants from temperate regions of North America and Europe. The type of overwintering organ is quite constant for a particular species and may serve as a useful diagnostic feature. The following summarizes the main types involved in overwintering or propagation (after Irmisch 1859; Sauvageau 1894; Raunkiaer 1903).

(i) **Vegetative parts not overwintering.** Here overwintering and propagation are entirely by seed. This is not a regular feature in any species of *Potamogeton*, but *Groenlandia*, which behaves essentially as an annual, is conspicuously different in this respect.

(ii) **Vegetative parts overwintering but not morphologically specialized.** This type is exemplified by *P. natans* in which the rhizomes persist through the winter without obvious morphological change. Erect shoots may or may not die back. Vegetative propagation is by isolation of branches when older parts of the plant die and become decomposed.

(iii) **Winter-buds developed on rhizomes.** Winter-buds are short, sometimes having the form of segments of the rhizome modified in various ways, and the leaves are always reduced. They may be restricted to the rhizome itself or arise as short lateral branches at the bases of the erect shoots (e.g. *P. alpinus*). These may serve as organs of propagation if they become detached.

Two main sub-groups may be recognized (after Raunkiaer (1895–9):

(a) *Coleogeton*-type (*P. filiformis*, *P. pectinatus*): stem swollen, tuber-like, the renewal shoot and parent shoot partly adnate (cf. Singh 1964).

(b) *Praelongus*-type (*P. alpinus*, *P. perfoliatus*, *P. praelongus* and other spp). Stem not tuber-like, but individual internodes enlarged to give a beaded appearance (e.g. *P. lucens*, Monoyer 1927).

(iv) **Winter-buds developed on aerial shoots.** (Sometimes additional buds may be developed on rhizome) In this type the rhizome may die at the end of the growing season so that the winter-buds, released in large numbers, may be the chief perennating as well as propagating organs. Two main types are recognized (after Raunkiaer (1895–9):

(a) *Pusillus*-type (*P. acutifolius*, *P. gemmiparus*, *P. obtusifolius*, and many other linear-leaved spp). The winter-buds are short lateral branches with short internodes, the leaves are reduced to their sheathing parts. The rhizome in this type does not normally persist. Sauvageau (1894) makes the important observation that, in this type, regrowth of the winter-bud is from the terminal bud which is only temporarily arrested during the overwintering period (cf. type (b) below).

(b) *Crispus*-type. The winter-buds are short lateral branches bearing leaves with reduced, rigid toothed blades. The rigidity is due to the presence of abundant starch rather than of sclerenchyma. The winter-bud varies in its appearance, even on a single individual, from short stout shoots with conspicuous leaves, to slender shoots with spine-like leaves. Sauvageau notes that regrowth from these shoots is always from a lateral branch so that growth of the shoot apex is permanently arrested. Regrowth may occur at any time of year without rest, so that the shoot becomes primarily a dispersal unit.

Modifications of this type involve the presence of enveloping scales which are equivalent to bladeless sheaths (*P. rufescens*). They may form only on underground rhizomes (*P. fluitans*).

VEGETATIVE ANATOMY
(Fig. 9.8)

Leaf blade

Varying greatly, even on a single individual. [The following account is very generalized, the most complete account is by Sauvageau (1891*a*).] **Squamules** as few as two per node in small linear-leaved spp (e.g. *P. pectinatus*) but usually numerous and forming an extended nodal palisade (e.g. Fig. 9.7 C–E).

Hairs absent except for minute marginal teeth, abundant in certain spp; the teeth mostly consisting of a single enlarged marginal cell; cells large and fairly conspicuous in *P. crispus*.

Apical pore present in all spp and arising in mature leaves by disorganization of apical epidermal cells, thus exposing group of tracheal elements terminating median vein. In some spp median vein somewhat recurved so that pore more or less abaxial. In some spp marginal fibrous strands extending as far as apical pore and providing something of a sheath for the apical tracheids.

Stomata present on upper surface of floating leaves and sometimes on distal submerged leaves, even of spp where floating leaves never develop.

Stomata paracytic (Fig. 9.8 B). **Epidermis** of uniform cells, thin-walled or with slightly thickened outer wall; cuticle thin except on floating leaves. Cells longitudinally extended or irregular and usually in distinct files (Fig. 9.8 B, C, K, L), chlorenchymatous in both submerged and floating leaves.

Among the great variety of leaf blades developed by *Potamogeton*, three main types may be distinguished (after Raunkiaer 1895–9) and examples of these are briefly described below. They are illustrated in Fig. 9.8. Diagnostic features which distinguish species with the same leaf types are mentioned.

(i) *Submerged, narrow, thick, rounded* (e.g. *P. pectinatus*) (Fig. 9.8 O, P). Distinguished by absence of a marginal wing, the blade more or less equivalent to the midrib region of the broader types of leaf (ii) and (iii). Veins 3 (–5). Mesophyll wholly lacunose, lacunae in 1–2 series between the veins and often with a series of smaller lacunae above and below the median vein (sometimes also the larger lateral veins). Number and arrangement of veins varying considerably according to width and thickness of leaf (cf. Fig. 9.8 J).

(ii) *Submerged, broad* (e.g. *P. lucens, P. epihydrus*) (Fig. 9.8 F–I). Corresponding to above but with a narrow to broad marginal wing, the lacunose midrib region broad and gradually merging with wing (e.g. *P. acutifolius*), or wing more abruptly delimited from midrib region (e.g. *P. pusillus*). Mesophyll of wing usually uniseriate between the lateral nerves (cf. Fig. 9.8 M).

(iii) *Floating, broad* (e.g. *P. natans*) (Fig. 9.8 A–E). Leaf with a thickened midrib and thick wing, often with a well developed petiole. Mesophyll of wing multiseriate with several adaxial palisade layers of anticlinally extended cells, compact except for deep substomatal chambers. Abaxial mesophyll with a reticulum of uniseriate partitions separating the lacunae.

Petiole of floating leaves thick, terete and with numerous lacunae separated by uniseriate partitions. Veins 3–7, more or less equidistant from each surface.

Apart from the general topography of leaves as outlined above there are additional features which provide systematically useful information. **Hypodermis** in lacunose part of leaf present or absent. Sauvageau (1891*a*) distinguished these two types developmentally according to the plane of first longitudinal divisions in the uniseriate mesophyll of the leaf primordium as follows:

(i) First divisions oblique, cutting off future hypodermal cells; lacunae developing subsequently and therefore not in direct contact with epidermis.

(ii) First divisions periclinal after lacunae initiated by enlargement of intercellular spaces; lacunae in direct contact with epidermis.

Fibrous strands either in midrib region and then usually at points of junction of longitudinal partitions with epidermis (Fig. 9.8 N); or in wing and more or less uniformly distributed between veins (Fig. 9.8 J, M). Strands formed by repeated longitudinal division of a single file of meosphyll cells. Marginal strand usually present and varying from narrow and inconspicuous to wide and forming a conspicuous rib; but never continuous into leaf sheath.

Veins as few as three in narrow linear leaves (midrib and two submarginal lateral veins), to many in broad leaves, the lateral veins all united apically with median vein; veins united basally so that usually only three enter stem centre except in wide leaves of larger spp with 5–7 veins continuous into central cylinder. Veins always collateral, sometimes partly or more or less completely sheathed by thick-walled cells (Fig. 9.8 I, N). Xylem lacuna well developed, in certain spp made up by enlargement of cells around proto- as well as metaxylem (according to Sauvageau 1891a). Phloem well developed with conspicuous sieve-tubes.

Transverse commissures frequently connecting longitudinal veins, the spacing affording useful diagnostic characters; in lacunose midrib region transverse veins restricted to transverse diaphragms of short, densely cytoplasmic cells.

Prophyll

Usually colourless and delicate, largely biseriate and composed solely of two epidermal layers. Vascular tissue reduced or veins largely represented by delicate fibrous strands, sometimes including narrow thin-walled cells as vestiges of conducting elements.

Rhizome scales

Usually colourless. Equivalent morphologically to bladeless sheaths but usually with several (2–5) mesophyll layers without well developed air-lacunae. Modified scales of rhizomatous winter-buds often including abundant starch. Vascular tissue reduced, usually represented by a few reduced veins; fibrous strands usually absent.

Ligular sheath

[The morphology and anatomy of the ligular sheath is of considerable diagnostic value; the following account is largely taken from that of Glück (1901), with anatomical information added by Sauvageau (1891a).]
Anatomically sheaths of two main kinds:

(i) Uniformly two-layered, consisting of epidermal layers without intervening mesophyll, the ligule then either wholly without vascular or fibrous tissue (e.g. *P. pauciflorus, P. pusillus, P. trichoides*) or bicarinate, each keel with a conspicuous vascular bundle together with numerous fibrous strands (e.g. *P. acutifolius, P. crispus, P. robbinsii*). Fibrous strands shown by Sauvageau to originate by division within the cells of the ventral (i.e. adaxial) epidermal layer.

(ii) Dorsally 3–5 layered and conspicuously bicarinate, the ventral wings 1–2 layered (e.g. *P. lucens, P. natans*).

Vascular system variously developed; either (i) absent, as in *P. pusillus*, the mechanical tissue represented by numerous strands of narrow fibres; (ii) represented by four vbs, the two larger of them occupying the keel, as in *P. obtusifolius*; (iii) represented by ten or more vbs, the two larger of them being situated in the keels and provided with well developed xylem. Keel bundles shown by Colomb (1887) and subsequently by numerous authors (e.g. Sauvageau 1891a; Chrysler 1907; Monoyer 1927) to be continuous basally with the larger lateral bundles of the blade which enter the stele.

Fibrous bundles sometimes absent (e.g. *P. crispus, P. fluitans, P. lucens*) or well developed (e.g. *P. obtusifolius, P. pusillus*). Fibrous bundles forming two more or less distinct systems, one ventral, in the wings of the sheath, the other dorsal and often anastomosing with a similar system in the adaxial side of the sheath. These two systems also frequently distinguished by their different behaviour at the node. Sauvageau distinguished two systems in the sheath, one anastomosing and with constituent strands including narrow, phloem-like cells, the other non-anastomosing and without such narrow cells. The former system perhaps to be regarded as a vestigial vascular plexus.

STEM
(Figs. 9.9–9.12)

[Anatomical characters of the stem are important diagnostically as was first established by Raunkiaer (1895–9, 1903) and subsequently elaborated by Hagström (1916) and Ogden (1943) (but see discussion under Taxonomic Notes). This does not overlook the observations of many authors that there is some variation throughout an individual, the most extreme differences being found between peduncle, rhizome, and turion anatomy. These are described separately later (p. 296). A standard level for comparison is provided by sections taken at the middle of the second or third internode below the spike of a flowering specimen and the following description, in general, refers to sections at this level (Figs. 9.9 and 9.10). Taxonomic writers who have used anatomy of the stem in diagnoses have all emphasized the surprising constancy of structural features in a genus otherwise noted for its polymorphism.

The following general account describes the stem of *P. praelongus*, one of the larger species, which has a relatively elaborate construction. Ways in which other species, of more simple construction, differ are described later.]

Stem of P. cf. praelongus (Fig. 9.9 A)

Hairs absent. Epidermis uniform, outer wall cutinized and somewhat thickened; epidermal cells in surface view more or less rectangular, longitudinally extended. **Hypodermis** 1–3 layered (Fig. 9.9 C). **Cortex** very wide, more or less uniformly lacunose (Fig. 9.9 D) with about five irregular series of lacunae separated by a reticulum of uniseriate longitudinal partitions becoming continuous at the nodes. Innermost 3–4 layers of cortex compact, small-celled. Cortical fibrous strands abundant within the angles at the junctions between longitudinal partitions and the epidermis, usually consisting wholly of narrow, thick-walled fibres with very narrow cell lumina. Larger strands in the middle region of the cortex including distinct but reduced vascular tissues (Fig. 9.9 E), often consisting of phloem alone. Fibrous strands always absent from the inner cortex.

Stele (central cylinder) relatively narrow, usually more or less rectangular in TS, the long axis at right angles to the plane of leaf insertion; outline often distinctly fluted (Fig. 9.9 B). **Endodermis** uniseriate, cell walls conspicuously but unevenly thickened at maturity, the cells appearing U-shaped in TS (Fig. 9.9 F). Pericycle 1–2 layered, indistinct but with frequent thick-walled cells. Conjunctive tissue largely thin-walled except for occasional thick-walled cells next to vbs. Vascular tissues represented by several (usually about 14) discrete vbs, each with a distinct phloem strand including wide sieve-tubes. Xylem usually represented by a wide xylem lacuna sometimes including the remnants of protoxylem elements (Fig. 9.9 B, F). Arrangement of vbs regular and related to three-dimensional distribution as described below, but essentially consisting of four median bundles, one on one side of the stele, three on the other, all in the plane of insertion of the leaves, together with a series of lateral bundles, 3–5 on each side of this plane.

Variation of stem anatomy within Potamogeton

[Features in which other species differ from the above example, and which are often diagnostically useful, involve cortical, endodermal and stelar tissues (e.g. Fig. 9.10). In general *P. praelongus* may be described as an elaborate 'type' to which most other species of *Potamogeton* can be referred as 'reductions.' This is discussed below.]

The following indicates the range of variation in microscopical features which appear to be diagnostically useful:

(i) **Epidermis and cortex.**

Epidermis: cuticular ridges or striations noted by Hagström (1916) in certain species. Outline shape and particularly the length of epidermal cells in surface view. Precise information lacking concerning the range of variation in epidermal cell length in a single individual or even in one internode.

Hypodermis: whether present (Fig. 9.10 E) or absent (Fig. 9.10 B) and, if present, whether one or more (up to four) layers thick. [This layer is described as a 'pseudohypodermis' by some taxonomic authors (e.g. Hagström, Ogden) in order to distinguish it from a 'hypodermis' formed of the inner layers of a multiseriate epidermis. Developmentally the stem epidermis in *Potamogeton* is uniseriate and for descriptive purposes the distinction which has been made is not necessary.]

Fibrous strands: (a) hypodermal strands either present (Fig. 9.10 B) or not (Fig. 9.10 E); (b) remaining cortical strands (interlacunar strands of earlier authors) either present or not (Fig. 9.9 A, D). [The number of strands and their distribution vary somewhat in different parts of an individual plant because the presence of cortical strands is related to some extent to the degree of development of the ligular sheath. At a standard comparative level, the feature appears to be constant.]

(ii) **Endodermis.** Consistently uniseriate but varying in ways summarized in the following key, the terminology used by Hagström being included in parenthesis:

A. Endodermis uniform, all cells alike ('uniform endodermis'),
    (1) Cells uniformly thin-walled ('non-stratified o-endodermis').
    (2) Cells uniformly thick-walled ('stratified o-endodermis') (e.g. Fig. 9.10 C, F).
    (3) Cells unequally thick-walled, U-shaped in TS ('U-endodermis'), and then either
        (a) cells ± isodiametric in TS ('common U-type') (e.g. Fig. 9.9 F).
        (b) cells laterally compressed ('gramineous-type').
B. Endodermis non-uniform, of two kinds of cells ('mixed endodermis'),
    (4) 'o-u endodermis,' i.e. an endodermis with predominantly U-shaped

cells except for either o-cells or relatively thin-walled U-shaped cells at intervals.
(5) 'o-o endodermis', i.e. an endodermis with predominantly o-shaped cells, but cell shape varying.

[Hagström notes that the type of endodermis can vary somewhat in different parts of a single internode.]

(iii) **Stele.** The elaborate arrangement of the vascular bundles in the larger species of *Potamogeton* is regarded by those authors who have studied their anatomy extensively (e.g. Chrysler, Hagström, Monoyer, Ogden, Raunkiaer, Schenck) as a 'prototype' from which the stelar arrangement of other spp can be derived by 'reduction,' the implication being that the reduction is a phylogenetic one. Whereas it is very convenient for descriptive purposes to compare the stelar anatomy of different species of *Potamogeton* in this way, it is misleading to emphasize the phylogenetic possibilities in the absence of other evidence, particularly that from stem development. This should be borne in mind when the subsequent account is read.

Information has been presented previously in the form of simplified TS diagrams (e.g. Hagström, Monoyer) and this has allowed the recognition of a number of 'types'. While this is a necessary and useful way of presenting information it tends to obscure what might be important differences between species which have the same type of stele. Since there is a more or less continuous range of variation, there may be difficulty in including certain species within one of a restricted range of types. This introduces a certain subjectivity into descriptive work.

For purposes of description the following types are used, corresponding essentially to those of Hagström and Ogden, but without using their nomenclature. Instead they are described simply by the approximate total number of vascular bundles (or apparent vascular bundles) visible in transverse section. Since vascular bundles are often represented by a phloem strand alone it may not always be possible to distinguish between a single strand and two closely adjacent but separate strands.

A. **Lateral bundles more than one on each side**

(i) *14-bundle type* ('Proto-type' of Ogden) (Fig. 9.11 A). This corresponds to the arrangement described above (p. 291) for *P. praelongus.* All vascular bundles may remain discrete. Additional lateral bundles may increase the total number to more than 14. The two small central bundles (not in contact with the endodermis) may be amphiphloic *(P. praelongus)* or more usually collateral (e.g. *P. linguatus*).

(ii) *Ten-bundle type* ('trio-type' of Ogden) (Fig. 9.11 B, C). This is recognized by the presence of a median bundle consisting of three separate phloem strands surrounding a central lacuna (the 'trio-bundle'). The trio bundle is interpreted as three separate vascular bundles which have become 'fused;' the opposed bundle is simple. There are usually three lateral bundles on each side, but this number is somewhat variable so that the total is not always ten. In some species the 'trio-bundle' is represented by a bicollateral bundle with two groups of phloem, one to the inside, the other to the outside of the xylem cavity (e.g. *P. nodosus*). This may be regarded as a situation arising by 'fusion' of the originally separate central phloem strands; it is described by Ogden as a distinct variant, as a 'trio-type with one patch of phloem on the inner face of the trio bundle.'

(iii) *Eight-bundle type* (Fig. 9.11 D). This is uncommon and represents a situation where the lateral vbs are still six but the median bundles are reduced to two, both of which have but one phloem strand so that a 'trio-bundle' is not obvious (e.g. *P. lucens, P. vaginatus*).

B. Lateral bundles only one on each side

(iv) *Four-bundle type* ('oblong-type' of Ogden) (Fig. 9.11 E). This is a simplification of the previous type by reduction of the lateral bundles to two, although on occasions there may be an additional lateral bundle so that the total may exceed four (e.g. *P. confervoides*).

In the more elaborate examples of this type the two median vbs have separate xylem lacunae ('oblong-type with two median bundles' of Ogden, e.g. *P. cheesemannii*). In the simpler examples the two median vbs have a common xylem lacuna (e.g. *P. obtusifolius*); this is essentially the arrangement described by Ogden as 'oblong-type with one median bundle.' Further reduction involves an arrangement whereby all vascular bundles have a common xylem lacuna so that the stele contains a central cavity surrounded by four distinct phloem strands (e.g. *P. rutilus, P. foliosus*). This type obviously includes quite a wide range and this variety may be encountered within a single species.

(v) *One-bundle type* ('one-bundle type' of Ogden) (Fig. 9.11 F, G). The stele now includes but a single xylem lacuna surrounded more or less completely by phloem in which individual strands cannot be recognized. This is the ultimate in reduction and is represented by a number of species with linear leaves and slender stems (e.g. *P. pusillus, P. pectinatus, P. trichoides, P. acutifolius*). The distinction between type (v) and the simplest examples of type (iv) may be somewhat arbitrary since it depends on the ability of an observer to recognize separate phloem strands.

Further variation in stelar anatomy, which is often very useful diagnostically, is brought about by the degree of development and distribution of sclerenchyma in the form of thick-walled conjunctive parenchyma. This may consist either of scattered cells only loosely associated with the vascular bundles, or of fibres largely associated with vascular bundles and with the appearance of irregular bundle sheaths, or of more dense masses of sclerenchyma which separate different bundles or groups of bundles from each other. In many species stelar sclerenchyma is wanting.

Course of the vascular bundles in the stem. (Largely after Chrysler, Hagström, Monoyer, Raunkiaer, Sauvageau, and Schenck; the pioneering observations of de Bary (1884) have mostly been superseded.) See summary in Fig. 9.12.

Two more or less separate systems exist: (i) stelar, (ii) cortical; the two are continuous only at the nodes via anastomosing strands which form a distinct nodal plexus. In the description of the vascular system of *Potamogeton* by earlier authors it has been assumed that the vascular bundles remain discrete at the nodes and can be traced through the nodal plexus. This has led to the description of apparently clear-cut patterns which can easily be represented by diagrams (notably by Monoyer). In nature, however, the situation may be less precise and only Chrysler makes any detailed reference to the nodal plexi. In addition, all earlier studies are apparently based on analyses of mature stems with no attempt to follow by serial sections the initiation of the vascular system as procambial strands in unextended parts of shoots.

(i) *Stelar system.* The following description refers to a rather generalized pattern found in the ten-bundle type, of which *P. natans* described by Monoyer provides a good example (Fig. 9.12 A). The stelar supply from the leaf is represented by only five bundles, one median (M), two large laterals (L) and two small laterals (m), the remainder of the leaf trace system becoming the cortical stem system. All five vascular bundles remain distinct in the first internode (hatched in Fig. 9.12 A, upper), the median forming the conspicuous simple median bundle, the two laterals interpolated between two existing laterals ('sympodial bundles,' to be described below). At the second node the median and two laterals become associated as the 'trio-bundle' and descend through a further internode in the median position, opposite the simple median bundle most recently inserted. At the next node below, the bundles anastomose and almost immediately divide to anastomose with two adjacent laterals (sympodial bundles) which have descended through two internodes in an unmodified state. The actual fate of the trio-bundle at its lower extremity does not seem very certain or constant when different accounts are compared and this is a point which has still to be examined developmentally.

On this basis the bundles visible in a TS of any one internode can be related to the trace systems of the two next highest leaves in a precise manner:

(a) Median and four laterals of leaf at next node immediately above (M_o, L_o, and m_o in Fig. 9.12 A).

(b) Median and two large laterals of leaf at second node above, forming 'trio-bundle' in internode below (M_1 and L_1 in Fig. 9.12 A).

(c) Four small bundles (S in Fig. 19.12 A). Small lateral leaf traces (m) unite with them two internodes below the level at which the leaf trace is inserted. These bundles were first recognized by de Bary (1884) on the basis of rather crude clearing techniques and he described them as 'cauline' bundles since they were considered to have no direct contact with leaves, in contrast to 'common' bundles which had such a contact. This distinction is obviously artificial. Monoyer (1927) referred to them as sympodial bundles and showed that they were continuous with leaf traces in the manner described above.

The situation is more complex than the simple description so far provided suggests, because of the presence of bud and root traces anastomosing with the nodal complex.

(a) Bud traces (dotted line in Fig. 9.12 A) have been represented as a pair of bundles, one diverging from each of the sympodial traces on the bud side of the stem, at the level where the lateral traces are inserted. Singh (1964) makes no mention of this configuration. Chrysler (1907) draws attention to the peculiar phloem tissue, with narrow thin-walled cells, in these bud traces at the node in certain *Potamogeton* species.

(b) Anastomoses between all four sympodial and lateral bundles occur at each node and these are connected by a further anastomosing reticulum with cortical bundles. Furthermore, the anatomy of traces entering the stele becomes more or less amphivasal at the node, as reported by Chrysler. Under these circumstances it can be questioned as to whether individual strands do retain their independence across the node. This problem needs further investigation.

(c) Roots on rhizomes and lower nodes are continuous via anastomosing traces with the nodal complex.

On the basis of the above outline it is possible to account three-dimensionally for the variation in stelar anatomy in *Potamogeton* which has led to the recognition of the 'types' described earlier (p. 293). Variation is the result of the following trends: (i) median and lateral bundles remaining discrete, without forming a 'trio-bundle'; (ii) additional petiolar lateral bundles entering the stele. This increases the number of bundles seen in a TS of the stele. The more common decrease (Fig. 9.12 B, C) depends on progressive elimination of sympodial bundles, fusion of bundles at the node of entry and finally aggregation of bundles (which may or may not be discrete) around the central xylem lacuna. The ultimate is reached in the one-bundle type, in which discrete strands are not identifiable in the stele. Variation in the extent of the cortical system adds further diversity; the cortical bundles may extend only a short distance in the internode (e.g. *P. robbinsii*).

Variations in stem anatomy in individual species. The above account refers to the anatomy of the distal part of erect stems of species grown in typical habitats. Several authors, from Raunkiaer onwards, have emphasized the relative constancy of anatomical features in sections from stems at standardized levels, and these provide such useful diagnostic characters. Nevertheless, both variation throughout a single individual (topographical) and between individuals of contrasted habitat (ecological) has been recorded and is outlined below.

Topographical variation.

(i) *Rhizome.* This may differ little from the erect stem although it is always terete, even when the aerial axis is flattened or bicanaliculate. Commonly, where the erect stem has cortical strands the rhizome lacks them (e.g. *P. pectinatus,* according to Singh 1964). Stelar tissues are often reduced owing to a reduction in leaf size, the branch traces then providing the major disruption of the stelar system. The endodermis may remain thin-walled.

(ii) *Turions.* Those on the underground parts are distinguished by ground tissues which become starch-filled; the lacunar system is less well developed than in erect stems, or even wholly absent. Either the cortex or the stele may be enlarged; the stelar system is often reduced in relation to the reduction in leaf size. Sauvageau (1894), Raunkiaer (1895–9), and Singh (1964) give detailed accounts of the anatomy of turions.

(iii) *Peduncle.* This differs strikingly from the vegetative stem in the presence of several (usually 4–5) separate vascular bundles which may or may not be surrounded by a common endodermis. A hypodermis may be developed even when it is absent from the stem and this may be related to the need for greater mechanical strength in an erect aerial axis. On the other hand cortical bundles may be absent even if they are present in the stem.

Information about the distribution in the peduncle of anatomical characters of diagnostic importance in distinguishing major sub-groups may be summarized as follows:

1. Vbs each with an individual endodermis; cortical bundles absent (except in *P. vaginatus*); hypodermis often absent. subgenus *Coleogeton* (Peduncle not erect; pollination at water surface.)
1A. Vbs without an endodermis, each with a well developed fibrous sheath; hypodermis always present (even if absent from vegetative stem); cortex thicker than in vegetative stem and with numerous narrow lacunae.
subgenus *Eupotamogeton*
(Peduncle erect; pollination above water surface.)

Ecological variation. Under extreme conditions the stem anatomy in *Potamogeton* may be considerably modified. An example is provided by Monoyer (1927), who illustrates a section of the stem of *P. natans* from a specimen collected at an altitude of 2700 m. This contrasts with the usual 14-bundle type of this species in having a four-bundle type stele with a common xylem lacuna, with the tracheids of the lateral bundles persistent.

Another source of variation not discussed by any previous author is that which takes place as the plant develops from a seedling.

ROOT
(Fig. 9.13)

[The following descriptive account is largely taken from Sauvageau (1889*c*), who undertook an investigation of the root anatomy of *Potamogeton* in order to demonstrate the incorrectness of the earlier belief that roots did not develop mechanical tissue (see also Schenck 1886*b*). Sclerenchyma is present either in the form of a thick-walled endodermis or a thick-walled endodermis plus sclerotic conjunctive parenchyma of the stele (e.g. Fig. 9.13 A), and even together with thickened cells of the inner cortex. In extreme examples, all stelar cells in old roots (except phloem elements) may be thick-walled. Sauvageau's observations led him to the conclusion that there was a correlation neither between root structure and taxonomic subdivision nor between root structure and degree of movement of surrounding water. The degree of sclerification varies even in roots arising at the same node, and is not correlated with root diameter although older parts of any one root are the more sclerotic.]

Epidermis thin-walled, uniform except for root-hairs arising at one end of a cell in certain roots. **Cortex** uniform, without specialized mechanical layers and not developing lacunae. **Endodermis** always uniseriate, at first thin-walled, but becoming thick-walled (wholly or in part) with age. Mature endodermal cells either all thickened, but often including passage cells (e.g. *P. rufescens*), or sometimes only those endodermal cells outside sieve-tubes becoming thickened (e.g. *P. polygonifolius*). Endodermal cells commonly with uniform wall thickening (O-shaped in TS; e.g. *P. lucens, P. pectinatus, P. plantagineus, P. polygonifolius, P. pusillus*); less commonly endodermal cells unequally thickened, appearing U-shaped in TS (e.g. *P. gramineus, P. natans*). **Stele** narrow (Fig. 9.13 A–E). **Pericycle** always uniseriate, cells either remaining thin-walled or becoming thick-walled if next to a sieve-tube. **Phloem** represented by sieve-tubes always included within pericycle, the sieve-tube usually associated with a distinct but narrow, persistently thin-walled companion cell derived from same pericyclic initial as sieve-tube. **Xylem** various, usually in wider steles represented by a fairly regular to irregular series of distinct tracheal files more or less alternating with the sieve-tubes but never in the pericycle (e.g. Fig. 9.13 B). Wider tracheal elements often associated with one or more narrower peripheral elements (e.g. *P. polygonifolius*), medulla either lacking central xylem elements (e.g. *P. polygonifolius*), or more usually with one or more central elements continuous with peripheral elements (e.g. Fig. 9.13 C). Narrower steles (e.g. in *P. gramineus, P. lucens*) with a wide central tracheal element surrounded by several smaller ones more or less alternating with sieve-tubes (e.g. Fig. 9.13 D); in narrowest steles (e.g. *P. pusillus*) xylem represented by a single wide central tracheal element (e.g. Fig. 9.13 E). [Schenck (1886*b*) describes this as an 'axile canal' in certain spp, but Sauvageau demonstrates that it has a distinct, but thin, cell wall.]

Conjunctive parenchyma often including thick-walled cells in mature roots, especially those cells next to sieve-tubes (but not companion cells); the mechanical tissue in *P. polygonifolius* further augmented by thickening of walls of inner cortical cells.

SECRETORY, STORAGE, AND CONDUCTING ELEMENTS

Oil

The epidermis of the ligular sheath and floating leaves in certain species (e.g. *P. lucens, P. praelongus*) has a shiny upper surface apparently related to the presence of large oil-drops in the epidermal (and sometimes mesophyll) cells between the veins, according to Lundström (1888). The oil appears early before the leaf matures. Lidforss (1898), who corrected certain observations of Lundström, says that the oil is in the vacuole (not the cytoplasm) and that the 'plastids' reported by Lundström to be the source of the oil are in fact calcium oxalate crystals. The oil is volatile since it disappears within three days when cut leaves are left to dry and Lidforss identified the oil as an aromatic aldehyde.

Xylem

Xylem is represented throughout the greater part of the leaf and stem by lacunae; the protoxylem tracheids sometimes persist in unextended organs, notably at the nodes in the stem. In the development of the xylem lacunae in the larger veins of the leaf in several larger species of *Potamogeton* (e.g. *P. natans*), Sauvageau (1889*c*) recorded that the lacunae are initiated by active enlargement and separation of cells surrounding the first-formed xylem tracheids (protoxylem) and not by any passive collapse of tracheal elements. At a later stage a second series of xylem elements is differentiated (metaxylem) and this leads to a further enlargement of the lacunae which are then obviously divided into two distinct regions. In other, smaller species no metaxylem is developed and the lacuna remains simple.

REPRODUCTIVE MORPHOLOGY

Inflorescences represented by terminal spikes on erect leafy shoots (Fig. 9.6 A, F), usually with a pair of sub-opposite foliage leaves at the base of the inflorescence, i.e. **inflorescence leaves,** commonly somewhat different in shape from normal adult foliage leaves and possibly important in supporting erect spikes. Inflorescence leaves each subtending a bud, one or both of which grow out as renewal shoots (with a basal prophyll) to continue the sympodium. Spikes sometimes produced sufficiently rapidly to be described as 'a panicle of spikes,' e.g. in *P. robbinsii* by Haynes (1978). Inflorescence axis (cf. p. 296) often somewhat thickened (Fig. 9.7 G). **Flowers** diffuse or congested in few (–3) whorls or spirally arranged (Fig. 9.7 G), the apex of the inflorescence ending blindly, never with a distinct terminal flower; distal flowers often somewhat imperfect. Flower usually described as ebracteate, but in some spp. subtended by (cf. *Ruppia*) an inconspicuous bract e.g. *P. richardsonii* (Posluszny and Sattler 1974*a*). Each flower (Fig. 9.7 H, J, K) consisting of four sepaloid perianth segments (tepals) opposite four stamens, the tepals at maturity appearing as outgrowths of the connective (Fig. 9.7 M, N); carpels usually four, each alternate with a tepal–stamen complex. Carpels each ovoid, with a short style and obscurely peltate stigma (Fig. 9.7 L); ovule solitary, bitegmic, more or less basal or lateral (Fig. 9.7 I). Number of parts slightly variable (e.g. sometimes in fives) but never to the same extent as in *Ruppia*.

Pollination either above the surface of the water, by insects or wind, or possibly at and even below the water surface.

Fruit (Fig. 9.7 O) an achene with soft outer layers and a hard endocarp provided with narrow sutures defining a narrow flap-like lid thrown off at germination (Lüpnitz 1969; Muenscher 1936). Fruits at first floating, often for long periods, by virtue of green spongy mesocarp, the flotation dependent on the continued photosynthesis carried out in these outer layers (i.e. in light), the seed eventually sinking because of rotting of soft outer layers. Germination inhibited by presence of seed coat tissues, as shown initially by Sauvageau (1894); more recently shown to be dependent on red–far-red interaction characteristic of the phytochrome system by Spence *et al.* (1971), providing an important experimental analysis of germination requirements in relation to the ecology of certain *Potamogeton* spp. Embryo curved, hypocotyl long, radicle distinct.

REPRODUCTIVE ANATOMY

(After Hegelmaier 1870; Uhl 1947; Sattler 1965; Posluszny and Sattler 1974*a*.)

Each flower is irrigated by a single bundle branching from the stele of the inflorescence axis. The flower trace divides to produce four bundles, each of which becomes a combined tepal–stamen trace and which forks to give a single trace to the stamen and one to the tepal. Within the broad part of the tepal, the trace branches freely without anastomosing, the main strand continues into the tip of the organ. The carpel traces are derived singly from the remainder of the flower trace; each divides below the carpel to produce a short ventral ovule trace and a longer dorsal stylar trace which extends to the stigma. The bract subtending the flower remains diminutive, inconspicuous, and without vascular tissue.

The above account is based on the description of several different species by several authors, but the vasculature seems very constant. The only significant variation is whether or not the carpel traces are derived from a temporarily reconstituted stele above the departure of the tepal–stamen traces.

TAXONOMIC AND MORPHOLOGICAL NOTES

A. Morphology

Morphology of the ligular sheath. Some difficulty exists in the comparative interpretation of the morphology of the sheathing structure associated with but free from the blade in many species of *Potamogeton*. In the juvenile stages of all Potamogetons the leaf has a tubular sheath and is easily equated with that of other monocotyledons. In many species this condition persists and the adult leaf may be described as having a sheathing base. The mouth of this may be extended as a tubular ligule, free of the blade, a condition not uncommon in other monocotyledons, e.g. Commelinaceae, Costaceae, and many Helobiae. In the remaining numerous species of *Potamogeton* there is a tubular structure in the leaf axil to greater or lesser extent free of the blade (Fig. 9.6 D), and it is reasonable to regard this as the ligular part of a sheath developed independently of the blade. This is the conclusion of Colomb and Sauvageau, and the one which is adopted here so that the organ is described as a 'ligular sheath,' or more generally and simply as the 'sheath.' Evidence for this interpretation is the frequent presence of intermediate types of leaf in which the ligule is progressively longer as the sheath becomes progressively free of the blade (as illustrated by Glück (1901) for *P. rufescens*). In addition, the keel bundles of the sheath show the same derivation from lateral traces to the blade, regardless of whether the sheath is attached or free.

Developmental study by Glück (1901) showed that the sheath arises as a single structure relatively late in the development of its associated blade as a cushion of tissue in the leaf axil. It grows to enclose the shoot within, including the bud subtended by its blade. In contrast Schalscha-Ehrenfeld (1940–1) shows that the sheath is an outgrowth of the leaf blade, with the normal position of a ligule.

In an alternative and more widely held interpretation (e.g. Haynes 1978) the organ is described as a 'stipule' or 'stipular sheath' and may even be regarded as two separate lateral stipules which are fused. The double nature of this structure is said to be indicated by the pair of prominent keel bundles which are usually present. However, there is no developmental evidence for a double structure and the keels, which are not present in all species, can be interpreted as resulting from the requirements of close packing, in much the same way as the bicarinate prophyll is developed in most groups of monocotyledons.

The further extension by Glück (1901) of this interpretation to encompass the leaf base in all monocotyledons, so that even the large woody one of palms becomes equivalent to 'two fused stipules' is quite unreal and reflects a too strongly dicotyledon-orientated bias towards the investigations of monocotyledons (see Tomlinson 1970). Paired stipules never occur in monocotyledons and only an over-strong acceptance of 'phylogenetic morphology' could ever have led to their 'discovery.'

Interpretation of morphology of reproductive parts. The 'flower' of *Potamogeton* has been subjected to a variety of interpretations on a comparative morphological basis. Discussion has centered around the nature of the tepals and the nature of the flower itself.

Opinion is divided as to whether the tepal is:

(i) an outgrowth of the stamen connective (the interpretation of numerous older authors, e.g. Irmisch 1851; Eichler 1875; Ascherson and Graebner 1907; as well as more recent ones, e.g. Markgraf 1936);

(ii) a perianth member which has become adnate to the connective (an interpretation held by others, e.g. Hegelmaier 1870; Čelakovský 1896; Graves 1908; Saunders 1937–9; Hutchinson 1959; Eames 1961). The organ itself is variously described as perianth segment, perigon segment, tepal, sepal . . . etc. depending on each author's use of these words.

Interesting light on the morphology of these organs is shed by the developmental studies of Hegelmaier (1870), Sattler (1965), Posluszny and Sattler (1974a). These authors show that the pairs of tepals and stamens originate in acropetal order as independent primordia which become 'fused' by meristematic activity at their common level of insertion in such a way that they come to have a common axis of indeterminate morphological status. Sattler points out that although this demonstrates the *developmental* independence of tepal and stamen it can throw no light on their *phylogenetic* derivation although it does render interpretation (i) untenable. Of interest is the observation that the opposite pairs of tepals originate successively, the two lateral ones first, the pairs of stamens having a similar sequence. This may be relevant to the further problem as to whether or not the individual groups of four stamens and four carpels each represents a flower. Kunth (1841) suggested that each group is an inflorescence consisting of four male flowers, each subtended by a 'bract' (the tepal) and four ebracteate female flowers, each represented by a naked carpel, the whole shoot system in the axil of a bract, where present. This view is supported by Uh (1947). Evidence for this interpretation might be that the tepals do stand opposite the stamens. A further modification of this concept is that of Miki (1935a), who interprets each of the male flowers as consisting of two fused stamens. It is possible that each stamen is, at a very early stage, represented by two separate primordia, as is suggested by Sattler, and this might be regarded as evidence for Miki's concept. Miki (1937) also suggested that the four carpels represented a single female flower surrounded by the four male flowers on the same 'simplified' inflorescence.

B. Taxonomy

Potamogeton is a notoriously difficult genus in which either to circumscribe spp or to recognize spp which have been circumscribed. There are a number of reasons (for a general summary see Taylor 1949). Firstly, the reproductive parts are very uniform and distinguishing features which use them are slight and quantitative. Secondly, species are widespread and vegetatively poly-morphic, the polymorphism being the result both of marked heterophylly and extreme ecological plasticity. There may be a range from linear, sub-merged to ovate, petiolate floating leaves all on a single shoot; the type of leaf may be further influenced by such factors as the depth at which the plants are rooted and the strength of water currents. Thirdly, numerous hybrids have been recorded. A fourth complicating factor is that specimens have been studied somewhat uncritically by amateurs. The consequent no-menclatural complexities have discouraged serious study and certainly the genus is not one which the beginner finds easy to comprehend. This is unfortu-nate because there are numerous problems of biological significance for which *Potamogeton* would provide an ideal experimental material (e.g. Spence *et al.* 1971; Anderson 1978).

There has been no cytotaxonomic survey of large population samples such as those carried out by Reese on *Ruppia* and *Zannichellia* and which have been so revealing. Cytological information is very limited (cf. Haynes 1978). Some early systematists did cultivate specimens of *Potamogeton* in what must constitute primitive transplant experiments. Pearsall (1930) has made field and experimental observations designed to determine the relative effects of light intensity, pH, and water currents on the morphology of certain *Pota-mogeton* species in the English Lake District.

Despite the voluminous literature on hybrids in this genus, no attempt seems to have been made to confirm the parentage of a hybrid by crossing experiments. Hagström, in fact, insisted that this was not necessary, since the hybrid could be identified solely by its morphologically and anatomically intermediate position between its two parents. The fact that this attitude led him to the discovery of hybrids between species with widely separated ranges, suggests that his conclusions may not be wholly reliable, as Fernald (1932) points out. Hagström primarily accepted hybrids on the basis of seed and pollen sterility, but even this may not be entirely acceptable, as suggested by Fernald for *P. robbinsii* which is a clear-cut species with low fruit set. Pearsall's observation that fruiting in *Potamogeton* may be an infrequent phenomenon dependent on exceptionally dry years is relevant to this problem. Nevertheless, that a hybrid can exist in an area of previous but not present overlap between its parental species is suggested by Dandy and Taylor (1946).

Classification of *Potamogeton* on anatomical grounds. All who have mono-graphed the genus have produced an elaborate subdivision. The first major subdivision of Kunth (1841) was amplified by Ascherson and Graebner in *Das Pflanzenreich* (1907), dealing with four sections and a large number of subsections. *Potamogeton densus* constitutes the sole species of a distinct fifth section. The characters used were largely features of leaf shape. Sauvageau (1889–94) produced much new anatomical information in a notable series of treatments which were not, however, taxonomic in scope. Raunkiaer (1895–9), on the basis of his treatment of European forms, broke new ground by using anatomical characters, mostly of the stem, in his diagnoses, since he considered these to be much more constant than many exomorphic features. He tentatively subdivided the species he treated into a number of 'groups.'

Hagström (1916) subsequently monographed the genus with a strong reliance on those anatomical characters established by Raunkiaer as diagnostically useful, largely incorporating Raunkiaer's groups. His subdivision involved an elaborate hierarchy of sections, subsections, series, and subseries within the two subgenera *Coleogeton* and *Eupotamogeton*. Although seemingly unwieldy at first sight since he provided no concise synopsis, Hagström's system has been followed (with slight additions) by subsequent monographers (e.g. Fernald 1932; Ogden 1943). Other authors have followed Raunkiaer in emphasizing features of microscopic vegetative anatomy in their diagnoses (e.g. Miki 1934*b,c*, 1935*a*).

Hagström's work was not without its critics (e.g. St. John 1925) and he has been accused of over-strong reliance on anatomy by Fernald, who in his own treatment makes no use of those microscopic diagnostic features which European authors had found so effective. Ogden (1943) takes the exact opposite view and makes enthusiastic use of anatomy. It may be significant that Fernald deals with the linear-leaved species, which may be relatively more uniform in their anatomy than the broad-leaved species, with their greater anatomical elaboration, investigated by Ogden.

Providing a synoptic key to this enormous literature has not been attempted, since it would be misleading without first-hand investigation of large numbers of reliably named examples. An initial problem would be to establish the range of variation within given taxa; comparison of different accounts of the same sp currently suggests this range is large. A better understanding of developmental processes in the construction of *Potamogeton* shoots is needed as essential background information.

MATERIAL EXAMINED (see Introductory Comments, p. 284)

Potamogeton cf. *capillaceus* Poir.; Harvard Pond, Petersham, Massachusetts; P. B. Tomlinson *s.n.*

P. cheesmanii A. Benn.; Huanua, New Zealand; P. B. Tomlinson *s.n.*

P. epihydrus Raf.; Harvard Pond, Petersham, Massachusetts; P. B. Tomlinson *s.n.* and Laurel Lake, Orange, Massachusetts; P. B. Tomlinson, 15.viii.67.

P. illinoensis Morong.; San Salvador, Bahama Islands; Gillis 8864.

P. cf. *natans* L.; No locality. Collection Jodrell Laboratory, Kew.

P. ochreatus Raoul.; Huanua, New Zealand; P. B. Tomlinson *s.n.*

P. pectinatus L.; Aldwark Bridge, Yorkshire; P. B. Tomlinson 7.vii.71 B.

P. cf. *perfoliatus* L.; Boroughbridge, Yorkshire; P. B. Tomlinson *s.n.* and Lake Champlain, New York; P. B. Tomlinson 8.vii.72 B.

P. zosteriformis Fernald; Lake Champlain, New York; P. B. Tomlinson 8.vii.72 A.

Potamogeton sp; Lake Champlain, New York; P. B. Tomlinson 8.vii.72 C

SPECIES REPORTED ON IN THE
LITERATURE—POTAMOGETONOIDEAE

(Information on *Groenlandia* has mostly been recorded as **Potamogeton densus** L.)

Anderson (1978): *Potamogeton nodosus*: leaf morphology.

Andersson (1888): *P. pectinatus*: vbs in young stem.

Andrei (1969): *P. gramineus*: leaf epidermis.

Andrei (1972): *P. gramineus*: leaf xylem.

Arber (1922*b*): *P. natans*: leaf.

Arber (1923): *P. natans, P.* sp: squamules.

Areschoug (1878): *P. natans*: leaf.

Bance (1946): *P. filiformis, P. pectinatus, P.* × *suecicus*: stem.

Borzi (1887–8): *P. crispus, P. natans*: developmental anatomy of root.

Campbell (1936): *P. pectinatus*: leaf, stem, turion.

Chauveaud (1897*b*): *P. natans*: sieve-tubes in root.

Chrysler (1907): *P. crispus, P. heterophyllus, P. hybridus, P. natans, P. nuttallii, P. pectinatus, P. pulcher, P. robbinsii*: course of vbs.

Colomb (1887): *P. natans, P. polygonifolius*: stipules.

Drawert (1938): *P. densus, P. obtusifolius, P. pectinatus*: leaf surface.

Esenbeck (1914): *P. alpinus, P. coloratus, P. fluitans, P. gramineus, P. natans, P. pectinatus, P. zizii*: morphology, leaf anatomy.

Falkenberg (1876): *P. crispus*: stem.

Fischer (1907): *P. acutifolius, P. alpinus, P. coloratus, P. crispus, P. densus, P. filiformis (= P. marinus), P. fluitans, P. gramineus, P. juncifolius, P. lucens, P. mucronatus, P. natans, P. nitens* (partly *P. perfoliatus* × *gramineus*), *P. obtusifolius, P. panormitanus, P. pectinatus, P. perfoliatus, P. polygonifolius, P. praelongus, P. pusillus, P. rutilis, P. trichoides, P. vaginatus, P. zizii, P. zosterifolius (= P. compressus)*, and the following hybrids: *P. cymbifolius (P. crispus* × *perfoliatus), P. decipiens (P. lucens* × *perfoliatus* and/or *P. lucens* × *praelongus), P. gessnacensis* (probably *P. natans* × *polygonifolius), P. noltei (P. lucens* × *natans), P. schreberi (P. fluitans* × *natans), P. spathulatus (P. alpinus* × *polygonifolius* and/ or *P. natans* × *polygonifolius), P. suecicus (P. filiformis* × *pectinatus)*: leaf, stem, and stipules.

Fontell (1908–9): (not seen; spp investigated according to Solereder and Meyer) *P. alpinus, P. compressus (= P. zosteraefolius), P. filiformis, P. gramineus, P. lucens, P. natans, P. obtusifolius, P. pectinatus* and var. *vaginatus, P. perfoliatus, P. praelongus, P. pusillus, P. zizii* and the following hybrids: *P. filiformis* × *vaginatus, P. gramineus* × *perfoliatus (= P. nitens), P. pectinatus* × *vaginatus*: vegetative anatomy.

François (1908): *P. natans, P. perfoliatus*: seedling.

Glück (1901): *Potamogeton* spp: stipules (pp. 49–51).

Graebner and Flahault (1908): *Potamogeton* spp: stem anatomy taken from other authors.

Haberlandt (1887): *P. natans*: stomata.

Hagström (1908): *P. dentatus, P. orientalis, P. parmatus, P. stylatus, P. ziziiformis*, and the following hybrids: *P. alpinus* × *perfoliatus (= P. prussicus), P. gramineus* × *nodosus (= P. argutulus)*: stem.

Hagström (1916): *P. acutifolius, P. alpinus, P. amblyophyllus, P. amplifolius, P. antaicus, P. apicalis, P. aschersonii, P. badioviridis, P. badius, P. berteroanus, P. biformis, P. bupleuroides, P. capensis, P. chamissoi, P. cheesemanii, P. coloratus, P. confervoides, P. conjugens, P. crispus, P. densus, P.*

dentatus, P. digynus, P. dimorphus, P. diversifolius, P. exiguus, P. ferrugineus, P. fibrosus, P. filiformis, P. foliosus, P. fragillimus, P. franchetii, P. fryeri, P. furcatus, P. gayii, P. gramineus, P. groenlandicus, P. hillii, P. hindostanicus, P. illinoensis, P. insulanus, P. javanicus, P. lacunatus, P. ligulatus, P. limosellifolius, P. linguatus, P. loculosus, P. lucens, P. maackianus, P. macrophylloides, P. malaianus, P. membranaceus, P. morongii, P. mucronatus, P. muricatus, P. natans, P. nodosus, P. nuttallii, P. oakesianus, P. obtusifolius, P. ochreatus, P. orientalis, P. oxyphyllus, P. pamiricus, P. panormitanus, P. pectinatus, P. perfoliatus, P. pleiophyllus, P. polygonifolius, P. polygonus, P. porrigens, P. praelongus, P. preussii, P. promontoricus, P. pseudopolygonus, P. pulcher, P. pusillus, P. quinquenervius, P. recurvatus, P. reduncus, P. repens, P. richardii, P. richardsonii, P. robbinsii, P. rostratus, P. rotundatus, P. rutilus, P. samariformis, P. sclerocarpus, P. semicoloratus, P. sessilifolius, P. spirilliformis, P. spoliatus, P. stagnorum, P. strictifolius (probably hybrid), *P. strictus, P. stylatus, P. subjavanicus, P. suboblongus, P. subretusus, P. subsibiricus, P. tepperi, P. tricarinatus, P. trichoides, P. ulei, P. vaginatus, P. vaseyi, P. zosterifolius,* and the following hybrids: *P. acutifolius* × *pusillus* (*P. sudermanicus*), *P. acutifolius* × *zosterifolius* (*P. bambergensis*), *P. alpinus* × *crispus* (*P. venustus*), *P. alpinus* × *gramineus* (*P. nericius*), *P. alpinus* × *natans* (*P. drucei*), *P. alpinus* × *praelongus* (*P. nerviger*), *P. alpinus* × *pusillus* (*P. lanceolatus*), *P. amplifolius* × *illinoensis* (*P. scoliophyllus*), *P. aschersonii* × *polygonus* (*P. attenuatus*), *P. bupleuroides* × *gramineus* (*P. subnitens*), *P. bupleuroides* × *pusillus* (*P. mysticus*), *P. coloratus* × *gramineus* (*P. billupsii*), *P. crispus* × *perfoliatus* (*P. cooperi*), *P. crispus* × *praelongus* (*P. undulatus*), *P. crispus* × *pusillus* (*P. bennettii*), *P. filiformis* × *pectinatus* (*P. suecicus*), *P. filiformis* × *vaginatus* (*P. fennicus*), *P. gramineus* × *illinoensis* (*P. deminutus*), *P. gramineus* × *lucens* (*P. zizii*), *P. gramineus* × *natans* (*P. sparganifolius*), *P. gramineus* × *perfoliatus* (*P. nitens*), *P. gramineus* × *polygonifolius* (*P. seemanii*), *P. gramineus* × *praelongus* (*P. navicularis*), *P. illinoensis* × *lucens* (*P. pseudolucens*), *P. illinoensis* × *nodosus* (*P. faxoni*), *P. lucens* × *natans* (*P. sterilis, P. fluitans*), *P. lucens* × *nodosus* (*P. subrufus*), *P. lucens* × *perfoliatus* (*P. decipiens*), *P. lucens* × *praelongus* (*P. babingtonii*), *P. natans* × *polygonifolius* (*P. gessnacensis*), *P. natans* × *trichoides* (*P. variifolius*), *P. pectinatus* × *vaginatus* (*P. bottnicus*), *P. perfoliatus* × *praelongus* (*P. cognatus*), *P. trichoides* × *zosterifolius* (*P. ripensis*): leaf and stem anatomy, with some notes on rhizomes and turions.

Kaul (1976*b*): *P. nodosus*: leaf.

Le Blanc (1912): *P. natans*: aerenchyma.

Lüpnitz (1969): *P. densus* (as *Groenlandia densa*), *P. crispus, P. lucens, P. natans, P. pectinatus, P. perfoliatus, P. trichoides*: root development.

Luhan (1957): *P. lucens*: rhizome.

Luther (1947): *P. filiformis, P. pectinatus*: leaf.

Ly Thi Ba *et al.* (1978): *P. lucens*: embryology.

Mayr (1915): *P. natans*: hydathodes.

Mer (1882): *P. rufescens*: stomata.

Mer (1886): *P. natans, P. rufescens*: stomata.

Miki (1934*b*): *P. biwaensis* (hybrid), *P. kamogawaensis* (hybrid), *P. maack-ianus*: leaf, stem.

Miki (1934*c*): *P. gramineus, P. maackianus, P. natans, P. nipponicus, P. pecti-natus, P. perfoliatus, P. praelongus, P. pusillus, P. sibericus*: leaf and stem anatomy illustrated.

Miki (1935*a*): *P. apertus* (hybrid), *P. panormitanus*: leaf, stem.

Monoyer (1927): *P. acutifolius, P. alpinus, P. compressus, P. crispus, P. densus, P. flabellatus, P. fluitans, P. gramineus, P. lucens, P. marinus, P. mucrona-tus, P. natans, P. nitens, P. obtusifolius, P. pectinatus, P. perfoliatus, P. plantagineus, P. polygonifolius, P. praelongus, P. pusillus, P. trichoides, P. zizii*: course of vbs.

Moore (1915): *P. crispus, P. pectinatus, P. robbinsii*: stem.

Murén (1934): *P. natans, P. perfoliatus*: root.

Myaemets (1979): *P. subretusus*: leaf.

Ogden (1943): *P. alpinus, P. amplifolius, P. crispus, P. gramineus, P. illinoensis, P. natans, P. nodosus, P. oakesianus, P. perfoliatus, P. polygonifolius, P. praelongus, P. pulcher, P. richardsonii* and the following hybrids: *P. alpinus* × *gramineus* (*P. nericius*), *P. alpinus* × *nodosus* (*P. subobtusus*), *P. berchtoldi* × *perfoliatus* var. *bupleuroides* (*P. mysticus*), *P. gramineus* × *perfoliatus* var. *bupleuroides* (? *P. nitens* sensu Morong), *P. illinoensis* × *P.* sp: key based on stem anatomy.

Ogden (1974): Brief notes on 29 spp, the following illustrated: *P. crispus, P. foliosus, P. gramineus, P. natans, P. pectinatus, P. perfoliatus, P. pul-cher, P. zosteriformis*.

Paliwal and Lavania (1979): *P. nodosus*: epidermis and stomata.

Pettitt and Jermy (1975): *P. natans, P. pectinatus*: pollen structure.

Porsch (1903): *P. natans*: stomata.

Posluszny and Sattler (1973): *P. densus*: floral development.

Posluszny and Sattler (1974*a*): *P. richardsonii*: floral development.

Priestley and North (1922): *P. perfoliatus*: stem endodermis.

Raunkiaer (1895–9): *P. acutifolius, P. alpinus, P. coloratus, P. crispus, P. densus, P. filiformis, P. gramineus, P. juncifolius, P. lucens, P. mucronatus, P. natans, P. obtusifolius, P. pectinatus, P. perfoliatus, P. polygonifolius, P. praelongus, P. pusillus,* P. rutilus, P. trichoides, P. zosterifolius*, and the following hybrids: *P. crispus* × *praelongus* (*P. undulatus*), *P. decipiens, P. gramineus* × *perfoliatus* (*P. nitens*), *P. lucens* × *natans* (*P. fluitans* auct.): leaf, stem; key based on stem anatomy.

Raunkiaer (1903): *P. acutifolius, P. alpinus, P. amplifolius, P. cheesemanii, P. claytonii, P. coloratus, P. confervoides, P. crispus, P. densus, P. drum-mondii, P. filiformis, P. fluitans, P. gramineus, P. hillii, P. hybridus, P. illinoensis, P. javanicus, P. lateralis, P. latifolius, P. linguatus, P. lucens, P. malaianus, P. mexicanus, P. miduhikimo, P. mucronatus, P. natans, P. oakesianus, P. obtusifolius, P. ochreatus, P. oxyphyllus, P. pauciflorus, P. pectinatus, P. perfoliatus, P. polygonifolius, P. polygonus, P. praelongus, P. pulcher, P. punctifolius, P. pusillus, P. robbinsii, P. rutilus, P. sclerocar-pus, P. spirillus, P. stenostachys, P. striatus, P. sumatranus, P.* ? *tenuifolius,*

---

* = *P. mucronatus*, according to Raunkiaer (1903).

P. tepperi, P. thunbergii, P. trichoides, P. vaginatus, P. vaseyi, P. zosterifolius and *P. lucens* × *natans* (one form of *P. fluitans*): leaf, stem.

Reinhardt (1897): *P. natans*: stomata.

Robards *et al.* (1976): *P. perfoliatus*: stem.

Sattler (1965): *P. richardsonii*: floral development.

Sauvageau (1889*c*): *P. amplifolius, P. crispus, P. densus, P. gramineus, P. lucens, P. microcarpus, P. natans, P. pectinatus, P. perfoliatus, P. plantagineus, P. polygonifolius, P. pusillus, P. robbinsii, P. rufescens, P. trichoides*: root.

Sauvageau (1890*c*): *P. densus* and notes on other spp: apical opening in leaf.

Sauvageau (1891*a*): *P. acutifolius, P. claytoni, P. compressus, P. crispus, P. densus, P. gramineus, P. lucens, P. marinus, P. natans, P. nitens, P. obtusifolius, P. pauciflorus, P. pectinatus, P. perfoliatus, P. plantagineus, P. polygonifolius, P. pusillus, P. robbinsii, P. rufescens, P. spirillus, P. trichoides, P. vaseyi*: leaf.

Sauvageau (1894): *P. acutifolius, P. crispus, P. lucens, P. natans, P. pectinatus, P. perfoliatus, P. pusillus, P. trichoides*: vegetative anatomy, including turion, with morphological notes on some other spp also.

Schalscha-Ehrenfeld (1940–1): *P. crispus*: developmental anatomy of shoot.

Scheifers (1877): (Not seen; spp investigated according to Solereder and Meyer) *P. acutifolius, P. compressus, P. crispus, P. densus, P. lucens, P. natans, P. perfoliatus, P. pusillus*: stem.

Schenck (1886*b*): *P. acutifolius, P. crispus* (and root), *P. densus* (and root), *P. lucens, P. natans* (root only), *P. nitens, P. pectinatus* (and root), *P. perfoliatus, P. pusillus*: leaf, stem.

Schilling (1894): *P. natans, P. rufescens*: mucilage.

Şerbănescu-Jitariu (1972*a*): *P. crispus, P. lucens, P. natans, P. pectinatus, P. perfoliatus, P. rutilus*: gynoecial morphology.

Şerbănescu-Jitariu (1976): *P. lucens, P. perfoliatus*: seed germination.

Shinobu (1952): *P. distinctus*: stomata with notes on other spp.

Singh (1964): *P. berchtoldi, P. crispus* (plus root), *P. epihydrus, P. indicus* (plus root), *P. natans, P. pectinatus, P. praelongus*: leaf, stem.

Souèges (1954): *P. natans*: developmental anatomy of shoot.

Spence *et al.* (1971): *P. richardii, P. schweinfurthii*: fruit morphology.

Streitberg (1954): *P. drucei, P. natans*: leaf.

Swamy and Parameswaran (1962*a*): *P. indicus*: embryology.

Symoens *et al.* (1979): *P. nodosus, P. thunbergii*: stem.

Tarnavschi and Nedelcu (1973): *P. crispus, P. fluitans, P. lucens, P. natans, P. pectinatus*: all parts.

Tichá (1964): *P. lucens, P. perfoliatus*: leaf.

van Tieghem and Douliot (1888): *P. natans*: developmental anatomy of root.

Tschermak-Woess and Hasitschka (1953): *P. densus, P. natans*: root-hairs.

Tutayukh and Arazov (1972): *P. natans* L.: all vegetative parts.

Uspenskij (1913): *P. perfoliatus*: leaf.

Wilson (1936): *P. densus*: root-hairs.

Ziegenspeck (1953–4): *P. alpinus, P. aschersonii, P. crispus, P. densus, P. fluitans, P. lucens, P. natans, P. pectinatus, P. perfoliatus, P. pusillus*: miscellaneous anatomical notes on stomata, leaf, etc.

308   POTAMOGETONACEAE

SIGNIFICANT LITERATURE—POTAMOGETONOIDEAE

Aalto (1970); Agardh (1852); Anderson (1978); Andersson (1888); Andrei
(1969, 1972); Arber (1922*a, b,* 1923); Areschoug (1878); Ascherson and Graebner (1907); Bance (1946); de Bary (1884); Bate-Smith (1968); Baumann
(1911); Behnke (1969); Bennett (1900–28, 1919); Berton (1978); van Beusekom
(1967); Bonnett and Millet (1970); Borzi (1887–8); Campbell (1936);
Čelakovský (1896); Chauveaud (1897*b*); Chrysler (1907); Clausen (1927);
Clifford (1970); Clos (1856); Cöster (1875); Colomb (1887); Cook (1908);
Cosson (1860); Costantin (1886*a, b*); Cranwell (1953); Dandy and Taylor
(1946); Davis (1966); Drawert (1938); Duval-Jouve (1873*b*); Eames (1961);
Eber (1934); Eichler (1875); Erdtman (1952); Esenbeck (1914); Falkenberg
(1876); Fernald (1932); Fischer (1901–4, 1907); Fontell (1908–9); François
(1908); Fryer (1886–97); Fryer and Bennett (1898–1915); Géneau de Lama-
rière (1906); Gibson (1905); Glück (1901, 1905, 1924); Goebel (1893, 1896);
Goffart (1900); Graebner and Flahault (1908); Gravis (1934); Gravis *et al.*
(1943); Gunning and Pate (1969); Gupta (1934); Haberlandt (1887); Hagström
(1906, 1908, 1916); Harada (1942); Haynes (1974, 1978); Hegelmaier (1870);
Hegnauer (1963); Helm (1934); Hofmeister (1861); Holferty (1901); Howard-
Williams *et al.* (1978); Hutchinson (1959); Irmisch (1851, 1858*b,* 1859, 1878);
Kaul (1976*b*); Klekowski and Beal (1965); Kudryashov (1964*a, b*); Kulesz-
anka (1934); Kunth (1841); Leavitt (1904); Le Blanc (1912); Lewin (1887);
Lidforss (1898); Lüpnitz (1969); Luhan (1957); Lundström (1888); Luther
(1947, 1951); Lyr and Streitberg (1955); Ly Thi Ba *et al.* (1978); Ly Thi
Ba and Guignard (1976); McClure (1970); Markgraf (1936); Mayr (1915);
Mer (1882, 1886); Meyer (1966); Miki (1934*b,* c, 1935*a,* 1937); Monoyer
(1926, 1927, 1929); Moore (1915); Muenscher (1936); Murén (1934); Myae-
mets (1979); Ogden (1943, 1974); Ohlendorf (1907); Ozimek *et al.* (1976);
Palamarev and Usunova (1969); Paliwal and Lavania (1979); Palmgren (1939);
Pearsall (1930); Peter *et al.* (1979); Pettitt and Jermy (1975); Philip (1936);
Porsch (1903, 1905); Porsild (1946); Posluszny and Sattler (1973, 1974*a*);
Priestley and North (1922); Ramati *et al.* (1973); Raunkiaer (1895–9, 1903);
Ravn (1894–5); Reinhardt (1897); Riede (1920); Robards *et al.* (1976); Rüter
(1918); Sahai and Sinha (1969); St. John (1925); Sattler (1965); Saunders
(1937–9); Sauvageau (1889*c,* 1890*c,* 1891*a,* 1894); Savich (1968); von Schalscha-Ehrenfeld (1940–1); Scheifers (1877); Schenck (1886*a, b*); Schilling (1894);
Schönherr (1976); Schürhoff (1926); Schuster (1910); Schwendener (1874,
1882); Şerbănescu-Jitariu (1972*a,* 1976); Sharpe and Denny (1976); Shinobu
(1952); Singh (1964, 1965*a*); Skottsberg (1913); Solereder and Meyer (1933);
Souèges (1940, 1954); Spence *et al.* (1971); Stebbins and Khush (1961); Streitberg (1954); Subramanyam (1962); Swamy and Parameswaran (1962*a*); Symoens *et al.* (1979); Tarnavschi and Nedelcu (1973); Taylor (1949); Tichá
(1964); van Tieghem and Douliot (1888); Tischler (1915); Tomlinson (1970);
Treub (1876); Tschermak-Woess and Hasitschka (1953); Tur (1976); Tutayukh and Arazov (1972); Uhl (1947); Uspenskij (1913); Weinrowsky (1899);
Wiegand (1898, 1899, 1900); Wilson (1936); Wiśniewska (1931); Ziegenspeck
(1953–4).

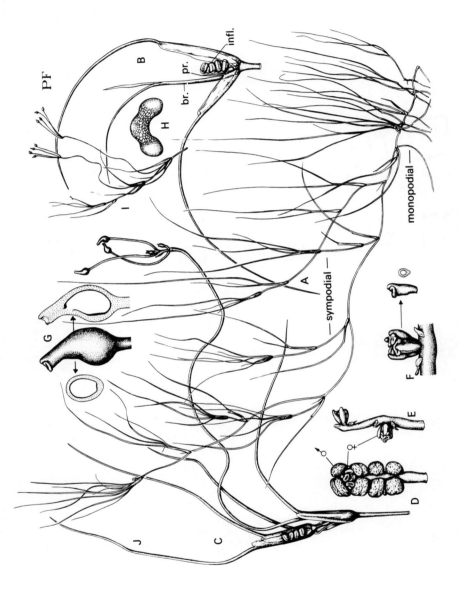

FIG. 9.1. POTAMOGETONACEAE. *Ruppia cirrhosa* (Petag.) Grande. Habit. (In brackish-water lakes, Fairchild Tropical Garden, Miami, P. B. Tomlinson *s.n.*)

A. Habit ($\times$ ⅓). Monopodial branching below with rooted nodes; sympodial branching above in association with flowering.

B. Young inflorescence (infl.) with floral leaves pulled apart; renewal shoot (br.) (with prophyll—pr.) in axil of lower floral leaf (to left) ($\times$ ⅗).

C. Two successive inflorescences, the lower extended one with developing fruits on extended fruit stalks (podogynes) ($\times$ ⅘).

D. Details of inflorescence ($\times$ 4).

E. Inflorescence with stamens removed ($\times$ 4).

F. Details of carpels with stamens removed ($\times$ 8); detail of single carpel to left.

G. Fruit ($\times$ 8); to right and left in LS and TS.

H. Single pollen grain ($\times$ *c.* 100).

I. Distal part of shoot with fruiting axes ($\times$ ⅓).

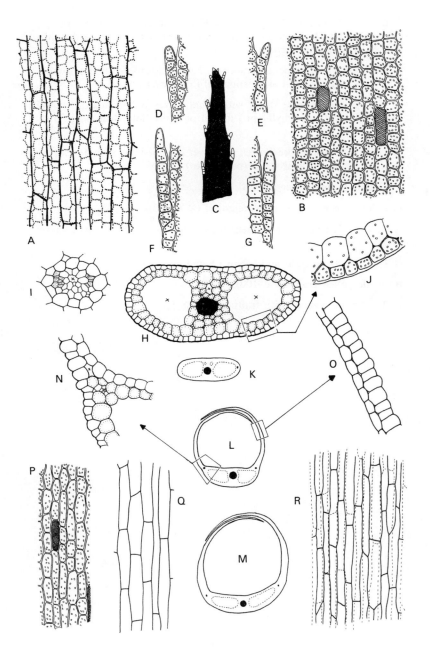

312

FIG. 9.2. POTAMOGETONACEAE. *Ruppia cirrhosa*. Leaf anatomy.

A. Lamina, surface view; subepidermal layer shown below dotted outline of epidermis (× 180).

B. Epidermis, surface view (× 180); tannin cells cross-hatched.

C. Apex of leaf, outline, with marginal hairs (× 65).

D–G. Hairs from margin of lamina, close to apex (× 180).

H–J. Lamina in TS about half-way between base and apex.
   H. (× 100); marginal bundles absent at this level in this leaf. I. Details of median vein from H (× 300); lateral group of phloem cells dotted. J. Details of surface layers from H (× 300).

K–O. TS leaf at successively lower levels.
   K. Lamina at a level where one marginal bundle is present (× 290). L. Distal part of sheath (× 290). M. Base of sheath (× 290). N. Enlargement of part of L, junction of sheath wing with its dorsal part showing lateral bundle (× 145). O. Enlargement of part of L, biseriate sheath wing (× 145).

P–Q. Epidermis of sheath in surface view, dorsal part half-way between base and apex of sheath (× 145).
   P. Abaxial epidermis. Q. Adaxial epidermis.

R. Surface view of biseriate sheath wing from within showing layers superposed (× 145).

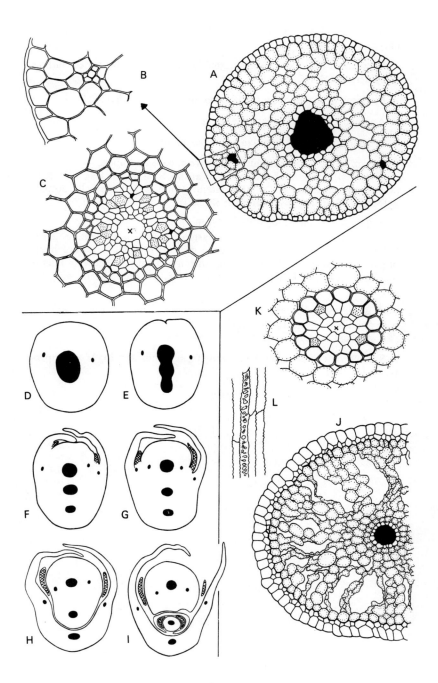

FIG. 9.3. POTAMOGETONACEAE. *Ruppia cirrhosa.* Stem, root anatomy.

A–I. Stem TS; A–C. stem internode; D–I node.

A. (× 110); vbs solid black.

B. Cortical bundle (× 300); enlargement of A.

C. Stele from A (× 300); sieve tubes stippled.

D–I. From Graves (1908). Successive sections through node, from below node upward showing relation between stem, leaf, and branch traces (× 300). Intravaginal scales (squamules) cross-hatched.

D. Normal internode appearance with two cortical bundles. E. Stele lobed. F. Leaf and branch trace split off from stele, one cortical bundle of next internode at right, originating blindly and independently of existing cortical bundle. G. Second cortical bundle evident at left. H. Leaf and intravaginal scales free of main axis, cortical bundles from internode below now lateral bundles of leaf sheath. I. Branch free of main axis with adaxial non-vasculated prophyll enclosing first pair of intravaginal scales on branch.

J–L. Root.

J. TS (× 110). K. TS stele (× 450); sieve-tubes dotted. L. Exodermal cells in surface view with characteristic plicate walls (× 110).

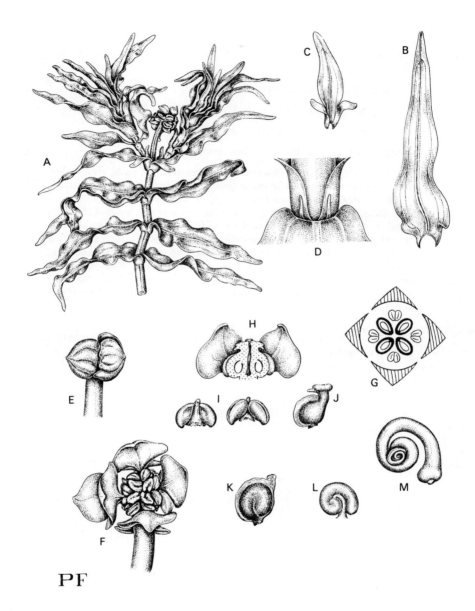

FIG. 9.4. POTAMOGETONACEAE. *Groenlandia densa* (L.) Fourr.

A. Distal part of flowering shoot ($\times \frac{3}{2}$) showing branching and leaf arrangment.

B. Single leaf ($\times$ 3).

C. Leaf ($\times$ 3) with auriculate sheath, below inflorescence ('inflorescence leaf').

D. Node with one leaf bent back ($\times \frac{9}{2}$) to show squamule pair.

E. Immature inflorescence from side ($\times$ 8) with two flower buds.

F. Inflorescence at anthesis from front ($\times$ 9).

G. Floral diagram of tetramerous flower.

H. Flower in LS ($\times$ 9).

I. Single stamen ($\times$ 12) from front and back, with tepal removed.

J. Carpel ($\times$ 12) from side.

K. Fruit ($\times \frac{9}{2}$).

L. Seed ($\times \frac{9}{2}$).

M. Isolated embryo ($\times$ 8).

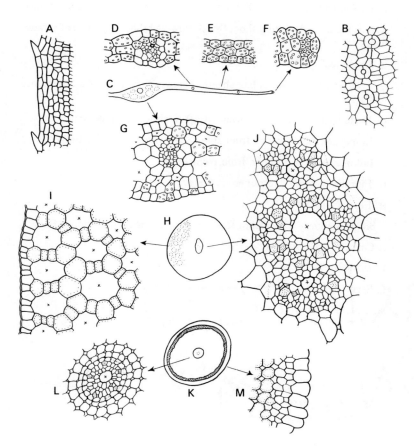

FIG. 9.5. POTAMOGETONACEAE. *Groenlandia densa.*

A–G. Leaf.

A. Leaf margin with prickle-hairs (× 120). B. Detail of leaf apex with stomata (× 225). C. TS lamina (× 18). D. Detail of lateral vein from lamina (× 225). E. Detail of triseriate lamina between vein (× 225). F. Detail of marginal fibrous strand (× 225). G. Detail of midrib region (× 160).

H–J. Stem.

H. TS (× 18), only a few lacunae indicated. I. Detail of cortex (× 120). J. Central cylinder (× 225).

K–M. Root.

K. TS (× 33), exodermis hatched. L. Stele (× 225). M. Surface layers (× 225).

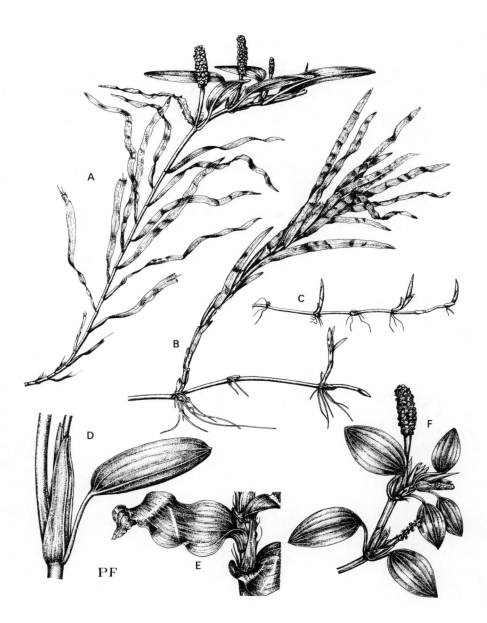

FIG. 9.6. POTAMOGETONACEAE. *Potamogeton* spp. Habit and vegetative morphology.

A–C. *Potamogeton epihydrus* Raf. (Harvard Pond, Petersham, Massachusetts; P. B. Tomlinson *s.n.*)

A. Distal shoot ($\times$ ½), with linear submerged leaves and broad floating leaves with terminal inflorescences. B. Proximal shoot ($\times$ ½), with rooted sympodially branched rhizomatous portion and developing erect shoots. C. Distal end of rhizome ($\times$ ½), showing regular alternation of 'sterile' and 'fertile' scales (see Fig. 9.7 A for detailed analysis).

D. *Potamogeton* sp (Lake Champlain, New York; P. B. Tomlinson 8.VII.72c). Ovate leaf ($\times$ ⅔) with ligular sheath, subtended branch with prophyll.

E. *Potamogeton* cf. *perfoliatus* L. (Boroughbridge, Yorkshire; P. B. Tomlinson *s.n.*) Portion of leafy shoot ($\times$ ⅔) showing sessile leaves, the ligular sheaths fraying into constituent fibrous vascular bundles.

F. *Potamogeton* cf. *capillaceus* Poir. (Harvard Pond, Petersham, Massachusetts; P. B. Tomlinson *s.n.*) Distal part of leafy shoot ($\times$ ⅔) with flowering sympodium.

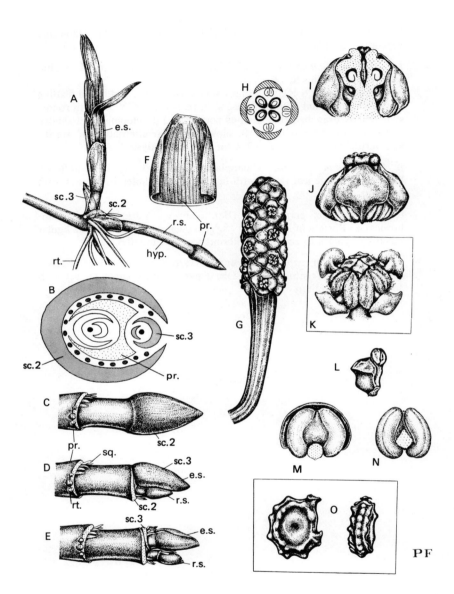

FIG. 9.7. POTAMOGETONACEAE. *Potamogeton* spp. Rhizome branching and reproductive morphology.

A–J, L–N. *P. epihydrus* Raf. (Harvard Pond, Petersham, Massachusetts, P. B. Tomlinson *s.n.*)

A–B. Vegetative branching.

A. Detail of rhizome apex from side (× ⅔) to show sympodial branching; erect shoot (e.s.) is terminal, evicted by renewal shoot (r.s.) in axil of scale leaf (sc.1). B. Diagram of sympodial construction at level of initiation of renewal shoot; sc.2 and sc.3 successive scale leaves, the lower subtending the renewal shoot (r.s.) with its prophyll (pr.). This becomes removed from the branch complex by extension of the hypopodium (hyp.).

C–E. Progressive dissection of apical bud shown in A (× 4).

C. Prophyll removed (shown in F). D. First scale leaf (sc.2) after prophyll removed to show axillary renewal bud (r.s.). E. Second scale leaf (sc.3) after prophyll, removed to show developing erect shoot (e.s.).

[Summary of annotations (in A–E) = e.s.—erect (terminal) shoot; hyp.—hypopodium of renewal shoot; pr.—prophyll or its scar; r.s.—renewal (lateral) shoot; rt.—adventitious root; sc.1—first scale leaf after prophyll; sc.2—second scale leaf after prophyll; sq.—squamules].

F. Prophyll (× 3) removed from terminal bud.

G–N. Floral morphology (all *P. epihydrus*, except K which is *P.* cf. *capillaceus*).

G. Inflorescence (× 3). H. Floral diagram. I. Single flower in LS (× 9). J. Single flower (× 9) before anthesis. K. Flower at anthesis (× 6) from the side. L. Single carpel (× 12). M. Detached stamen-scale complex (× 9) from within (scar of attachment stippled). N. Detached stamen (× 9) from outside, with scale removed, its scar stippled.

O. *P.* cf. *capillaceus.* Achene (× 24) lateral and dorsal views.

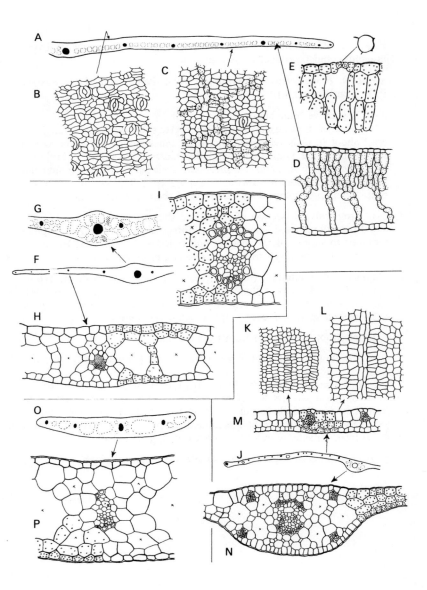

FIG. 9.8. POTAMOGETONACEAE. Leaf anatomy. Series showing progressive reduction in relation to submergence.

A–E. *P.* cf. *natans*. Floating broad leaf.
   A. TS half lamina ($\times$ 7), veins shown solid black. B. Adaxial epidermis, surface view ($\times$ 120). C. Abaxial epidermis, surface view ($\times$ 120), outline of mesophyll partitions dotted. D. Portion of lamina ($\times$ 100) to show palisade and lacunose abaxial mesophyll with uniseriate partitions. E. Detail of upper surface ($\times$ 200) with stoma and substomatal chamber, thin-walled guard cell—inset.

F–I. *P.* cf. *epihydrus*, submerged broad leaf.
   F. TS ($\times$ 12), half lamina (incomplete); veins—solid black. G. Detail of midvein region ($\times$ 20); veins—solid black. H. Detail of blade halfway between margin and midvein ($\times$ 100). I. Detail of main lateral vein ($\times$ 200).

J–N. *P.* cf. *zosteriformis*, submerged narrow leaf.
   J. TS half lamina ($\times$ 20), fibre bundles—solid black. K. Abaxial epidermis, surface view ($\times$ 120). L. Adaxial epidermis, surface view ($\times$ 120). M. Detail of blade halfway between midrib and margin ($\times$ 120). N. Midrib region ($\times$ 120).

O–P. *P.* cf. *pectinatus*, submerged linear leaf.
   O. TS lamina ($\times$ 40), veins—solid black. P. Detail of midvein ($\times$ 150).

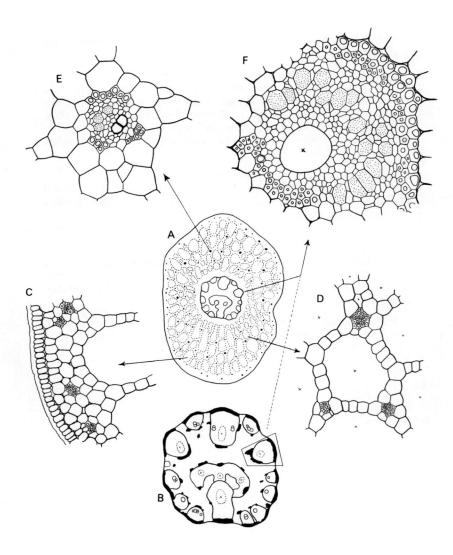

FIG. 9.9. POTAMOGETONACEAE. *Potamogeton* cf. *P. praelongus.* Stem anatomy (in TS).

A. Internode diagrammatic ($\times$ 12).

B. Details of stele ($\times$ 35); sclerenchyma—solid black; protoxylem lacuna—dotted outline in vbs directly continuous into leaves at next two nodes above; solid outline—tracheids with thin walls in vbs continuous into leaves at higher nodes.

C. Detail of peripheral tissues ($\times$ 100).

D. Cortical lacunar system ($\times$ 100).

E. Detail of cortical vb ($\times$ 260).

F. Peripheral vb of central cylinder ($\times$ 150), from position indicated in B.

Sieve-tubes—stippled (in C and D).

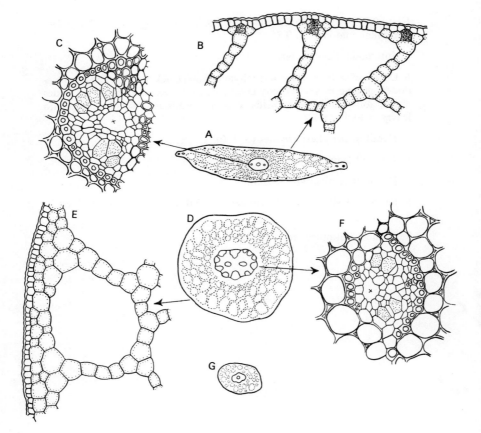

FIG. 9.10. POTAMOGETONACEAE. *Potamogeton* spp. Aspects of stem anatomy, in TS.

A–C. *P.* cf. *zosteriformis.*
  A. Internode, diagrammatic ($\times$ 12). B. Detail of surface layers ($\times$ 100). C. Detail of one lateral stelar vb ($\times$ 100).

D–F. *P.* cf. *pulcher.*
  D. Internode diagrammatic ($\times$ 12). E. Detail of surface layers ($\times$ 100). F. Detail of one peripheral vb of central cylinder ($\times$ 100).

G. *P.* cf. *pectinatus,* internode diagrammatic ($\times$ 12).

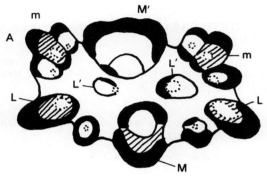

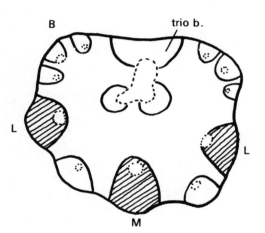

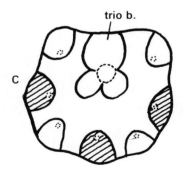

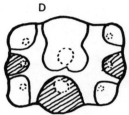

330

FIG. 9.11. POTAMOGETONACEAE. *Potamogeton*, stem anatomy (central cylinder) (after Monoyer 1927); all × 60, to show progressive 'reduction' in stelar complexity. Each figure has the same orientation with the median (M), lateral (L), and marginal (m) bundles inserted at the node immediately above hatched; sclerenchyma black. The 'trio-bundle' (triob.) where it is present, which represents the median portion of the leaf trace system inserted at the second node above the level of section (M'), is thus orientated toward the upper side of the figure.

A. *P. plantagineus.*

B. *P. fluitans.*

C. *P. perfoliatus.*

D. *P. praelongus.*

E. *P. compressus.*

F. *P. acutifolius.*

G. *P. pusillus.*

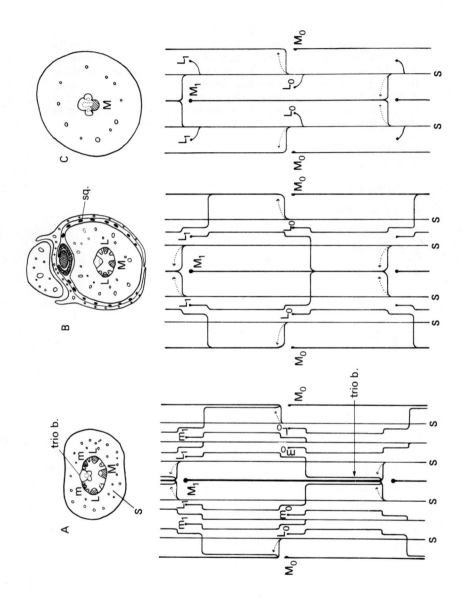

332

FIG. 9.12. POTAMOGETONACEAE. *Potamogeton*, course of vascular bundles in stem, to illustrate Monoyer's concepts. (After Monoyer 1927.)

A. *P. natans. Above.* TS stem internode (× 15). *Below.* Course of vbs, simplified. The stele is represented as if slit longitudinally along one orthostichy and rolled out flat. Cortical vbs, stipular traces and nodal anastomoses not included; branch traces dotted; trio b.—'trio bundle.'

B. *P. lucens. Above.* TS stem just above node (× 9) including petiole and stipular sheath of leaf inserted at next node below; axillary bud with prophyll cross-hatched; squamules (sq.)—solid black. *Below.* Course of vbs simplified as in A.

C. *P. pectinatus. Above.* TS vigorous rhizome (× 5). *Below.* Course of vbs simplified as in A. The laterals (L) fuse with S at the node of insertion. M, L, m—median, lateral, and, where present, marginal traces from leaf inserted at next node above, hatched in the upper figures which all have the same orientation. The corresponding traces in the lower figures are shown for two successive nodes, labelled Mo, Lo, mo and M, L, m.

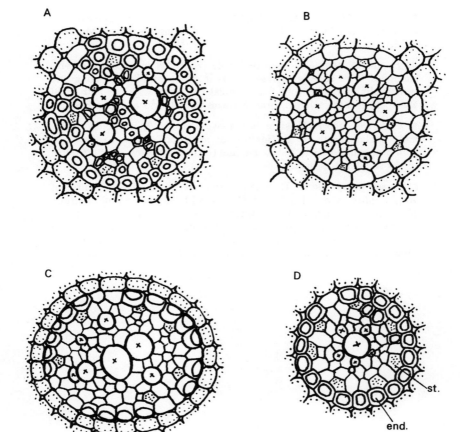

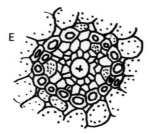

FIG. 9.13. POTAMOGETONACEAE. *Potamogeton,* root anatomy (stele) (after Sauvageau 1889*a*); all × 430. Chosen to represent a progressive 'reduction' in stelar complexity.

A. *P. plantagineus.*

B. *P. polygonifolius.*

C. *P. natans.*

D. *P. lucens* (in text, not *P. gramineus* as in Sauvageau's legend).

E. *P. pusillus.*

end.—endodermis; s.t.—sieve-tubes (pericyclic) stippled; x.—xylem tracheid.

Family 10
ZANNICHELLIACEAE (Dumortier 1829)
(i.e. in the sense of Takhtajan but not Hutchinson)
(Figs. 10.1–10.9 and Plates 13–14)

SUMMARY

THIS is a natural family consisting of four genera (Tomlinson and Posluszny 1976) and about ten species. They occur inland in fresh but mainly in coastal brackish waters, although never truly marine, and have a limited tolerance for saline water (e.g. Onnis and Mazzanti 1971). *Zannichellia* L. is widely distributed in tropical and temperate latitudes; *Althenia* Petit occurs in the western Mediterranean and North Africa (den Hartog 1975); records for South Africa may refer to introductions. The remaining two genera are restricted to the southern hemisphere: *Vleisia*[1] Tomlinson and Posluszny to the Cape Province in South Africa; *Lepilaena* Drumm. ex Harv. to Australia and New Zealand. The plants are often ephemeral, probably behaving as annuals in many localities. They are always wholly submerged, filamentous and individually inconspicuous, with very narrow linear leaves (1–3 mm wide, up to 10 cm long) on much-branched stems; sometimes making conspicuous mats of vegetation (e.g. *Zannichellia*) but never to the same extent as most sea grasses. The habit is diffuse, but the plants usually consist of a creeping, sympodially branched (less commonly monopodially branched) rhizome rooted at the nodes in the substrate (mud) and giving rise to more or less erect, much-branched leafy and flower-bearing shoots. The plants are usually monoecious, and the flowers terminal in complex sympodial aggregates. The male flower consists of a single stamen, the anther having from 1–6 pairs of microsporangia, naked (*Vleisia, Zannichellia*) or with a minute basal three-lobed structure (*Althenia, Lepilaena*); the pollen is globose. The female flower is composed of one or more (up to eight) carpels on a common axis, surrounded either by a closed membranous cupule or three separate scales; styles variable. The solitary ovule is bitegmic, pendulous and anatropous. The fruit is an achene protected by the inner persistent sclerotic layer of the fruit wall. Endosperm is absent, and the cotyledon circinnately coiled.

Anatomically the family is very reduced. Stomata are absent, except from some races of *Zannichellia*; the chief photosynthetic tissue is the epidermis; the leaf mesophyll includes one or more air-lacunae on each side of the solitary midvein. Fibres and mechanical tissue are little developed. Stems and roots are essentially protostelic, and the xylem is represented by a central lacuna with tracheids mainly restricted to the nodes. Tannin and crystals are scarcely developed.

---

[1] (*Vleisia* Tomlinson and Posluszny = *Pseudalthenia* Nakai; den Hartog, 1980)

336

FAMILY DESCRIPTION

Vegetative Morphology

Leaves distichous, sometimes sub-opposite, in pseudowhorls of 3 (–4) in *Zannichellia.* Plane of distichy often changing by 90° from parent to branch axis. Leaves of two main kinds: (i) foliage leaves, (ii) scale leaves representing the bladeless sheathing portion of a foliage leaf. Foliage leaves in all genera except *Zannichellia* with basal, membranous, open but often overwrapping sheaths, extended beyond the insertion of the blades as ligule-like, often auriculate structure. Length of ligule commonly increasing progressively in the distal direction for different leaves on a single shoot, apparently in a quite regular pattern. Foliage leaf in *Zannichellia* consisting of a narrow blade quite free from the associated sheath. Apex of foliage leaf rounded, pointed but sometimes truncate or three-toothed in *Lepilaena*; apical pore absent but stomata, possibly with same function, sometimes present in *Zannichellia.* Basal **rhizome scales** membranous; biseriate, without vascular tissue but usually subtending a bud. **Prophyll** a biseriate, non-vasculated scale leaf, at first enclosing younger organs but itself apparently never subtending a branch. **Squamules** filiform, inconspicuous, usually in pairs at each node, inserted laterally, sometimes more irregularly distributed in association with reproductive parts. **Roots** always unbranched, arising singly or in groups at lower nodes, provided with a distinct, often conspicuous coleorhiza.

Vegetative Anatomy

Hairs absent from stem and leaf, except for the paired squamules. Marginal teeth at leaf apex recorded for *Lepilaena australis*. **Epidermis** very uniform, cells thin-walled, chlorophyllous (Fig. 10.8 B, C), cubical or tabular. **Stomata** absent except from some races of *Zannichellia* and then present only in apex of foliage leaf blade, leading to a peculiar kind of heterophylly. **Foliage leaf** always with a single median vascular bundle (Figs. 10.8 A, G, I and 10.9 A, C), surrounded by a thin-walled (occasionally thick-walled) uniseriate endodermis in *Lepilaena bilocularis*. Submarginal fibres (Figs. 10.8 B and 10.9 B, D) consistently present in all genera except certain races of *Zannichellia* (Fig. 10.8 H, J). Hypodermal fibrous strands otherwise present only in *Lepilaena preissii* (Sauvageau 1891*a*). **Mesophyll** either irregularly lacunose (Fig. 10.9 B, D) or, in *Vleisia* and *Zannichellia*, with one–several more or less distinct and large lacunae, on each side of the midvein (Fig. 10.8 A, G, I). Leaf blade in *Vleisia* including a narrow submarginal strand of elongated cells representing a conducting strand. Sheathing **leaf base** colourless, with a biseriate margin, including fibrous strands in *Althenia* and *Lepilaena*. **Prophyll** (and other scale leaves) equivalent to a bladeless sheath and represented simply by its biseriate wing, without any dorsal thickening or vascular tissue but sometimes with fibres (Fig. 10.9 F, G).

Stem (Figs. 10.8 D–F and 10.9 H, I) with a narrow central stele surrounded by a thin-walled endodermis and a wide lacunose cortex.

Course of vascular bundles in stem very simple, said by Monoyer (1927) to have the most reduced vascular system of any aquatic monocotyledonous

family, with traces to the lateral organs diverging directly from the stele of
the parent axis. Cortical vascular system absent; submarginal and hypodermal
strands, when present in the leaf blade, continuous into the leaf sheath, but
never extending into stem. Submarginal vascular strand in the leaf of *Vleisia*
united with median leaf trace at node.

Root (Figs. 10.8 L, M and 10.9 J) with a narrow stele and a uniseriate,
thin-walled endodermis; epidermal cells thin-walled, large, and in submerged
roots with conspicuous root-hairs arising from short trichoblasts. **Exodermis**
1–2-layered, usually consisting of narrow, slightly thick-walled, sometimes
lignified cells. Vascular tissues reduced throughout. **Tracheids** absent from
most mature organs, but present in young developing organs, and in mature
stems and leaves replaced by a xylem lacuna. Tracheids persistent in unextended stem nodes. Tracheal elements absent from root or at most represented
by partially differentiated cells with unthickened walls and devoid of contents
when mature.

REPRODUCTIVE MORPHOLOGY

Plants monoecious (or possibly dioecious in species of *Lepilaena*). 'Flowers'
(see p. 345) associated with modified, often reduced leaves ('bracts') to form
complex, sympodially branched 'inflorescences' resembling a cincinnus, the
latter either axillary or terminating the segments of the vegetative sympodium
(Figs. 10.1, 10.2, 10.3, 10.5, and 10.7). Flowers themselves always terminal,
the male usually on the axis that initiates the cincinnus (except in *Zannichellia*). Female flowers (one or more) terminal on branches of higher orders
and therefore often more numerous than male flowers in monoecious species.
Male flower either naked (*Vleisia* (Fig. 10.2 D), *Zannichellia* (Fig. 10.1 C,
H)), or with a short three-lobed 'perianth' (*Althenia* (Fig. 10.5 F), *Lepilaena*
(Fig. 10.7 H, I)) consisting of a single stamen on a short (but often late-
elongating) pedicel. Anther made up of one or more bisporangiate units (often
interpreted as representing more than one fused stamen), the connective sometimes extended distally into a short appendage. Connective in *Vleisia* with
two minute vestigial lateral appendages. Anther dehiscing by longitudinal
slits. Pollen always globose. **Female flower** of one in *Vleisia* (Fig. 10.2 E–H),
1–3 in *Althenia* (Fig. 10.5 E) and *Lepilaena* (Fig. 10.7 B–E), or (1–) 4–5
(–8) in *Zannichellia* (Fig. 10.1 C–G), separate short-stalked, usually asymmetrical carpels on a common pedicel. Carpels surrounded by a membranous,
biseriate 'perianth' consisting of either a closed tube (*Vleisia, Zannichellia*)
or three separate segments (*Althenia, Lepilaena*), the segments sometimes
bifid and always situated opposite the carpels. **Styles** of individual carpels
either short (*Zannichellia*; Fig. 10.1 G) or long (*Althenia*; Fig. 10.5 E), the
stigmas enlarged, peltate, or irregularly funnel-shaped; feathery in *Lepilaena
bilocularis* (Fig. 10.7 B, D). **Ovule** single, bitegmic, anatropous and pendulous,
the micropyle eventually directed basally.

Fruit sometimes becoming long-stalked, dimorphic in this respect in *Vleisia*.
Pericarp two-layered, the soft outer layers becoming eroded, the sclerotic
inner layers persisting. Fruit sometimes dehiscent into two unequal valves
(*Althenia, Vleisia*) via a longitudinal parenchymatous suture in the sclerotic

layer. Fruit wall made warty by protuberances from the inner layers in *Vleisia*. Fruit with a dorsal (sometimes also ventral) warted ridge in *Zannichellia* (Fig. 10.1 I), sometimes tuberculate in *Lepilaena*. Seed without endosperm. Cotyledon circinnately coiled in a characteristic manner above the swollen hypocotyl (Fig. 10.1 J); germination epigeal.

REPRODUCTIVE ANATOMY

Vascular system of the flower, where studied, apparently very simple and offering few clues concerning morphology of the various parts (Uhl 1947; Posluszny and Tomlinson 1977; cf. however, Singh 1965a). Vascular system of the floral axis consisting of a single acropetally developing vb (Posluszny and Sattler 1976b), supplying stamens and carpels. Prophylls and tepals unvasculated. Stamens always with a single vascular bundle in the connective. Each carpel in *Zannichellia* supplied by a single trace, the trace dividing in the stalk to produce a short dorsal and a longer single ventral bundle, the ventral bundle becoming the ovule trace directly. Ventral bundle finally becoming sclerotic and conspicuous in the fruit.

TAXONOMIC AND MORPHOLOGICAL NOTES FOR THE FAMILY

(Further discussion is given under the descriptions of individual genera.)

Stamen morphology

As pointed out by Reinecke (1964) the description of stamen morphology, especially in *Zannichellia palustris*, is confused. This is sometimes because of variability within a single population, but more confusion arises because no standard descriptive terminology has been used (theca, anther sac, locule, lobe, sporangium being used indiscriminately). Furthermore, there is an inherent tendency to interpret the stamen not as a simple but as a compound structure (see Ascherson and Graebner (1907) in their key to *Althenia*). Comparative anatomy shows that the basic construction and development of the stamen unit is uniform throughout the family and there certainly seem to have been evolutionary changes in the male reproductive capacity of the plant. Phylogenetic interpretations of the stamen as a compound structure merely confuse the issue, since there is no developmental evidence to support them. Singh (1965a) claimed that the existence of a double xylem strand in certain stamens of *Zannichellia* with eight microsporangia supported his interpretation of it as a double structure.

Anatomically the constructional unit is an anther sac containing two *microsporangia*: Each microsporangium has the normal endothecium and tapetum of a terrestrial plant. At maturity the wall separating the two microsporangia breaks down to form a common locule in which the mature pollen is retained. The pollen is released through longitudinal slits, and this may be taken to indicate that these plants, which all have hydrophilous pollination (Roze 1887), may nevertheless have had an aerophilous ancestor. Confusion in description could arise if the common locule formed by the breakdown of the two microsporangia was not interpreted correctly.

On this basis, the stamen construction in Zannichelliaceae may be summarized as follows:

Althenia, one unit (single anther sac, two microsporangia).
Lepilaena bilocularis, two units (two anther sacs, four microsporangia).
Lepilaena (other species, e.g. *L. cylindrocarpa*), six units (six anther sacs, 12 microsporangia).
Vleisia, four units (four anther sacs, eight microsporangia).
Zannichellia, two units (two anther sacs, four microsporangia) in the majority of forms (Reese 1967); four units (four anther sacs, eight microsporangia) in ssp *pedicellata* (Reese 1967); rarely a single unit (one anther sac, two microsporangia)

Further anatomical surveys of this family should take into consideration the populational variability within a single species (e.g. *Althenia*) especially in view of the correlation established by Reese (1967) between anatomy and cytology in races of *Zannichellia*.

Vegetative anatomy

Despite the extreme reduction, diagnostic features of leaf anatomy present themselves and are used to a limited degree in the subsequent keys and descriptions. Taxa seem to be separable on the presence or absence of apical 'stomata,' fibre distribution in blade mesophyll and leaf sheath, continuity or otherwise of leaf blade hypodermis and the extent of air-lacunae. *Vleisia* is distinguished by the presence of a submarginal strand of elongated cells which can be interpreted as a vascular strand, although its constituent cells are little differentiated. However, it appears to be homologous with the marginal vascular strand of the *Ruppia* leaf and is continuous with the median leaf trace at the node.

Relationship of the family

Although they have sometimes been associated with the group of genera here treated as the family Cymodoceaceae, the Zannichelliaceae are certainly a discrete group and should be so recognized. They differ from the Cymodoceaceae in features which have been summarized in the account of that family (see p. 401).

The taxonomic position of the Zannichelliaceae partly depends on how the morphology of the reproductive parts is interpreted, and this question is not yet entirely resolved. It seems clear, however, that the family is loosely allied to *Potamogeton*, *Ruppia*, and *Najas*, all of which are plants with slender filiform axes and reduced flowers strongly disposed to become, or which are actually, hydrophilous and therefore submerged. It is usual to regard the Zannichelliaceae as standing towards the end of this reduction series, although comparisons have usually been made on the basis of information relating only to *Zannichellia* itself (e.g. Vijayaraghavan and Kumari 1974). However, the flower aggregations in the Zannichelliaceae are so complex in their morphology that it is not easy to make part for part comparisons, so that contrasting interpretations all based on reasonable premises can co-exist. The best way to deal with these uncertainties is to regard the family as representing one line of elaboration within the Alismatidae, without close relatives or obvious ancestors.

MATERIAL EXAMINED

The same material as reported on by Posluszny and Tomlinson (1977) has been used and is listed under the separate generic descriptions.

DELIMITATION OF AND KEY TO THE GENERA

Ascherson and Graebner (1907) subdivided the family into two genera each with two subgenera, but it now seems clear that four separate genera are best recognized (Tomlinson and Posluszny 1976). Since the literature on the family is large and scattered, the following synoptic key is included, representing the distillation of this large literature together with the results of detailed morphological study (Posluszny and Tomlinson 1977), and the addition of anatomical data.

1. Blade of distal leaves without attached sheath, but accompanied by a membranous 'ligular sheath' (Fig. 10.1 B). Leaf blade with a continuous hypodermis and a large air-lacuna each side of the midvein; marginal strands often absent (Fig. 10.8 G, H, I, J). Creeping rhizome not well differentiated. Erect shoots sympodial, the branch complex associated with two flowers, one male and one female; female flower lowest on branch unit (Fig. 10.1 A). Male flower lacking a perianth, the anther usually 4– (but sometimes 8–) sporangiate (Fig. 10.1 H). Female flower with a cupular membranous perianth, (1–) 4–5 (–8) carpellate (Fig. 10.1 F, G). Fruit usually with a dorsal (sometimes also a ventral) warted ridge (Fig. 10.1 I) *Zannichellia* (one cosmopolitan species of considerable cytological complexity).

1a. Blade of distal leaves with attached sheath; marginal strands always present. Sympodial creeping rhizome well differentiated. Flowers in sympodial groups, not in fixed numbers per group (Fig. 10.3). Lowest flower within inflorescence unit always male   2

2. Leaf blades up to 2 mm wide with a submarginal 'vascular strand' and oblique transverse strands continuous with midvein. Hypodermis uniseriate, continuous, air-lacunae well developed. Leaf sheath without fibres. Male flower naked, 8–sporangiate, with a pair of vestigial lateral appendages on the connective (Plate 13A.4). Female flower a single carpel enclosed by a tubular membranous perianth (Fig. 10.2 E). Fruit uniformly warty   *Vleisia* (one species, South Africa).

2a. Leaf blade normally only 1 mm wide, usually with marginal fibres but without submarginal 'vascular strands' or transverse commissures; hypodermis discontinuous, air-lacunae obscure. Male flower with a short three-lobed perianth, 2 to 12–sporangiate. Female flowers with (1–) 3 carpels enclosed by three separate tepals. Fruits smooth   3

3a. Male flower 2–sporangiate (Fig. 10.5 F), carpels each with a long style and peltate stigma (Fig. 10.5 E). Fruit dehiscent. Leaf fibres sometimes absent   *Althenia* (one or two species, Mediterranean).

3b. Male flower 2 to 12–sporangiate, carpels each with a short style and
funnel-shaped or feathery stigma (Fig. 10.7 D). Fruit not reportedly
dehiscent. Leaf fibres usually well developed (e.g. Fig. 10.9 A, B).
Lepilaena (three to four species, Australasia).

GENERIC DESCRIPTIONS

The above account summarizes observations for the family as a whole;
in subsequent paragraphs features which refer to individual genera are emphasized, with notes on species within genera and a taxonomic discussion.

ZANNICHELLIA L. (sensu Tomlinson and Posluszny)
(Figs. 10.1 and 10.8 G–M)

Cosmopolitan in fresh or brackish water. Usually treated as a single species
but numerous infraspecific taxa have been described which themselves include
several cytological races (e.g. Reese 1963, 1967).

VEGETATIVE MORPHOLOGY

Habit and inflorescence

Erect shoots arising from a creeping rhizome, apparently without the regularity of *Althenia*; first leaf on each branch a scale-like prophyll. Roots usually
in pairs at lower nodes. Erect shoots showing sympodial branching of a
regularity and complexity not found in other genera, resulting in apparent
whorls of 3 (–4) leaves associated with a male and female flower (Fig. 10.1
A), the pseudowhorl separated by a long internode usually with a prophyll
towards its base.

Leaf morphology distinct within the family, blade of the foliage leaf free
from the open or closed tubular sheathing base, described as a ligular sheath
(cf. certain species of *Potamogeton*). Each ligular sheath enclosing younger
parts.

For purposes of description the structure of each unit of the distal sympodium may be represented as follows (Fig. 10.1 Aa) based the developmental
studies of Posluszny and Sattler (1976b: (i) two basal scales, the lowest the
prophyll (L_1), often separated from structures above and below by long internodes; (ii) two subopposite foliage leaves (leaves L_2 and L_3) each with its
separate but associated ligular sheath; (iii) a terminal female flower. In the
axil of leaf L_2, a short, precociously developed branch with two leaves, a
prophyll and a foliage leaf (leaf L'), the axis ending in a male flower. In
the axil of leaf L_3, a renewal shoot capable of repeating the construction of
the sympodium. A bud in the axil of the foliage leaf L' on the male branch
likewise capable of growing into a branch of the sympodium and repeating
the unit structure. Stamen sometimes becoming separated from the rest of
the complex by elongation of the internode above it (cf. Fig. 10.1 B–D, E).
From this it is evident that the three leaves of each nodal complex are separated (morphologically) by internodes (leaves L_2 and L_3 have been shown

to arise in sequence by Campbell 1897; Posluszny and Sattler 1976b) and that L' belongs to a branch one order higher than those from which leaves L_2 and L_3 are produced. The system also involves a change of 90° in the plane of distichy from one pseudowhorl to another (Fig. 10.1 Ab). Campbell (1897) indicates that leaf L' may also be at right angles to L_2 and L_3. The above arrangement is far from constant, some of the variation observed is illustrated in Fig. 10.1 B–E.

VEGETATIVE ANATOMY

Leaf blade usually distinguished by two well developed lacunae, one on each side of the midvein (e.g. Fig. 10.8 G, H). Submarginal fibrous strands variously developed in a way determined by leaf size in relation to chromosome number (Reese 1967). Fibrous strand wholly absent from ssp *repens*, represented by a few narrow cells in ssp *pedicellata*, relatively well developed in ssp *major*. Stomata, mostly chlorophyll-free and with a permanently open pore, present apically in some ssp of *Z. palustris*.

[Burgemeister (1968) has studied the complex way in which apical stomata occur and this has led to the recognition of a microscopic heterophylly. Leaves on an individual plant may be distinguished microscopically by the shape and width of their apices as either 'pointed' or 'blunt' tipped. Stomata are present only on the blunt-tipped leaves in ssp *palustris* and *pedicellata*, but they are wholly absent from both kinds of leaf in ssp *major*. At a standard level (175 μm) behind the leaf apex the three subspecies may be distinguished by a constant difference in blade width which is obscured by the greater variation in blade width at lower levels. The influence of certain environmental factors such as light intensity and photoperiod as well as the effect of growth substances on the distribution of these different leaf types was studied in great detail. From these studies it was concluded that because the three races responded differently to the experimental treatments they differed from each other in their physiological make up as well as in their chromosomes. Burgemeister also showed that the cuticle of the stomata-bearing region lacks the specialized properties of the rest of the cuticle of the whole leaf and, therefore, he equated the narrow stomata-bearing zone with the rudimentary remains of the original 'land-leaf.']

REPRODUCTIVE MORPHOLOGY

Flowers

Male. Perianth absent. Stamen naked (Fig. 10.1 H), usually with two anther sacs and four microsporangia (as in *Lepilaena bilocularis*), but very rarely with a single anther sac and two microsporangia. Connective prolonged into a blunt appendage. Stamens with four anther sacs and eight microsporangia sometimes present. This last situation is interpreted by Singh (1965a) as two stamens fused back to back because he considered that the connective trace in some examples had two, and not a single, xylem strand.

Female. A group of (1–) 2–5 (–8) carpels collectively surrounded by a membranous, tubular perianth (Fig. 10.1 F, G). Each carpel asymmetric, short-stalked, with a funnel-shaped stigma on a short style. Vascular system a feebly developed ventral strand extending into the base of the ovule, the dorsal extension recorded by Uhl not having been seen by other workers.

Fruit

Short-stalked, the pronounced dorsal ridge with numerous protuberances from the persistent inner wall of the pericarp (Fig. 10.1 I).

TAXONOMIC AND MORPHOLOGICAL NOTES

Zannichellia, in the present restricted sense, is generally treated as a single cosmopolitan species *Z. palustris* L., within which many infraspecific taxa have been described, especially in western Europe where the plant has been most extensively collected (e.g. Fischer 1907). The situation has been greatly illuminated by the detailed cytotaxonomic survey carried out by Reese (1963, 1967) on *Zannichellia* in northwest Europe. He has also made a careful assessment of morphological and anatomical features that have been used traditionally in subdividing the species. Reese concludes that his population samples represent three distinct forms ('Sippe') and that these include several cytological races. This chromosomal variation influences quantitative characters considerably and Reese concludes that the most constant morphological differentiation is in fruit shape and size. Trends in other features demonstrated by measurement of large samples tend to be obscured by the wide range of variation. The conclusion is that *Z. palustris* in northwest Europe is represented by a group of 'sibling species.' These are equated with types recognized by classical methods as:

Zannichellia palustris L.
    ssp. *palustris* = '*repens*' of Reese (2n = 24, 34)
    ssp. *pedicellata* (Wahlenb. et Rosén) Archangeli = '*pedicellata*' of Reese (2n = 24, 36)
    ssp. *polycarpa* (Nolte) F. Richter = '*major*' (including '*polycarpa*') of Reese (2n = 32)

The study by Reese represents a model for the kind of work needed on these polymorphic water plants and it is hoped that populations of *Zannichellia* throughout its range can be examined in the same detail.

Morphology

The morphological interpretation of the sympodial unit in *Zannichellia,* given above, originated with Irmisch (1851) and has been supported (or accepted) by numerous subsequent authors (e.g. Schumann 1892; Reese 1963; and most recently Posluszny and Sattler 1976b). Alternative explanations have been offered, however. On the basis of developmental observations, Campbell (1897) regarded branching within the unit as involving an equal dichotomy of the shoot apex, so that the femal flower and the axis that initiates the renewal shoot (in the axil of leaf L$_3$ in the above description) originate simultaneously. In a similar way Campbell regarded the meristem of the male branch (in the axil of leaf L_2) as also branching dichotomously to produce a renewal shoot together with the male flower. Posluszny and Sattler (1976b) do not support an interpretation of branching in either meristem as being dichotomous.

Floral morphology

The cupular sheath surrounding the group of carpels is variously interpreted. It is looked upon as 'perianth' if the group of carpels is collectively regarded as a 'flower.' This is the opinion of Magnus (1870, 1894). Ascherson & Graebner (1907) and Wettstein (1935). Alternatively if the group of carpels is regarded as an inflorescence, then the cupular sheath is considered to be an 'involucre' or tubular bract, homologous with the ligular sheath associated with the foliage leaf, and it is pictured as surrounding a group of naked, unicarpellate flowers. This is the opinion of Campbell (1897) and Uhl (1947). Developmental observation (Posluszny and Sattler 1976b) does not resolve the problem; these authors suggest the cupule may have a morphological status intermediate between phyllome and emergence.

MATERIAL EXAMINED

Zannichellia palustris L.; Havelock North, New Zealand; P. B. Tomlinson, 14-I-72A; all parts. Isle of Grain, England; all parts. Boomer Marsh, Tasmania; W. M. Curtis s.n.; all parts. (No attempt has been made to identify these further as to subspecies)

VLEISIA[1] Tomlinson and Posluszny
(Figs. 10.2, 10.3, and 10.8 A–F and Plate 13 A)

A monotypic genus consisting of an annual sp in seasonal vleis in Cape Province, South Africa.

VEGETATIVE MORPHOLOGY

Rhizomatous with erect shoots, length probably determined by the depth of the water. Axis basally sympodial, the lower internodes on each segment rhizomatous and comparable to those of *Althenia* but probably not as regular, distally monopodial with axillary sympodial flower complexes (Reinecke 1964; Posluszny and Tomlinson 1977). Prophyll on renewal shoot inserted basally and so serving to distinguish branch from continuing main axis. Roots restricted to rhizomes or lower nodes, solitary or in groups of up to six, coleorhiza short.

Leaves

Foliage leaf with narrow blade up to 6.5 cm long and an open sheathing base; ligule variously developed (Fig. 10.2 B). Leaves associated with flowers (bracts) usually with v. short blades or with blades scarcely developed (e.g.

---

[1] (= *Pseudalthenia* Nakai; see der Hartog, 1980)

Fig. 10.2 I). Apex of well developed blade either rounded or apiculate and slightly pointed (Fig. 10.2 A). Stomata absent.

VEGETATIVE ANATOMY

Leaf mesophyll with a continuous hypodermis and several air-lacunae on each side of midrib (Fig. 10.8 A–C), with transverse diaphragms at long intervals. Margin including a fibrous strand and a vascular strand, represented by narrow, elongated cells (Fig. 10.8 B), continuous with midvein via obliquely ascending transverse veins. Submarginal vein continuous with midvein at the node.

REPRODUCTIVE MORPHOLOGY

Inflorescence (Plate 13A)

Flowers terminal on main axis or on short lateral axes, the inflorescence variously monopodial or sympodial (cf. Fig. 10.3 A, C). Each unit consisting of a single male and a single (rarely more) female flower, the whole unit at first enclosed by the basal tubular prophyll; prophyll later caused to split by the enlargement of younger organs. Inflorescence complex made up of three successive sympodial units shown in Fig. 10.2 C, D, the first two each with a prophyll and a reduced foliage leaf (bract), each branch axis of a higher order originating in the axil of this bract (never of the prophyll). First-order axis ending in a male flower, second-order axis ending in a female flower, the third-order axis either aborted but capable of producing a further inflorescence unit, or even proliferating as a vegetative branch. Bract (second leaf) on each axis inserted at right angles to its prophyll, plane of distichy changing regularly (Fig. 10.3 B, D). A determinate complex with a terminal female flower shown in Fig. 10.3 A, B.

Male flower. A single stamen with four pairs of microsporangia (interpreted by Reinecke (1964) as two fused bisporangiate anthers); elongating at anthesis on a short naked pedicel (Fig. 10.2 C, D). Connective with a rounded apical knob and two small outgrowths on the opposite side.

Female flower. On a short pedicel elongating in fruit; the single carpel surrounded by a tubular membranous perianth. Style short, ending in an asymmetrical funnel-shaped, ephemeral stigma (Fig. 10.2 E–H).

Fruit

Somewhat dimorphic; those on lower nodes with elongated, positively geotropic stalks, up to 4 cm long and burying the ripe fruit; those on upper nodes remaining short-stalked. Inner, thick-walled persistent layers of pericarp with warty protuberances on dorsal and ventral edges; pericarp possibly dehiscent.

TAXONOMIC NOTES

Vleisia, when first described, was included in *Zannichellia* by Ascherson and Graebner (1907) by virtue of its naked male flower, and tubular female perianth and warty fruit, However, in other characters, summarized in the generic key (on p. 341), it is very different and its recognition as a distinct genus is justified (Tomlinson and Posluszny 1976).

MATERIAL EXAMINED

Vleisia aschersoniana (Graebner) Tomlinson and Posluszny; Loc River, South Africa. P. Reineke s.n.; all parts.

ALTHENIA Petit
(Figs. 10.4, 10.5, and 10.9 c–j and Plate 13 b)

One or two species in the Mediterranean region (den Hartog 1975; Onnis 1969). Possibly introduced into Southern Africa (Schonland 1924; Obermeyer 1966*b*). Shoots dimorphic, with creeping sympodial rhizome producing erect leafy and flowering shoots (Prillieux 1864).

VEGETATIVE MORPHOLOGY

Habit

Leaves regularly distichous in lower parts of the plant, but becoming less uniformly so in association with flowers. Each segment of sympodium including two basal scale leaves separated by long internodes (Fig. 10.4 A, B), the first scale leaf (prophyll) consequently well separated from the parent axis. Renewal segments originating in axil of second scale leaf of previous unit (Fig. 10.4 C). Third leaf a foliage leaf inserted subopposite the second scale leaf; the unit ending in a series of 3–4 further foliage leaves (total number of leaves on each unit then 5–6). Lowest foliage leaf sometimes subtending a branch developing as the first unit of another sympodium (and so leading to proliferation of rhizome). Upper foliage leaves subtending leafy branches (Fig. 10.5 B, C), usually of determinate growth, here referred to as 'inflorescences,' the main axis itself ending in one such inflorescence. Roots borne only at second node of each unit (i.e. node with subopposite scale and foliage leaf), singly or in groups of 3(–5); the first (prophyllar) node always rootless.

[The overall habit (Fig. 10.4 A) thus suggests a seemingly continuous rhizomatous axis consisting of alternate rootless nodes, each bearing a single scale, and root-bearing nodes, each with a pair of leaves (one a scale, the other a foliage leaf), the erect leafy shoots originating at the root-bearing nodes.

This description provides a very formal analysis, further detailed examination may show a greater plasticity in growth expression.]

Leaves

(i) Prophyllar scale (Fig. 10.9 F, G) and rhizome scales with characters of the family. (ii) Foliage leaves with distinct auriculate ligule-like extensions of their sheaths the extensions being most pronounced in distal leaves (Fig. 10.4 D). In inflorescence leaves, blade short and scarcely exceeding the length of the sheath. Apex of blade pointed, lacking an apical pore and stomata.

Vegetative Anatomy (Fig. 10.9 c–j)

Exhibiting the features of the family, but with differences between certain forms (see Taxonomic Notes below). Mesophyll (Fig. 10.9 c–e) irregularly lacunose and devoid of a continuous uniseriate hypodermis.

Reproductive Morphology

Inflorescence (Fig. 10.5, Plate 13 b)

Distal branches of limited growth, all ending in a flower, possibly proliferating vegetatively under certain circumstances. Construction variable. The generalized scheme, represented in Fig. 10.5 A suggests the following simple rules of construction: (i) All axes ending in a flower, the main axis always in a male flower, the remaining axes usually in a female flower but a rare male flower also occurring on axes of lower orders; (ii) Each axis bearing 0–3 leaves (some ultimate axes devoid of leaves, others with three, but most commonly with one, the system then resembling a simple cincinnus); (iii) A prophyllar scale may be present, e.g. on the lowest first-order branch, but otherwise without a leaf in the prophyllar position; [This means that in most branches the first leaf is not adaxial but abaxial, being situated on the same side as the parent subtending leaf. It is reasonable to interpret such branches, as did Prillieux, as 'lacking a prophyll';] (iv) Distichous leaf arrangement tending to become irregular on distal parts of the plant, possibly because of displacement as a result of their becoming closely packed during the maturation of fruits on the lower branches.

Male flowers. Terminal on main axis (less commonly lateral axes) on a long pedicel, with a three-lobed, inconspicuous perianth; anther bisporangiate, the connective not produced distally (Fig. 10.5 F).

Female flowers. Terminal on lateral axes of inflorescence sympodium, each flower (Fig. 10.5 E) attached to a short pedicel, and consisting of three carpels in the axils of three corresponding perianth segments (bracts according to Prillieux). Style distinct, stigma peltate. Pedicel and style elongating with age.

Fruit

Endocarp splitting into two unequal valves along an oblique line of dehiscence, the style persisting on the larger valve.

Taxonomic Notes

Two forms of this genus have been described either as distinct sp or ssp. *Althenia filiformis* Petit, the type species, is widely distributed, and has been contrasted with *A. barrandonii* Duval-Jouve, a more robust form with a localized distribution near Montpellier in southern France (A. *filiformis* ssp *filiformis* Aschers. et Graebner, and ssp *barrandonii* (Duval-Jouve) Aschers. et Graebner). They can be distinguished in their leaf anatomy according to Sauvageau (1891) as follows:

1. Submarginal fibrous strand of leaf blade well developed, additional fibrous strands on inner surface of sheath; endodermal sheath of leaf vein thick-walled *A. barrandonii.*

1a. Fibrous strands absent from blade and sheath; endodermal sheath of leaf vein thin-walled *A. filiformis.*

In this sense the material described by Prillieux would correspond to *A. barrandonii.*

In view of the work of Reese (1963, 1967) on *Zannichellia*, the cytological examination of these forms would be worthwhile.

MATERIAL EXAMINED

Althenia filiformis Petit ssp *barrandonii* (Duval-Jouve) Aschers. and Graebner; Mares des Onglous, near Sète (or Cette), Hérault, France; L. G. de Solignac *s.n.*; all parts.

LEPILAENA Drumm. ex Harv.
(Figs. 10.6, 10.7, and 10.9 A, B and Plate 14)

Four poorly understood species growing in fresh or brackish water in southern Australia, Tasmania, and New Zealand (Aston 1973).

VEGETATIVE MORPHOLOGY

Shoots probably dimorphic, as in *Althenia*. Creeping, rooted rhizome apparently branching sympodially and giving rise to erect lateral branches at each node, the rhizomatous axis thicker than the erect stems (Fig. 10.6 A, B). Lower part of erect stem unbranched, but branched distally (Fig. 10.6 C) to produce short determinate flowering shoots (inflorescences). Foliage leaves with a ligular extension of the basal sheath, much as in *Althenia* (Fig. 10.6 D, E). Leaf apex truncate or sometimes more or less three-toothed (Fig. 10.6 F); rounded and with marginal teeth in *L. australis*.

VEGETATIVE ANATOMY

Exhibiting the characters of the family, but mesophyll fibres arranged in three alternate ways in the leaf blade:

(1) *L. bilocularis*, fibres exclusively submarginal, in a slender inconspicuous strand; (2) *L. cylindrocarpa*, fibres exclusively submarginal, in a wide, conspicuous strand (Fig. 10.9 A, B); (3) *L. preissii* (Sauvageau 1891a), submarginal fibrous strand conspicuous and accompanied by additional irregular hypodermal strands throughout mesophyll.

REPRODUCTIVE MORPHOLOGY

Inflorescence (e.g. Fig. 10.7 A, Plate 14)

Unisexual (plants possibly dioecious), each axis ending in a male or female flower (Fig. 10.7 A). Inflorescence axis bearing usually one (male) or two (female) pairs of subopposite, reduced leaves. These subtending (i) further flowers, (ii) further inflorescence units, or (iii) vegetative branches which may abort. Some inflorescence branches each with a basal scale-like prophyll; others lacking this basal scale, as in *Althenia*. Developmental details illustrated for *L. bilocularis* (Plate 14 A) and *L. cylindrocarpa* (Plate 14 B).

Male flower. Anther with either two pairs of microsporangia (*L. bilocularis*, Fig. 10.7 H, I), or six pairs of microsporangia (remaining species), on a short pedicel with a short basal unequally three-lobed perianth, the method of overlapping of the lobes not constant (e.g. Fig. 10.7 I). Connective extended into a short point in *L. bilocularis* (Fig. 10.7 F–I).

Female flower. Each including (1–)3 carpels, each carpel subtended by a perianth segment (Fig. 10.7 B–E); segments simple (e.g. *L. australis*) or bifid (e.g. *L. cylindrocarpa*). If carpel solitary, then perianth very short. Style long or short, the stigma expanded and either funnel-shaped (e.g. *L. cylindrocarpa*) or feathery (e.g. *L. bilocularis*; Fig. 10.7 C). Pedicel either not or scarcely elongating in fruit (e.g. Fig. 10.7 J), or much extended (e.g. *L. australis*).

TAXONOMIC AND MORPHOLOGICAL NOTES

The morphology of this genus is in need of detailed study because descriptions even of types remain very confused. The best account is by Aston (1973). Some attempt at clarification is provided by Posluszny and Tomlinson (1976) in their study of floral development. The earlier revision by Ascherson and Graebner (1907) was characterized by certain inconsistencies in their descriptions. For example, they distinguish *Lepilaena* from *Althenia* by stamens with six instead of two thecae. However, *L. bilocularis*, the New Zealand species, has two thecae. As Reinecke (1964) points out, the descriptions of anthers in this genus (and even the family as a whole) are not clear despite the fact that they are diagnostically useful. The tendency of authors to interpret the single anther of several species as consisting of two or three 'fused stamens' according to the total number of pairs of microsporangia adds to the confusion, although there is no convincing evidence for this interpretation.

MATERIAL EXAMINED

Lepilaena bilocularis Kirk; Kaituna Lagoon, Canterbury, New Zealand; L. B. Moore; duplicate of CHR 184514. Lake Ellesmere, Canterbury, New Zealand; J. Clarke; duplicate of CHR 184515: all parts.
L. cylindrocarpa (Koern.) Benth.; Boomer Marsh, Tasmania; W. M. Curtis: all parts.

SPECIES REPORTED ON IN THE LITERATURE

Burgemeister (1968): *Zannichellia palustris*: leaf.
Campbell (1897): *Z. palustris*: all parts.
Chrysler (1907): *Z. palustris*: vascular system of stem.
Esenbeck (1914): *Z. palustris*: note on stomata.
Graebner and Flahault (1908): *Z. palustris*: general morphology.

Hochreutiner (1896): *Z. palustris*: all parts.
Irmisch (1851): *Z. palustris*: morphology.
Lüpnitz (1969): *Z. palustris*: root development.
Monoyer (1927): *Z. palustris*: vascular system of stem.
Pettitt and Jermy (1975): *Z. palustris*: pollen structure.
Posluszny and Sattler (1976*b*): *Z. palustris*: floral development.
Posluszny and Tomlinson (1977): *Althenia filiformis, Lepilaena bilocularis, L. cylindrocarpa, Zannichellia palustris.*
Prillieux (1864): *Althenia filiformis*: all parts.
Reese (1963): *Zannichellia palustris:* leaf.
Reinecke (1964): *Vleisia aschersoniana* (as *Zannichellia aschersoniana*): general morphology.
Sauvageau (1891*a*): *Althenia barrandonii, A. filiformis, Lepilaena preissii, Zannichellia lingulata, Z. pedicellata, Z. repens*: leaf, root.
Schenck (1886*b*): *Zannichellia palustris*: leaf, root.
Şerbănescu (1980): *Z. palustris, Z. prodanii*: stem, leaf.
Şerbănescu-Jitariu (1972*b*): *Z. palustris*: floral anatomy.
Singh (1965*a*): *Z. palustris*: floral anatomy.
Uhl (1947): *Z. palustris*: floral morphology and anatomy.

SIGNIFICANT LITERATURE—ZANNICHELLIACEAE

Arber (1920); Ascherson and Graebner (1907); Aston (1973); van Beusekom (1967); Burgemeister (1968); Campbell (1897); Chrysler (1907); Eber (1934); Esenbeck (1914); Fischer (1907); Graebner and Flahault (1908); Gravis (1934); den Hartog (1975); Hisinger (1887); Hochreutiner (1896); Irmisch (1851); Jönsson (1881); Lakshmanan (1965); Lüpnitz (1969); McClure (1970); Magnus (1870, 1871); Markgraf (1936); Monoyer (1927); Obermeyer (1966*b*); Ohlendorf (1907); Onnis (1969); Onnis and Mazzanti (1971); Pettitt and Jermy (1975); Posluszny and Sattler (1976*b*); Posluszny and Tomlinson (1977); Prillieux (1864); Raunkiaer (1895–9); Reese (1963, 1967); Reinecke (1964); Roze (1887); Sauvageau (1891*a*); Schenck (1886*a, b*); Schonland (1924); Schumann (1892); Şerbănescu (1980); Şerbănescu-Jitariu (1972*b*); Singh (1965*a*); Solereder and Meyer (1933); Souèges (1954); Tomlinson and Posluszny (1976); Uhl (1947); Venkatesh (1952); Vijayaraghavan and Kumari (1974); Warming (1873); Wettstein (1935); Wille (1882).

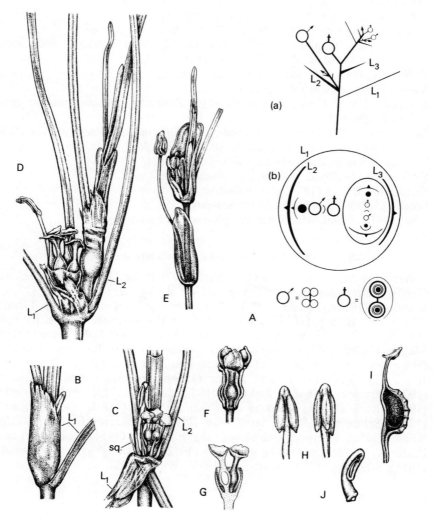

PF

Fɪɢ. 10.1. ZANNICHELLIACEAE. *Zannichellia palustris* L. Reproductive morphology. (Havelock North, New Zealand, P. B. Tomlinson 14.i.72 *A*.)

A. Interpretative diagram of flower complex (after Posluszny and Sattler 1976). (a) Two successive bisexual complexes in side view, the parts shown in one plane. (b) Same in plan view to show actual orientation of parts. L_1, L_2, L_3 = foliage leaves of a single pseudowhorl, L_1 = tubular prophyll. Axes ending in an arrow continue the development of the shoot system.

B–E. Flower complexes at different stages.
   B. Young flower complex ($\times$ 9) enclosed by ligular sheath of leaf L_1 (= prophyll of renewal shoot). C. Somewhat older complex ($\times$ 9) with L_1 pulled back to show male and female 'flowers'; sq. = squamules. D. Older complex ($\times$ 9) with renewal shoot well developed. E. Shoot ($\times$ 4) with renewal shoot well developed.

F–J. Details of reproductive organs.
   F. Female 'flower' ($\times$ 9) from the side. G. Same in LS ($\times$ 9). H. Stamen ($\times$ 12) from opposite sides. I. Young fruit ($\times$ 4). J. Embryo ($\times$ 9).

F IG. 10.2. ZANNICHELLIACEAE. *Vleisia aschersoniana* (Graebn.) Tomlinson and Posluszny. Vegetative and reproductive morphology. (Loc River, South Africa; P. Reineke *s. n.*)

A. Branching of erect shoot below flowering level ($\times$ 3); branch in axil of leaf to right, with its prophyll-scale pr.

B. Details of leaf morphology ($\times$ 3); (*above*) ligulate leaf bases of two foliage leaves showing different degrees of development of the sheath; (*below*) leaf tip.

C–D. Male flowering branch (cf. Fig. 10.3 C, D).
    C. Young branch ($\times$ 3) with unextended stamens. D. Older branch ($\times$ 3) with extended stamens, inset single stamen ($\times$ 6).

E–I. Female flowering branch (cf. Fig. 10.3 A, B).
    E. Axis with solitary terminal female 'flower' ($\times$ 3). F. Older axis ($\times$ 6) with second lateral flower just visible. G. Single 'flower' with enveloping leaf ($\times$ 3). H. 'Flower' in LS ($\times$ 9), enclosed by tubular spathe-like leaf I. Two scale-leaves ($\times$ 3) associated with female 'flowers.'

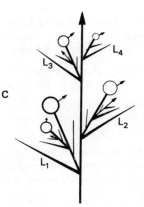

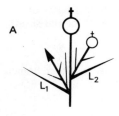

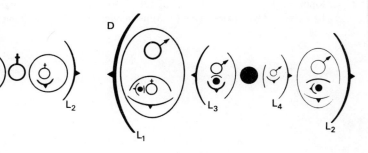

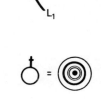

FIG. 10.3. ZANNICHELLIACEAE. *Vleisia aschersoniana.*

A. Diagram of representative determinate (female) shoot from the side with parts shown in one plane.

B. The same in plan view to show actual orientation of parts.

C. Diagram of representative indeterminate (bisexual) shoot from the side with parts shown in one plane.

D. The same in plan view to show actual orientation of parts. L_1, L_2, L_3, L_4 = foliage leaves on main axis subtending flower-bearing complexes. Axes ending in an arrow continue the development of the shoot system.

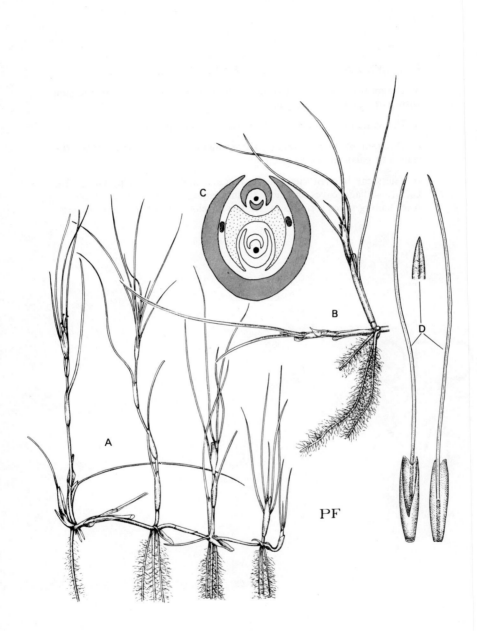

FIG. 10.4. ZANNICHELLIACEAE. *Althenia filiformis* Petit. Vegetative morphology. (Mares des Onglous, Sète, Hérault, France. L. G. Solignac *s.n.*).

A. Habit of rooted rhizomatous portion (× 2), the rhizome a sympodium, each erect shoot the evicted terminal shoot.

B. Detail (× 3) of terminal shoot (to left) showing its precocious development before it turns erect.

C. Diagram of leaf arrangement; rhizome scales-hatched; prophyll-stippled.

D. Single foliage leaf from front and back (× 4), with a leaf tip inset (× 18).

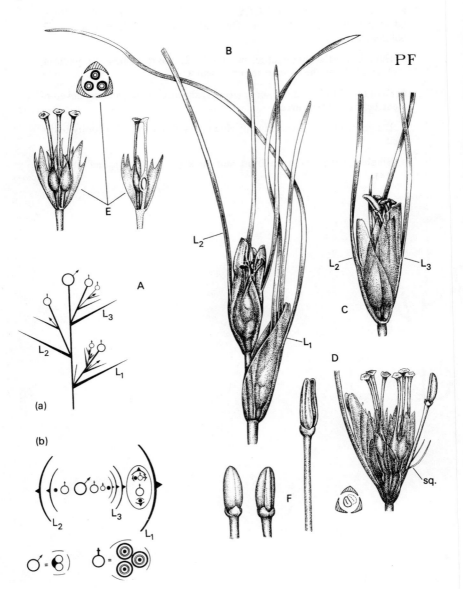

FIG. 10.5. ZANNICHELLIACEAE. *Althenia filiformis* Petit. (Same locality as Fig. 10.4.) Reproductive morphology.

A. Interpretative diagrams of arrangement of parts (see text). (a) diagram of representative flowering shoot from the side with leaves shown in one plane; (b) plan diagrams of flowering shoot, showing actual orientation of parts. L_1, L_2, L_3 = foliage leaves on main axis, variously subtending flowers or flower complexes. Axes ending in arrows continue the development of the shoot system.

B–F. Whole flower complexes and their dissections.

B. Two successive flower complexes ($\times$ 6) from the side, corresponding to Aa. C. Terminal flower complex ($\times$ 6); stamen and styles protruding. D. Older flower complex ($\times$ 6) with two successive units at similar developmental stages. E. Female 'flower' ($\times$ 6), to left from the side, to the right in LS, with floral diagram above. F. Male 'flower' ($\times$ 12), three different views, the one to the right with the pedicel extended and anther dehisced, floral diagram to the right.

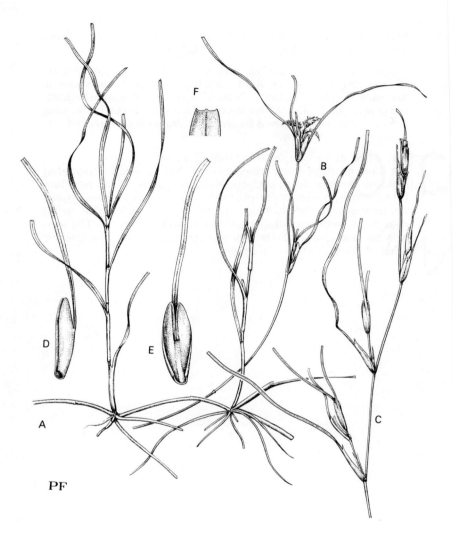

PF

FIG. 10.6. ZANNICHELLIACEAE. *Lepilaena bilocularis* Kirk. Vegetative morphology. (Kaituna Lagoon, Canterbury, New Zealand; L. B. Moore CHR 184514.)

A. Part of basal rooted rhizomatous system (× 2), with two developing erect shoots.

B. Distal part of erect shoot (× 2), with female flowers initiating reproductive cycle.

C. Distal part of erect shoot (× 2) with young male flowers, the prophylls of the two lowest branches exposed.

D–E. Foliage leaf (× 6) from outside (D) and inside (E) respectively.

F. Leaf tip (× 35).

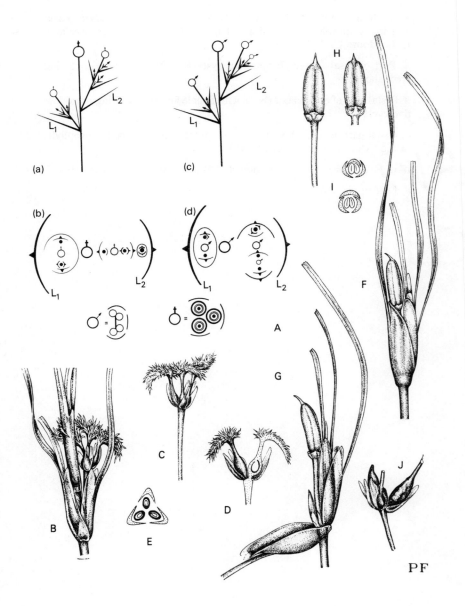

FIG. 10.7. ZANNICHELLIACEAE. *Lepilaena bilocularis* Kirk. (Same locality as Fig. 10.6.) Reproductive morphology.

A. Interpretative diagrams of arrangement of parts (see text).

(a) Diagram of female flowering shoot from the side with leaves shown in one plane.

(b) Same in plan view showing actual orientation of parts.

(c) and (d) Same for representative male flowering shoot.

L_1, L_2 = foliage leaves on main axis. Axes ending in arrows continue the development of the shoot system.

B–E. Female inflorescence, corresponding to Aa–b above.

B. Inflorescence from the side ($\times$ 6). C. Single female 'flower' ($\times$ 6). D. Female 'flower' in LS ($\times$ 6). E. Floral diagram of female 'flower.'

F–I. Male inflorescence, corresponding to Ac–d above.

F. Inflorescence from the side ($\times$ 6). G. Same ($\times$ 6) with lower leaf pulled back to expose its subtended branch. H. Male 'flower' from two sides ($\times$ 6). I. Floral diagram of male 'flower' with two alternative arrangements of subtending scales.

J. Young fruits ($\times$ 6).

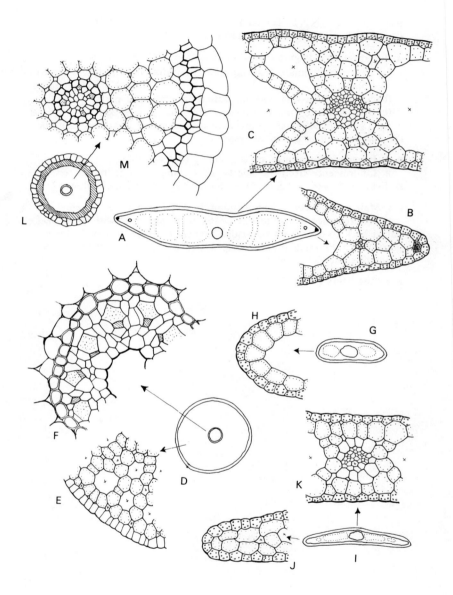

FIG. 10.8. ZANNICHELLIACEAE. Vegetative anatomy.

A–F. *Vleisia aschersoniana.*
A. TS lamina (× 40). B. Detail of leaf margin (× 150). C. Detail of midvein (× 150). D. TS stem (× 20). E. TS surface layers (× 80). F. TS stele (× 150).

G–M. *Zannichellia palustris.* (G, H, Boomer Marsh, Tasmania; I,J. Havelock North, New Zealand.).
G. TS lamina (× 40). H. Detail of leaf margin (× 150). I. TS lamina (× 40). J. Detail of leaf margin (× 150). K. Detail of midvien (× 150). L. TS root (× 80). M. Detail including stele (× 150).

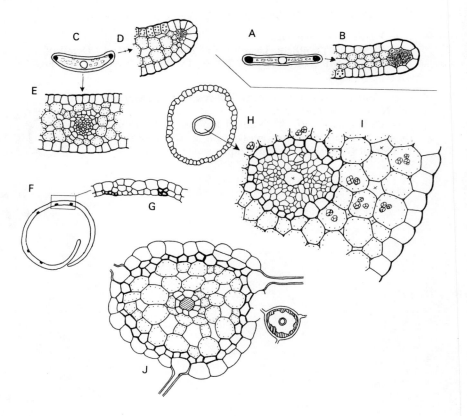

FIG. 10.9. ZANNICHELLIACEAE. Vegetative anatomy.

A–B. *Lepilaena cylindrocarpa*; leaf anatomy.
  A. TS lamina (× 40). B. Detail of leaf margin (× 150).

C–J. *Althenia filiformis*; leaf, stem, and root anatomy.
  C. TS lamina (× 40). D. Detail of leaf margin (× 150). E. TS midrib
(× 150). F. Outline TS of prophyll (× 40). G. Detail of median part of
prophyll to show fibres (× 15). H. TS stem outline (× 20). I. Detail of
stele (× 300). J. TS root with only base of root hairs shown, inset
(× 10), in detail (× 100) the stele shown hatched.

Family 11
POSIDONIACEAE (Lotsy 1911)
(Figs. 11.1–11.3 and Plate 3E)

SUMMARY

THIS monogeneric family of marine herbs consists of at least five species: *Posidonia oceanica* (= *P. caulinii*) is Mediterranean, the remaining four species are Australian (Cambridge and Kuo 1979). They grow in coastal waters to a depth of 50 m. The rhizomes are monopodial, irregularly branched, with strap-shaped to almost terete ligulate leaves, and the older rhizomes are clothed with persistent fibrous remains of leaf bases (Fig. 11.1 A). The axillary inflorescences are racemose and bear leafy bracts. The ultimate units are spike-like with several lateral and apparently naked flowers. Each flower has three stamens and a single carpel (Fig. 11.2 A–F). Each stamen has two bilocular thecae attached to a broad connective. The pollen is filamentous. The carpel is ellipsoid, with a short, irregularly-lobed style and a solitary, orthotropous, pendulous ovary. The fruit has a spongy pericarp but dehisces to release the elongated seed.

Sixteen-twenty squamules are present at each node (Fig. 11.1 D, F, *sq.*). The epidermis is photosynthetic; hairs and stomata are absent. Fibres are well developed as a subepidermal system of strands in the lamina, and they become aggregated into wide, lignified, long-persisting strands towards the base of the leaf sheath. Fibre strands are also present in the cortex of the rhizome. Air-lacunae are developed in the mesophyll of the leaf, as well as in the cortex of stem and root. Tannin cells are abundant in all parts, but especially in the epidermis.

FAMILY DESCRIPTION
VEGETATIVE MORPHOLOGY

Axis

Rhizome ± horizontal, bearing foliage leaves, monopodially but irregularly branched (Fig. 11.1 B); most foliage leaves axillant to a prophyllate bud Fig. 11.1 G); the bud either remaining undeveloped (Fig. 11.1 I), or occasionally becoming a precocious vegetative branch (Fig. 11.1 F, H) repeating the structure of the parent rhizome. Bud alternatively developed as an erect flattened shoot with a terminal, branched inflorescence. Rhizome thick (up to 1 cm diameter) with short internodes in *P. oceanica*, but often slender with internodes elongated at intervals in *P. australis*, and especially *P. ostenfeldii*; but the periodicity of internodal extension in the last two spp not known. Branch sometimes adnate to parent axis in *P. ostenfeldii*.

Leaf

Foliage leaves transversely distichous (Fig. 11.1 E) on the rhizome, each leaf having an open sheathing base and a distal strap-shaped blade, with a narrow ligule marking the junction of the sheathing base and leaf blade (Fig. 11.1 c). Wings of the sheath widest at the leaf base, overwrapping somewhat where inserted on the stem. Blade eventually becoming detached in the region of the ligule, the persistent sheath decomposing and so leaving only the constituent fibres. Leaf apex rounded, without either a pore or marginal teeth. Veins all united apically with a marginal commissure (Fig. 11.1 E) except for a minute prolongation of the median vein beyond the commissure. **Prophyll** (Fig. 11.1 G) undifferentiated, with several (7–10) veins.

VEGETATIVE ANATOMY

Leaf

(i) **Lamina** (Figs. 11.2 G–I and 11.3 A–F). **Hairs** absent. Apical pore not developed. **Cuticle** thin, penetrating somewhat between the anticlinal walls. **Epidermis** (Fig. 11.3 B) uniformly chlorenchymatous, consisting of regular longitudinal files of mostly rectangular, longitudinally-extended cells, but cells sometimes more or less square in surface view. Anticlinal walls apparently slightly sinuous in surface view in *P. ostenfeldii*. Outer wall of the epidermal cells thickened (Fig. 11.3 D, E), the remaining walls slightly thicker than those of the mesophyll cells. Many, often most, epidermal cells including brown tanniniferous deposits but otherwise uniform. Subepidermal cell layers often compact (Figs. 11.2 H, I and 11.3 A). **Mesophyll** homogenous, composed of wide, elongated cylindrical cells with truncate ends, and containing a few large chloroplasts. Inner mesophyll layers in *P. ostenfeldii* fairly compact and not developing pronounced air-lacunae; in *P. australis* and *P. oceanica* commonly forming a reticulum of loose cells enclosing irregular longitudinal air-lacunae. **Air-lacunae** in *P. oceanica* relatively narrow, hexagonal in TS and each usually enclosed by six mesophyll cells; those in *P. australis* wider (Fig. 11.3 A), polygonal and each enclosed by more than six mesophyll cells. Individual lacunae of relatively limited longitudinal extent, continuity from one lacuna to another resulting from the irregular distribution of mesophyll cells. **Transverse diaphragms** largely represented by transverse veins embedded in narrow, mostly biseriate transverse partitions of small compact, densely cytoplasmic mesophyll cells, separated by narrow intercellular spaces; diaphragms never extending to epidermis; occasional partitions without vascular tissue observed.

Fibres (Figs. 11.2 G–I and 11.3 A, D, E) abundant in small strands; unlignified (except for the middle lamellae, according to Sauvageau 1890a, 1891a). Fibrous strands mostly hypodermal or subhypodermal in the compact outermost mesophyll layers (Fig. 11.3 A, D, E), most abundant in the leaf margin, and sometimes also abundant in the mesophyll below the median vein. **Vascular bundles** forming a single series of veins (Figs. 11.2 G and 11.3 c) up to 17 in *P. oceanica*, or as few as five in *P. ostenfeldii*; independent of and equidistant from each surface. Median vein wider than lateral veins, but not forming a distinct midrib (Fig. 11.2 H); enclosed by an indistinct 1–2 seriate outer sheath of elongated, slightly lignified compact cells, the cell contents appearing mucilaginous in fresh sections. Sheath cells described by

Sauvageau (1891a) as an endodermis and the adjacent layer as a pericycle but simply a single lignified layer in *P. australis* according to Kuo (1978). Vascular elements enclosed by an indistinct layer of conjunctive parenchyma. Phloem including several wide sieve-tubes. Xylem represented either by a single narrow tracheid with inconspicuous wall thickenings or, more usually, by a distinct lacuna. Lateral veins similar to midvein but narrower (Fig. 11.3 F).

Transverse commissures frequently connecting the longitudinal veins, the spacing varying in different spp; conspicuous to the naked eye due to the dense tannin deposits in the associated mesophyll. Transverse veins sheathed by a single layer of slightly thick-walled cells, these cells conspicuously tiered in surface views of cleared leaves, suggesting the development of each tier from a longitudinal file of cells in the mesophyll of the leaf primordium. Vascular elements including a single file of short tracheids (see above).

(ii) **Leaf margin.** Rounded in TS. Marginal epidermal cells usually narrow and thick-walled, especially in *P. ostenfeldii.* Epidermal cells often without tannin, and so conspicuously different from those elsewhere. Marginal mesophyll compact, including numerous fibrous strands, air-lacunae absent.

(iii) **Ligule.** Short, curved (Fig. 11.1 C), originating from the epidermis according to Weber (1956); at maturity biseriate distally, three-layered at insertion. Ligular region of the leaf marked by a close aggregation of transverse diaphragms. Mesophyll at the level of the ligule consisting of short, thick-walled cells but not constituting a distinct abscission layer. Blade usually becoming detached above this level in old leaves. Sheath tissues below the ligule, especially the fibres, tending to be lignified. Tissues at the distal end of the blade not, or less, lignified, this possibly accounting for the ephemeral nature of the blade and persistence of its sheath, as pointed out by Sauvageau (1891a) and Kuo (1978).

(iv) **Leaf sheath.** Differing from the blade by having a scarious, non-vasculated margin; overwrapping below. Sheath mesophyll less lacunose than that of the blade; but more lacunose in *P. australis* than in *P. oceanica* according to Sauvageau (1891). Sheath wing narrowed gradually to a biseriate margin, without vascular tissue. Hypodermal **fibrous strands** of the lamina continued into the sheath, initially forming a more or less continuous hypodermal fibrous layer, most of them extending into the marginal wings on the abaxial side. In *P. oceanica* fibres of the abaxial surface near the sheath insertion becoming aggregated into several massive lignified fibrous strands separated from the epidermis by 2–3 compact mesophyll layers; sheath fibres persistent on old rhizomes. Adaxial fibrous strands in other *Posidonia* spp less well developed, with fewer conspicuous subepidermal fibrous strands but including numerous narrow fibrous strands scattered throughout the mesophyll and embedded in the longitudinal parenchyma surrounding the air-lacunae. In *P. australis* Kuo (1978) described fibres as discontinuous from leaf sheath into the blade, this and their lignification accounting for the early loss of the blade.

Stem

(i) **Rhizome** (Fig. 11.3 H–K and Plate 3 E). Shoot apex flattened in the plane of insertion of leaves, the rhizome remaining elliptical in TS (Fig.

11.3 H) and markedly dorsiventral (Weber 1956). Epidermis thin-walled, tanniniferous. **Cortex** wide in proportion to narrow central stele (Fig. 11.3 H). Peripheral cortical cells compact, slightly thick-walled or even collenchymatous, the outermost 4–5 layers becoming lignified to form an indistinct peripheral mechanical layer. Middle and inner cortex lacunose (Fig. 11.3 K), including numerous and often massive lignified fibrous strands (Fig. 11.3 H–K). Individual fibres thicker-walled and with wider lumina than those of leaves (Fig. 11.3 J). Cortical vascular system inconspicuous, as described below. **Stele** narrow, irregularly lobed or stellate in transverse outline (Plate 3 E) and delimited from the small-celled inner cortex by an **endodermis** with well developed Casparian thickenings (Fig. 11.3 I). Endodermis discontinuous at leaf trace insertions. Stele consisting of several indistinct vbs representing the fused median traces of successive leaves, together with lateral strands associated with root insertions, the separate leaf traces remaining recognizable for some distance below their exsertion into a leaf.

Each vascular strand (Fig. 11.3 I) represented by a phloem strand with numerous wide conspicuous sieve-tubes, the xylem on the inner side represented at most by an indistinct series of short, narrow tracheids with scalariform-reticulate wall thickenings, but xylem usually undifferentiated or represented by a lacuna. Centre of the stele occupied by an irregularly fluted pith of thick-walled, lignified parenchyma.

(ii) **Inflorescence axis.** More leaf-like than stem-like in its anatomy. Surface layers as in the leaf, with numerous hypodermal fibrous strands. Cortex lacunose. Vascular system not distinctly differentiated into a cortical system and a stele, but with numerous independent vbs, most closely aggregated at the base of the axis.

(iii) **Course of vascular bundles in rhizome** (after Monoyer 1927, based on *P. oceanica*). Vascular bundles arranged in two discrete series: (a) A complex of **cortical vbs** arranged in positions corresponding to the planes of insertion of the leaves, and representing the fused lateral bundles of successive leaves, one cortical stem bundle corresponding to each lateral leaf bundle. Each lateral leaf vein extending independently downwards in the cortex through one internode below its insertion before uniting with a corresponding cortical bundle. Vascular bundles in the cortex not directly continuous with those in the stele; (b) **Stelar vbs,** continuous with the median veins of the leaves. Each median vein below its insertion passing obliquely downward through cortex over a distance of one internode, before being inserted in the stele in a median position. Median vein remaining recognizable as a discrete bundle for a further internode, before splitting into two separate lateral strands. Lateral strands then moving tangentially to opposite sides of the stele, the split strands still remaining recognizable for another two internodes before uniting with a lateral 'sympodial' strand (*faisceaux sympodique* of Monoyer).

[If the vascular strands are always strictly developed according to this scheme, at any one level in the internode the stele would include seven vascular bundles or at least be seven-lobed. The cortex would include the basal continuation of the lateral strands leading to two successive leaves, together with a median leaf trace passing across the cortex on its way to the stele. The distribution of vascular tissues in rhizomes which have been examined by me does not coincide precisely with Monoyer's formalized scheme. Furthermore, Monoyer's analysis does not describe the insertion of branch and root traces. In my own investigation vegetative branching was indicated by sections of rhizomes including two separate steles, reflecting the tendency for branch adnation, but this has not been studied more completely. Root traces apparently unite with the lateral 'sympodial bundle' of the stele. Larger fibrous strands of the leaf sheath and stem cortex are also apparently continuous, thereby accounting for the persistence of the fibrous part of leaf remains. Albergoni *et al.* (1978) interpret the vascular system of *P. oceanica* as 'polystelic' since they consider the cortical bundles to be 'lateral steles,' largely because they have an individual endodermis and are amphicribal. The outermost leaf trace is excluded because it lacks these anatomical features. Comparison with other Helobiae suggests that this elaborate interpretation is not necessary.

Developmental studies of the vascular system of this genus are much needed, taking into account the rhizome architecture and variation in internode length.]

Root (Fig. 11.3 G). Roots branched, arising in two alternating series from each side of stem, towards its lower surface, each root corresponding to a leaf according to Weber (1956). In spp with elongated internodes the roots more obviously inserted in pairs at a node. Root-hairs infrequent, but recorded by Sauvageau (1889*b*). **Epidermis** thin-walled; exodermis several-layered and forming a rigid mechanical region of thick-walled compact cells below the epidermis. **Cortex** (Fig. 11.3 G) wide; outer layers within the exodermis compact, cells wide; inner cortex including a well developed system of radial air-lacunae delimited by usually uniseriate radial plates of cells. Innermost cortex compact, small-celled, cells in regular concentric and radial series. **Endodermis** with distinct Casparian thickenings, the cells radially coincident with innermost cortical cells. Endodermal cells in old roots and even the inner cortical cells opposite the phloem often becoming slightly and uniformly thickened and lignified. **Stele** narrow. Pericycle indistinct. Xylem represented by up to 15 peripheral strands, each with 2–5 narrow tracheal elements with inconspicuous, mostly scalariform-reticulate wall thickenings. Alternatively xylem elements not differentiated. Phloem strands alternating with the xylem, each including a single sieve-tube situated in the pericycle. Medulla wide; cells becoming thick-walled in old roots.

SECRETORY, STORAGE, AND CONDUCTING ELEMENTS

Tannin abundant in all parts in otherwise unmodified cells (e.g. Figs. 11.2 H and 11.3 I, K, *cross-hatched*) as dense brown deposits; common and often v. abundant in the epidermis, locally abundant around transverse veins of leaf, often in short longitudinal series of cells. Tanniniferous contents sometimes appearing foam-like, as noted by Solereder and Meyer (1933). Epidermis of axis and old leaves apparently becoming wholly tanniniferous. The high tannin content supported by the histochemical observations of Kuo (1978) and Kuo and Cambridge (1978).

Starch abundant as small, mostly simple, spherical or ellipsoidal grains in the cortex of the rhizome; less commonly in the mesophyll of the leaf.

Xylem: tracheary elements poorly differentiated in all parts as narrow cells with inconspicuous spiral or scalariform-reticulate wall thickenings. Wall thickenings often not evident in TS and the tracheary element then represented by a narrow lacuna, as in the longitudinal veins of the leaf.

Phloem: sieve-tubes of the leaf, and wide conspicuous sieve-tubes of the rhizome provided with simple transverse sieve-plates.

REPRODUCTIVE MORPHOLOGY

Inflorescences arising as erect, lateral, apparently axillary, branches from the rhizome. Bracts resembling foliage leaves but with their blades progressively more reduced towards the distal end of the axis (Fig. 11.2 A). Branching racemose throughout, the bracts each subtending a prophyllate branch, each branch with two (rarely more) further short bracts (shortly bladed leaves or bladeless sheaths) below the terminal series of flowers. Ultimate units of the inflorescence essentially spicate, and consisting of a 1–7-flowered axis, apparently always ending in a terminal sterile appendage (never a terminal flower; Fig. 11.2 B, arrow). Inflorescence of *P. oceanica* much more congested than that of *P. australis* and *P. ostenfeldii,* and the parts more difficult to see.

Flowers (Fig. 11.2 F) perfect (occasionally male by abortion), each with three stamens and a single carpel. Stamens each with a broad connective, the connective extending into a long appendage in *P. oceanica* (Fig. 11.2 C, D). Each connective with two bisporangiate thecae, apparently attached abaxially, the thecae dehiscing longitudinally and extrorsely. Carpel ellipsoid, style short, stigma irregularly lobed (Fig. 11.2 B). Ovule solitary, orthotropous, and pendulous from the upper part of the ovary (Fig. 11.2 E). Fruit ovoid with a fleshy and more or less spongy pericarp, initially detached and free-floating, but eventually dehiscing at one end by longitudinal slits to release the single seed. Embryo with an enlarged hypocotyl. Seedling with (*P. oceanica*), or without (*P. australis, P. ostenfeldii*), a well developed primary root.

ECONOMIC USES

Detached leaf blades and rhizome segments of *P. oceanica* in the Mediterranean become rolled into small balls by wave action, with their persistent fibres enmeshed. These objects (*boules feutrées, pelotes de mer, aegagropiles de mer, Posidonian-balles, See-ballen, Pilae marinae,* etc.) are frequently washed ashore. They have been produced artificially—in a washing machine (Cannon 1979)!. *Posidonia australis* fibres are lignified and have been examined as a potential source of textile or paper fibre (Winterbottom 1917). The fact that the Australian spp of *Posidonia* also make these seaballs is independent of their somewhat different fibre structure and distribution. The major economic importance of *Posidonia* is indirect, by virtue of the habitat it provides for many marine animals and its ability to prevent submarine erosion. However, massive meadows are sensitive to pollution and there is ecological concern for the ability of *Posidonia* to recover from man-made destructive processes.

Taxonomic and Morphological Notes

Characters which may be used to distinguish *Posidonia* species in the vegetative state include leaf morphology and rhizome organization, but there is so much phenotypic variation that it is difficult to provide a precise diagnosis. Among the characters that can be used are blade width and thickness (becoming subterete in some forms of *P. ostenfeldii*), frequency of transverse veins, number of longitudinal veins per leaf, shape and dimensions of epidermal cells, frequency and distribution of fibrous strands, and the extent to which air-lacunae are developed. Other diagnostic characters are the shape of leaf margin in TS, the shape of the ligule and sheath auricles, and the distribution of fibres in the leaf sheath. Cambridge and Kuo (1979) show that the three spp of *Posidonia* in the vicinity of Perth, Western Australia (two of them new) can be distinguished by the ultrastructure of the epidermal cells. It would be useful to establish consistently different and easily recognizable vegetative characters for the separation of species since *Posidonia* seems to flower rarely. However, if vegetative characters are to be used, it will be necessary to establish their reliability by examining a wide range of material.

Inflorescence and floral morphology

Den Hartog's suggestion (1970*a*) that the inflorescence of *Posidonia* is cymose in its construction seems incorrect. I have confirmed Ostenfeld's (1916) analysis in *P. oceanica* and *P. australis*. The presence of a terminal sterile continuation of each ultimate flower-bearing axis supports a racemose interpretation, because there is no terminal flower.

Attention has been drawn to scarious downwardly directed outgrowths below each flower, which surround the base of the flowers to varying degrees. Not all observers, however, have recorded these structures, possibly because they seem to originate at a late stage in flower development, e.g. at anthesis. The most obvious interpretation is that these outgrowths represent the subtending bract of the flower, as suggested by Ferguson-Wood (1959). This author goes on to complicate the issue by referring to 'bracteoles.' The presence of bracts below flowers would make the spicate structure a true raceme. Markgraf (1936) and Eckardt (1964) both describe these structures as appendages of the connective. Den Hartog (1970*a*) is of the opinion that the scales arise from the place of insertion of the flower and not the connective, and he interprets them as a group of fused squamules, even though there is no evidence that squamules ever become fused in this way in the Helobiae.

Information from microscopic anatomy and particularly development is badly needed to provide useful interpretative evidence.

Material Examined

Posidonia australis Hook.f.; Spencer's Gulf, S. Australia; no collector's name (113), Jodrell Slide Collection; leaf, stem; dried material. V. I. Cheadle

CA-361 and CA 366; all vegetative parts. Perth, W. Australia; M. C. Cambridge *s.n.* 10.v.77; all vegetative parts.

P. oceanica (L.) Delile; Marine Biological Station, Naples, Italy; M. Délépine *s.n.*; all vegetative parts. Station Marine d'Endoumé, Marseille, France; M. L. Vivien *s.n.* 13.iii.74; flowers.

P. ostenfeldii den Hartog; Perth, W. Australia; M. C. Cambridge *s.n.* 10.v.77; all vegetative parts.

SPECIES REPORTED ON IN THE LITERATURE

Albergoni *et al.* (1978): *Posidonia oceanica*: leaf, rhizome, root.

Cambridge and Kuo (1979): *P. angustifolia, P. sinuosa*: leaf.

Duval-Jouve (1873*b*): *P. oceanica* (as *P. caulinii*): diaphragms in stem.

Grenier (1860): *P. oceanica* (as *P. caulinii*): general morphology.

Kuo (1978): *P. australis*: anatomy and ultrastructure of leaf.

Kuo and Cambridge (1978): *P. australis*: rhizome, root.

Monoyer (1927): *P. oceanica*: stem.

Ostenfeld (1916): *P. australis*: leaf; *Posidonia* sp [?*P. ostenfeldii* den Hartog]: leaf.

Pottier (1934): *P. oceanica*: developmental anatomy.

Sauvageau (1889*b*): *P. oceanica* (as *P. caulinii*): root.

Sauvageau (1890*a*, 1891*a*): *P. oceanica* (as *P. caulinii*): leaf.

van Tieghem and Douliot (1888): *P. oceanica* (as *P. caulinii*): origin of lateral roots.

Weber (1956): *P. oceanica* (as *P. caulinii*): general morphology.

SIGNIFICANT LITERATURE—POSIDONIACEAE

Albergoni *et al.* (1978); Cambridge and Kuo (1979); Cannon (1979); Duval-Jouve (1873*b*); Eckardt (1964); Ferguson-Wood (1959); Gessner (1968); Grenier (1860); den Hartog (1970*a*); Kuo (1978); Kuo and Cambridge (1978); Magnus (1871); Markgraf (1936); Monoyer (1927); Ostenfeld (1916); Pottier (1934); Sauvageau (1889*b*, 1890*a*, 1891*a*); Solereder and Meyer (1933); van Tieghem and Douliot (1888); Weber (1956); Winterbottom (1917).

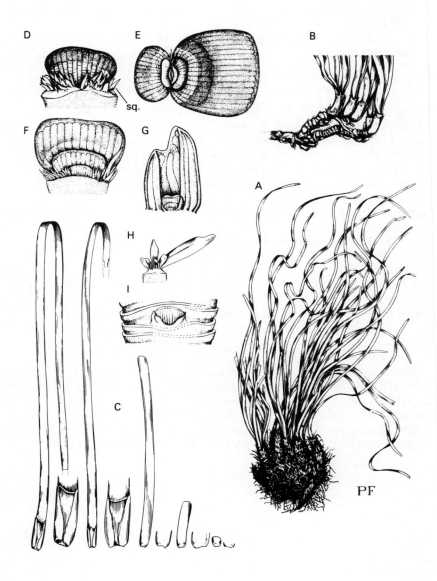

sq.

PF

378

POSIDONIACEAE 379

FIG. 11.1. POSIDONIACEAE. *Posidonia oceanica* (L.) Delile (from fluid preserved material, two Mediterranean collections).

A. Habit (× ½), of specimen with congested rhizomes.

B. Portion of branched rhizome with fibrous remains of older leaf bases all removed (× ½), to show roots restricted to lower rhizome surface and irregular distribution of branches.

C. Leaf series from a single shoot to show basipetal maturation, each leaf (× ¼) with inset (× ½) of leaf base.

D–I. Details of bud construction and branch initiation (× 4), older leaves removed.
   D. Bud from side to show squamules (sq.). E. Bud from above with older primordia laid back. F. Early stage in branch development, from side, subtending leaf removed, leaf behind belongs to parent shoot. G. Dormant or aborted branch with well developed prophyll. H. Semi-diagrammatic representation to show precocious development of a vigorous branch. I. Aborted or dormant bud with small prophyll; produced irregularly along rhizome.

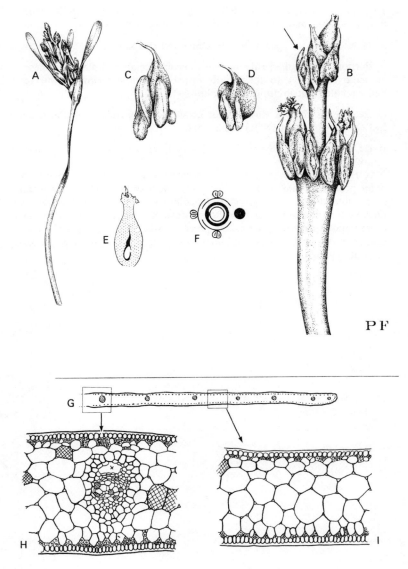

PF

Fɪɢ. 11.2. POSIDONIACEAE. *Posidonia oceanica* (L.) Delile.

ᴀ–ꜰ. Reproductive morphology. (Marseille, at a depth of 6 m. M. L. Vivien *s.n.*)

ᴀ. Entire lateral inflorescence ($\times \frac{1}{2}$). ʙ. Determinate flower axis with bracts removed, arrow indicates sterile extension of main axis ($\times$ 3). ᴄ. Single stamen from outside, with dehiscing anthers ($\times$ 3). ᴅ. Stamen with anthers removed to show expanded connective and distal extension ($\times$ 3). ᴇ. Carpel in LS ($\times$ 3). ꜰ. Floral diagram. In this material there is no subtending 'collar.'

ɢ–ɪ. Leaf anatomy (after Tomlinson 1980).

ɢ. Diagrammatic TS of lamina ($\times$ 16). ʜ. Midvein ($\times$ 150). ɪ. Lateral portion of lamina mesophyll ($\times$ 150).

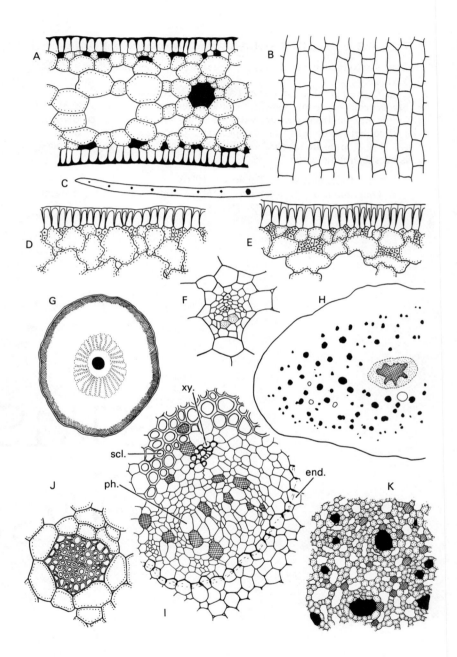

382

FIG. 11.3. POSIDONIACEAE. *Posidonia australis* and *P. oceanica*, vegetative anatomy.

A–F. Lamina.
A, B. *P. australis.* A. TS lateral part of lamina half-way between base and apex (× 150). Hypodermal fibres and vein—solid black. B. Lamina epidermis, surface view (× 310). Cell contents not shown. C, D. *P. ocean-ica.* C. TS one half of lamina, including median vein (× 12). D. TS abaxial surface layers of lamina (× 270). E, F. *P. australis.* E. TS abaxial surface layers of lamina (× 270). F. Enlargement of lateral vein from A; sieve-tubes stippled (× 310).

G. Root of *P. australis.* TS (× 18). Stele—solid black; air-lacunae—dotted outline; peripheral sclerenchyma—cross-hatched.

H–K. Rhizome of *P. australis.*
H. TS (× 12). Cortical fibrous strands—solid black; cortical vbs and median vb—outlined; endodermis—dotted outline; vascular tissue (cf. Plate 3 E)—stippled; medullary sclerenchyma—cross-hatched. I. Details of single 'vascular bundle' included within one groove of fluted sclerotic medulla as in H, TS (× 150). Tannin cells—cross-hatched; xy.—xylem; ph.—phloem; scl.—sclerenchyma of medulla; end.—endodermis. J. Corti-cal fibrous bundle in TS (× 150). K. Lacunose cortex in TS (× 35). Cortical fibrous strands—solid black; tannin cells—cross-hatched.

Family 12
CYMODOCEACEAE (N. Taylor 1909)
(Figs. 12.1–12.8)

SUMMARY

THIS is a natural and widely distributed family of submerged, wholly marine plants (seagrasses) which includes five genera and about 20 species (den Hartog 1970a). These are largely tropical, except for *Amphibolis* from warm temperate southern Australia and *Cymodocea nodosa* from the Mediterranean. The plants are perennials, and they often form extensive stands (submarine mead-. ows), with a more or less clear distinction between creeping (or rhizomatous) and erect axes. The rhizome is either (i) leafy, herbaceous, and monopodially branched (*Cymodocea, Halodule*, e.g. Fig. 12.1 A); (ii) scale-bearing and monopodially branched (*Syringodium*; Fig. 12.2 A, B); or (iii) distinctly woody, scale-bearing and ± sympodially branched (*Amphibolis, Thalassodendron*). The growth habit and mode of branching are described in detail below for individual genera. The leaves are always distichous, of three main types: (a) ligulate foliage leaves with an open sheath, (b) scales (= bladeless sheaths) represented either by prophylls or rhizome scales, (c) bracts, i.e. leaves with reduced blades, associated with flowers in certain genera. The leaf blade is terete in *Syringodium*, but otherwise linear and then narrow (e.g. *Halodule*) to broad (e.g. *Amphibolis*), with from as few as three to as many as 25 veins. Apical pores sometimes develop by the irregular breakdown of leaf tissue (e.g. *Halodule*). The shape of the leaf apex provides diagnostic characters for most species. The roots are branched in all genera except *Halodule*. There are 2–several squamules at each node.

The plants are dioecious. The flowers are naked and usually solitary and terminal at the ends of short erect shoots, or on branches of erect shoots, but they form distinct cymose inflorescences in *Syringodium* (Tomlinson and Posluszny 1978). Male flowers are either subsessile or stalked, the stalk elongating at anthesis. Each male flower has two tetrasporangiate anthers standing back-to-back and partly or completely fused. The anthers in *Halodule* are attached at different heights. The pollen is filamentous (Ducker *et al.* 1978; Yamashita 1976a). The female flower is sessile or shortly stalked, the stalk never elongating, and each flower has two free carpels, each carpel having either a simple style (*Halodule*) or a style divided into two (three in *Amphibolis*) slender stigmas. The solitary bitegmic ovule is ± pendulous. The fruit is always indehiscent, and either unspecialized with a stony endocarp (*Cymodocea, Halodule, Syringodium*) or viviparous and biologically specialized (*Amphibolis, Thalassodendron*). The single seed has no endosperm, and the embryo is often very specialized.

Features of microscopic anatomy which distinguish this family are: (i) the absence of stomata and hairs, except for marginal teeth of leaf apex; (ii) the presence of tannin ('secretory') cells, which are especially abundant in the epidermis of leaf blades; (iii) the veins of the leaf blade often have a specialized sheath; (iv) aerenchyma is developed in the leaf blade; (v) a system of cortical vascular bundles in the stem in which the xylem is usually reduced and represented by xylem lacunae, or at most by narrow tracheids (e.g. stem of *Thalassodendron*). The root anatomy is also reduced, and sometimes without tracheids.

KEY TO GENERA, BASED MAINLY ON VEGETATIVE MORPHOLOGY AND ANATOMY

The five genera and most species are readily distinguished by differences in their vegetative morphology and anatomy (as well as in their reproductive morphology). This is indicated in the following diagnostic key (and subsequent keys). The individual genera are described separately in the next section. Determination usually depends to a large extent on vegetative features since the plants are not readily found with flowers and fruits. The flowers are poorly known for some species.

I. Rhizome herbaceous, monopodial with foliage or scale leaves; erect stems short (< 5 cm, except in *Syringodium*); leaf blade usually narrow (5 mm or less, except in *C. serrulata*); sheath of leaf veins not thick-walled; tannin cells of leaf epidermis enlarged. Anthers stalked   1

    1A. Rhizome with scale leaves, without short-shoots, leaf blade terete, mesophyll without fibrous strands, flowers in conspicuous, much-branched, cymose inflorescences terminating erect shoots
Syringodium Kützing

    1B. Rhizome with foliage leaves; sometimes with distinct short-shoots, never forming much-branched inflorescences   2

    2A. Leaf blade narrow, usually < 3 mm wide; veins three; leaf apex usually tridentate, without conspicuous marginal teeth; mesophyll without fibrous strands; roots unbranched. Style single, anthers attached at unequal heights.   *Halodule* Endlicher

    2B. Leaf blade wider (usually > 3 mm wide, up to 9 mm wide in *C. serrulata*); veins 7–17; leaf apex rounded, with distinct, often conspicuous, marginal teeth; roots branched. Styles two, anthers attached at the same height   *Cymodocea* König

II. Rhizome woody, with well developed sclerotic outer cortex, scale-bearing, sympodial; erect stems long, the foliage leaves in terminal tufts at the ends of naked axes; leaf blade broad (5–10 mm or more); sheath of leaf veins thick-walled, mesophyll without fibrous strands; leaf epidermis without tannin cells. Anthers sessile   3

    3A. Rhizome with erect branches and roots at intervals of four internodes; leaf apex rounded with conspicuous coarse teeth; epidermis without sinuous anticlinal walls. Flowers enclosed by 3–4 leafy bracts; styles two; fruit unappendaged, ± viviparous, with one enlarged inner bract
Thalassodendron den Hartog

    3B. Rhizome with erect branches at irregular intervals, roots at every node; leaf apex usually bidentate, broadly emarginate; epidermis with sinuous anticlinal walls. Flowers not enclosed by leafy bracts, styles three; fruit with four comb-like outgrowths, viviparous

Amphibolis Agardh

DESCRIPTIONS OF INDIVIDUAL GENERA

GROUP 1.   *Halodule, Cymodocea, Syringodium*

HALODULE Endlicher

A genus of possibly six very similar species (den Hartog 1964, 1970*a, b*) distinguished largely by vegetative characters, with one group of species in the Old World (*H. ciliata, pinifolia, uninervis*), the other in the New World (*H. beaudettei, bermudensis*), the remaining species (*H. wrightii*) having an unusual distribution in both the Old and New Worlds.

VEGETATIVE MORPHOLOGY

Growth monopodial with a dominant creeping main **axis** (rhizome) bearing foliage leaves and producing lateral branches but without consistent differentiation between long- and short-shoots (Fig. 12.1A, B). Rhizome extension produced by elongation of the internode below the youngest expanded leaf. Branches arising in the axils of foliage leaves borne on the main axis, each branch beginning with an adaxial scale leaf (prophyll, Fig. 12.1 B, C). Branches initially short, erect, but older ones growing out to establish a new main axis as a branch of the old one; internode length very variable and habit irregular. Much-branched rhizomes with numerous short erect internodes common.

Leaves distichous, the plane of distichy vertical. Each leaf with a narrow flattened blade up to 15 cm long and a shorter open sheath with opposite overwrapping margins. Mouth of the sheath with two lateral auricles; ligule short, few-celled (Fig. 12.1 E). Leaf blade narrow, 1–3 mm wide. Leaf apex 1–3 dentate (e.g. Fig. 12.1 F), rounded or emarginate, the outline said to be a major diagnostic feature (den Hartog 1970*d*) but central tissues breaking down to leave an indistinct pore. Squamules always two (Fig. 12.1 D). **Roots** 1–2 (–5) at nodes on rhizome, always unbranched (Fig. 12.1 A), each root with a distinct coleorhiza, root apex sometimes becoming tuberous.

[Anomalous specimens with much-branched stems and series of short, swollen internodes resembling strings of beads are infected by a parasitic fungus (*Plasmodiophora diplantherae*) which attacks several species (den Hartog 1965).]

VEGETATIVE ANATOMY

Leaf

Lamina (Fig. 12.7 A–D)

Leaf blade with three veins (Fig. 12.7 A), the median vein ending in a group of tracheids more or less continuous with the apical pore. Marginal veins not anastomosing with the median vein and usually ending blindly in marginal lobes (if present). Apical cells in the region of the median vein losing their cytoplasmic contents, becoming transparent and eroded to produce a distinct invagination. Microscopic marginal teeth often frequent at the leaf apex, each represented by an outgrowth of a few marginal epidermal cells. **Epidermis** of elongated cells with abundant chloroplasts, outer wall thickened, anticlinal walls somewhat sinuous in surface view (Fig. 12.7 B). Epidermal cells uniform except for numerous elongated secretory cells protruding somewhat into the mesophyll (Fig. 12.7 C, D). **Mesophyll** including 1 (–2) hypodermal layers and from 1–6 large air-canals separated by uniseriate longitudinal partitions between the median and lateral veins. The number of canals dependent on the width of the leaf. Canals segmented by perforated diaphragms. Fibrous strands absent. Mesophyll towards leaf apex reduced to two layers of cells separated by intercellular spaces; median vein and its xylem lacuna flattened. **Median vein** (Fig. 12.7 D) within a large partition including several narrow air-lacunae, the vein rounded in TS and surrounded by 2–3 layers of parenchyma with well developed intercellular spaces. Outermost layer of the bundle sheath differentiated as a slightly thick-walled endodermoid layer with the radial walls lignified. Phloem separated from the xylem by 1–3 layers of narrow cells. Xylem with a conspicuous lacuna, often including the remains of tracheids, separated from the endodermoid layer by a single layer of cells. **Marginal veins** (Fig. 12.7 C) each separated from the epidermis by 2–3 collenchymatous layers; endodermoid layer indistinct, xylem lacuna present.

Leaf sheath

Colourless, distinguished from the blade by its biseriate wing and adaxial epidermis of large cells frequently interrupted by secretory cells.

Prophyll

Open. colourless, biseriate throughout and without vascular or fibrous tissue. Apex rounded.

Stem

Epidermis with relatively large cells and frequent secretory cells, outer walls becoming thickened with age. **Cortex** lacunose, outermost layers compact, somewhat collenchymatous in thicker stems. Cortical bundles two, one on each side of the stele, together with several rudimentary strands without vascular tissue. **Endodermis** becoming thick-walled with age but remaining unlignified. **Stele** lozenge-shaped in TS, resembling that of *C. nodosa*, i.e. with a central xylem lacuna and four separate phloem strands. Pericyclic layer of thickened cells somewhat larger than those of the endodermis.

Course of vascular bundles. [The central stele bifurcates at the level of branch insertion, one strand going to each axis. The pair of large cortical vbs divides twice, one pair of derivatives becomes the lateral strand of the leaf, the other pair the cortical bundles of the branch. There is no direct connection between cortical and central vbs. Root traces connect directly to the stele.]

Root

Unbranched, *c.* 0.5 mm diameter (Fig. 12.8 G). **Epidermis** (Fig. 12.8 I) thin-walled, commonly with numerous root-hairs, each with a swollen base extending into the hypodermis somewhat. **Exodermis** uniseriate, thin-walled. **Cortex** lacunose, the outermost layer compact. **Endodermis** conspicuous, cells slightly thick-walled. **Stele** narrow (Fig. 12.8 H), scarcely 50 μm diameter. **Pericycle** uniseriate, including two diametrically opposed sieve-tubes, the centre of the stele occupied by 3–4 thin-walled cells, apparently representing undifferentiated tracheary elements.

Secretory elements

Tanniniferous secretory cells abundant in epidermis of leaf and stem but apparently absent from internal tissues.

REPRODUCTIVE MORPHOLOGY

Flower solitary, terminal on a short erect shoot, naked except for a foliage leaf immediately below, but the flowering shoot often extended by a lateral branch arising in the axil of the leaf next but one below the flower (Fig. 12.1 G, J). Male flower (Fig. 12.1 I–K) consisting of two bithecate (tetrasporangiate) anthers standing back-to-back, joined below, one slightly above the other (Fig. 12.1 K), pedicel elongating 1–2 cm at maturity (Fig. 12.1 I). Female flower (Fig. 12.1 G, H) more or less sessile, with two free carpels each with a long style (fig. 12.1 H). Fruit ovoid, somewhat flattened, with a short subapical beak, pericarp stony.

REPRODUCTIVE ANATOMY

Male flower

Anatomy of the axis supporting the stamens similar to that of the vegetative shoot; cortical vbs ending blindly a little below level of insertion of lowest pair of thecae. Renewal shoot inserted in the axil of leaf below the stamen and provided with a branch trace attachment resembling that of vegetative branch.

Female flower

Anatomy not investigated.

TAXONOMIC AND MORPHOLOGICAL NOTES

The genus *Halodule* (synonym *Diplanthera*) had been traditionally interpreted as consisting of two species, one from the New World (*H. wrightii* Aschers.) the other from the Old World (*H. uninervis* (Forsk.) Aschers.), until the critical work of Miki (1932). The more recent and more complete studies by den Hartog (1964, 1970*a*, *b*) demonstrated that this simple and convenient geographical segregation obscured a possibly more complex situation, although there still tends to be a marked disjunction between the species from the eastern and western hemispheres respectively. Different species are distinguished largely by differences in vegetative morphology, particularly the outline of the leaf apex. Since this is a feature which varies considerably on a single plant, and also because the apex of each leaf undergoes pronounced changes during ontogeny, the criteria used are difficult to apply and have been criticized (Phillips 1967; Phillips *et al.* 1974).

Since early accounts of the anatomy of *Halodule* (notably that by Sauvageau 1890*b*, 1891*a*) were based on the classical two species concept of the genus, the identity of material described in the literature is sometimes uncertain, especially as Sauvageau refers to size differences between different 'varieties' which must reflect the more complex situation revealed by den Hartog's revision. The above generalized account therefore refers to species within *Halodule* quite indiscriminately. Much more extensive work is required to establish whether the species of *Halodule* can, in fact, be distinguished by other microscopic features of their vegetative anatomy. For an extended discussion of these problems see den Hartog (1964).

MATERIAL EXAMINED

Halodule beaudettei (den Hartog) den Hartog; Matheson Hammock Wading Beach, Dade County, Florida; P. B. Tomlinson, numerous collections; all parts.

H. pinifolia (Miki) den Hartog; Mbutha (Buca) Bay, Vanua Levu, Fiji; P. B. Tomlinson 30.iv.69; all parts.

CYMODOCEA König

A very natural genus of four species with a wide but disjunct distribution: *C. rotundata* and *C. serrulata* (Indo-Pacific); *C. nodosa* (Mediterranean as far as Mauritania in W. Africa); *C. angustata* (western Australia). Most information relates to *C. nodosa*, but differences in other species are indicated where appropriate.

VEGETATIVE MORPHOLOGY

Habit

Seedling at first vertical (for three years) before becoming abruptly horizontal. Growth habit diffuse and somewhat irregular (Fig. 12.4 A), but essentially developing via a monopodial rhizome with distichously arranged foliage leaves, some leaves subtending a more or less erect leafy short-shoot. In *C. nodosa* internode length of rhizome variable and related to seasonal periodicity of growth; a segment of a few long internodes produced during a period of vigorous growth in spring alternating with a segment with several short internodes produced during late summer and forming an indistinct terminal bud. Regular alternation of segments allowing annual increments to be determined, with records by Bornet (1864) of shoots up to 12 years old. Erect branches usually restricted to nodes separated by long internodes, i.e. branches absent from or infrequent on short segments. Proliferation of the rhizome usually by extension of a lateral shoot as a long-shoot.

Periodicity of growth not obvious in other (tropical) spp, which tend to have a more regular and clear-cut distinction between long-shoots (rhizomes) and short-shoots.

Leaves

Somewhat dimorphic; shorter and with shorter blades when separated by short internodes, i.e. the dimorphism usually correlated with axis dimorphism. Each leaf with a short open, but overwrapping sheath, the mouth of the sheath auriculate (Fig. 12.4 C), the ligule inconspicuous and originating by division of three horizontal series of epidermal cells; blade up to 30 cm long, width a specific diagnostic character. Leaf apex rounded (Fig. 12.4 B; Fig. 12.6 I–N) with either microscopic marginal outgrowths (e.g. *C. nodosa*; Fig. 12.6 K, N; *C. rotundata*, Fig. 12.6 J, M) or conspicuous marginal teeth (e.g. *C. serrulata*; Fig. 12.6 I, L). First leaf on each branch a scale-like membranous prophyll with 1–5 veins. **Squamules** usually ten (sometimes more) in two groups of five laterally at each node.

VEGETATIVE ANATOMY

Leaf (Fig. 12.6)

Blade. Apical pore absent. **Hairs** absent except for irregular microscopic to macroscopic marginal teeth (Fig. 12.6, L–N). **Epidermis** with outer wall thickened; cells chlorenchymatous, uniform except for numerous elongated tannin cells, either uniformly scattered (e.g. Fig. 12.6 D) or confined to marginal areas (e.g. Fig. 12.6 F, H), the tannin cells deeper than normal epidermal cells and protruding into the mesophyll. **Mesophyll** including a uniseriate hypodermis forming a roof and floor to the longitudinal air-canals, the latter being separated laterally from each other by either veins or uniseriate partitions 3–4 cells high (e.g. Fig. 12.6 A–C). Number of canals determined by width of leaf, but usually two between each pair of veins (e.g. Fig. 12.6 A, E), more in wider leaves of *C. serrulata* (Fig. 12.6 G).

Veins. 7–17 depending on the sp, including a larger median vein. All veins joined at the apex by a marginal commissure, the median vein said by den Hartog (1970*a*) to be continued beyond the commissure in *C. angustata*. Transverse veins inconspicuous, rather infrequent. Parenchyma of bundle sheath including up to six narrow longitudinal air-lacunae (Fig. 12.6 B, C). **Fibres** in discrete strands more or less restricted to the hypodermis above and below the veins (Fig. 12.6 B, C). Marginal fibrous strand conspicuous, usually single (e.g. Fig. 12.6 F) but several in *C. serrulata* (Fig. 12.6 H). Larger veins each with a distinct uniseriate, sometimes thick-walled and slightly lignified endodermoid sheath together with a less distinct pericycle-like layer. Xylem represented either by a single tracheid with reticulate wall thickenings, or, more usually, by a narrow lacuna; phloem strand indistinct. [Sauvageau (1891*a*) described changes in the construction of the mesophyll relating to its progressively more shallow profile in a distal direction which suggests that there may be two or three series of canals at the leaf base, but I have not confirmed this and the subject requires reinvestigation. Sauvageau also described the mesophyll as originating from two cell layers, which again I have not confirmed.]

Leaf sheath. Persistent after detachment of the blade in earliest formed leaves of each annual increment in *C. nodosa* (Bornet 1864). Sheath resembling the blade in anatomy, but dorsiventrally differentiated, the adaxial epidermis thin-walled; narrow wing biseriate marginally but with numerous secretory cells. Fibrous strands absent marginally, but often present above and below the veins.

Prophyll. Colourless, three-veined in *C. nodosa*, the veins united apically below the rounded tip; mesophyll 1–3-layered but without air-lacunae.

Stem.

Epidermis uniform, outer wall thickened and sometimes becoming lignified; cuticle well developed. **Cortex** wide; outer layers compact, small-celled; gradual transition to lacunose middle cortex with numerous wide air-lacunae separated by uniseriate partitions; the lacunae segmented by transverse diaphragms only at the nodes. Inner cortex compact, cells becoming thick-walled with age but remaining unlignified. Cortical fibres present only in *C. serrulata* as conspicuous unlignified strands in the outer cortex, but independent of the epidermis (Fig. 12.8 A, B). Secretory cells conspicuous in the outer cortex of all species except *C. serrulata*. Cortical vbs numerous (12–33), fewest in shortest and narrowest internodes, in 1–2 irregular concentric series; each vb somewhat radially extended and surrounded by 1–2 layers of narrow, slightly thick-walled cells, with a conspicuous phloem strand and usually a narrow xylem lacuna. **Endodermis** distinguishable with difficulty from the slightly thick-walled, narrow cells of the inner cortex. **Stele** narrow, more or less lozenge-shaped in TS. Xylem represented by a single wide central lacuna surrounded by conjuctive tissue including four distinct phloem strands, each phloem strand with 1–4 conspicuous and often quite wide *(C. serrulata)* sieve-tubes.

Course of vascular bundles in the stem (after Monoyer 1927)

[The stelar arrangement evident in TS is interpreted as four separate vascular bundles (represented by their independent phloem strands) with a common xylem strand (represented by the xylem lacuna). The lateral veins of the leaf are continuous into the stem as cortical bundles; the smaller (marginal) bundles of the leaf extend two internodes within the stem, the larger bundles of the leaf (those immediately on each side of the median vein) extend up to six internodes before ending blindly. Monoyer's account is somewhat confusing because he says that the identity of the cortical traces is lost but his diagrams suggest that the trace system to individual leaves (and branches) can be recognized in the stem. Anastomosing of cortical traces is likely to occur. The cortical traces of branches are continuous with the cortical traces of the main axis.

The median leaf traces continue directly into the stele, where they unite with the large stelar phloem bundle on the adjacent side of the stem ('foliar trace' of Monoyer). The remaining pair of stelar bundles (those not in the plane of leaf insertion; 'sympodial traces' of Monoyer) are not directly continuous with the leaves but are continuous with the 'foliar traces' via an anastomosing strand at each node. Root traces arise from the stele and are continuous with both types of stelar bundle. Branch steles unite directly with the stele of the main axis.]

Root.

1 (–2) at rhizome nodes towards lower side of stem. Coleorhiza distinct but short. **Epidermis** including numerous root-hairs, but often lost from older

roots. **Exodermis** uniseriate, cell walls sometimes developing uneven thickenings and appearing U-shaped in TS. **Cortex** wide, outer cortex compact, up
to 12 cells wide, becoming enlarged, thick-walled, and lignified in older roots
of *C. serrulata*; in *C. nodosa* and *C. rotundata* middle cortex developing
lacunae separated by radial partitions one cell wide. Inner cortex of compact,
regularly concentric layers of narrow, somewhat thick-walled cells. **Endodermis** uniseriate, often indistinct but becoming somewhat thickened in *C. serrulata*. **Stele** narrow. **Pericycle** uniseriate, including 4–5 (–8) sieve-tubes, each
with associated companion cells. Xylem represented by indistinctly differentiated cells described by Sauvageau (1889) as 'vascular canals,' alternating
with phloem strands but within the pericycle. [The vascular canals may represent undifferentiated files of tracheids or cavities formed by the breakdown
of walls between contiguous cells.] Tracheids with lignified walls recorded
in *C. nodosa* by Bornet at proximal end of roots, i.e. near their insertion.

Secretory and Conducting Elements

Tannin cells: present in all parts but most conspicuous in the epidermis
of the leaf blade; each cell enlarged, with refractive contents and slightly
thickened walls.

Phloem: sieve-tubes with simple sieve-plates in all parts.

Xylem reduced, in leaf and stem represented largely by lacunae formed
by breakdown of original protoxylem elements, the elements persisting at
the nodes. Xylem lacunae stated by Sauvageau (1890*a*) to be separated from
the bundle sheath by a 'pericycle' in the median vein of the leaf, but not in
the laterals.

Reproductive Morphology

Flowers naked, terminal on lateral (erect) shoots. **Male flower** with a stalk
elongating considerably (to 10 cm) at anthesis, and with two anthers united
back-to-back at the same level, each anther tetrasporangiate and with a short
apical process. **Female flower** terminal, sessile or shortly stalked, the stalk
not elongating, each flower of two free carpels, each carpel with a simple
pendulous orthotropous ovule. Style divided into two slender stigmas up to
3 cm long. Fruit smooth or somewhat warty.

Key to the Species

Although very similar, the species of *Cymodocea* may be distinguished
vegetatively, as follows (see also den Hartog 1970*b*):

   I.   Leaf narrow (< 4 mm wide), marginal teeth rather obscure, leaf mar-
       gins with a single fibrous strand, rhizome without cortical fibres   1
  II.   Leaf broader (> 4 mm wide), marginal teeth prominent, leaf margin
       with two or more fibrous strands, rhizome with or without cortical
       fibres   2
  1A.   Veins 7–9, marginal teeth obscure (Fig. 12.6 K, N)   *C. nodosa*

1B. Veins 9–15, marginal teeth relatively conspicuous (Fig. 12.6
     J, M).   *C. rotundata*
2A. Veins 13–17, blade up to 9 mm wide, marginal teeth numerous,
     prominent, closely set (Fig. 12.6 I, L); mesophyll lacunae deep;
     cortical fibres present in stem   *C. serrulata*
2B. Veins 9–13, blade up to 6 mm wide, marginal teeth prominent
     but widely spaced; mesophyll lacunae shallow; cortical fibres
     absent from stem   *C. angustata*

Comparative information about the morphology and anatomy of reproductive parts is still to be sought and much desired.

MATERIAL EXAMINED

Cymodocea nodosa (Ucria) Aschers.; Khereddine, Tunisia; L. G. de Solignac
     10.iv.69; all parts.
C. rotundata Ehrenb. & Hempr. ex Aschers.; Tarauma Beach, Port Moresby,
     New Guinea; P. B. Tomlinson 31.x.74 M; all parts.
C. serrulata (R. Br.) Aschers. & Magnus; Townsville, Queensland; W. R.
     Birch *s.n.*; all vegetative parts.

SYRINGODIUM Kützing

A genus consisting of two distinct but evidently closely related species, one in the Caribbean (*S. filiforme*), the other widely distributed in the Indo-Pacific (*S. isoetifolium*).

VEGETATIVE MORPHOLOGY (Fig. 12.2 A, B)

Habit

Axes dimorphic. Rhizome monopodial, scale-bearing, with erect leafy shoots arising from the axils of the scales on the horizontal rhizome (Fig. 12.2 B). Rhizome with uniformly long internodes (Fig. 12.2 A). Erect shoots with short internodes, usually unbranched and terminating in an extended inflorescence. Branching of the rhizome taking place by modification of an erect shoot, usually as a response to damage. Roots branched, 1–3 at each node. Foliage leaf with a terete blade, the mouth of the sheath auriculate and ligulate, the blade usually abscissing from the sheath. Leaf apex somewhat flattened with a narrow toothed wing (Fig. 12.2 A, inset). Leaf sheath open, squamules two at each node.

VEGETATIVE ANATOMY

Leaf (Fig. 12.7 I–N)

[The following description refers to basal foliage leaves, and not to bracts of inflorescence]. **Epidermis** (Fig. 12.7 L, M) uniform, including numerous secretory cells, epidermal cells ± isodiametric in surface view. **Mesophyll** with 2–3 hypodermal layers surrounding a region of 5–7 longitudinal air-canals separated from one another by uniseriate partitions (Fig. 12.7 I, J, N). Air-canals with transverse perforated diaphragms. Fibrous bundles absent. **Midvein** (Fig. 12.7 J, K), prominent, surrounded by 6–10 large cells to which longitudinal partitions attached. Bundle sheath of narrow, slightly thick-walled cells, the outermost layer differentiated as an endodermis. Xylem lacuna and phloem strand conspicuous. **Lateral veins** two in *S. filiforme* (Fig. 12.7 N), establishing the sagittal plane of the leaf; lateral veins 7–12 in *S. isoetifolium* (Fig. 12.7 I), radially oriented in a ring within the inner layers of the outer mesophyll; endodermis not differentiated, vascular tissues as in midvein.

Leaf sheath

Dorsiventral with 1–3 series of air-canals on the dorsal side; air-canals absent from the narrow wings, margins of sheath biseriate. Lateral veins five in *S. filiforme*, the marginal pair ending at the blade insertion; lateral veins 7–10 in *S. isoetifolium*, all but the marginal two continuous into the blade.

Stem (rhizome) (Fig. 12.3 A–F)

Epidermis small-celled, slightly thick-walled, with numerous large to very large secretory cells (Fig. 12.3 B). **Cortex** with outer compact, thick-walled layers, gradually passing over to thinner-walled inner layers; including up to 15 narrow cortical vbs (Fig. 12.3 A, D); middle cortex lacunose (Fig. 12.3 C) with occasional transverse diaphragms (Fig. 12.3 E). Inner cortex compact, small-celled, often thick-walled, enclosing the narrow stele. **Endodermis** indistinct, slightly thick-walled. **Stele** (Fig. 12.3 F) elongated in TS, with a wide central xylem lacuna, the phloem represented by two conspicuous strands, with wide sieve-tubes, inserted in the plane of leaf distichy.

[The **aerial stem** differs from the rhizome only in having fewer cortical vascular bundles and in the somewhat thinner walls of the outer cortical vascular bundles.]

Course of vascular bundles in stem (after Chrysler 1907)

[Cortical vbs originate from lateral traces of the leaf, but they are connected via a branch with the median leaf trace at the node. The median trace is connected to the central cylinder, which is interpreted as consisting of two vbs inserted in the plane of leaf distichy. Tracheary tissue fills the xylem lacuna at the node.]

Root (Fig. 12.3 G–I)

Branched. **Epidermis** of wide, thin-walled cells, with frequent root-hairs, other cells becoming tanniniferous and even thick-walled (Fig. 12.3 I) with age. **Exodermis** uniseriate, thick-walled, small-celled, the cells below the root-hairs somewhat wider than those elsewhere. **Cortex** with 3–5 outer layers of compact, slightly thick-walled cells. Middle cortex becoming somewhat irregularly lacunose (Fig. 12.3 G) on account of the separation and the partial collapse of certain radial files of cells. **Endodermis** conspicuous, the cells becoming uniformly thick-walled with age. **Stele** (Fig. 12.3 H) narrow; conducting tissues represented by sieve-tubes in the pericycle; central cells narrow, undifferentiated except for four somewhat wider, thin-walled cells next to pericycle and alternating with sieve-tubes, the thin-walled cells probably representing vestigial tracheary elements.

Tannin cells

Common in the epidermis, especially of the leaf, as enlarged cells with a narrow surface exposure (Fig. 12.7 L, M).

REPRODUCTIVE MORPHOLOGY

Flowers in distinct **inflorescences** with long basal internodes (Fig. 12.2 A) terminating erect shoots (Tomlinson and Posluszny 1978). **Bracts** on the inflorescence resembling the foliage leaves but with shorter sheaths. Blades becoming progressively shorter in the more distal bracts, the blades of the most distal bracts being shorter than the sheaths. Branching of the inflorescence racemose below, cymose above, often appearing dichasial by branching from the axils of two subopposite bracts. Each distal unit with a terminal flower and a pair of bracts, a renewal shoot arising in the axil of one or both of the bracts (Fig. 12.2 F, H). Renewal shoots absent from ultimate flowers. **Male flower** (Fig. 12.2 C–F) shortly stalked, with two anthers back-to-back and fused below, lacking the apical processes of *Cymodocea*, but with an obscure ridge below the anthers, initiated early but soon concealed (Tomlinson and Posluszny 1978). **Female flower** (Fig. 12.2 H–I) of two free carpels each with two styles and a single pendulous ovary. Fruit (Fig. 12.2 J–K) with a stony pericarp, ellipsoidal, smooth and with a short beak.

REPRODUCTIVE ANATOMY

[The following information is taken from Tomlinson and Posluszny (1978). The vascular system of the floral axes corresponds closely to that of vegetative axes, with a central axial system and 4–6 cortical strands. The median traces of the leaves, bract, prophylls, and of the branch axes diverge from the central system: the lateral traces of the bracts arise from the cortical system. The central strand of the male flower ends blindly as a mass of tracheary tissue just above the blind-ending cortical system; the ridge of the male flower remains unvasculated. The central strand of the female flower divides to produce one trace to each carpel, each carpel trace itself producing a dorsal (ovular) trace and a ventral trace ending blindly in the base of the style.]

KEY TO SPECIES

The two species, though very similar, may most readily be distinguished by the following feature of the leaf blade, as has been pointed out by a number of authors (e.g. Magnus 1872; Sauvageau 1890*a*; den Hartog 1970*a*; Tomlinson 1980):

1. Leaf blade with 7–10 (–15) narrow sub-peripheral lateral veins (Fig. 12.7 I)   *S. isoetifolium*
1A. Leaf blade with two large sub-peripheral lateral veins (Fig. 12.7 N)   *S. filiforme*

MATERIAL EXAMINED

Syringodium filiforme Kütz.; Matheson Hammock Wading Beach, Miami, Florida; P. B. Tomlinson, several collections; all parts.
S. isoetifolium (Aschers.) Dandy; Singapore; P. B. Tomlinson *s.n.*

GROUP 2. *Amphibolis, Thalassodendron*

AMPHIBOLIS Agardh

A genus of two species restricted to temperate Australia, including Tasmania (Ducker *et al.* 1977). It is closely related to *Thalassodendron* but distinguished by its less robust and less precise habit and by conspicuous differences in floral morphology.

VEGETATIVE MORPHOLOGY

Axes dimorphic (Fig. 12.4 D, E). Horizontal rhizome woody, scale-bearing, sympodial, and producing erect, frequently-branched stems ending in tufts of relatively short leaves. Erect shoots produced only at intervals of 3–7 internodes, less regularly than in *Thalassodendron*. Rhizome proliferating by the development of second-order branches from the bases of erect shoots. Branches often adnate to the parent axis. Internodes of the rhizome uniformly long except at places of branching; internodes long at bases of erect shoots, progressively shorter in distal parts. Roots branched; one or two roots at every node. Foliage leaf short (Fig. 12.4 F) with a very short (*c.* 1 cm) compressed sheath narrowed basally; the mouth of the sheath auriculate and with a pronounced ligule. Leaf apex usually with two pronounced marginal teeth but first leaves on a branch with more rounded apices, the leaf shape possibly determined by seasonal periodicity of growth.

VEGETATIVE ANATOMY

Leaf

Blade with *c.* 20 veins (Fig. 12.7 E), the latter joined by frequent, usually oblique commissures. **Epidermis** of blade uniform, without tannin cells, chlorenchymatous; anticlinal walls distinctly sinuous in surface view. **Mesophyll** towards the base of the blade with shallow **air-lacunae,** the latter separated by shallow partitions 1–2 cells high. Lacunae without transverse diaphragms (except rarely in *A. griffithii*) other than the thick plates of cells supporting the obliquely transverse anastomosing veins. Lacunae absent distally, mesophyll then two-layered. Fibrous strands absent. Tannin cells common in mesophyll. Bundle sheath of veins well developed, thick-walled, often 2–3 layered. Marginal veins conspicuous, embedded in a massive sclerenchyma sheath and sometimes represented by two separate vbs. Marginal veins of the leaf sheath inconspicuous, not continuous into blade.

Stem (Fig. 12.8 c–f)

Anatomy of erect and horizontal stems very similar. **Epidermis** lignified with a thick cuticle, hypodermis 1-layered completely tanniniferous (Fig. 12.8 E). **Cortex:** outermost cortical layers compact, up to 12 cells wide, becoming sclerotic at an early age and eventually uniformly thick-walled and lignified. Secretory cells absent from the compact cortex. Middle cortex regularly lacunose (Fig. 12.8 F), most obviously so in horizontal stems; cell walls slightly to markedly thickened. Cortical bundles 6–9 in horizontal and 4–6 in erect stems; arranged in two lateral groups at the periphery of the lacunose layer. Cells of the bundle sheaths of the cortical vbs thick-walled (Fig. 12.8 D) and becoming lignified with age. Inner cortex sclerotic 7–10 cells wide, forming a complete investment to the stele (Fig. 12.8 c). **Stele** narrow and usually with four distinct phloem groups surrounding a central xylem lacuna, more or less as in *Cymodocea nodosa*.

Root

Epidermis shallow, ± uniformly thick-walled; without root-hairs in material examined. **Exodermis** uniseriate, thick-walled. Outer **cortex** of 6–7 compact layers of thick-walled, lignified cells, outermost 1–2 layers tanniniferous. Inner cortex reticulately lacunose, including occasional tannin cells. **Endodermis** uniseriate, uniformly thick-walled, lignified. **Stele** narrow, conducting tissues little differentiated.

Reproductive Morphology

Flowers solitary at the ends of short lateral leafy branches on erect shoots. Male flowers shortly stalked, each with two anthers standing back-to-back and fused below, and crowned by 2–3 branched appendages. Female flower subsessile with two free carpels, each with three slender styles. Mature fruit with one ripe ovary (the other aborting), developing four spreading toothed lobes as extensions of the pericarp. Germination viviparous (Black 1913).

Key to Species

The two species of *Amphibolis* are scarcely distinguishable anatomically, but the following vegetative features (mainly after Ducker *et al.* 1977) serve to separate them:

1A.   Plants robust with shoots containing up to 12 leaves, leaves with short (< 14 mm) sheaths, non-overlapping and not clasping the stem; blade broad, much twisted (to 90°); ligules and auricles of sheath large and confluent; roots solitary; axis with 8–10 cortical vbs

A. antarctica

1B. Plants slender, with shoots containing up to five leaves; leaves tightly clasping the stem with long ($<$ 23 mm) overlapping sheaths; blade narrow, little twisted (to 60°); ligules and auricles scarcely developed; roots in pairs; axis with 5–6 cortical vbs   *A. griffithii*

MATERIAL EXAMINED

Amphibolis antarctica (Labill.) Sonder et Aschers.; Cape Patterson, Victoria; P. B. Tomlinson 1.v.77A; all vegetative parts.
A. griffithii (J. M. Black) den Hartog; Safety Bay, Perth, Western Australia; P. B. Tomlinson 7.v.77; all vegetative parts.

THALASSODENDRON Den Hartog

A genus of two species, one (*T. ciliatum* (Forsk.) den Hartog) quite widely but discontinuously distributed from East Africa to Eastern Malaysia, the other (*T. pachyrhizum* den Hartog) in Western Australia.

VEGETATIVE MORPHOLOGY

Habit

Vegetative axes dimorphic (Fig. 12.4 G), with a sharp distinction between (i) creeping rhizomes bearing scale leaves and roots and (ii) erect, sparsely branched axes bearing foliage leaves and no roots (Tomlinson 1974). Branching of rhizome periodic and **sympodial,** producing an erect axis at each fourth internode and commonly a resting bud at the first node of the erect shoot. Rhizome proliferation initiated from the first node of the erect shoot (Fig. 12.4 I). Erect axes monopodial, diffusely but infrequently branched to produce further leafy shoots, or distally, to produce short lateral flowering shoots. Internodes on erect shoots initially long, but progressively shorter in a distal direction; older shoots becoming naked below by loss of lower leaves (Fig. 12.4 H). Roots branched, usually 2–6 at regions of rhizome branching.

Leaves

Foliage leaf short and broad, with a flattened open sheath, narrowed below, the sheath margins overwrapping at the node; mouth of sheath scarcely auriculate but markedly ligulate. Blade broad (*c.* 10 mm) with numerous (up to 25) veins; apex rounded with numerous close-set, prominent teeth. Rhizome scales without any blade vestige, deciduous.

VEGETATIVE ANATOMY

Leaf (Fig. 12.7 F–H)

Teeth on leaf margin conspicuous apically, each consisting of elongated cells with thick walls, the cell lumen often almost occluded. **Epidermis** with outer cell walls thickened, anticlinal walls scarcely sinuous in surface view.

Mesophyll shallow, with 1–2 hypodermal layers above and below the air-lacunae in the thicker parts of the leaf. Air-lacunae in a single series separated from one another by shallow uniseriate partitions. Up to six lacunae present between each adjacent pair of veins near the middle of the lamina, but lacunae absent between the 4–6 marginal veins and towards the leaf apex. Lacunae including occasional transverse diaphragms. Fibrous strands absent; tanniniferous secretory cells abundant (Fig. 12.7 G, H). **Veins** 17–25, united below the leaf apex, the median vein not prolonged beyond the apical commissure. Transverse veins frequent, often rather oblique, situated within transverse diaphragms. Median vein largest with a distinct 1 (–2)-layered sheath of wide, somewhat thick-walled, lignified cells (Fig. 12.7 G). Xylem represented by a lacuna, always separated from the sheath cells by a single layer of thin-walled cells; phloem strand well developed. Lateral veins similar but smaller, the xylem lacuna contiguous with sheath cells (Fig. 12.7 H).

Leaf sheath

Epidermis shallow, cells thin-walled on the ventral side. Mesophyll uniform, shallow, without lacunae. Veins each with a massive fibrous sheath on the abaxial side, continuous with the epidermis and forming a prominent rib; veins fewer distally, those in the wing represented by vascular tissues with an associated strand of fibres.

Rhizome scale

Abaxial epidermal cells with thickened and lignified walls. Median vein well developed and with a thick-walled sheath; lateral veins about 12 on each side of median vein, represented by strands of fibres more or less equidistant from each epidermis, those closest to the median vein being thickest.

Stem

(i) **Rhizome. Epidermis** tanniniferous, becoming lignified with age. Outer **cortex** compact with an outer mechanical layer of early-maturing tanniniferous cells, and an inner layer remaining persistently meristematic but eventually producing a lignified thick-walled zone. Both regions becoming equally thick-walled and less distinct with age. Mechanical layer interrupted by the passage of leaf traces. Middle cortex lacunose, with narrow air-canals occasionally traversed by uniseriate diaphragms; tannin cells common at the intersection of canal partitions. Cortical vascular system including up to 30 cortical vbs in an irregular circle; each vb with a 1–3-layered sheath of thick-walled lignified cells, a xylem lacuna (sometimes including one or more tracheids) and a strand of phloem with 1–3 sieve-tubes. Inner cortex at first thin-walled, but maturing as a massive, pitted, sclerotic layer by thickening and lignification of the endodermis and adjacent cortical layers. Inner and outer cortex thus constituting the chief mechanical tissue of these woody rhizomes.

Endodermis not, or poorly, differentiated. **Stele** essentially amphiphloic, the phloem forming an irregular peripheral cylinder with conspicuous, wide sieve-tubes; central tissue including a single wide lacuna, or, sometimes, persistent tracheids with spiral-reticulate thickenings.

(ii) **Erect shoot.** Narrow, differing from the rhizome only in having fewer (5–12) cortical vbs arranged in two groups to the left and right of the plane of insertion of the leaves, the number of vbs sometimes increased at the nodes by bifurcation. **Starch** common in cortex.

Root

Branched, associated exclusively with branch-bearing regions of rhizome. **Epidermis** small-celled, outer wall slightly thickened; root hairs infrequent as v. large cells. Outer **cortex** compact, consisting of up to ten layers of slightly thick-walled, lignified cells. Inner cortex with radial lacunae formed by the collapse of cortical cells. Innermost cortical layer and **endodermis** becoming uniformly thick-walled with age. Conducting tissues undifferentiated. **Tannin cells** common.

REPRODUCTIVE MORPHOLOGY

Flowers (Fig. 12.5) solitary and terminal on short lateral shoots, subsessile and enclosed by 3–4 leaves (Fig. 12.5 A–C, F, I) with reduced blades (bracts) red-pigmented at maturity. Inner bract in male flower a reduced scale, resembling a prophyll. Male flower of two anthers completely fused back-to-back, each anther with a short terminal, sometimes bilobed appendage (Fig. 12.5 F–I). Pollen filiform (Ducker *et al.* 1978). Female flower with two free uniovulate carpels, each with two slender styles, the latter shortly united below (Fig. 12.5 D). Ovules anatropous, inserted sub-basally (Fig. 12.5 E). Fruit made up of 1 (–2) fertilized carpels plus the fleshy innermost bract. Germination tending to be viviparous to produce a free-floating seedling.

No information is available concerning floral anatomy.

TAXONOMIC NOTES

The above account refers to the more common species *T. ciliatum*; the other species has not been investigated anatomically.

MATERIAL EXAMINED

Thalassodendron ciliatum (Forsk.) den Hartog; Nairobi, Kenya; Moorjani *s.n.*; all parts.

TAXONOMIC NOTES FOR THE FAMILY

Following Takhtajan (1966), the Cymodoceaceae are here treated as a distinct (and isolated) family and correspond to the tribe Cymodoceae of Ascherson and Graebner (1907) and the subfamily Cymodoceoideae of den Hartog (1970*a*). They are distinguished as a group by their wholly marine habitat and morphological features of vegetative parts and male and female flowers. The association of this group with *Zannichellia* and related genera, as in Hutchinson's Zannichelliaceae, is inappropriate; contrasting features are listed in Table 12.1.

TABLE 12.1

Comparison between Cymodoceaceae and Zannichelliaceae (s.s.)

Character	Cymodoceaceae	Zannichelliaceae
Habitat	marine	fresh or brackish water
Axis	± dimorphic, rhizome clearly differentiated	obscurely dimorphic, rhizome not always clearly differentiated
Roots	branched (except in *Halodule*)	unbranched
Leaves	often broad, with numerous veins	always with 1 (–3) veins
Ligule	specialized outgrowth	extension of mouth of sheath
Prophyll	usually vasculated	not vasculated
Squamules per node	two or more	always two
Stomata	absent	rarely present (*Zannichellia*)
Fibrous mechanical tissue	commonly well developed	uncommon, little developed
Cortical vascular system	well developed	usually absent
Tannin cells	frequent, specialized in epidermis	rare, not specialized in epidermis
Sex distribution	dioecious	usually monoecious
Floral axis	usually determinate, without sympodial proliferation	indeterminate by sympodial proliferation
Flower	naked	with an envelope
Male 'flower'	always two fused anthers (two tetrasporangiate structures)	anther single (1–4 bisporangiate structures)
Pollen	filamentous	globose
Female 'flower'	bicarpellate	1–8 carpellate
Styles	1, 2, or 3, long	single
Stigma	not enlarged, linear	enlarged, feathery or funnel-shaped
Fruit	sometimes viviparous	never viviparous
Cotyledon	obscure, short	conspicuous, circinnately coiled

Within the Cymodoceaceae the regrouping of taxa by den Hartog (1970*a*) in which the old genus *Cymodocea* (s.l.) is fragmented is a welcome and necessary advance since it produces a much more natural arrangement. In comparison with the arrangement of Ascherson and Graebner we now have:

	Ascherson and Graebner (1907)	den Hartog (1970*a*)	
Genus	*Diplanthera* (= *Halodule*)	*Halodule*	
Genus	*Cymodocea* (s.l.)		Group 1
	subgenus *Phycagrostis*	*Cymodocea* (s.s.)	
	subgenus *Phycoschoenus*	*Syringodium*	
	subgenus *Amphibolis*		
	sp *C. ciliata*	*Thalassodendron*	Group 2
	C. antarctica	*Amphibolis*	

On the basis of the anatomical and morphological evidence presented earlier, it is evident that the genera of the first group are more closely related to each other than they are to those of the second group. *Halodule* quite clearly belongs with the first group, indeed in habit it closely resembles the smaller species of *Cymodocea*.

BIOLOGICAL NOTES

Because of their habitat detailed knowledge of the seagrasses is very deficient. Morphological and functional aspects of the flowers and fruits in these plants are very poorly understood, but there is increasing interest, especially in aspects of pollination (e.g. Ducker *et al.* 1977; Ducker and Knox 1976, 1978; Yamashita 1976*a*). The suggestion that such plants have the C_4 carbon cycle has been made (Doohan and Newcomb 1976) although their leaf anatomy is distinctive. Of particular interest is the biological specialization of the embryo, fruit, and seedling in relation to the specialized problems of establishment under the sea in a substrate subject to wave action (cf. Black 1913; Bornet 1864).

Cymodocea nodosa

The embryo is described by Bornet (1864) as having a reduced radicle adpressed to an enlarged, partially vasculated structure which is interpreted either as an expansion of the radicle itself (Bornet) or as the adnate hypocotyl (Ascherson 1882). The embryo of *Syringodium* (Fig. 12.2 K) seems similar. In this peculiar morphological development the seedling of *Thalassia* is recalled (see p. 161).

Amphibolis

The fruit consists usually of only one fertilized carpel, the other aborting. Its most distinctive feature is the presence of four distal unequal spreading lobes arising at a late stage from the outer wall. The lobes become rigid, owing to the persistence of the inner fibrous layers of the pericarp and the lobes then serve as a grapnel apparatus which assists to anchor the detached seedling. For details of this organ see Ascherson (1882), Black (1913), den Hartog (1970*a*), Ostenfeld (1916), Tepper (1882).

The embryo is described by den Hartog (1970*a*) as having a long cotyledon, a short axis and no radicle (primary root). It begins its development whilst still on the parent plant, breaking through the hard ovary wall and producing a series of reduced leaves with progressively better-developed blades. The seedling is detached only when it reaches a length of about 8 cm.

Thalassodendron

The male and female flowers each have four enveloping bracts, but only in the female do the two innermost bracts continue to develop after anthesis. The innermost bract eventually encloses the carpel(s) and forms a false fruit. Both carpels may develop but normally only one is fertile. Germination begins while the fruit is still attached. The pericarp splits to reveal the plumule; the cotyledon itself is reduced and apparently specialized as an aril-like structure which covers the split in the pericarp. Early leaves are bladeless and adventitious roots arise on the first extended internode. No radicle (primary root) is developed. The seedling is eventually released from the enveloping bract of the false fruit. For details see Isaac (1969) and den Hartog (1970*a*).

It is of general biological interest that in these last two genera seed development is partly or wholly viviparous and that establishment is made by a seedling rather than a seed, with specialized grapnel-like organs developed in *Amphibolis*.

SPECIES REPORTED ON IN THE LITERATURE

Arber (1923): *Syringodium isoetifolium* (as *Cymodocea isoetifolia*): squamules.

Ascherson (1882): *Amphibolis antarctica* (as *Cymodocea antarctica*): general morphology.

Black (1913): *Amphibolis antarctica* (as *Pectinella antarctica*): floral morphology, fruit biology.

Bornet (1864): *Cymodocea nodosa* (as *Phucagrostis major*): general anatomy and morphology.

Chrysler (1907): *Syringodium filiforme* (as *Cymodocea manatorum*): course of vbs in stem.

Doohan and Newcomb (1976): *Cymodocea rotundata, C. serrulata*: leaf ultrastructure.

Duchartre (1872): *Cymodocea nodosa* (as *C. aequorea*): general anatomy and morphology.

Ducker *et al.* (1977): *Amphibolis antarctica, A. griffithii*: all vegetative parts.

Ducker *et al.* (1978): *Amphibolis antarctica, Thalassodendron ciliatum*: pollen development.

Feldmann (1936): *Halodule beaudettei* (as *Diplanthera wrightii*), *Syringodium filiforme* (as *Cymodocea manatorum*): morphology.

Feldmann (1938): *Halodule beaudettei* (as *Diplanthera wrightii*), *H. uninervis* (as *Diplanthera uninervis*): leaf.

Gravis (1934): *Cymodocea nodosa*: leaf traces.

den Hartog (1964): *Halodule* spp: detailed taxonomic account, diagnostic features of leaf apex.

den Hartog (1970*a*): taxonomic account, including detailed morphological description of each sp; anatomical details largely derived from the literature.

Isaac (1969): *Thalassodendron ciliatum* (as *Cymodocea ciliata*): floral biology and seedling development.

Kay (1971): *Cymodocea serrulata, Thalassodendron ciliatum*: floral morphology.

Lüpnitz (1969): *Halodule beaudettei*: root development.

Magnus (1871): *Amphibolis antarctica* (as *Cymodocea antarctica*), *Cymodocea nodosa, C. rotundata, C. serrulata, Syringodium isoetifolium* (as *Cymodocea isoetifolia*), *Thalassodendron ciliatum* (as *Cymodocea ciliata*): brief notes on leaf and stem anatomy.

Magnus (1872): *Cymodocea nodosa, C. serrulata, Syringodium filiforme* (as *Cymodocea manatorum*), *S. isoetifolium* (as *Cymodocea isoetifolia*): mucilage cells.

Miki (1932): *Cymodocea rotundata, C. serrulata, Halodule pinifolia* (as *Diplanthera pinifolia*), *H. uninervis* (as *Diplanthera uninervis*): leaf and stem.

Miki (1934*a*): Same spp as 1932 paper, plus *Syringodium isoetifolium* (as *Cymodocea isoetifolia*): leaf and stem anatomy; *Halodule pinifolia* (as *Diplanthera pinifolia*): floral morphology.

Monoyer (1927): *Cymodocea nodosa*: course of vbs in stem (information on other spp taken from Sauvageau 1891*b*).

Ostenfeld (1916): *Amphibolis antarctica* (as *Cymodocea antarctica*); seedling morphology; *Cymodocea angustata, Syringodium filiforme* (as *Cymodocea manatorum*), *S. isoetifolium* (as *Cymodocea isoetifolia*): morphology, leaf anatomy.

Pettitt and Jermy (1975): *Halodule wrightii*: pollen structure.

Phillips (1960*a, b*): *Halodule beaudettei* (as *Diplanthera wrightii*), *H. uninervis* (as *Diplanthera uninervis*): leaf morphology and anatomy.

Pottier (1934): *Cymodocea nodosa*: developmental anatomy of leaf, stem, and root.

Sauvageau (1889*b*): *Cymodocea nodosa* (as *C. aequorea*): root.

Sauvageau (1890*a*): *Amphibolis antarctica* (as *Cymodocea antarctica*), *Cymodocea nodosa* (as *C. aequorea*), *C. rotundata, C. serrulata, Syringodium filiforme* (as *Cymodocea manatorum*), *S. isoetifolium* (as *Cymodocea isoetifolia*), *Thalassodendron ciliatum* (as *Cymodocea ciliata*): leaf.

Sauvageau (1890*b*): *Halodule uninervis, H. wrightii*: leaf. (For a discussion of the possible identity of plants under these names see den Hartog 1964.)

Sauvageau (1891*a*): Same spp as 1890*a* paper, plus *Halodule uninervis, H. wrightii*: leaf.

Sauvageau (1891*b*): Same spp as 1891*a* paper: stem.

Sauvageau (1891*c*): Same spp as 1891*a* paper: root.

Schenck (1886*b*): *Cymodocea nodosa* (as *C. aequorea*): leaf.

Tepper (1882): *Amphibolis antarctica* (as *Cymodocea antarctica*): morphology and dispersal.

van Tieghem and Douliot (1888): *Cymodocea nodosa* (as *C. aequorea*): origin of lateral roots.

Tomlinson and Posluszny (1978): *Syringodium filiforme*: floral development and anatomy.

Yamashita (1976*a*): *Halodule pinifolia, H. uninervis*: pollen development.

Significant Literature—Cymodoceaceae

Arber (1923); Ascherson (1882); Ascherson and Graebner (1907); van Beusekom (1967); Black (1913); Bornet (1864); Chrysler (1907); Clifford (1970); Cohen (1939); Doohan and Newcomb (1976); Duchartre (1872); Ducker *et al.* (1977); Ducker and Knox (1976, 1978); Ducker *et al.* (1978); Feldmann (1936, 1938); Gessner (1968); Gravis (1934); Gravis *et al.* (1943); den Hartog (1960, 1964, 1965, 1970*a, b*); Irmisch (1851); Isaac (1969); Kay (1971); Lakshmanan (1968); Lüpnitz (1969), McClure (1970); Magnus (1871, 1872); Markgraf (1936); Miki (1932, 1934*a*); Monoyer (1927); Ostenfeld (1916); Pettitt

(1976); Pettitt and Jermy (1975); Phillips (1960*a*, *b*, 1967); Phillips *et al.* (1974); Pottier (1934); Sauvageau (1889*b*, 1890*a*, *b*, *c*, 1891*a*, *b*, *c*); Schenck (1886*a*, *b*); Solereder and Meyer (1933); Subramanyam (1962); Tepper (1882); van Tieghem and Douliot (1888); Tomlinson (1974*a*, 1980); Tomlinson and Posluszny (1978); Yamashita (1976*a*).

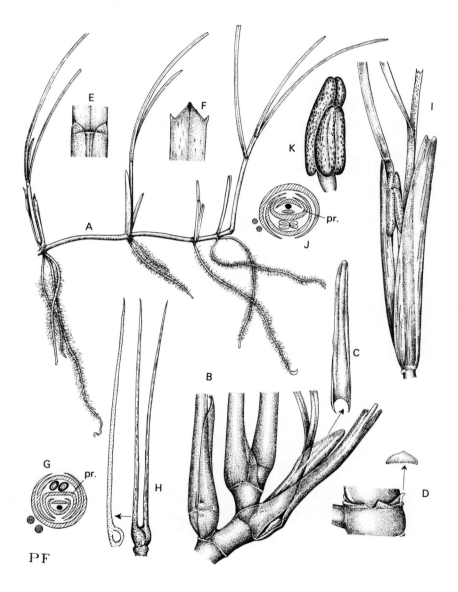

FIG. 12.1. CYMODOCEACEAE. *Halodule* spp. Vegetative morphology (A–F), based on *H. beaudettei* (den Hartog) den Hartog, Matheson Hammock Wading Beach, Florida, P. B. Tomlinson, mixed collections. Reproductive morphology (G–K), based on *H. pinifolia* (Miki) den Hartog, Suva, Fiji, P. B. Tomlinson *s.n.*

A. Habit of monopodial shoot (× 1) with long internodes (some foliage leaves have lost their blades).

B. Apex of shoot (× 3) with congested internodes, to show monopodial branching.

C. Detail of prophyll from B (× 3).

D. Node with foliage leaves detached (× 12) to show squamules (inset, × 12); base of adventitious root to left.

E. Mouth of sheathing base of foliage leaf (× 12).

F. Leaf apex (× 12).

G, H. Female 'flower.'
    G. Diagram of female 'flower,' terminal on an erect shoot (pr., prophyll of continuation shoot, adventitious roots to left, cross-hatched). H. Female 'flower' (× 4), with pair of carpels, one in LS to left.

I–K. Male 'flower.'
    I. Male flowering shoot (× 4) with emerging anthers. J. Diagram of male 'flower' terminal on erect shoot (pr.—prophyll of continuation shoot, adventitious roots to left, cross-hatched). K. Detail of unequal anthers of male 'flower' (× 9).

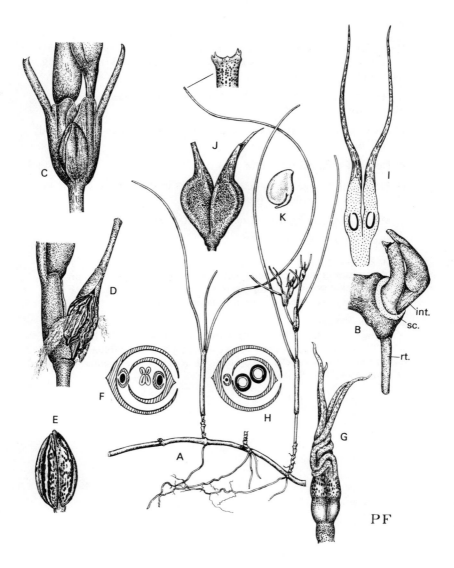

int.

sc.

rt.

PF

408

FIG. 12.2. CYMODOCEACEAE. *Syringodium filiforme* Kütz. Vegetative and reproductive morphology. (Matheson Hammock Wading Beach, Miami, Florida, P. B. Tomlinson, mixed collections). (Partly after Tomlinson and Posluszny 1978.)

A. Habit of female plant ($\times$ ½) showing dimorphic axis system; inset detail of leaf apex ($\times$ 3).

B. Rhizome apex ($\times$ 3), with enveloping scale removed (stippled scar, sc), showing adventitious root (rt.), axillary shoot (future erect shoot) and continuing apex with next internode (int.) beginning to extend.

C. Part of male inflorescence ($\times$ 3) with terminal 'flower' enclosed by two bracts each subtending a lateral continuation shoot.

D. Male 'flower' ($\times$ 6) at anthesis rupturing enveloping bract; pollen filamentous.

E. Single male 'flower' (solitary anther) ($\times$ 3).

F. Floral diagram of arrangement of parts in C, with two continuation shoots.

G. Immature female 'flower' ($\times$ 4), with paired carpels and crumpled styles.

H. Floral diagram of female 'flower' with one continuation shoot.

I. LS female 'flower' ($\times$ 4).

J. Young fruits ($\times$ 3).

K. Embryo ($\times$ 3).

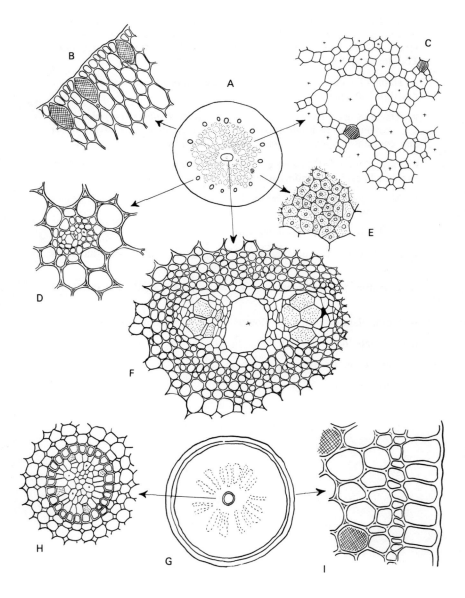

410

FIG. 12.3. CYMODOCEACEAE. *Syringodium filiforme.*

A–F. Rhizome.
   A. Diagrammatic TS rhizome ($\times$ 10). B. Detail of surface layers, epidermal tannin cells cross-hatched. C. Detail of lacunose middle cortex ($\times$ 50), tannin cells cross-hatched. D. Cortical vb ($\times$ 80). E. Transverse diaphragm from cortical air-canal, in surface view ($\times$ 200). F. Vascular cylinder ($\times$ 180); endodermis is indistinct as inner layer of thick-walled sheathing cells.

   G–I. Root. All TS.
   G. Root, diagrammatic TS ($\times$ 35). H. Stele ($\times$ 200), putative thin-walled undifferentiated tracheary elements—x. I. Details of surface layers ($\times$ 200) without differentiated root-hairs.

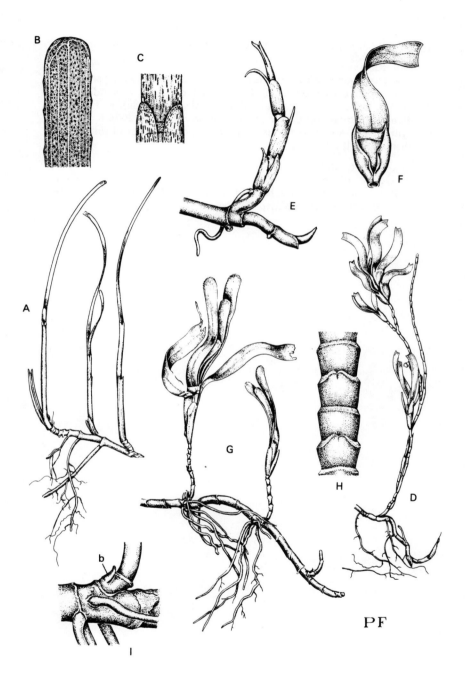

FIG. 12.4. CYMODOCEACEAE. Vegetative morphology in *Cymodocea*, *Amphibolis*, and *Thalassodendron*.

A–C. *Cymodocea nodosa.*
  A. Part of shoot system (× ½) with indistinct long- and short-shoots.
B. Leaf tip (× 2). C. Mouth of sheath with ligule (× 2).

D–F. *Amphibolis antarctica.*
  D. Part of dimorphic shoot system (× ½) with scale-bearing rhizome and branched erect shoot. E. Detail of rhizome apex (× 1) showing sympodial construction, erect shoot is original terminal shoot, new rhizome apex is lateral shoot. F. Foliage leaf (× 1).

G–I. *Thalassodendron ciliatum.*
  G. Habit (× 1) showing dimorphic shoot system with regularly spaced erect shoots. H. Portion of erect stem (× 5) showing regular leaf scars. I. Detail of branching (× 3⁄2), erect axis is main axis evicted by lateral rhizome renewal shoot, with a resting bud (b) at first node of erect shoot. shoot. (A, E, G, and I after Tomlinson 1974.)

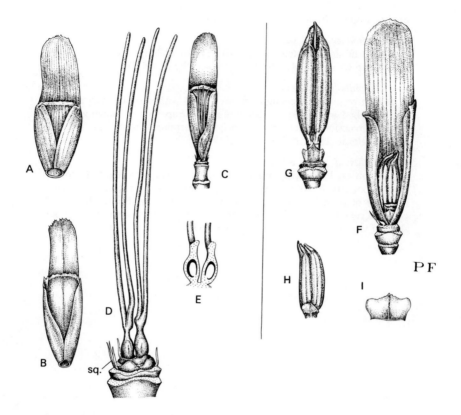

FIG. 12.5. CYMODOCEACEAE. *Thalassodendron ciliatum* (Forsk.) den Hartog. Reproductive morphology.

A–E. Female inflorescence.
A. Outer bract of 'inflorescence' (× 1). B. Middle bract of 'inflorescence' (× 1). C. Inner bract (× 1), immediately enclosing female 'flower'. D. Female 'flower' (× 4), with all three bracts removed (bract scars stippled), showing pair of carpels; sq., squamules. E. Carpel pair in LS, styles omitted (× 4).

F–I. Male inflorescence.
F. Male inflorescence (× 3) with three outer bracts removed, their scars stippled. G. Male inflorescence (× 3) with four outer bracts removed showing 'flower' and innermost scale-bract. H. Male 'flower' (× 3) with innermost scale bract. I. Inner bract (× 6) detached.

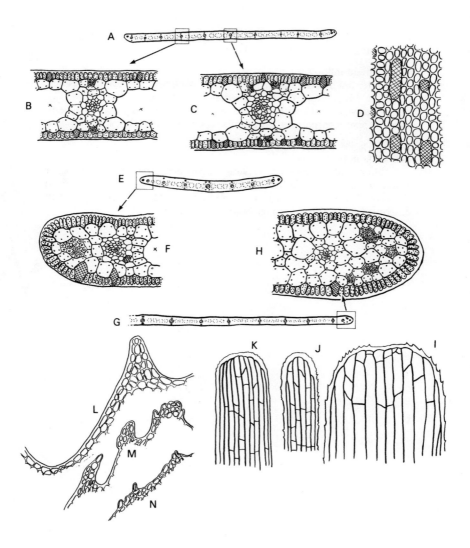

FIG. 12.6. CYMODOCEACEAE. *Cymodocea.* Leaf anatomy. (After Tomlinson 1980.)

A–D. *C. rotundata.*
A. Lamina in TS (× 17). B. Detail of lateral vein (× 150). C. Detail of midvein (× 150). D. Epidermis in surface view (× 300).

E–F. *C. nodosa.*
E. Lamina TS (× 17). F. Detail of leaf margin (× 150).
G–H. *C. serrulata.*
G. Lamina TS (× 17). H. Detail of leaf margin (× 150).

I–N. Leaf apex (× 6) and details of diagnostic apical teeth (× 150).
I and L. *C. serrulata.* J and M. *C. rotundata.* K and N. *C. nodosa.* Tannin cells cross-hatched in B, C, D, F, and H.

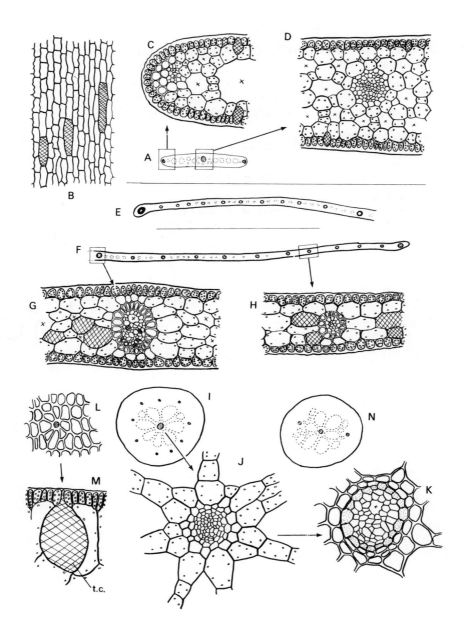

FIG. 12.7. CYMODOCEACEAE. Leaf anatomy. (After Tomlinson 1980.)

A–D. *Halodule beaudettei.*
   A. Lamina TS (× 17). B. Epidermis in surface view (× 300). C. Detail of leaf margin (× 150). D. Detail of midvein (× 150).

E. *Amphibolis antarctica*, half lamina TS (× 17).

F–H. *Thalassodendron ciliatum.*
   F. Half lamina TS (× 17). G. Detail of midvein (× 150). H. Detail of lateral vein (× 150).

I–M. *Syringodium*
   I. Lamina (*S. isoetifolium*) TS (× 17) to show diagnostic multiple lateral veins. J. Detail of midvein (× 150). K. Midvein (× 300), endodermis stippled. L. Epidermis in surface view (× 300) with surface exposure of tannin cell (cross-hatched). M. Same in TS to show enlarged tannin cell (t.c.) below epidermis.

N. Lamina (*S. filiforme*) TS (× 150) to show diagnostic pairs of lateral veins. Tannin cells cross-hatched in B, C, D, G, H.

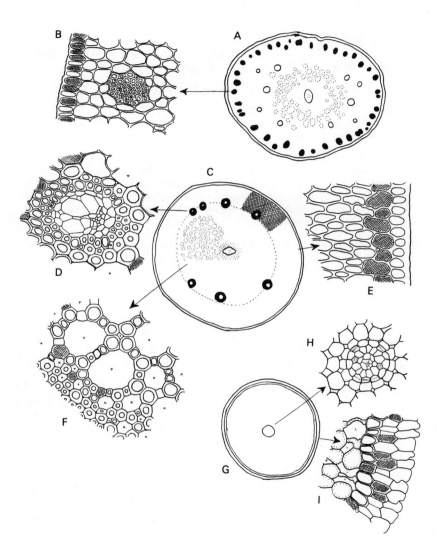

FIG. 12.8. CYMODOCEACEAE. Stem and root.

A–B. *Cymodocea serrulata.* Rhizome.

A. Diagrammatic TS (× 15); cortical fibrous bundles solid black. B. Surface layers (× 100) including a single fibrous bundle; epidermal cells mainly tanniniferous—cross-hatched.

C–F. *Amphibolis griffithii.* Rhizome.

C. Diagrammatic TS (× 15) only lacunae on one sector shown, outer sclerenchyma shown cross-hatched on only one sector, sclerenchyma of cortical vbs—solid black. D. Single cortical vb (× 200). E. Surface layers (× 200); hypodermal tanniniferous layer—cross-hatched. F. Lacunose cortex (× 100) with thick-walled cells.

G–I. *Halodule beaudettei.* Root.

G. Diagrammatic TS (× 60). H. Stele (× 200), diarch with thin-walled central cells surrounding a narrow xylem lacuna—x. I. Surface layers (× 200), thin-walled epidermis without root-hairs at this level; tannin cells—cross-hatched.

Family 13
ZOSTERACEAE (Dumortier 1829)
(Figs. 13.1–13.7 and Plates 1A, B and 6)

SUMMARY

THIS natural family consists of three entirely marine genera, but the plants sometimes penetrate into brackish-water habitats. The family occurs almost entirely in temperate latitudes of the northern and southern hemispheres (den Hartog 1970a). *Zostera* L. (including two distinct subgenera, *Zostera* and *Zosterella*) is widely distributed; *Phyllospadix* Hook. is restricted to the shores surrounding the north Pacific; *Heterozostera* den Hartog occurs in Australia and South America. The plants typically have a creeping, often extended perennial rhizome (congested in *Phyllospadix*), giving rise to longer or shorter, sometimes annual, erect shoots with linear leaves. The erect shoots become secondarily rooted and form new rhizomes in *Heterozostera*. The branched rhizome is obscurely monopodial. The branch axis is frequently adnate to the parent axis (e.g. Fig. 13.2 C, D and Plate 6).

The leaves are distichous and of two main kinds: (i) foliage leaves with a narrow strap-shaped blade; (ii) scale leaves, usually developed only as prophylls at the base of the branch. Foliage leaves that are associated with reproductive organs often become modified as bracts or 'spathes.' The basal sheath of the foliage leaf is closed in subgenus *Zostera* (Fig. 13.1 E), but in other taxa it is open, with overwrapping margins. The mouth of the sheath is often auriculate and there is always a distinct ligule (e.g. Fig. 13.1 G). The leaf apex is rounded, but it becomes emarginate by the dissolution of apical cells to form an apical pore. The apex in *Phyllospadix* has marginal 'fin cells.' Roots are unbranched and arise in 2–4 regular linear series, at, or close to, the node (e.g. Fig. 13.4 B), with from one to many roots in each series.

The reproductive parts consist of flattened spike-like axes ('spadices'), each bearing flowers on one flattened side (e.g. Fig. 13.1 I). The whole spadix is enclosed within the sheath of the leaf ('spathe') inserted below, but it sometimes becomes exserted in fruit. The plants are either monoecious (*Heterozostera* and *Zostera*) or dioecious (*Phyllospadix*). The male 'flower' has two free bilocular anthers, dehiscent longitudinally. The pollen is filamentous. The female 'flower' is an ellipsoidal or crescent-shaped carpel with a short style and two stigmas. The solitary ovule is pendulous, orthotropic, and bitegmic. Additional structures associated with flowers are (i) a narrow flap of tissue connnecting the base of each pair of anthers; (ii) retinacules ('retinaculae'), i.e. marginal outgrowths (Fig. 13.2 G) which more or less enclose each pair of stamens. Retinacules are, however, absent from *Zostera* subgenus *Zostera*. The fruit is either ovoid to ellipsoidal and with a scarious pericarp, or else crescent-shaped and with a soft ephemeral exocarp and a hard fibrous endocarp. The seed lacks endosperm and the embryo has an enlarged hypocotyl (Taylor 1957). The seedling is without a radicle (primary root) (Yamashita 1973); the morphology of the embryo and seedling of *Zostera capensis* is misinterpreted by Edgecumbe (1980).

The Zosteraceae are anatomically reduced. They lack stomata and have a chlorophyllous leaf epidermis. There is a virtual absence of lignin, but tannin (polyphenols) are plentiful. The leaves have well developed hypodermal fibres. Protoxylem lacunae are well developed. The vascular system of the axis is more or less protostelic and amphiphloic, and cortical vascular bundles are present. Tracheary elements of the roots are incompletely differentiated.

KEY TO GENERA

The anatomy as well as the morphology of the vegetative parts is useful in the diagnosis of genera. The following key summarizes the information for major taxa.

1A. Dioecious. Rhizome with thick, congested, often flattened internodes, commonly dimorphic and with long- and short-shoots, these obviously monopodial; sclerenchyma well developed but intercellular fibre bundles absent. Roots short, thick, often inserted throughout node. Leaf blade thick, somewhat coriaceous and often with marginal 'fin-cells.' Leaf sheaths persisting as fibrous masses
Phyllospadix (Figs. 13.4 and 13.5).

1B. Monoecious. Rhizome usually with elongated, slender, terete internodes, itself not dimorphic (but usually with distinction between horizontal and erect axes); sclerenchyma not well developed but intercellular fibre bundles common. Roots long, thin, usually restricted to nodes. Leaf blade thin, margin without fin-cells. Leaf sheath ± deciduous as a unit 2.

2A. Erect shoots long, wiry, containing abundant phenolics, branched and commonly rooting distally to form new rhizomes; axes with more than two (4–12) cortical vbs, fibrous strands distributed throughout cortex, commonly associated with stele *Heterozostera** (Fig. 13.3)

2B. Erect shoots herbaceous, not wiry, not commonly rooting distally; axes with only two cortical vbs, fibrous strand restricted to periphery of cortex 3 (*Zostera*)

3A. Leaf sheath closed, rupturing with age, leaf blade wide with (3–) 7 (–11) veins. Erect shoots terminating in an inflorescence. Retinacules absent (or at most singly in association with lower flowers). Cortical fibrous bundles of rhizome in outermost layers,† commonly next to epidermis *Zostera* subgenus *Zostera* (Fig. 13.1)
(*Alega* of some authors)

---

* Den Hartog (1970a) distinguishes *Heterozostera* by its sympodial rhizome, but examination of a large population does not support this. The growth habit in the family is seemingly not consistent, and would repay further detailed investigation, especially with reference to seasonal variations in architecture which is ecologically important (Harrison 1979). The disposition of spadices on erect shoots also needs more complete analysis; to what extent the distal axis complexes are determinate (cymose) is not reliably known.

† This distinction, originating with Setchell (1933), is subjective and seems difficult to apply when large samples are examined (cf. Fig. 13.7 C, I).

3B. Leaf sheath open, abscissing or shredding with age, leaf blade narrow with 3 (–5) veins (lateral veins often inconspicuous). Erect shoots not terminating in an inflorescence. Retinacules present. Cortical fibrous bundles of rhizome less superficial, not usually next to epidermis

Zostera subgenus Zosterella (Fig. 13.2)

GENERIC DESCRIPTIONS

The subsequent account describes the individual genera, but only with reference to the better known species. A general discussion of reproductive morphology is given later (p. 432).

ZOSTERA L. subgenus *Zostera* (Fig. 13.1)

This subgenus includes four species (*Z. asiatica, Z. caespitosa, Z. caulescens, Z. marina*) with a distribution in the northern Atlantic and Pacific. The following account refers to *Z. marina* L. which has been studied extensively; ways in which other species may be distinguished are given later.

VEGETATIVE MORPHOLOGY

Habit (Fig. 13.1)

Creeping rhizome monopodial, each leaf subtending a short erect lateral branch of determinate growth. Lowest nodes on erect shoots usually remaining unbranched, although aborted lateral buds present. Distal nodes with adnate branches, the ultimate units of the erect shoots being 'inflorescences.' Each branch with a basal prophyll (Fig. 13.1 D), the prophyll itself often with a bud capable of development. Adnation of the branch to the main axis due to late elongation of the axis below the originally axillary branch, in extreme cases the mature branch becoming separated by an entire internode from the leaf subtending it (Bugnon 1963). [The development processes involved in this change are illustrated in Plate 6.]

Leaf

Leaf blade up to 120 cm long, 2–12 mm wide, the leaves of the main axes being longer and wider than the leaves on lateral axes. Leaf sheath closed (Fig. 13.1 E, F) but becoming ruptured with age. Auricles distinct (Fig. 13.1 G); ligule short, a specialized outgrowth of the adaxial surface of the leaf. Veins 5–11 per blade, the lateral ones joining each other and the midvein below the rounded apex (Fig. 13.1 G); frequent conspicuous transverse commissures connecting the longitudinal veins. Squamules 2–4 (Fig. 13.1 B–D) (not consistently four as suggested by Setchell 1929).

Root

Roots in two (sometimes four) lateral groups at each node with 6–8 (up to 12) roots initiated in each group, but typically rather few developed (Fig. 13.1 B, C).

VEGETATIVE ANATOMY

Leaf blade (Fig. 13.6 D–F)

Apex initially rounded (Fig. 13.1 G), but distal cells becoming eroded to produce a shallow apical pore more or less continuous with the distal extension of the median vein. **Cuticle** thin; **stomata** absent. **Epidermis** assimilatory, uniform; outer epidermal wall somewhat thicker than remaining walls. Cells more or less cubical or somewhat extended longitudinally, with abundant chloroplasts. Chloroplasts appearing fewer in mesophyll layers. **Hypodermis** uniseriate (less commonly biseriate). **Mesophyll** including uniseriate parenchymatous girders alternating with longitudinal air-lacunae (Fig. 13.6 F), the lacunae segmented at intervals by transverse, mostly uniseriate diaphragms of small, densely cytoplasmic cells. Every three (at margin) to five (towards midvein) series of uniseriate girders separating a more massive girder including a vein, the total number of veins varying from 5–11 according to the overall width of the leaf (e.g. Fig. 13.6 D). Strands of unlignified, mostly hypodermal **fibres** frequent near both surfaces, most commonly above and below each girder, numerous at leaf margin. Similar fibrous strands adjacent to the larger veins. Leaf margin including a massive fibrous strand. **Veins** towards the abaxial surface, median larger than laterals; vbs without any distinct continuous sheath but demarcated somewhat by associated fibres (Fig. 13.6 E). Phloem relatively well developed. Xylem reduced and represented by a wide lacuna when mature, the 1-(2–3) files of tracheids present in the developing leaf becoming ruptured and disappearing with expansion, but those in the leaf apex persisting. Transverse vascular commissures frequent, each occupying a 2–3-layered transverse diaphragm.

Leaf blade thinner distally, largely by a decrease in the height of the girders, the air-lacunae becoming progressively shallower. Leaf apex including only two mesophyll layers (equivalent to two hypodermal layers). Fibrous strands absent close to the leaf apex except for those associated with the median vein.

Leaf sheath

Tubular, closed. **Epidermis** of thin-walled, colourless cells. **Mesophyll** resembling that in the blade but fibrous strands more numerous. Ventral region including only two mesophyll layers; becoming even thinner distally and then composed solely of the two epidermal layers but with intervening fibres. **Ligule** two- or three-layered basally, with 2–4 closely set transverse diaphragms marking its level of insertion. Abscission of the leaf blade taking place immediately above the level of insertion of the ligule and assisted by decomposition (gelatinization) of the cell walls.

Prophyll

Delicate colourless organ enclosing the base of the stem; transparent and easily overlooked. Adaxial face ending in a short rounded point, representing the vestigial blade. Blade somewhat more pronounced in vegetative (lower) than in floral (upper) prophylls. **Mesophyll** including several layers of cells and fibres between each epidermis but no lacunae. **Veins** 5–7; the median prolonged beyond the apical commissure but not continuous into an apical pore.

Leaf development

Described in some detail by Sauvageau (1890*a*, 1891*a*) on the basis of the examination of cleared leaves and evidently identical in all respects with the type described on page 191 for *Thalassia* (cf. Roth (1961) for *Z. noltii*).

Horizontal stem

Epidermis with outer walls thickened. **Cortex** very wide, peripheral layers compact; cells narrow, the walls with somewhat collenchymatous thickening. Fibrous strands abundant in the outer cortex; (Fig. 13.7 I), often hypodermal, but progressively well developed towards the base of each internode and absent from the node. Middle cortex regularly lacunose. Internode including two narrow cortical vbs, one on each side of the stem and inserted perpendicular to the plane of distichy of the leaves (cf. Fig. 13.7 A). Cortical vbs somewhat variable in their anatomy, but usually with a wide xylem lacuna (cf. Fig. 13.7 B) and 1–2 phloem strands, the phloem enveloped by an endodermis. **Stele** narrow but surrounded by a thin-walled, distinct **endodermis,** the latter becoming wholly lignified with age. Vascular tissues of stele represented by a wide central xylem lacuna surrounded by ± four large but often indistinct phloem strands (cf. Fig. 13.7 G). In more regularly constructed steles each large phloem strand alternating with a much narrower strand of phloem associated with a narrow xylem lacuna. This arrangement can be interpreted as consisting of eight alternate large and small vbs, the four larger ones with a common xylem cavity. Each larger phloem strand with a narrow strand of parenchyma representing obliterated protophloem.

[According to Sauvageau (1891*b*), these two types of stelar bundle differ developmentally in their time of origin. The large 'first-order bundles' differentiate before internodal elongation is complete and the xylem at first includes one or more narrow tracheal elements. The smaller 'second-order bundles' differentiate later, when the internode has almost finished elongation, and the small lacunae are never represented by distinct tracheal elements. First- and second-order lacunae are continuous at the nodes via 1–2 large, radially extended cells with spirally or reticulately thickened and lignified walls (see description under 'Course of vascular bundles' below).

Adnation of the branch to the main axis is not evident anatomically, because the internode below the branch insertion elongates after the vascular system is established.]

Erect stem and inflorescence axis

Often reduced compared with the horizontal stem. Second-order bundles partially or wholly absent. Branch trace often visible in the internode some distance below the level of branch exsertion.

Course of vascular bundles in stem

This has been described in a somewhat stylized fashion by Chrysler (1907) and Monoyer (1927). In their accounts the strands of conducting tissue (mostly phloem) within the stele are interpreted as separate vascular bundles. The cortical system of the stem is derived from lateral bundles on each side of the median bundle, which unite together in the node. The collective bundle is continuous through the internode below and joins the incoming lateral trace of the leaf below. The cortical system is continuous with the central cylinder via a short connection at the node. The median vascular bundle of the leaf continues directly into the stele, and extends downwards through two internodes as the 'median' of the large stelar bundle, splitting above the incoming trace at the next leaf below on the same orthostichy to unite with the 'lateral' of the large bundles. Since this 'lateral' bundle is not directly continuous with a leaf it is interpreted as a 'sympodial stelar bundle' by Monoyer. The branch trace diverges from the median at more or less one internode above the subtending leaf so that adnation of the branch vascular system is complete anatomically as well as morphologically. Fibrous bundles of the outer cortex are continuous with those of the leaf sheath. The root traces are continuous with the stele via a single common bundle at the node.

Preliminary examination of vascular development suggests that the system is less precise and less formalized than the above account indicates.

Root

In 2–4 series at the node, with up to 12 roots in each series; the system sometimes reduced to a single root on either side of stem.

Individual roots unbranched, coleorhiza absent. **Epidermis** large-celled. Root-hairs arising from distinct elongated initials with slightly thickened walls and dense protoplasm; these initials protruding somewhat deeper into the cortex than the remaining undifferentiated cells (Fig. 13.7 κ). Outer cortex differentiated into a single-layered **exodermis** of narrow, hypodermal, thick-walled cells, and a wider zone of somewhat collenchymatous cells, 1–3 cells wide. Middle cortex wide, lacunose; inner cortical cells regularly arranged. **Endodermis** uniseriate; cells uniformly thickened with lignified middle lamellae. Stele (Fig. 13.7 J); conducting tissues represented by 4–5 pericyclic sieve-tubes, each with a single companion cell on the inside (probably derived from same mother cell as sieve-tube). No tracheal elements differentiated; centre of stele occupied by 3–9 undifferentiated cells.

Conducting elements

Xylem represented largely by lacunae developed by dissolution of first-formed tracheids, the elements persisting only at the nodes. In the stem, according to Sauvageau (1891*d*), xylem lacunae formed by resorption of the elements before the acquisition of any wall thickening.

TAXONOMIC NOTES FOR ZOSTERA SUBGENUS ZOSTERA

The remaining three species of this subgenus, which are all Asiatic, are distinguished from each other by combinations of characters which include variations in: the number of leaf veins; the shape of the leaf apex; the degree of persistence of the leaf sheath; the construction of erect shoots; the texture of seeds (Miki 1932, 1933; den Hartog 1970a). Of interest is *Z. caespitosa* which is described as having short internodes and consequently a tufted habit which recalls that of *Phyllospadix*. *Zostera caulescens* is distinguished from the other 3 species by the fact that the erect axis ends in a leafy vegetative shoot rather than in an inflorescence.

MATERIAL EXAMINED

Zostera marina L.; Isle de Groix, Brittany, France; F. Hallé and P. B. Tomlinson 8.ix.71. Plum Island, Massachusetts; P. B. Tomlinson 2.ix.73B; all parts. Mole Harbour, Admiralty Island, Alaska; D. Gregory 23.viii.79; vegetative parts.

ZOSTERA L. subgenus *Zosterella* (Fig. 13.2)

This subgenus includes eight species which have a wider total distribution than the subgenus *Zostera*, as they occur in Europe, Asia, East Africa, and Australasia. Certain of these species extend further into the tropics than any other members of the family (den Hartog 1970a). The most extensively studied species is *Z. noltii* (*Z. nana* of most workers).

VEGETATIVE MORPHOLOGY

Species differing from *Z. marina* in their method of branch construction, but published information confused. Aspects of shoot constriction in *Z. novazelandica* shown in Fig. 13.2. Adnation between branch and parent axis indicated in Fig. 13.2 c, but the branch meristem sometimes abortive (Fig. 13.2 D). Roots commonly only two at each node (Fig. 13.2 A, B).

VEGETATIVE ANATOMY

Leaf blade

As in *Z. marina* but narrow and with only 3–5 veins (Fig. 13.6 B, C), apex becoming more deeply emarginate, especially with age (e.g. Fig. 13.2 E). Hypodermal layer between girders always uniseriate. Fibrous strands always present above and below the girders except in the region of the midvein and at the margin, but never forming a conspicuous marginal strand (Fig. 13.2 E).

Leaf sheath

Always open, margins overwrapping (Fig. 13.2 E).

Prophyll and spathe (cf. Fig. 13.2 c, pr)

Anatomy reduced to the same extent, compared with the foliage leaf, as in *Z. marina*.

Stem

Said to be distinguished from that of *Z. marina* by the fibrous bundles in the outer cortex being well separated from the epidermis (Fig. 13.7 C), and never occupying a hypodermal position (Setchell 1933). Cortical vbs two, as in *Z. marina*. [The general features illustrated in Fig. 13.7 A–C are similar to those of *Z. marina*.]

TAXONOMIC NOTES FOR ZOSTERA SUBGENUS ZOSTERELLA

The species of the subgenus *Zosterella* are distinguished from one another by similar variations in morphological characteristics to those by which subgenus *Zostera* is subdivided (den Hartog 1970*a*) (see p. 428). There is insufficient evidence to show whether they may also be diagnosed by features of vegetative anatomy.

MATERIAL EXAMINED

Zostera noltii Hornem.; Isle de Groix, Brittany, France; P. B. Tomlinson 20.viii.74; all vegetative parts.

Z. muelleri Irmisch ex Aschers.; Westernport Bay, Victoria, Australia; B. Clough 11.i.74; all parts.

Z. novazelandica Setchell; Akaroa, New Zealand; CHR 168130, Lucy B. Moore. Kawakawa Bay, Auckland, New Zealand; P. B. Tomlinson 23.iii.69; all parts.

HETEROZOSTERA den Hartog

The one species in this monotypic genus, *H. tasmanica* (Martens ex Aschers.) den Hartog, is distributed in Australia and South America (Chile). Its absence from New Zealand is rather surprising.

MORPHOLOGY

Shoot dimorphism pronounced. **Rhizome** often looped, branching not consistently sympodial as suggested by den Hartog (1970*a*). Either sympodial, the rhizome axis becoming directly an erect shoot (Fig. 13.3 A), with a renewal shoot commonly developed near the base of the erect shoot; or monopodial (Fig. 13.3 B) with branches at every node, sometimes developing into either a branch of the rhizome, or an erect shoot. Branches adnate to the parent axis for one internode, with a prominent prophyll as in *Zostera*; frequently aborted. Internodes of the rhizome long (e.g. Fig. 13.3 B) or short (Fig. 13.3 A). **Erect shoots** up to 1 m long, determinate or indeterminate, thin, wiry, but unlignified, with long internodes. Basal branches usually aborted. Distal branches sometimes developed on long axes meeting the substrate, the branches rooting and forming a new rhizome (Fig. 13.3 C), thereby proliferating the shoot system. **Roots** usually in pairs at each node, not forming clusters as in other genera (Fig. 13.3 B). **Leaves** with an open sheath, ligulate, the mouth of the sheath auriculate (Fig. 13.3 E); leaf apex usually distinctly emarginate. **Spathe** with a somewhat inflated sheath (Fig. 13.3 F).

Inflorescences consisting of cymose complexes of spadices, with pronounced adnation between branch and axis.

VEGETATIVE ANATOMY

Leaf blade

With three veins; not distinguishable from *Zostera* subgenus *Zosterella*.

Stem (Fig. 13.7 D–H)

Similar to *Zostera*, but fibrous strands distributed throughout the cortex and often in contact with the epidermis (Sauvageau 1891*d*); especially common near the cortical vbs. Cortex including (2) –6 (–12) vbs in two series, one on each side of stele (Fig. 13.7 D).

TAXONOMIC NOTES FOR HETEROZOSTERA

An important diagnostic feature of this taxon, namely its ability to branch in a proliferative manner from the distal parts of the erect shoots, has not been carefully described. The genus is distinctive because of its wiry erect stems and numerous cortical vbs.

MATERIAL EXAMINED

Heterozostera tasmanica (Martens ex Aschers.) den Hartog; Westernport Bay, Victoria Australia; B. Clough, I.A. Staff, P. B. Tomlinson 10.xi.74; all parts.

PHYLLOSPADIX Hook.

There are five species of this genus in Pacific North America and Japan. They are ecologically distinct amongst seagrasses because they are frequently attached directly to rocks to form mat-like colonies. They are distinguished from the other Zosteraceae by their coarse habit which results from the rhizome having congested internodes, and also by the longer leaves and the short roots with adhesive root-hairs (Fig. 13.4 C).

VEGETATIVE MORPHOLOGY

Creeping axis monopodial (Dudley 1893), covered with persistent fibrous remains of old leaf sheaths. Internodes usually 1 cm long or less, but up to 5 mm wide. Branches supra-axillary, inserted shortly below the next node above the axillant leaf. Most laterals of limited growth function as short-shoots (e.g. Fig. 13.4 A). Roots in a double row attached to the internode on the side of the stem opposite a branch; with up to five roots per row (Fig. 13.4 B, C). Rhizome usually orientated so as to make the plane of distichy of the leaves horizontal, the leaves growing erect, and the roots downward.

Leaves as much as 2 m long, with a relatively short sheath (up to 50 cm long) and a much longer blade. Leaf sheath open, with overwrapping wings, auriculate at the mouth of the sheath. Ligule distinct. Leaf apex with apical pore, as in *Zostera.*

VEGETATIVE ANATOMY

Leaf blade

Lamina often thick, becoming ± elliptic in transverse outline (Fig. 13.6 G). **Hairs** absent but anticlinally extended marginal cells ('fin-cells') frequently developed, forming a more or less complete narrow, sometimes discontinuous and then denticulate, distal fringe to the blade. The 'fin-cells' at the proximal part of the blade largely extensions of the cell wall. Leaf apex rounded (Fig. 13.4 D) or slightly emarginate. Cuticle thin. **Epidermis** of uniform cells, more or less isodiametric in surface view, but anticlinally extended, 2–4 times as high as wide in TS, the cells including abundant chloroplasts; outer epidermal wall quite thick. **Mesophyll** (Fig. 13.6 J) very slightly dorsiventral, including a single series of lacunae, but lacunae occasionally in two series near the midvein. Lacunae separated from one another by uniseriate girders, but girders becoming biseriate in the thicker proximal part of the lamina. Lacunae 3–4 on each side of the midvein but up to 15 in wider blades; none outside the marginal vein. Hypodermal layers forming roof and floor of lacunae up to four cells deep, the outermost layer of cells being the smallest. Mesophyll cells with few chloroplasts. **Fibrous strands** abundant especially at the leaf margin, always hypodermal and forming a discontinuous series (Fig. 13.6 H, J); individual fibres narrow, unlignified.

Veins 3–5 (Fig. 13.6 G, I), the median somewhat larger than the laterals; each with an indistinct 1–2-layered sheath of narrow cells with slightly thickened walls surrounding the phloem but not the xylem, but without an endodermis. Xylem represented by an inconspicuous xylem lacuna (Fig. 13.6 H). Fibres adjacent to the veins in the proximal part of the blade.

Leaf sheath

Colourless. Overwrapping wing biseriate, the abaxial layer large-celled. Thickened dorsal region including several (up to four) series of lacunae, the hypodermal tissues up to five cells deep on abaxial side, shallower adaxially. Hypodermal **fibrous strands** below both surfaces in the dorsal region, but absent below the adaxial surface laterally; present in the wings, and becoming massive in the leaf base; persistent after the decay of older leaves. Veins five basally, three distally, as in the blade but here with associated fibres sometimes forming an almost continuous sheath around the marginal veins and continuous with the hypodermal fibres.

Stem (Rhizome)

Epidermis of uniform cells, outer walls becoming much thickened. Cortex with a distinct wide **mechanical layer** of narrow compact cells with slightly thickened, unlignified walls; mechanical layer most obvious on the lower (flattened) side of rhizome. **Cortex** otherwise of uniform parenchyma cells, but somewhat lacunose towards the stele. The cortical vascular system represented by 2 (–3) narrow vbs, each surrounded by a distinct endodermis, the two bundles being in a plane perpendicular to that of the leaf insertion (Fig. 13.7 L). **Endodermis** distinct. **Stele** including wide sieve-tubes and tracheids with spiral or somewhat reticulate wall thickenings, the tracheids scattered or in small groups (Fig. 13.7 M). The regular conducting system reported for other members of this family may be represented by four protophloem strands, the two larger ones perpendicular to the plane of leaf distichy, and the two smaller ones in the plane of leaf distichy, but this is not obvious in wider axes with congested internodes.

Root

Epidermis thin-walled, or, at most, with the outer wall a little thickened. Root-hairs short, thick-walled and frequently lobed distally where in contact with the substrate. **Exodermis** small-celled and thin-walled, but lignified. **Cortex** with an outer mechanical layer of unlignified thick-walled cells, up to nine cells wide. Middle cortex uniform, not lacunose, the cells being in regular radial and concentric files. **Endodermis** thin-walled with well developed Casparian strips. **Stele** with polyarch sieve-tubes within the pericycle, the xylem elements apparently undifferentiated.

TAXONOMIC NOTES FOR PHYLLOSPADIX

According to den Hartog (1970*a*) the North American species (*P. torreyi*, *P. scouleri*) are distinguished from the Alaskan and Japanese species (*P. serrulatus*, *P. iwatensis*, *P. japonicus*) by differences in the degree of branching of the reproductive shoots. There are also differences in the number of roots on the internode (usually many, up to ten in the former group, few, usually two, in the latter). The number of leaf veins in the first species is often five or seven as compared with three in the last three species. The species are still insufficiently well known anatomically to decide whether they can be circumscribed by anatomical features of their vegetative parts. They seem more variable than species of *Zostera*.

MATERIAL EXAMINED

Phyllospadix scouleri Hook.; Admiralty Pt., Whidbey Island, Washington; R. C. Phillips 14.v.75. Cattle Point, San Juan Island, Washington; R. C. Phillips 20.iii.75. Mussell Point, Pacific Grove, California; I. A. Abbott 12356; all parts except male flowers.

REPRODUCTIVE MORPHOLOGY FOR THE FAMILY

Inflorescences indistinctly circumscribed, but essentially consisting of terminal flattened, flower-bearing axes (spadices), the spadices usually in complexes at the ends of higher order branches on erect shoots (Fig. 13.5 A). Axis of the spadix basally 'adnate' to the parent stem. Each branch of the inflorescence with a conspicuous basal membranous **prophyll** (e.g. *Hetero-zostera*), but this kind of prophyll never subtending a bud. Inflorescence enclosed by a second leaf ('spathe'), with a narrow (3–7-veined) short blade and a somewhat inflated, always open sheath (e.g. Figs. 13.1 H, 13.2 F and 13.5 E). Successive spathes in a complex with their sheaths overwrapping alternately to left and right. Inflorescence axis with a few to many (up to 20) pairs of male and female 'flowers.' Inflorescence axis in *monoecious taxa* (Figs. 13.1 I and 13.2 G), bearing two series of carpels alternating with paired stamens, the stamens of each pair connected to one another by a shallow longitudinal flap of tissue. **Stamen** pair consisting of two bilocular sessile anthers (Figs. 13.1 J and 13.2 H), dehiscing longitudinally. **Carpel** (Figs. 13.1 K, 13.2 I, J, and 13.5 D) v. asymmetric; naked, short-stalked with a sagittate base. Ovule solitary, bitegmic, more or less pendulous from the ventral side. Style short with two longer or shorter, simple stigmas (e.g. Fig. 13.2 I). Axis either naked (at least distally in *Zostera* subgenus *Zostera* (Fig. 13.1 I), or in all other taxa (and occasionally in association with basal flowers even of the subgenus *Zostera*) a narrow scale (retinacule) on the margin of the axis opposite each pair of stamens, at first covering them but later reflexed (Figs. 13.2 G and 13.5 B, C). Retinacules of considerable diagnostic significance, but morphologically puzzling (see below).

In *dioecious taxa* (*Phyllospadix*), sterile organs in each sex represented by staminodia (Fig. 13.5 C) and pistillodia, the retinacules being opposite the staminodia or stamen pairs.

Fruit an enlargement of the carpels, the style either persistent (Figs. 13.1 L and 13.2 J) or non-persistent (Fig. 13.5 E). **Seed** ellipsoidal (Fig. 13.1 L). Embryo complex (Taylor 1957) and said to lack a primary root (Yamashita 1973). Germination described by Taylor (1957).

REPRODUCTIVE ANATOMY FOR THE FAMILY
(mainly after Uhl 1947)

Zostera marina. Axis with three longitudinal bundles, one median and one in each margin, the longitudinal veins connected below each pistil and stamen pair by a complete, oblique commissure. Vascular supply to the carpel a single trace from the commissure; this trace remaining unbranched and ending at the base of the ovule. Vascular supply to the stamen pair provided directly from the commissure which is enlarged below them.

Phyllospadix scouleri (female only). Differences from *Zostera marina*: trace to the carpel diverging from the median vb without forming a complete commissure across the axis. Carpel trace branched to produce a dorsal trace and a ventral trace ending below the ovule.

Vascular supply to the retinacule (when present) represented by a marginal branch from the stamen commissure; the branch remaining undivided within the retinacule itself.

Interpretation of reproductive parts

Two main interpretations may be recognized:
(i) The pair of anther sacs and a carpel collectively represent a perfect flower. One group of authors interprets this perfect flower as composed of

a single carpel and a pair of stamens, the reproductive organs from the opposite sides of the axis (e.g. Eichler 1875; Irmisch 1851; Bornet 1864). Ascherson and Graebner (1907) interpret the perfect flower in precisely the same way except that the parts are thought to arise on the same side of the axis. Wettstein (1935) considers the two anther lobes belong to different stamens.

(ii) According to the second interpretation each individual organ represents a separate unisexual flower. This is the opinion of such authors as Bentham and Hooker (1883), Kunth (1841), and Hutchinson (1959), who all appear to regard the two anther lobes as separate stamens from separate flowers. However, it seems more reasonable to follow den Hartog (1970a) and treat the anther lobes as the separate thecae of a single stamen.

Retinacules

These organs, which it must be remembered are partly absent from one group of species, have been variously interpreted. One concept, e.g. that of Ascherson and Graebner (1907), treats them as bracts, Bentham and Hooker (1883) regard them as a reduced perianth. Another very different view is that of Markgraf (1936) and Eckardt (1964) who consider the retinacules to be appendages of the connective of a stamen represented by the anther pair.

The vascular anatomy offers no ready answer to which of these interpretations is correct because the veins irrigate the various organs in the simplest and most direct manner possible. Developmental evidence is no more helpful and some of it is even conflicting (cf. Grönland 1851; Hofmeister 1852; with de Lanessan 1875). The conclusion must be that in the absence of a thorough developmental and detailed comparative study the morphology of the zosteraceous 'spadix' remains uninterpretable in the light of present concepts.

GENERAL NOTES

The Zosteraceae provide a good example of a clearly circumscribed group whose affinities with other helobian families is obscure. This is because the arrangement of their reproductive organs is unlike that of any other angiosperm family and, despite considerable discussion, no acceptable interpretation of them has been offered. Clearly this high degree of morphological reduction reflects considerable biological specialization, which remains little investigated.

ECONOMIC USES

Zostera species were among the first taxa to indicate the importance of aquatic plants in biological food chains and of their consequent economic significance. The 'wasting disease' of eel-grass, apparently caused by a fungus, led to a decline in the population of marine organisms and water-fowl, which fed extensively on these marine beds in Western Europe (Blackburn 1934; Tutin 1938; den Hartog 1970a). Another very different economic fact is that *Zostera marina* has been used directly as a packing material. In the Gulf of California the fruit of *Z. marina* was at one time harvested by the Seri

Indians of Sonora and appears to have been an important food source (Felger and Moser 1973; Felger *et al.* 1980).

GENERA AND SPECIES REPORTED ON IN THE LITERATURE

Areschoug (1878): *Zostera marina*: leaf.

Barnabas *et al.* (1977): *Z. capensis*: leaf ultrastructure.

Blackburn (1934): *Z. marina, Z. noltii* (as *Z. nana*): cytology.

Butcher (1933): *Z. marina* and vars., *Z. noltii* (as *Z. nana*): leaf venation, stem.

Chrysler (1907): *Phyllospadix scouleri, Zostera marina*: stem.

Clavaud (1878): *Zostera marina*: pollination.

Dahlgren (1939): *Z. marina*: embryology.

Duchartre (1872): *Z. marina, Z. noltii* (as *Z. nana*): all parts.

Dudley (1893): *Phyllospadix scouleri, P. torreyi*: general anatomy and morphology.

Duval-Jouve (1873*a*): *Zostera marina, Z. noltii* (as *Z. nana*): leaf anatomy, general morphology, embryology.

Duval-Jouve (1873*b*): *Z. marina, Z. nodosa*: diaphragms.

Eber (1934): *Z. marina*: floral morphology.

Engler (1879): *Z. marina*: reproductive morphology.

Falkenberg (1876): *Z. marina*: stem.

Flahault (1908): *Z. marina, Z. noltii* (as *Z. nana*): general anatomy and morphology.

Goffart (1900): *Z. marina*: apical opening in leaf.

Gravis (1934): *Z. marina*: leaf traces.

Grönland (1851): *Z. marina*: leaf anatomy, reproductive morphology.

Hofmeister (1852, 1861): *Z. marina, Z. noltii* (as *Z. minor*): reproductive morphology.

Jensen (1889): *Z. angustifolia, Z. marina*: germination.

Lanessan (1875): *Z. marina, Z. noltii* (as *Z. nana*): flower and fruit development.

Lüpnitz (1969): *Z. marina*: root development.

Magnus (1870): *Z. marina, Z. noltii* (as *Z. nana*): leaf and stem.

Markgraf (1936): *Phyllospadix scouleri, P. torreyi, Zostera marina, Z. noltii* (as *Z. nana*): floral morphology.

Miki (1932): *Zostera asiatica, Z. caespitosa, Z. caulescens, Z. marina*: leaf anatomy, general morphology.

Miki (1933): Same spp. as 1932 paper plus *Phyllospadix iwatensis, P. japonicus, Z. noltii* (as *Z. nana*): general anatomy and morphology.

Monoyer (1927): *Zostera noltii* (as *Z. nana*): stem anatomy, with brief comparative notes on other spp.

Ohlendorf (1907): *Z. marina*: fruit.

Pettitt and Jermy (1975): *Z. marina*: pollen structure.

Raunkiaer (1895–9): *Z. marina*: root, stem, leaf, fruit.

Rosenberg (1901*a*): *Z. marina*: note on root anatomy; embryology.

Roth (1961): *Z. noltii* (as *Z. nana*): leaf development.

Sauvageau (1889*b*): *Z. marina, Z. noltii* (as *Z. nana*): root.

Sauvageau (1890*a*, 1891*a*): *Heterozostera tasmanica* (as *Zostera tasmanica*), *Phyllospadix scouleri*, *P. torreyi*, *Zostera capricorni*, *Z. marina*, *Z. muelleri*, *Z. noltii* (as *Z. nana*): leaf anatomy and development.

Sauvageau (1890*b*): *Phyllospadix torreyi*: leaf.

Sauvageau (1890*c*): *Phyllospadix*, *Zostera marina*, *Z. muelleri*, *Z. noltii* (as *Z. nana*): apical opening in leaf.

Sauvageau (1891*d*): *Heterozostera tasmanica* (as *Zostera tasmanica*), *Zostera capricorni*, *Z. marina*, *Z. muelleri*, *Z. noltii* (as *Z. nana*): stem.

Schenck (1886*b*): *Zostera marina*, *Z. noltii* (as *Z. nana*): leaf.

Schilling (1894): *Z. marina*: mucilage.

Şerbănescu-Jitariu (1974): *Z. marina*: floral anatomy.

Setchell (1929): *Z. marina*: germination, seedling and general morphology.

Setchell (1933): brief notes on diagnostic features in 11 spp of *Zostera* (including *Heterozostera*).

Setchell (1934): *Heterozostera tasmanica* (as *Zostera tasmanica*): stem.

Taylor (1957): *Z. marina*: embryology, germination, seedling anatomy.

Tutin (1938): *Z. marina*: with notes on *Z. hornemanniana*: germination, general morphology.

Uhl (1947): *Phyllospadix scouleri* (female only), *Zostera marina*: floral anatomy.

Warming (1890): *Zostera marina*: general morphology.

Yamashita (1973): *Z. japonica*: embryo development.

SIGNIFICANT LITERATURE—ZOSTERACEAE

Areschoug (1878); Ascherson and Graebner (1907); Barbour and Radosevich (1979); Barnabas *et al.* (1977); Bate-Smith (1968); Bentham and Hooker (1883); van Beusekom (1967); Blackburn (1934); Bornet (1864); Boyd (1932); Bugnon (1963); Butcher (1933); Chrysler (1907); Clavaud (1878); Clifford (1970); Dahlgren (1939); Davis (1966); Drawert (1938); Drysdale and Barbour (1975); Duchartre (1872); Dudley (1893, 1894); Duval-Jouve (1873*a*,*b*); Eber (1934); Eckardt (1964); Edgcumbe (1980); Eichler (1875); Engler (1879); Erdtman (1952); Falkenberg (1876); Felger and Moser (1973); Felger *et al.* (1980); Flahault (1908); van Fleet (1942); Gessner (1968); Goffart (1900); Gravis (1934); Grönland (1851); den Hartog (1970*a*); Hofmeister (1852, 1861); Hutchinson (1959); Irmisch (1851, 1859*b*); Jacobs and Williams (1980); Jensen (1889); Kunth (1841); de Lanessan (1875); Lüpnitz (1969); McClure (1970); Magnus (1871); Markgraf (1936); Miki (1932, 1933); Monoyer (1927); Ohlendorf (1907); Pate and Gunning (1969); Pettitt and Jermy (1975); Raunkiaer (1895–9); Rosenberg (1901*a*, *b*); Roth (1961); Sauvageau (1889*b*, 1890*a*, *b*, *c*, 1891*a*, *d*); Schenck (1886*a*, *b*); Schilling (1894); Şerbănescu-Jitariu (1974); Setchell (1929, 1933, 1934); Solereder and Meyer (1933); Taylor (1957); Tutin (1938); Uhl (1947); Warming (1890); von Wettstein (1935); Yamashita (1973).

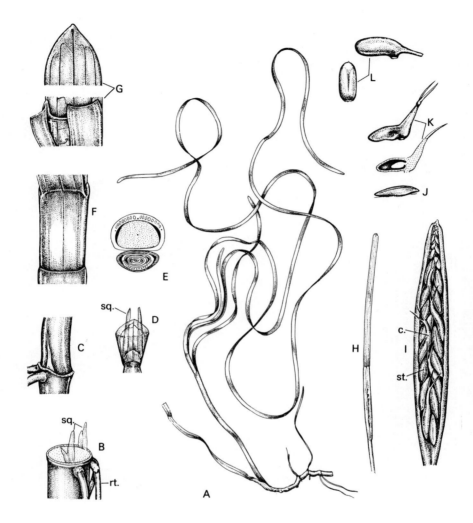

PF

FIG. 13.1. ZOSTERACEAE. *Zostera marina* L. Vegetative and reproductive morphology. (Isle de Groix, France, F. Hallé and P. B. Tomlinson
8.ix.71.)

A. Habit ($\times$ $\frac{1}{4}$).

B. Stem cut off above node to show squamules (sq.) and roots (rt.)
($\times$ 3).

C. Detail of node with leaf removed ($\times$ 3).

D. Prophyll with enclosed branch removed to show squamules (sq.)
($\times$ 4).

E. Diagram of node in TS to show position of prophyll (hatched) of
adnate branch.

F. Leaf sheath from dorsal side ($\times$ 3).

G. Details of foliage leaf; (below) mouth of sheath split open to show
ligule; (above) apex of blade ($\times$ 3).

H. Bract (spathe) enclosing inflorescence ($\times$ $\frac{1}{2}$).

I. Inflorescence (spadix) with spathe removed ($\times$ 3) to show alternate
stamens (st.) and carpels (c.).

J. Stamen ($\times$ 3).

K. Female flower ($\times$ 3) in side view (above) and LS (below).

L. Embryo ($\times$ 3) from side and from above.

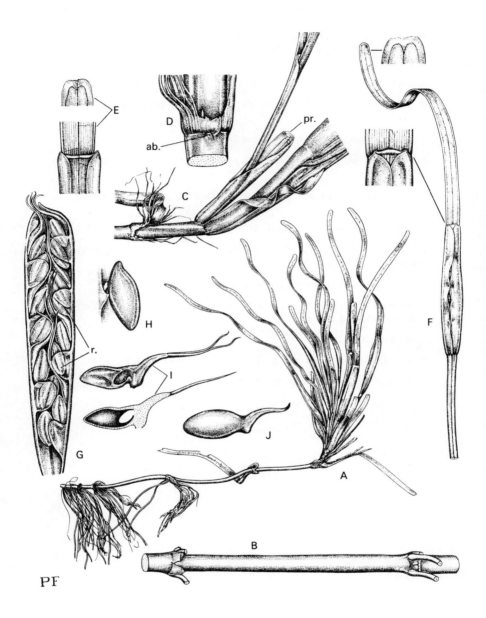

Fɪɢ. 13.2. ZOSTERACEAE. *Zostera novazelandica* Setchell. Vegetative and reproductive morphology. (Akaroa, New Zealand, Lucy B. Moore, CHR 168130.)

A. Portion of leafy shoot ($\times$ 1) to show variation in internode length.

B. Internode and two nodes with leaves removed to show distichy and root insertion ($\times$ 3).

C. Detail of stem to show insertion of adnate branch with prophyll (pr.) ($\times$ 2).

D. Node from ventral side with aborted bud (a.b.) ($\times$ 9).

E. Details of foliage leaf ($\times$ 6); (below) apex of sheath with ligule; (above) apex of blade.

F. Spathe enclosing inflorescence ($\times$ 2). Insets: details of ligule (below) and apex (above) ($\times$ 6).

G. Inflorescence (spadix) with spathe removed, r.—retinacule ($\times$ 6).

H. Details of stamen ($\times$ 12).

I. Female flower ($\times$ 12) from side and in LS.

J. Fruit ($\times$ 6).

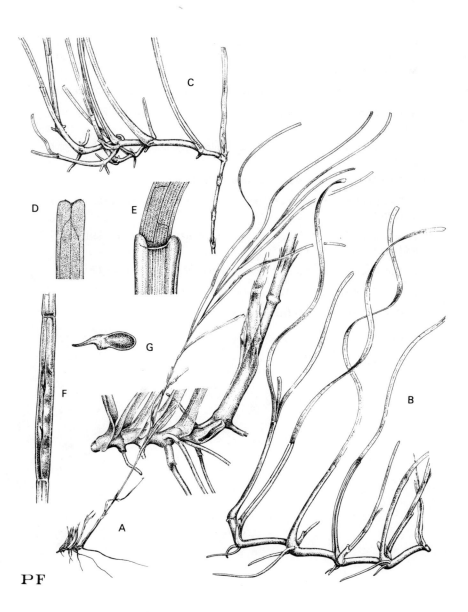

PF

FIG. 13.3. ZOSTERACEAE. *Heterozostera tasmanica* (Martens ex Aschers.) den Hartog. Vegetative and reproductive morphology. (Westernport Bay, Victoria, P. B. Tomlinson, B. Clough, Ian Staff, 10.xi.74.)

A. Habit ($\times$ ½), with underlying details of rhizome suggesting sympodial construction ($\times$ 3).

B. Apical portion of rhizome suggesting monopodial construction ($\times$ 1).

C. Distal portion of erect shoot ($\times$ 1) with leaf remains and new rhizomatous branch showing unique method of vegetative spread.

D. Leaf apex ($\times$ 4).

E. Mouth of leaf sheath with ligule ($\times$ 4).

F. Fruiting spadix ($\times$ 3) enclosed by sheathing base of bract ('spathe').

G. Young fruit ($\times$ 4).

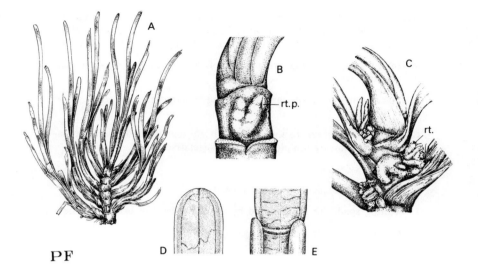

PF

FIG. 13.4. ZOSTERACEAE. *Phyllospadix scouleri* Hook. Vegetative morphology. (A, D, E from Cattle Point, San Juan Island, Washington R. C. Phillips *s.n.* B, C from Mussel Point, Pacific Grove, California, 1–5' tide level in north channel, I. A. Abbott 12356.)

A. Habit from above with long- and short-shoots (× ½).

B. Detail of stem from below with older leaves detached to show unextended primordia (rt. p.) of root complex (× 3).

C. Rhizome from below to show anchoring roots (rt.) (× 3/2).

D. Leaf apex (× 4).

E. Insertion of blade on stem, with ligule (× 4).

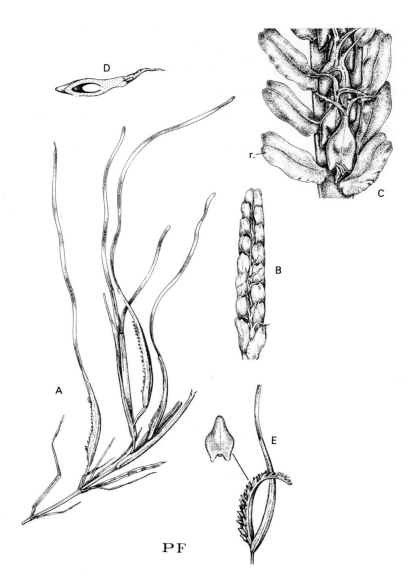

PF

446

FIG. 13.5. ZOSTERACEAE. *Phyllospadix scouleri* Hook. (In its branched inflorescence this resembles *P. torreyi* S. Watson.) Reproductive morphology. (Mussel Point, Pacific Grove, California. 1–5' tide level in north channel. I. A. Abbott 12356.)

A. Erect shoot with fruiting spadices ($\times$ $\frac{1}{2}$).

B. Female spadix with spathe removed ($\times$ $\frac{3}{2}$).

C. Part of female spadix with retinacules (r.) spread to show female flowers ($\times$ 3).

D. LS single female flower ($\times$ 3).

E. Fruiting spadix ($\times$ $\frac{1}{2}$). Inset: single fruit ($\times$ $\frac{3}{2}$).

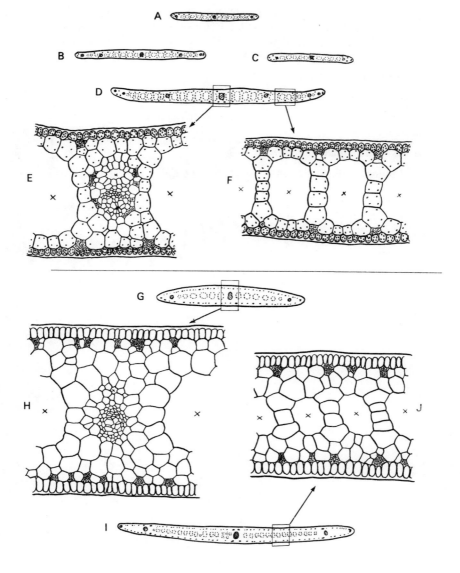

FIG. 13.6. ZOSTERACEAE. Leaf anatomy, all TS. (After Tomlinson 1980.)

A. *Heterozostera tasmanica* ($\times$ 15).

B, C. *Zostera* subgenus *Zosterella*.
   B. *Z. muelleri* ($\times$ 15). C. *Z. noltii* ($\times$ 15).

D–F. *Zostera* subgenus *Zostera (Z. marina)*.
   D. Entire lamina ($\times$ 15). E. Detail of midvein ($\times$ 15). F. Detail of lamina air-lacunae ($\times$ 150).

G, H. *Phyllospadix scouleri.*
   G. Lamina ($\times$ 15). H. Detail of midvein ($\times$ 150).

I, J. *Phyllospadix torreyi.*
   I. Lamina ($\times$ 15). J. Detail of lamina air-lacunae ($\times$ 150).

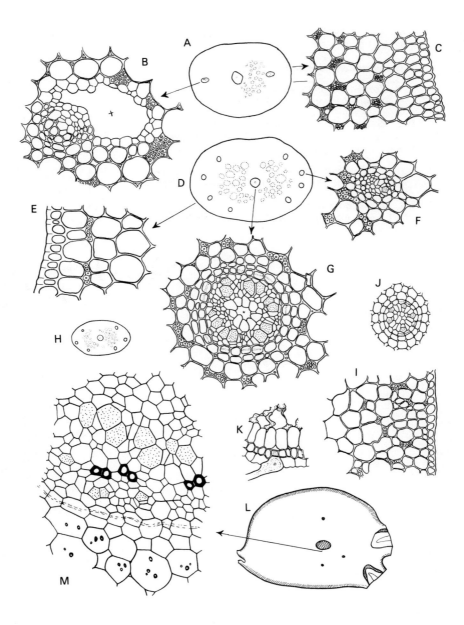

FIG. 13.7. ZOSTERACEAE. Stem and root anatomy.

A–C. *Zostera muelleri,* rhizome TS.

A. Diagrammatic (× 10) with two cortical vbs, only part of lacunose cortex shown (dotted outlines); fibre bundles not shown. B. Detail of single cortical vb with wide xylem lacuna (× 200). C. Surface layers (× 100) with intercellular fibre bundles.

D–H. *Heterozostera tasmanica,* rhizome and erect stem TS.

D. Diagrammatic TS of rhizome (× 15) with eight cortical vbs, fibre bundles not shown. E. Surface layers (× 200) with intercellular fibre bundles. F. Single cortical vb (× 200). G. Details of stele, with central xylem lacuna (x.) and peripheral phloem; intercellular fibre bundles in inner cortex. H. Erect stem (× 15), same organization as D, but different size.

I–K. *Zostera marina,* stem and mature root TS.

I. Surface layers of rhizome (× 100) to compare with C. J. Root stele (× 200) xylem represented by a few central thin-walled cells. K. Root, surface layers (× 200) to show a root-hair with subjacent modified exodermal cell.

L, M. *Phyllospadix scouleri,* rhizome TS.

L. Diagrammatic TS of rhizome (× 6), suberized surface layers hatched. M. Detail of outer layers of central vascular cylinder, with thick-walled tracheary elements.

Family 14
NAJADACEAE (A. L. de Jussieu 1789)
(Figs. 14.1–14.3)

SUMMARY

THIS is a monogeneric (*Najas* L.), isolated but cosmopolitan family of
c. 40 species of wholly submerged, often annual, aquatics in still or slowly
moving fresh or brackish water (Rendle 1899, 1901; Haynes 1977). The plants
have slender, filamentous much-branched axes rooted below but readily detached and forming extensive mats (Figs. 14.1 A and 14.2 A). The leaves
are in sub-opposite pairs or, more usually, pseudowhorls of three. They are
linear and narrow, with a single midvein, while the base is sheathing and
often auriculate. The leaf margin is minutely to conspicuously toothed and
there are conspicuous toothed emergences on the stem and leaf in *N. marina.*
Prophylls are absent. The plants are either dioecious (subgenus *Eunajas*) or
monoecious (subgenus *Caulinia*), and the flowers are associated with leaf
whorls and appear to be more or less axillary, but they develop by precocious
bifurcation of a lateral meristem. The very reduced flowers are either naked,
or they may have one or two cupulate structures. The male flower has a
single, sessile or shortly pedicellate, one- or four-sporangiate anther. The
female flower consists of a single, sessile, flask-shaped 'carpel,' with a solitary
basal anatropous ovule and 2–4 styles. The large seeds have a sculptured
testa, and are released by the rupture or decay of the enclosing parts. The
embryo is straight and endosperm absent. There are paired squamules at
each leaf insertion, sometimes with toothed margins.

Anatomically the plants are very reduced, uniform and with virtually no
mechanical tissue. Stomata and hairs are absent except for the usually marginal unicellular 'prickle hairs' (Fig. 14.3 D), the leaf is bi- or triseriate; the
stem cortex is typically lacunose. The vascular system of the stem and root
is a protostele containing a central protoxylem lacuna surrounded by a cylinder of phloem. A cortical vascular system is absent. Tannin and crystals
are little developed.

FAMILY DESCRIPTION

VEGETATIVE MORPHOLOGY

(Magnus 1869, 1870; Campbell 1897; Posluszny and Sattler 1976*a*.) **Leaves**
originating in sub-opposite pairs, but usually appearing as whorls of three
(Figs. 14.1 B and 14.2 B, C); the phyllotaxis essentially spirodistichous, only
the internodes between the pseudowhorls of three extended. Pseudowhorl
of three made up of two leaves (leaves 1 and 2) on main axis and basal
leaf (leaf 3) of vegetative branch subtended by the lower leaf (1) of the pair.
Upper leaf (2) of the pair always empty. Lateral meristem in axil of leaf 1
developing precociously and usually immediately bifurcating in the lateral

452

plane of the subtending leaf to produce (i) a floral meristem, (ii) a vegetative meristem which normally develops to repeat the pattern of branching of the parent, its first leaf becoming a normal foliage leaf, leaf 3 of the associated pseudowhorl (Figs. 14.1 c and 14.2 E). Lateral vegetative meristem sometimes suppressed, but never developing as a dormant or overwintering bud. Unbranched precocious meristems also produced. [Magnus (1869, 1870) interprets the floral meristem (i) as occupying the site of the first leaf on the vegetative branch, but it is easier to regard it as terminal on a lateral axis and becoming evicted by a branch arising precociously in the axil of leaf 3.] **Roots** produced singly at lower nodes (Fig. 14.2 A).

VEGETATIVE ANATOMY

Leaf

Linear with a single midvein, the base often with a toothed sheathing auriculate expansion (Fig. 14.2 D), the expansion least pronounced in leaf 1 of each pseudowhorl (Fig. 14.1 D). Apical pore absent. **Hairs** represented by unicellular prickle hairs on leaf margin (Fig. 14.3 D), the hair sometimes on an enlarged base; in *N. marina* the hairs terminating a prominent outgrowth or emergence (Fig. 14.3 I) with similar outgrowths above and below midrib and diffusely distributed along the stem (Fig. 14.2 B–D).

Leaf **blade** (Fig. 14.3 A) usually biseriate (except for midrib) and consisting of two epidermal layers with occasional narrow internal intercellular spaces; leaf blade in *N. gracillima*, *N.* leichhardtii, *N. major*, *N. marina*, triseriate, with two epidermal layers and a single (sometimes more) mesophyll layer (e.g. Fig. 14.3 G). Cuticle thin. **Epidermis** thin-walled, or the outer wall slightly thickened, chlorenchymatous; epidermal cells rectangular in surface view (Fig. 14.3 C), those of adaxial surface somewhat the larger. **Mesophyll** in *N. marina* of elongated chlorenchymatous cells enclosing narrow intercellular spaces. Air-lacunae otherwise represented by two or more fairly wide canals in the thickened midrib region (e.g. Fig. 14.3 A), the lacunae never traversed by diaphragms; lacunae well developed in *N. ancistrocarpa*, *N. foveolata*, and *N. minor* in the thickened blade (Miki 1935a, b).

Mechanical tissue represented at most by occasional epidermal fibres in midrib region (e.g. *N. graminea*, *N. leichhardtii*) or by slight wall thickening of marginal 2–3 cell files (Fig. 14.3 B; *N. microdon*). **Midrib** prominent, single vb sheathed by a mesophyll layer (Fig. 14.3 A–G); vascular tissue reduced, tracheids largely replaced by a protoxylem lacuna at maturity surrounded by enlarged radiating cells (Fig. 14.3 H).

Stem (Fig. 14.3 E, F)

Simple and protostelic in construction. **Cuticle** thin, stomata and hairs absent. **Epidermis** usually large-celled, thin-walled, chlorophyllous, but smaller-celled than hypodermis in subgenus *Eunajas* according to Miki (1935a, b). **Cortex** wide, including 1–2 peripheral compact layers, the number of these hypodermal layers specifically diagnostic, according to Miki (1935a, b); e.g. one in *N. ancistrocarpa, tenuicaulis,* two in *N. major*, *N. oguraensis*, but rhizomatous part of stem sometimes with one more hypodermal layer than distal stem (e.g. *N. gracillima*). Lacunose middle region of cortex with several enlarged longitudinally extended intercellular spaces forming a single series of longitudinal air-canals (Fig. 14.3 E). Innermost cortical layers compact and sometimes becoming slightly thick-walled. **Endodermis** thin-walled, or at most outer wall becoming somewhat thickened in older, basal stem parts. **Stele** (Fig. 14.3 F) including a wide protoxylem lacuna surrounded by a regular uniseriate layer of radially extended cells with the wall adjacent to the cavity slightly thickened. Phloem represented by a cylinder of sieve-tubes, often in contact with endodermis. Mechanical and cortical vascular system not developed.

Nodal anatomy (Singh 1965*b*; Uhl 1947): stele enlarged at a branch-bearing nodal complex, and each leaf of a pseudowhorl supplied in sequence by a single leaf trace; the remaining stele bifurcating, one fork continuing the vascular system of the main axis, the other fork splitting again immediately to supply (i) the vegetative bud, (ii) flower. **Xylem** at node represented by a mass of short narrow tracheary elements. Ground tissue cells somewhat thickened; air-lacunae not continuous across nodal complex.

Root

Unbranched, restricted to lower nodes. Corresponding closely in anatomy to stem, but with numerous root-hairs.

SECRETORY, STORAGE, AND CONDUCTING ELEMENTS

Crystals: never conspicuously developed.
Tannin: little developed.
Starch: common in the inner cortical layers of the stem.
Phloem: with simple sieve-plates in all parts.
Xylem: vessels absent. Tracheary elements represented largely by a wide protoxylem lacuna, except at nodes. [The record by Caspary (1858) of 'vessels' at the base of the funiculus in *N. flexilis* should not be interpreted in the modern use of the term 'vessel.']

REPRODUCTIVE MORPHOLOGY AND ANATOMY

Male flower

Represented by a single ± sessile stamen enclosed by two non-vasculated cupular structures, the inner one ('perianth') closely adnate to the stamen and typically bilobed apically, the outer one ('spathe') typically with irregular apical teeth (Fig. 14.1 E). Outer cupule absent from *N. graminea*. Stamens with one or four sporangia, the separating walls breaking down at maturity; endothecium absent (Venkatesh 1956). Anther extending at maturity, irregularly rupturing the cupules and releasing the pollen via an apical split. Pollen trinucleate. Male flower supplied by a single vb ending below the solitary sporangia or extending slightly into the connective of the tetrasporangiate anther (Singh 1965*b*).

Female flower

Represented by a solitary, usually naked ovary but the ovary in Rendle's section Spathaceae enveloped by a 'spathe' comparable to the outer envelope of the male flower. Ovary (Fig. 14.2 G) consisting of a basal bitegmic anatropous ovule (Fig. 14.2 H) enclosed by a cupulate 'carpel' with a short style extended into 2 (–3) short linear stigmas (sometimes with additional short teeth interpreted as 'barren stigmas'). Female flower with a single vb, entering the funiculus and remaining undifferentiated at the base of the nucellus.

Fruit

Pericarp initially succulent, becoming dry and adherent to the seed (Fig. 14.2 F); enclosed by the persistent spathe, when present. Seed ellipsoidal (Fig. 14.2 I). Testa hard, and presenting the following diagnostic characters, e.g. testa multilayered in *N. marina*, the outer layer eventually becoming eroded; three-layered in the other spp with additional diagnostic differences shown by the sculpturing of the cells of the outer layer (Haynes 1979). Embryo straight (except presumably in the curved seed of *N. ancistrocarpa*). Germination epigeal, the cotyledon developing before the radicle.

Floral development

[The several studies (e.g. those by Magnus, Rendle, Campbell, and more recently Posluszny and Sattler) all agree that the reproductive organs arise terminally and that the ovule is terminal on the axis, with the 'spathe,' 'perianth,' and 'carpel' wall all originating as separate annular outgrowths below and subsequent to ovule initiation. The outgrowths become enveloping organs by the upward extension of the annular primordium.]

TAXONOMIC AND MORPHOLOGICAL NOTES

The Najadaceae are anatomically and morphologically very reduced and despite their simple structure their relationships with the other Helobiae have been much discussed. Their relationship with the other Helobiae is most obviously supported by their similar embryos (Rendle 1901). Some authors (e.g. Campbell 1897; Rendle 1901) have considered the Najadaceae to be primitively simple. However, most investigators have regarded them as extremely reduced forms and have tried to relate them to families with a similar vegetative organization (notably the Zannichelliaceae). Miki (1937) considered they could have originated from submerged Hydrocharitaceae. The specialized sieve-tubes do not suggest that the family is truly primitive.

The morphological nature of the floral parts has occasioned much discussion (e.g. Swamy and Lakshmanan 1962*b*; see also summary by Posluszny and Sattler 1976*a*). The most straightforward morphological interpretation of the flower corresponds to the description given above, indicating that the male flower has a separate cupular perianth and spathe, while the female flower is usually naked, but sometimes has a spathe. Earlier and sometimes more elaborate interpretations have been discussed by Rendle (1899). Comparative morphology supports the interpretation of the outer cupule as a modified leaf, since de Wilde (1961) has found transitional structures between cupules and amplexicaul leaf bases when examining large samples. Multiple stigmas might be taken to suggest that the female flower had a multicarpellary origin. However, there is a wide range of stigma morphology in Zannichelliaceae. De Wilde concludes that the spathe is less useful as a sectional character than is traditionally assumed.

Magnus (1894) adopted an extreme interpretation, which assumes that *Najas* is gymnospermous, because he equated the inner and outer integuments of the ovule in the female flower with the two cupules of the male flower, the carpel being homologized with the cupulate structure surrounding the flower complex of *Zannichellia palustris* on the basis of the similar development of the two organs.

Posluszny and Sattler (1976*a*), essentially invert this argument and equate the two cupules of the male flower to the integuments of the ovule, but in morphogenetic terms rather than in terms of strict homology. They further conclude that the gynoecium of *Najas* is acarpellate.

It should be noted that in floral development, differentiation of floral envelopes follows rather than precedes initiation of the reproductive organs (ovule and stamen) and that the flowers have an origin via an apical bifurcation comparable to that in Zannichelliaceae and many other Helobiae.

The taxonomic isolation of *Najas* is well justified, especially in view of the distinctive method of ovule initiation and the controversial nature of its reproductive parts. Most botanists will find it convenient to accept the genus as a highly specialized member of the Helobiae in the absence of more decisive evidence. Re-investigation of the family, especially using modern analytic methods is much needed, an example is the use of scanning electron microscopy in the examination of testa sculpturing (Haynes 1979). *Najas* is not a difficult plant to collect!

Anatomical information has been used in the identification of *Najas* spp, notably by Miki (1935*a*, *b*) who provided a key to the eight Japanese species he studied, using features of leaf anatomy (number of marginal teeth) and stem anatomy (discreteness of epidermis, number of hypodermal layers) as well as features of vegetative and reproductive morphology. It seems that the triseriate (three-layered) leaf is not restricted to either of the two subgenera (e.g. it occurs in *N. marina* and *N. gracillima*).

A note on shoot morphology

There is some confusion in the literature regarding the interpretation of the nodal complex which, when fully developed, is represented by a pseudo-whorl of three leaves, a lateral vegetative branch, and a flower. All authors who have examined the morphology of *Najas* in detail follow Magnus (1869) and are agreed that the third leaf (leaf 3) of each pseudowhorl is a basal leaf on the lateral shoot subtended by leaf 1. Magnus (1869, 1870, 1894) and Rendle (1899) indicate that there are two leaves at the base of this lateral shoot, of which one is reduced to a small scale of microscopical size bearing a bud in its axil. In flower-bearing pseudowhorls the scale and bud are said to be replaced by a flower. This interpretation is repeated by de Wilde (1961). However, neither Campbell (1897) nor Posluszny and Sattler (1976*a*) illustrate or refer to such a basal scale. It is possible that earlier authors confused a squamule for a scale since there is an extra squamule at the base of the flower. The statement by Singh (1965*b*) is even more ambiguous, because he implies that staminate and pistillate flowers occupy different positions. His interpretation is based on the disposition of traces to appendages. Since the morphology of *Najas* is complex, the point should be resolvable by further comparative study.

MATERIAL EXAMINED

(Vegetative parts of all spp.)

Najas flexilis (Wild.) Rostk. & Schmidt; Lake Champlain, New York; P. B. Tomlinson 6.viii.72.

N. gracillima (A.Br.) Magnus; Old Mill, Fitchburg, Massachusetts; P. B. Tomlinson *s.n.*

N. guadalupensis (Spreng.) Magnus; Fairchild Tropical Garden, Florida; P. B. Tomlinson *s.n.*

N. marina L.; Fairchild Tropical Garden, Florida; P. B. Tomlinson *s.n.*

SPECIES REPORTED ON IN THE LITERATURE

Bailey (1884): *Najas graminea* var. *delilei*: leaf, stem, root.

Campbell (1897): *N. flexilis, N. graminea*: all parts.

Caspary (1858): *N. minor*: root; *N. flexilis*: ovule.

Chauveaud (1897*b*): *Najas* sp: root.

Chrysler (1907): *N. flexilis, N. marina*: stem vasculature.

François (1908): *N. major*: seedling anatomy.

Magnus (1870): Extensive morphological and anatomical notes on numerous spp.

Magnus (1883): *N. graminea*: leaf.

Miki (1935*a, b*): *N. ancistrocarpa, N. foveolata, N. gracillima, N. graminea, N. major, N. minor, N. oguraensis, N. tenuicaulis*: leaf, stem.

Nedelcu (1972): *N. marina, N. minor*: stem.

Ogden (1974): *N. flexilis*: leaf, stem.

Pettitt and Jermy (1975): *N. flexilis*: pollen structure.

Posluszny and Sattler (1976*a*): *N. flexilis*: floral development.

Sattler and Gifford (1967): *N. guadalupensis*: shoot apex.

Sauvageau (1888): *N. major*: root.

Sauvageau (1889*a*): *N. major, N. minor*: root.

Schenck (1886*b*): *N. flexilis, N. major, N. minor*: leaf, stem.

Singh (1965*b*): *N. flexilis, N. graminea, N. guadalupensis*: nodal and floral anatomy.

Swamy and Lakshmanan (1962*b*): *N. graminea, N. lacerata*: floral morphology and embryology.

Tarnavschi and Nedelcu (1973): *N. minor*: leaf, stem.

van Tieghem (1870–1) *N. major*: root.

Uhl (1947): *N. gracillima, N. marina* (brief comparison with other N. American spp). nodal and floral anatomy.

Venkatesh (1956) *N. graminea, N. palustris*: anther.

SIGNIFICANT LITERATURE—NAJADACEAE

Arber (1921); Bailey (1884); Campbell (1897); Caspary (1858); Chauveaud (1897b); Chrysler (1907); François (1908); Haynes (1977, 1979); Jönsson (1883–4); Kirchner (1908); Leitgeb (1857); Magnus (1869, 1870, 1883, 1889, 1894); Miki (1935a, b, 1937); Nedelcu (1972); Ogden (1974); Ohlendorf (1907); Pettitt and Jermy (1975); Posluszny and Sattler (1976a); Raunkiaer (1895–9); Rendle (1899, 1901); Roze (1892); Sattler and Gifford (1967); Sauvageau (1888, 1889a); Schenck (1886a, b); Singh (1965b); Solereder and Meyer (1933); Swamy and Lakshmanan (1962b); Tarnavschi and Nedelcu (1973); van Tieghem (1870–1); Uhl (1947); Venkatesh (1956); Wilde (1961).

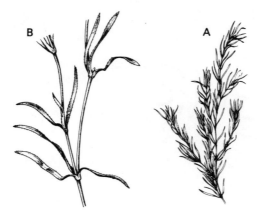

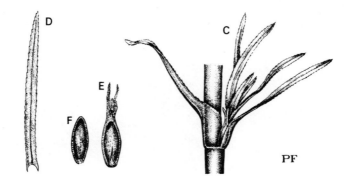

FIG. 14.1. NAJADACEAE. *Najas guadalupensis* (Spreng.) Magnus. (Fresh water, Fairchild Tropical Garden, Florida; P. B. Tomlinson *s.n.*)

A. Distal part of branched leafy shoot ($\times$ $\frac{1}{2}$).

B. Details of shoot to show leaf arrangement and branching ($\times$ 1).

C. Details of branch-bearing node with subtending leaf removed to show male flower and associated vegetative branch ($\times$ 2).

D. Basal leaf of branch, without expanded base ($\times$ 2).

E. Male flower ($\times$ 2).

F. Male flower with 'spathe' removed to show enveloping perianth ($\times$ 2).

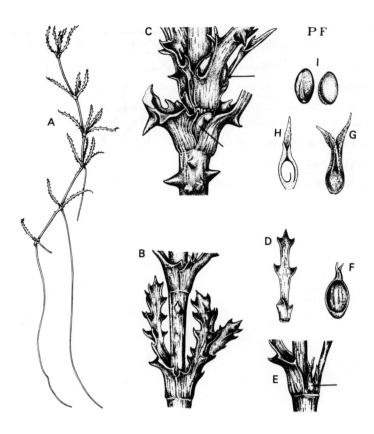

Fɪɢ. 14.2. NAJADACEAE. *Najas marina* L. (Brackish water, Fairchild Tropical Garden, Florida; P. B. Tomlinson *s.n.*)

A. Rhizomatous portion of shoot with roots at lower nodes ($\times$ ½).

B. Details of leaf arrangement ($\times$ 2).

C. Shoot details ($\times$ 6); two nodes (lower leaf to left with blade cut off) to show female flowers in leaf axils.

D. Leaf from base of branch ($\times$ 4).

E. Node ($\times$ 4) with lower leaf removed to show female flower paired with vegetative branch.

F. Fruit ($\times$ 2).

G. Female flower ($\times$ 4).

H. Female flower in LS ($\times$ 4).

I. Seed ($\times$ 4), from two aspects.

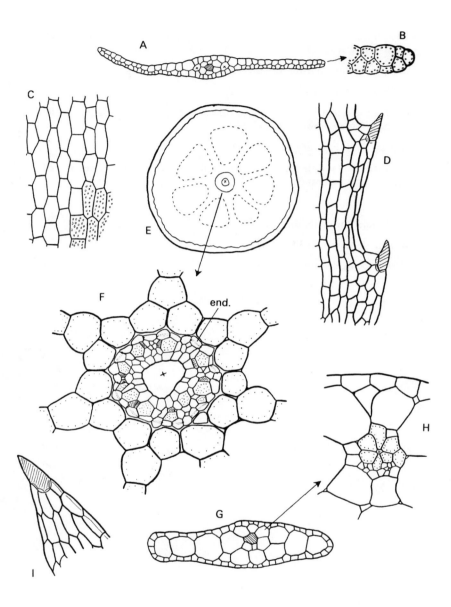

FIG. 14.3. NAJADACEAE. Stem and leaf.

A–F. *Najas guadalupensis.*
   A. Lamina TS ($\times$ 40). B. Detail of leaf margin ($\times$ 140). C. Adaxial
epidermis in surface view ($\times$ 70), chloroplasts indicated in only a few
cells. D. Leaf margin with prickle-hairs in surface view ($\times$ 20). E. Stem
TS ($\times$ 30). F. Stele from E ($\times$ 140), sieve-tubes stippled; companion
cells—hatched; end.—endodermis

G–I. *Najas marina.*
   G. Lamina TS ($\times$ 30). H. Detail of midvein from G ($\times$ 140), cells
surrounding protoxylem lacunae—dotted. I. Marginal emergence with
terminal prickle-hair ($\times$ 70).

Family 15
TRIURIDACEAE (Lindley 1846)
(Triuraceae of Gardner 1843)
(Fig. 15.1 and Plate 15)

SUMMARY

THIS is a small, very natural and almost entirely tropical family of saprophytic herbs (Schmucker 1959) which grow in forest litter, rotten wood, or rarely termite nests. There are seven genera and about 80 species (Giesen 1938). The diminutive plants are usually 10–15 cm tall, sometimes smaller, but reach a height of 1.4 m in *Sciaphila purpurea*. The erect stem is slender, scaly, glabrous, and it arises from an irregular slender rhizomatous system (Fig. 15.1 A). The plants contain no chlorophyll but are often coloured pink, purple, or brown. The leaves are always reduced, scale-like and each is provided with a single median vascular strand.

The flowers are typically small, inconspicuous, and unisexual. They are also often markedly dimorphic, and the plants are usually monoecious. The flower parts are variable, but often trimerous. The perianth is not more than slightly differentiated, and the tepals commonly bear appendages or glands. The anthers are typically in a single free or united series, and situated opposite the tepals. They are sessile, with extrorse, usually horizontal dehiscence, and sometimes also bear appendages. Staminodia may be present. The numerous carpels are free, and each has a lateral or basal style and a single ovule. The fruit is follicular and the seeds are endospermous.

Anatomically the family is reduced, in relation to its nutritional status. Stomata are absent. Stem vascular bundles are always in a single cylinder, often associated with sclerenchyma and rather obscure. The roots include endotrophic mycorrhizal hyphae in enlarged cells of the cortex, and the root stele is narrow and commonly with a sclerotic pericycle. Vascular elements are narrow and obscure. The phloem elements are thick-walled, and vessels have not been observed with certainty.

FAMILY DESCRIPTION

VEGETATIVE MORPHOLOGY

Rhizome irregularly branched, slender or somewhat swollen and congested at the base of the erect shoots. Erect shoots apparently arising by sympodial growth, often clustered, usually themselves unbranched and terminating in a simple racemose inflorescence (Fig. 15.1 A). Branching of erect shoots usually restricted to their bases, or with distal branches arising from dormant buds as a response to damage. Prophyll usually absent (Fig. 15.1 c). Branching sometimes a regular feature of distal parts, producing a somewhat corymbose inflorescence.

Leaves spirally arranged (whorled in *Hyalisma*), scale-like, triangular, scarcely 1–2 mm long, commonly subtending minute suppressed buds (Fig.

466

15.1 B–D). **Roots** solitary or in pairs at rhizome nodes (Fig. 15.1 F, G), slender, branched, and usually conspicuously hairy in contrast to glabrous rhizome; stele sclerotic and persistent.

VEGETATIVE ANATOMY

Leaf

Hairs absent. **Epidermis:** cells thin-walled, tabular. **Stomata** absent. **Mesophyll** colourless, 1–3 cells deep enclosing a median vein with 1–2 files of tracheids. Bracts identical with vegetative scale leaves, but sometimes longer and narrower. Scales enveloping and protecting younger parts in developing shoots (Fig. 15.1 D).

Stem (Plate 15 B–E)

Cuticle thin, often slightly ridged or papillose. **Epidermis** slightly thick-walled and usually lignified. **Stomata** absent. Cortex either undifferentiated (Plate 15 B, C) or consisting of 1–3 peripheral cell layers with thick, lignified walls forming a mechanical sheath (Plate 15 E). **Endodermis** not observed. **Stele** including a single irregular cylinder of narrow vbs anastomosing tangentially, the vbs either contiguous with cortical sclerenchyma (Plate 15 D, E) or enclosed in a continuous sclerotic ring (Plate 15 B, C). Individual vbs with narrow endarch xylem, in TS distinguished with difficulty from adjacent sclerenchyma (Plate 15 B, D). Phloem elements obscure, becoming thick-walled in older stems.

Basal stem (or **rhizome**) typically with a relatively wider stele, with or without associated sclerenchyma, the vbs sometimes scarcely distinguishable in TS. Fungal mycorrhizae absent.

Root (Plate 15 A)

Epidermis usually with conspicuous root hairs, the latter either arising from single superficial cells, or appearing as enlarged cells with inflated bases and enclosed by a cylinder of surrounding cells, the root hair then usually septate. **Exodermis:** cells small, often slightly thick-walled but never markedly differentiated. Middle and inner cortex consisting of 1–3 cell layers, with abundant mycorrhizal hyphae in one or more layers (Plate 15 F). **Endodermis** of slightly thick-walled cells, usually scarcely lignified, always narrow, each cell coincident radially with an enlarged inner cortical cell (Plate 15 A). **Pericycle** variable, the cells sometimes becoming thick-walled and sclerotic; uniseriate but frequently appearing biseriate in TS owing to cells with oblique overlapping end walls. **Vascular tissue** reduced and obscure, with one or two xylem poles and obscure phloem elements, the latter becoming thick-walled. In simplest roots, stele including a single central xylem element, the vascular structure of a normal root scarcely evident.

SECRETORY, STORAGE, AND CONDUCTING ELEMENTS

Crystals; not observed.
Tannin; occasional as in superficial cells of rhizome.

Phloem elements; obscure, but simple sieve-plates observed in *Andruris vitiensis.*

Tracheary elements; long, v. narrow with scalariform or pitted lateral walls; end walls in the stem and root oblique and possibly obscurely perforated.

REPRODUCTIVE MORPHOLOGY

Flowers (e.g. Fig. 15.1 I–N) usually unisexual (sometimes perfect); plants typically monoecious, the basal part of the inflorescence female, the distal part male; rarely dioecious (*Hexuris, Triuris*). Flowers usually lateral, in simple racemes (Fig. 15.1 H), each flower usually on a long pedicel in the axil of a distal scale leaf (bract), the pedicel with neither prophyll nor bracteoles except rarely in *Seychellaria.* Flowers in *Triuris* terminal, solitary or in a few-flowered sympodium. **Perianth** of several (3–8, most commonly six) imbricate members, sometimes in two indistinct and slightly dimorphic series; tepals typically pointed apically, sometimes gland-tipped (Fig. 15.1 I) or with one or more apical filaments, the inner surface sometimes roughly papillose (as in many *Sciaphila* spp) or glandular (Fig. 15.1 H). **Male flower** with 2–6 anthers, the number fairly constant for a given taxon, usually in a single series and opposite the tepals (alternate with them in *Triuris*). **Anthers** sessile, sometimes shortly stalked and again sometimes united to form a central column. Anthers two; three- or four-sporangiate (Fig. 15.1 K); dehiscence always extrose, usually transverse, but longitudinal in *Triuris* and *Hexuris.* **Staminodia** present as filamentous (rarely glandular) organs alternating with fertile stamens (male flowers of *Seychellaria*) or in female flowers of some *Sciaphila* species. Connective in fertile stamens of *Andruris* (Fig. 15.1 I, J) and *Seychellaria madagascariensis* extended into a filamentous organ somewhat resembling a staminode, but could also be interpreted as a pistillode. Central cushion of tissue in male flowers of *Sciaphila* interpreted by Wirz (1910) as a pistillode.

Female flower (Fig. 15.1 M, N) with the perianth usually unelaborated compared with the male flower; **carpels** numerous, free, each with a lateral or even basal style ending in a conspicuous, sometimes papillose stigma. **Ovule** obscurely bitegmic, solitary, basal, anatropous. **Fruit** a one-seeded follicle (Fig. 15.1 O, P), usually splitting longitudinally but indehiscent in *Soridium.* Seed endospermous in *Sciaphila* sp, described by Wirz (1910) as wedge-shaped, undifferentiated and attached to the wall of the embryo sac by a two-celled suspensor. Germination never described.

REPRODUCTIVE ANATOMY

[Little information is available, the most detailed work is by Wirz (1910) who deals especially with embryology. The following notes are based largely on observation of microtome sections of flowers and flower buds of *Andruris vitiensis*, but with additional information summarized from the literature.]

Male flowers

Pedicel with a central strand of vascular tissue, including scattered sieve-tubes and short tracheids, and ending in the receptacle (Plate 15 H). Each stamen without vascular tissue, but supplied simply from the pedicel strand by a mass of small cells with dense cytoplasm, the nuclei being irregularly lobed or even fragmented. [These specialized cells do not enter the filament so there is no vascular tissue within the stamen itself.]

Glands of petal tips in *Andruris vitiensis* consisting of 2–3 layers of large, densely cytoplasmic cells enclosing an irregular central cavity (Plate 15 H).

Female flower

Pedicel, as in male flower, with a central strand of vascular tissue, but here with more obvious, discrete, peripheral vascular strands (Plate 15 G). Each carpel supplied by a branch from the central strand continuous as a series of elongated but undifferentiated cells in the funiculus. Carpel and ovule without differentiated vascular tissue.

Carpel morphology

In carpel ontogeny (Poulsen 1906) ovule initiated on the floral axis and rapidly enclosed by the developing carpel wall. Subsequent closure on the ventral side due to the morphological apex of the carpel growing over and behind the ovule, leaving a narrow ventral aperture (acropyle). Style arising from an extension of the adaxial side of the carpel. Ovule becoming inverted by chalazal extension and asymmetric growth at a later ontogenetic stage.

Ovular morphology and function

[Some uncertainty has existed concerning the number of integuments in the ovules of the Triuridaceae. Van Tieghem (1911) accepted early reports that the ovules were unitegmic, but Poulsen (1906) and especially Wirz (1910) demonstrated that two integuments develop, the inner closing the micropyle completely and fusing with the nucellus. Wirz found that the uniseriate nucellus was absorbed along with the inner integument during the development of the embryo, the seed coat arising from the outer integument alone.

The flowers of Triuridaceae show so many peculiar features that their further intensive study, using modern histological techniques, is much to be desired. Wirz (1910) could find no evidence for fertilization of the embryo sacs he studied and suspected apomixis.]

Taxonomic and Morphological Notes

This is a very natural and easily identified group whose segregation as a discrete family or even order is justified. It has been associated with the similarly saprophytic Petrosaviaceae in the order Triuridales and included in the Alismatidae by Cronquist (1968). Anatomical evidence essentially supports the first but not the second association. The affinities of the Triuridaceae have been discussed by several authors (e.g. Engler 1889*b*; Johow 1889; Poulsen 1906; van Tieghem 1911; Wirz 1910), but not always on the basis of correct information (see above). This topic has been amply summarized by Giesen (1938). It is generally accepted that the family has helobial affinities, largely because of its apocarpous gynoecium, but there are sufficient differences in ovule organization, endosperm, and embryo to exclude it. Affinities with the Liliaceae have been pointed out. Erdtman (1952) suggested that the pollen grains of *Andruris vitiensis* were of a unique type, without any counterpart in a number of apocarpous monocotyledons (Alismataceae, Burmanniaceae) or dicotyledons (Ranunculaceae). However, the limited cytological data do not conflict with information available for putatively related families (Green and Solbrig 1966; Ohga and Sinotô 1924). It should be noted that the fundamental place of the family in the plant kingdom remains uncertain; it is not known if Triuridaceae have a single cotyledon because their embryo is undifferentiated and germination is not recorded. They are considered to be monocotyledonous because of their predominantly trimerous flowers.

Anatomically the family is so reduced in relation to its saprophytic existence that useful evidence for its affinities are lacking. Stem anatomy, with a single ring of vbs, certainly could be interpreted as dicotyledonous, as could the essentially monarch or diarch root stele. The absence of stomata parallels the condition found in a number of submerged aquatics, but the Triuridaceae lack any intercellular space system and they are quite sclerotic. Reduction here is presumably influenced by different physiological needs so that there is no anatomical evidence to support the inclusion of the Triuridales in the Alismatidae.

Insufficient information is available to comment on diagnostic anatomical features at the generic level within the family. Current evidence suggests anatomy is very uniform.

Many species seem to be differentiated on quantitative characters which may vary in a single population. Most species are known from few collections, some only from the type collection.

BIOLOGY

There is a dearth of information about the biology and development of the Triuridaceae because of their relative insignificance in tropical communities. Our knowledge of the family is based almost entirely on the study of herbarium or fluid-preserved specimens and quite old literature. Information about habit could be obtained by growing plants from seed, which never seems to have been attempted. Flowering sequence needs careful study. Most species would appear to mature flowers in acropetal order, the basal flowers being female and the apical flowers male, but it is possible that there may be a second basipetal wave when lower dormant buds expand as a second generation of flowers. There is no information on pollination biology. There is evidence for apomixis.

MATERIAL EXAMINED

Andruris vitiensis (A. C. Smith) Giesen; Mt. Korombamba, Suva, Fiji; P. B. Tomlinson 21.iv.69 D; all parts.
Sciaphila africana A. Chev.; Ghana; Enti & Hall 36286 aerial parts.
Sciaphila cf. *albescens* Benth.; Saül, French Guiana; P. B. Tomlinson 21.ii.72; all parts.
Sciaphila sp; Papua New Guinea; D. W. Bierhorst 269; all parts.

Species Reported on in the Literature

Engler (1909): *Sciaphila*: stem.
Fiebrig (1922): *Triuris mycoides*: stem, root.
Giesen (1938): summary of older literature.
Janse (1897): *Sciaphila tenella*: root.
Johow (1889): *Sciaphila schwackeana*: root, inflorescence axis, embryology.
Larsen (1963): *Sciaphila thaidanica*: root.
Malme (1896): *Triuris lutea*: all parts.
Milanez and Meira (1943): *Triuris alata*: all parts.
Poulsen (1886): *Sciaphila caudata*: root, stem, flower.
Poulsen (1890): *Triuris major*: root, stem, flower.
Poulsen (1906): *Sciaphila nana*: floral development.
van Tieghem (1911): ovule morphology.
Wirz (1910): *Sciaphila* sp: embryology and floral development.

Significant Literature—Triuridaceae

Cronquist (1968); Engler (1889*b*, 1909); Erdtman (1952); Fiebrig (1922); Gardner (1845); Giesen (1938); Green and Solbrig (1966); Janse (1897); Johow (1889); Larsen (1963); Malme (1896); Milanez and Meira (1943); Ohga and Sinotô (1924, 1932); Poulsen (1886, 1890, 1906); Schmucker (1959); van Tieghem (1911); Wirz (1910).

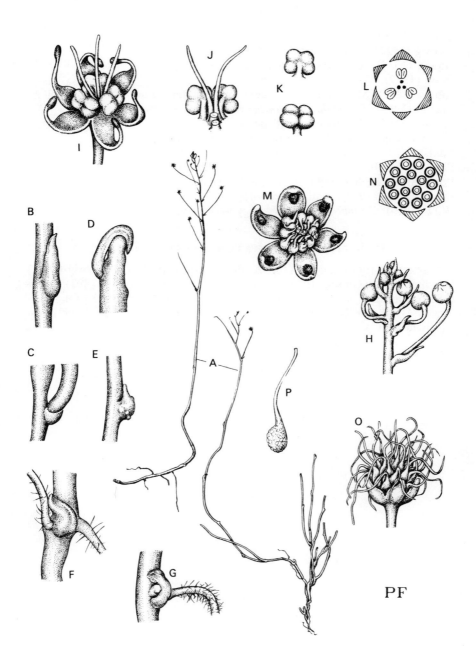

FIG. 15.1 TRIURIDACEAE. *Andruris vitiensis* (A. C. Smith) Giesen. (Mt. Korombamba, Suva, Fiji; P. B. Tomlinson 21.iv.69 *D*.)

A. Habit (× 1), to left a single flowering axis, to right rhizomatous system producing several erect shoots.

B–G. Details of vegetative morphology (all × 6).
  B. Scale leaf. C. Scale leaf subtending lateral branch. D. Scale leaf overarching leaf apex. E. Scale leaf with developing root primordia. F. Scale leaf and adventitious roots. G. Scale leaf, adventitious root and undeveloped axillary bud.

H. Distal part of inflorescence with mostly male flowers (× 6).

I–L. Male flower (× 16).
  I. From above. J. Pair of anthers from the back to show the corresponding extended connective or pistillode; other anther and pistillode cut off. K. Stamen from back (above) and front (below). L. Floral diagram.

M, N. Female flower (× 16).
  M. From above. N. Floral diagram.

O, P. Fruit (× 16).
  O. Young fruiting head. P. Single mature achene.

Family 16
PETROSAVIACEAE (Hutchinson 1934)
(Figs. 16.1–16.2 and Plate 16)

SUMMARY

THIS is a family of colourless saprophytic herbs with endophytic mycorrhiza in the root cortex. It includes the genus *Petrosavia** Becc. (two spp, *P. sakuraii* (Makino) J. J. Sm. ex Van Steenis and *P. stellaris* Becc. = *P. borneensis* Becc.) and probably *Protolirion paradoxum* Ridley, ranging from the Malay Peninsula through Borneo to southern China and Japan. The diminutive plants (to 15 cm high) grow on the forest floor or sometimes in drier localities. The erect shoots bear spirally-arranged, often overlapping scale leaves which arise sympodially from a branched subterranean rhizome system. The inflorescence is a simple raceme (Fig. 16.1 A). The flowers are on long pedicels, with basal, lateral bracteoles; they are bisexual and actinomorphic, with six colourless perianth segments in two distinct series. There are six stamens and three (rarely more) carpels, each with numerous ovules attached to the inner angle of the locule. The fruits are dehiscent.

Anatomically the family is reduced in relation to its saprophytic existence. Hairs and squamules are absent, but stomata without subsidiary cells occur on the leaves. The axis is sclerotic with abundant lignin, aerenchyma is not developed, and the vascular tissue is reduced, that of the stem including a single cylinder of poorly differentiated strands contiguous with sclerenchyma. Vessels probably do not occur.

FAMILY DESCRIPTION

VEGETATIVE MORPHOLOGY AND ANATOMY

Erect stems (Fig. 16.1 A) arising sympodially from the axils of basal scale leaves of existing stems. Apex of the renewal shoot with a uniseriate tunica layer (Stant 1970).

Leaf (Fig. 16.2 A–D and Plate 16 C)

Scale-like, up to 5 mm long and 2 mm wide, undifferentiated, shallowly V-shaped in TS (Fig. 16.2 D and Plate 16 C) with a broad insertion (Fig. 16.2 G). **Hairs** absent. **Epidermis** (Fig. 16.1 A) of elongated, irregularly rectangular or rhombohedral cells, the cell files indistinct; outer wall of the epidermis slightly thickened and cuticularized. **Stomata** (Fig. 16.2 A, B) restricted to abaxial epidermis, distally and in midrib region, without subsidiary cells, the guard cells with two ± equal thickened ledges. **Mesophyll** undifferentiated, consisting of 1–4 layers of rounded parenchyma cells and only small intercellu-

---

* Synonomy in this family seems confused; the status of *Protolirion sakurai* (Makino) Dandy and *Petrosavia miyoshia-sakuraii* Makino needs re-examination (see Sterling 1978).

474

lar spaces. **Midrib** including one central bundle of intermingled xylem and phloem elements, the bundle sheath at most represented by an irregular or discontinuous ring of lignified cells (Fig. 16.1 C). **Xylem** consisting of thin-walled spirally-thickened tracheids, and the **phloem** of narrow sieve-tubes with oblique sieve-plates.

Groom (1895) recorded raphides in mucilaginous mesophyll cells between the midvein and the abaxial epidermis, but crystals not observed in material examined here.

Stem (flower-bearing scape; Fig. 16.2 G and Plate 16 B)

Up to 1 mm in diameter. **Epidermis** of slightly thickened, lignified and cuticularized cells. **Stomata** occasional. **Cortex** consisting of 3–6 undifferenti-ated layers of elongated pitted and slightly lignified cells; subepidermal layer compact; cells of the inner layers more rounded with narrow intercellular spaces. **Stele** delimited externally by an **endodermoid layer** of unequally thick-ened cells, U-shaped in TS, the wall thickenings stratified. Endodermoid layer sometimes multiseriate and interrupted by thick-walled cells resembling 'pas-sage cells'. **Stele** (Plate 16 B) including an outer sheath 4–6 cells wide, next to endodermoid layer, of thick-walled sclerenchyma. Sheath interrupted only by departing leaf traces (Fig. 16.2 G). Vbs represented by a cylinder of 7–14 strands attached to the inner face of the sclerenchyma sheath, the size and number of the strands varying because of tangential anastomoses. Vascu-lar tissues obscurely differentiated and intermingled. **Xylem** elements with scalariform pitting on part of the wall, but protoxylem elements with thin cellulose spiral thickenings, the elements frequently disintegrating to form irregular cavities. **Phloem** cells thin-walled, with transverse to slightly oblique sieve-plates. Centre of stele occupied by a wide-celled and usually continuous pith, the cell walls thickened, lignified, and pitted.

Basal stem (rhizome)

Up to 1.5 mm in diameter, and organized in the same way as the scape; sometimes slightly eccentric, the cortex becoming thick-walled and lignified. Sclerenchyma sheath well developed and the vascular cylinder with more extensive conducting tissue than in the scape, forming a discontinuous ring.

Root (Fig. 16.2 E, F and Plate 16 A)

Branched, fine but few, up to 0.5 mm in diameter. **Epidermis** thin walled; root hairs not observed. **Exodermis** uniseriate, the radial and outer tangential walls thickened and lignosuberized. **Cortex** consisting of 4–6 layers of large thin-walled parenchyma cells including **mycorrhizal hyphae** (Plate 16 A). **Stele** narrow, delimited by an **endodermis** of elongated cells with unequal, stratified wall thickenings appearing U-shaped in TS (Fig. 16.2 F). **Pericycle** narrow, cells thin-walled. **Vascular tissue** poorly differentiated, including four groups of narrow, thin-walled phloem cells; xylem occupying the centre of the stele, its elements with bordered pits, those in the centre somewhat like vessel elements because of distinct elongated scalariform pitting on oblique end walls.

Secretory, Storage, and Conducting Elements

Crystals not observed, apart from the unconfirmed report of raphides by Groom (1895).

Starch not observed.

Xylem: vessel elements not known with certainty for this family (Stant 1970); xylem elements in the roots with scalariform pitting restricted to distinct long oblique end walls, but without any evidence for the dissolution of the pit membranes.

Other cell types

The wide range of cell types and their size in this plant have been described in detail by Stant (1970), who worked with macerated material. It is difficult to classify the cell types precisely because they form a continuous integrading series.

Reproductive Morphology and Anatomy

Flowers (Fig. 16.1 F–I) in simple distal, corymb-like racemes terminating the erect shoots (Fig. 16.1 A). Each flower on a long pedicel with one (sometimes more) basal but lateral bracteole. **Perianth** biseriate, the three outer much narrower than the three inner tepals (Fig. 16.1 B, F, H). **Stamens** six ('partly inferior' according to Groom 1895), each attached basally on a tepal and each with a slender filament; the tepals and stamens partly adherent to the carpels (Fig. 16.1 G). Anthers ovate, dorsifixed, and introrse, the connective slightly pointed. **Carpels** three (sometimes more), free in the upper two-thirds, each with a short style and indistinctly bifid hairy stigma; septal nectaries well developed according to Groom (1895) and illustrated by Sterling (1978). **Ovules** numerous, ± anatropous, bitegmic, attached to the inner angle of the carpels. Fruiting carpels spreading (Fig. 16.1 C), opening by an adaxial slit to release the numerous seeds (Fig. 16.1 D, E). Seeds with a spongy covering, ribbed.

[Groom (1895) shows a series of diagrams of the flower in TS. According to Sterling (1978) three amphiphloic dorsal vascular bundles depart from the central cylinder of the pedicel. These each divide tangentially to give off two centrifugal traces: a tepallary vascular bundle on the outside supplying the outer tepals ('sepals') and a staminal vascular bundle between the tepallary and dorsal bundles, supplying the antesepalous stamens. The inner tepals ('petals') and antepetalous stamens are supplied in a similar way so that all these appendages receive one bundle. The carpels are supplied by six vascular bundles in three pairs, described as placental bundles, each carpel receiving two which both reach the style (sometimes fusing over the locular apex), the ventral bundle supplying the ovules via branches.]

Taxonomic Notes

The genus *Petrosavia* has been of uncertain taxonomic position since it was first described by Beccari (1871) who placed it in his Melanthaceae, a group later included in the Liliaceae. This position was essentially accepted by Bentham and Hooker (1883) and in successive editions of Engler and Prantl (e.g. Eckardt 1964). Groom (1895) regarded his genus *Protolirion*, which he separated from *Petrosavia*, as a link between Liliaceae and Triuridaceae; he also stressed the occurrence of septal nectaries. Hutchinson (1959) departed considerably from this scheme, raising the genus to family level and associating it with Scheuchzeriaceae in his Alismatales. The justification for this seems to be the similarly trilocular apocarpous ovary with two (*Scheuchzeria*) or many (*Petrosavia*) ovules in each carpel, attached to the adaxial surface. However, in its essentially hydrophytic habit, *Scheuchzeria* differs considerably from *Petrosavia*.

Stant (1970) has pointed out the considerable anatomical similarity between Triuridaceae and Petrosaviaceae and comments that she 'would have no hesitation in placing *Petrosavia* in the Triuridaceae on the basis of anatomical evidence.' However, since both families are saprophytic, convergence might account for the similarity, leading to this opinion. Anatomical investigation of other monocotyledonous families which include saprophytic members, e.g. Burmanniaceae, Corsiaceae, and Thismiaceae, might throw further light on this problem, as would more extensive knowledge of the range of variation in *Petrosavia* itself.

For the present it seems most logical to take the structural evidence at its face value and place the Petrosaviaceae and Triuridaceae next to each other, since neither can be accommodated more conveniently in any other linear sequence and their association with other apocarpous groups is justifiable. A comparison between the Scheuchzeriaceae, Petrosaviaceae and Triuridaceae is given in Table 16.1 (p. 478).

Further information concerning pollen morphology, floral development, and other microscopic details of the reproductive parts would help to clarify the relationships among these groups.

MATERIAL EXAMINED

Petrosavia stellaris Becc.; Mt. Kinabalu, Sabah; J. & H. S. Clemens 32454; also Chew, Corner, & Stainton 601; all parts (see Stant 1970).

SPECIES REPORTED ON IN THE LITERATURE

Beccari (1871): *Petrosavia stellaris*: general morphology.
Groom (1895): '*Protolirion paradoxum*' all parts.
Stant (1970): *Petrosavia stellaris*: all vegetative parts.
Sterling (1978): *Petrosavia miyoshia-sakuraii*; *Protolirion sakuraii*: floral anatomy.
Watanabe (1944): *Petrosavia miyoshia-sakuraii*: all parts.

SIGNIFICANT LITERATURE—PETROSAVIACEAE

Beccari (1871); Bentham and Hooker (1883); Eckardt (1964); Erdtman (1952); Groom (1895); Hutchinson (1959); Stant (1970); Sterling (1978); Watanabe (1944).

TABLE 16.1

Comparison of three monocotyledonous families with putative affinities

Character	Scheuchzeriaceae	Petrosaviaceae	Triuridaceae
No. of taxa	one sp	2 (–3) spp	*c.* 80 spp
Nutrition	green, autotropic	colourless, sapro-phytic	colourless, sapro-phytic
Range habitat	circum-boreal, marshes	tropical, subtropical, forests	tropical, forests
Leaves	linear, well developed	scale-like	scale-like
Aerenchyma	well developed	absent	absent
Sclerenchyma, lignin	little developed	well developed	well developed
Crystals	present	absent	absent
Stem anatomy	vbs scattered	vbs annular	vbs annular
Hairs	present at nodes	absent	absent
Stomata	tetracytic	anomocytic	absent
Vessels	possibly present	possibly absent	possibly absent
Infloresence	racemose	racemose	racemose
Bracteoles	absent	present	absent
Flowers	perfect	perfect	usually unisexual
Perianth	uniseriate, trimerous	biseriate, trimerous	± uniseriate, dimer-ous or trimerous
Stamens	six, stalked	six, stalked	2–6, usually sessile
Dehiscence	extrorse by longitudi-nal slits	introrse by longitu-dinal slits	extrorse by transverse slits
Carpels	three	three	many
Ovules/carpel	two	many	one
Placentation	basal	adaxial	basal
Fruit	dehiscent	dehiscent	usually dehiscent
Endosperm	absent	absent	absent

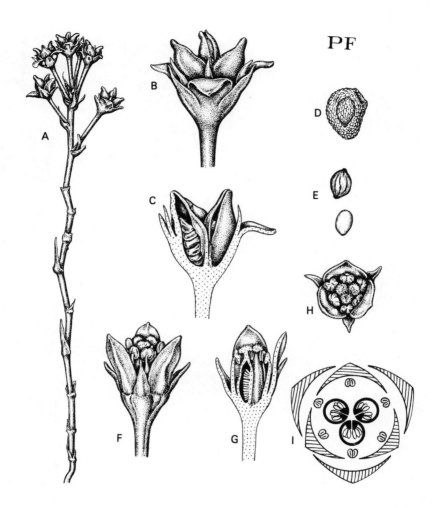

PF

FIG. 16.1. PETROSAVIACEAE. *Petrosavia stellaris* Becc. Fruits and flowers. (Mt. Kinabalu, Sabah, Chew, Corner & Stainton, 601.)

A–E. Fruiting material.

A. Distal portion with mature infructescence ($\times \frac{3}{2}$). B. Detail of mature fruits ($\times$ 6). C. Fruiting head in LS ($\times$ 6). D. Single seed ($\times$ 15). E. Seed ($\times$ 15) with testa removed; below, embryo.

F–G. Flowering material ($\times$ 6).

F. Flower from the side. G. Flower in LS H. Flower from above. I. Floral diagram.

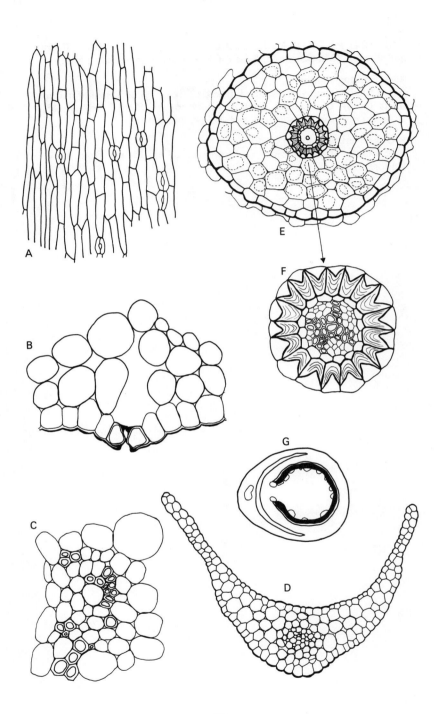

FIG. 16.2 PETROSAVIACEAE. (After Stant 1970.) *Petrosavia stellaris* Becc. Mt. Kinabalu, Sabah.

A–D. Leaf.

A. Abaxial epidermis in surface view with stomata (× 36). B. TS stoma from abaxial epidermis below midrib (× 110). C. TS vb from midrib, with scattered sheathing fibres (× 110). D. TS lamina (× 36).

E–F. Root.

E. TS (× 36), position of mycorrhizal hyphae in dotted outline. F. Stele (× 110), delimited by massive endodermis with U-shaped wall thickenings. G. TS lower part of stem at attachment of sheathing base of scale leaf (× 8) with leaf gap in sclerenchyma cylinder.

BIBLIOGRAPHY

AALTO, M. (1970). Potamogetonaceae fruits. I. Recent and subfossil endocarps of the Fennoscandian species. *Acta bot. fenn.* **88**, 1–85.

AFZELIUS, K. (1920). Einige Beobachtungen über die Samenentwicklung der Aponogetonaceae. *Svensk bot. Tidskr.* **14**, 168–75.

AGARDH, J. G. (1852) [1853]. Rotknölar hos *Potamogeton pectinatus. Öfvers. kongl. Vet.-Akad. Förh.* 29–31.

AGRAWAL, J. S. (1952). The embryology of *Lilaea subulata* H. B. K. with a discussion on its systematic position. *Phytomorphology* **2**, 15–29.

ALBERGONI, F. G., BASSO, B., and TEDESCO, G. (1978). Considérations sur l'anatomie de *Posidonia oceanica* (Zosteraceae). *Pl. Syst. Evol.* **130**, 191–201.

ANCIBOR, E. (1979). Systematic anatomy of vegetative organs of the Hydrocharitaceae. *Bot. J. Linn. Soc.* **78**, 237–66 (+ microfiche).

ANDERSON, L. W. J. (1978). Abscisic acid induces formation of floating leaves in the heterophyllous aquatic angiosperm *Potamogeton nodosus. Science, N.Y.* **201**, 1135–8.

ANDERSSON, S. (1888). Om de primära kärlsträngarnes utveckling hos monokotyledonerna. *Bih. svenska Vet.-Akad. Handl.* **13** (12), 23 pp.

ANDREI, M. (1969). Cercetări anatomice comparative asupra epidermei unor plante acvatice. *Comun. Bot. Soc. Ştiinţ. biol. Rep. soc. România* **10**, 35–49. (German summary.)

—— (1972), Cercetări anatomice comparative asupra ţesutului conducător lemnos la cîteva specii acvatice şi temporar acvatice. [Vergleichende anatomische Untersuchungen über das Holzleitgewebe bei einigen Wasser- und zeitweiligen Wasserarten.] *Lucr. Grăd. bot. Bucuresti* (1970–1), 153–71. [Rumanian, German summary.]

ARBER, A. (1914). On root development in *Stratiotes aloides* L. with special reference to the occurrence of amitosis in an embryonic tissue. *Proc. Camb. Phil. Soc.* **17**, 369–79.

—— (1918a). Further notes on intrafascicular cambium in monocotyledons. *Ann. Bot., Lond.* **32**, 87–9.

—— (1918b). The phyllode theory of the monocotyledonous leaf, with special reference to anatomical evidence. *Ann. Bot., Lond.* **32**, 465–501.

—— (1919). On heterophylly in water plants. *Am. Nat.* **53**, 272–8.

—— (1920). *Water plants. A study of aquatic angiosperms.* Cambridge University Press, Cambridge. [Reprint 1963 J. Cramer Weinheim Historiae Naturalis Classica. Tomus xxiii.]

—— (1921). Leaves of the Helobieae. *Bot. Gaz.* **72**, 31–8.

—— (1922a). Studies on intrafascicular cambium in monocotyledons. V. *Ann. Bot., Lond.* **36**, 251–6.

—— (1922b). On the nature of the 'blade' in certain monocotyledonous leaves. *Ann. Bot., Lond.* **36**, 329–51.

—— (1923). On the 'squamulae intravaginales' of the Helobieae. *Ann. Bot., Lond.* **37**, 31–41.

—— (1924). Leaves of *Triglochin. Bot. Gaz.* **77**, 50–62.

—— (1925a). *Monocotyledons. A morphological study.* Cambridge University Press.

—— (1925b). On the "squamulae intravaginales" of the Alismataceae and Butomaceae. *Ann. Bot., Lond.* **39**, 169–73.

—— (1940). Studies in flower structure VI. On the residual vascular tissue in the apices of reproductive shoots, with special reference to *Lilaea* and *Amherstia. Ann. Bot., Lond.* n.s. **4**, 617–27.

ARESCHOUG, F. W. C. (1878). *Jemförande undersökningar öfver bladets anatomi*, 242 pp. Lund.

ARGUE, C. L. (1973). The pollen of *Limnocharis flava* Buch., *Hydrocleis nymphoides* (Willd.) Buch., and *Tenagocharis latifolia* (Don) Buch. (Limnocharitaceae). *Grana* **13**, 108–12.

ARZT, T. (1937). Die Kutikula bei einigen Pteridophyten, Gymnospermen, und Monokotyledonen. *Ber. dt. bot. Ges.* **55**, 437–64.

ASCHERSON, P. (1882). Die vegetative Vermehrung einer australischen Seegrasart, der *Cymodocea antarctica* (Labill.) Endl. *Sber. bot. Ver. Prov. Brandenburg:* in *Verh. bot. Ver. Prov. Brandenburg* **24**, 28–33.

—— and GRAEBNER, P. (1907). Potamogetonaceae. In *Das Pflanzenreich* (ed. A. Engler) Vol. 4, 11, Heft 31: pp. 1–184, W. Engelmann, Leipzig.

—— and GÜRKE, M. (1889). Hydrocharitaceae. In *Die natürlichen Pflanzenfamilien* II (ed. A. Engler and K. Prantl) Vol 1, pp. 238–58.

ASTON, H. (1973). *Aquatic plants of Australia.* Melbourne University Press.

AYENSU, E. S. (1972). *Dioscoreales.* Vol. VI, *Anatomy of the monocotyledons* (ed. C. R. Metcalfe). Clarendon Press, Oxford.

BAILEY, C. (1884). Notes on the structure, the occurrence in Lancashire, and the source of origin, of *Naias graminea* Delile var. *delilei* Magnus. *J. Bot., Lond.* **22**, 305–33.

BALFOUR, B. (1879). On the genus *Halophila. Trans. Proc. Bot. Soc. Edinb.* **13**, 290–343.

BANCE, H. M. (1946). A comparative account of the structure of *Potamogeton filiformis* Pers. and *P. pectinatus* L. in relation to the identity of a supposed hybrid of these species. *Trans. Bot. Soc. Edinb.* **34**, 361–7.

BARBOUR, M. G. and RADOSEVICH, S. R. (1979). ^{14}C uptake by the marine angiosperm *Phyllospadix scouleri. Am. J. Bot.* **66**, 301–6.

BARNABAS, A. D., BUTLER, V., and STEINKE, T. D. (1977). *Zostera capensis* Setchell. I. Observations on the fine structure of the leaf epidermis. *Z. Pflanzenphysiol.* **85**, 417–27.

BARY, A. DE (1884). *Comparative anatomy of the vegetative organs of the phanerogams and ferns.* Clarendon Press, Oxford.

BATE-SMITH, E. C. (1968). The phenolic constituents of plants and their taxonomic significance. II. Monocotyledons. *J. Linn. Soc. (Bot.)* **60**, 325–56.

BAUMANN, E. (1911). *Die Vegetation des Untersees (Bodensee). Eine floristischkritische und biologische Studie.* Inaug.-Diss. Univ. Zürich, Stuttgart. (Also *Arch. Hydrobiol. Planktonkunde*, Suppl. I.)

BECCARI, O. (1871). *Petrosavia*, nuovo genere di piante parassite della famiglia delle Melanthaceae. *Nuovo G. bot. ital.* **3**, 7–11.

BECKEROWA, Z. (1934). Zytologische Untersuchungen an den Trichocysten von *Stratiotes aloides* L. *Acta Soc. bot. pol.* **11**, 347–66.

BEHNKE, H.-D. (1969). Die Siebröhren-Plastiden der Monocotyledonen. Vergleichende Untersuchungen über Feinbau und Verbreitung eines charakteristischen Plastidentypus. *Planta* **84**, 174–84.

BELL, A. D. and TOMLINSON, P. B. (1980). Adaptive architecture in rhizomatous plants. *Bot. J. Linn. Soc.* **80**, 125–60.

BENEDICT, C. R. and SCOTT, J. R. (1976). Photosynthetic carbon metabolism of a marine grass. *Pl. Physiol.* **57**, 876–80.

BENNETT, A. (1900–28). Notes on *Potamogeton. J. Bot., Lond.* **38**, 125–30 (1900); **39**, 198–201 (1901); **40**, 145–9 (1902); **42**, 69–77 (1904); **45**, 373–7 (1907); **46**, 160–3 (1908); **65**, 113–6 (1927); **66**, 102–4 (1928).

—— (1914). *Hydrilla verticillata* Casp. in England. *J. Bot., Lond.* **52**, 257–58.

—— (1919). Notes on Dr. Hagström's "Critical researches on *Potamogeton*, 1916". *Trans. Bot. Soc. Edinb.* **27**, 315–26.

BENTHAM, G. and HOOKER, J. D. (1883). *Genera plantarum 3*, pp. 1017. London.

BERTON, A. (1978). Contribution de l'anatomie à la détermination des *Potamogeton. Monde Plantes* **73**, no. 393, 3–4.

BEUSEKOM, C. F. VAN (1967). Ueber einiger Apiose-vorkommnisse bei den Helobiae. *Phytochemistry* **6**, 573–6.

BIEBL, R. (1951). Freilandbeobachtungen an Spaltöffnungen von *Hydrocharis, Nymphaea* und *Nuphar. Verh. Zool.-Bot. Ges. Wien* **92**, 249–53.

BIRCH, W. R. (1974). The unusual epidermis of the marine angiosperm *Halophila* Thou. *Flora, Jena* **163**, 410–14.

BJÖRKQVIST, I. (1967). Studies in *Alisma* L. I. Distribution, variation and germination. *Opera Bot., Lund* No. **17**, 1–128.

—— (1968). Studies in *Alisma* L. II. Chromosome studies, crossing experiments and taxonomy. *Opera Bot., Lund* No. **19**, 1–138.

BLACK, J. M. (1913). The flowering and fruiting of *Pectinella antarctica* (*Cymodocea antarctica*). *Trans. Proc. R. Soc. S. Aust.* **37**, 1–5.

BLACKBURN, K. B. (1934). Wasting disease of *Zostera marina. Nature, Lond.* **134**, 738.

BLASS, J. (1890). Untersuchungen über die physiologische Bedeutung des Siebtheils der Gefässbündel. *Jb. wiss. Bot.* **22**, 253–92.

BLOEDEL, C. A. and HIRSCH, A. M. (1979). Developmental studies of the leaves of *Sagittaria latifolia* and their relationship to the leaf-base theory of monocotyledonous leaf morphology. *Can. J. Bot.* **57**, 420–34.

BÖHMKER, H. (1917). Beiträge zur Kenntnis der floralen und extrafloralen Nektarien. *Beih. bot. Zbl.* **33**, (1) 169–247.

BONNETT, B. and MILLET, B. (1970). Problèmes soulevés par la structure morphologique de la tige du *Potamogeton densus* L. *Annls scient. Univ. Besançon, Bot.*, sér. **3** (7), 32–7.

BORNET, E. (1864). Recherches sur le *Phucagrostis major* Cavol. *Annls Sci. nat., Bot.*, sér. 5, **1**, 5–51.

BORZI, A. (1887–8). Formazione delle radici laterali nelle Monocotiledoni. *Malpighia* **1**, 391– 413, 541–50 (1887); **2**, 52–85, 394–402, 477–506 (1888).

BOUTARD, B., BOUILLANT, M.-L., CHOPIN, J., and LEBRETON, P. (1973). Chimiotaxinomie flavonique des Fluviales. *Biochem. Syst.* **1**, 133–40.

BOYD, L. (1932). Monocotylous seedlings. *Trans. bot. Soc. Edinb.* **31**, 1–224.

BOYSEN-JENSEN, P. (1959). Untersuchungen über Determination und Differenzierung 6. Über den Aufbau des Zellwandmusters des Blattes von *Helodea densa*. *Biol. Meddel. Kgl. Danske Vidensk. Selskab.* **23**, 1–33.

BRISTOW, J. M. (1975). The structure and function of roots in aquatic vascular plants. *The development and function of roots* (ed. J. G. Torrey and D. T. Clarkson) pp. 221–36. Academic Press, New York.

BRUGGEN, H. W. E. VAN (1968). Revision of the genus *Aponogeton* I. The species of Madagascar. *Blumea* **16**, 243–63.

—— (1969). Revision of the genus *Aponogeton* III. The species of Australia. *Blumea* **17**, 121– 37.

—— (1970). Revision of the genus *Aponogeton* (Aponogetonaceae) IV. The species of Asia and Malesia. *Blumea* **18**, 457–86.

—— (1973). Revision of the genus *Aponogeton* (Aponogetonaceae). The species of Africa VI. *Bull. Jard. Bot. Nat. Belg.* **43**, 193–233.

BRUNAUD, A. (1976). Ramification chez les Hydrocharitaceae. I. Ontogénie du système des pousses. *Rev. gén. Bot.* **83**, 397–413.

—— (1977). Ramification chez les Hydrocharitaceae. II. Organisation des rameaux latéraux. *Rev. gén. Bot.* **84**, 137–57.

BRUNKENER, L. (1975). Beiträge zur Kenntnis der frühen Mikrosporangienentwicklung der Angiospermen. *Svensk bot. Tidskr.* **69**, 1–27.

BUCHENAU, F. (1857). Über die Blüthenentwicklung von *Alisma* und *Butomus*. *Flora, Jena* **40**, 241–54.

—— (1872). Eigenthümlicher Bau der Blattspitze von *Scheuchzeria palustris* L. *Bot. Ztg.* **30**, 139.

—— (1882). Beiträge zur Kenntniss der Butomaceen, Alismaceen und Juncaginaceen. *Bot. Jb.* **2**, 465–510.

—— (1889). Alismataceae. In *Die natürlichen Pflanzenfamilien* (ed. A. Engler and K. Prantl) Vol. 2 (1), pp. 227–32.

—— (1903*a*). Alismataceae. In *Das Pflanzenreich* (ed. A. Engler) Vol. **4** (15), 66 pp. Wilhelm Englemann, Leipzig.

—— (1903*b*). Butomaceae. In *Das Pflanzenreich* (ed. A. Engler) Vol. **4** (16), 12 pp. Wilhelm Englemann, Leipzig.

—— (1903*c*). Scheuchzeriaceae. In *Das Pflanzenreich* (ed. A. Engler) Vol. **4** (14), 20 pp. Wilhelm Englemann, Leipzig.

—— and HIERONYMUS, G. (1889). Juncaginaceae. In *Die natürlichen Pflanzenfamilien* (ed. A. Engler and K. Prantl) Vol. 2 (1), pp. 222–7.

BUGNON, F. (1963). La notion de concrescence congénitale et la cas de bourgeons "extra-axillaires" du *Zostera marina* L. *Bull. Soc. bot. Fr.* **110**, 92–101.

BUGNON, F. and JOFFRIN, G. (1962). Recherches sur la ramification de la pousse chez le *Vallisneria spiralis* L. *Mém. Soc. bot. Fr.* 1962, 61–72.

—— (1963). Ramification de la pousse chez l'*Hydrocharis morsus-ranae* L.; comparaison avec le cas du *Vallisneria spiralis* L. *Bull. Soc. bot. Fr.* **110**, 34–42.

BURGEMEISTER, H. (1968). Entwicklungsphysiologische Untersuchungen zur Heterophyllie und Stomatabildung bei *Zannichellia palustris* L. *Beitr. Biol. Pfl.* **44**, 67–121.

BURGER, W. C. (1978). The Piperales and the Monocots. Alternate hypotheses for the origin of monocotyledonous flowers. *Bot. Rev.* **43**, 346–93.

BUTCHER, R. W. (1933) [1934]. Notes on the variation of the British species of *Zostera*. *Bot. Soc. Exchange Club Brit. Isles, Rep.* 1933, **10**, 592–7.

CAMBRIDGE, M. L. and KUO, J. (1979). Two new species of seagrasses from Australia, *Posidonia sinuosa* and *P. angustifolia* (Posidoniaceae). *Aquat. Bot.* **6**, 307–28.

CAMPBELL, D. H. (1897). A morphological study of *Naias* and *Zannichellia. Proc. Calif. Acad. Sci. (Bot.)*, Ser. 3, **1**, 1–70.

—— (1898). Development of the flower and embryo in *Lilaea subulata* H.B.K. *Ann. Bot.* **12**, 1–28.

CAMPBELL, G. K. G. (1936). The anatomy of *Potamogeton pectinatus. Trans. bot. Soc. Edinb.* **32**, 179–86.

CANNON, J. F. M. (1979). An experimental investigation of *Posidonia* balls. *Aquat. Bot.* **6**, 407–10.

CARBONELL, C. S. and ARRILLAGA, B. R. (1959). Sobre la relacion anatomica de las ootecas de *Marellia remipes* Uvarov (Orthoptera, Acrid., Pauliniidae) con las hojas de su planta huesped, y su posible significacion fisiologica. *Rev. Soc. Uruguaya Ent.* **3**, 45–56.

CARIGNAN, R. and KALFF, J. (1980). Phosphorus sources for aquatic weeds: water or sediments? *Science*, N.Y. **207**, 987–9.

CARTER, S. (1960). Alismataceae. *Flora of tropical East Africa*. Crown Agents, London.

CASPARY, R. (1857). Note sur la division de la famille des Hydrocharidées, proposées par M. Chatin. *Bull. Soc. bot. Fr.* **4**, 98–101.

—— (1858). Die Hydrilleen (Anacharideen Endl.) *Jb. wiss. Bot.* **1**, 377–513. (Translated as: Les Hydrillés (Anacharidées Endl.) *Annls. Sci. nat., Bot.*, Sér. 4, **9**, 323–96.)

—— (1861). Über das Vorkommen der *Hydrilla verticillata* Casp. in Preussen, die Blüthe derselben in Preussen und Pommern und das Wachsthum ihres Stammes. *Verh. XXXV Vers. dtsch. Naturf. Ärzte Königsberg* 1860, 293–310.

CASTELL, C. P. (1935). The water soldier. *School Nature Study* **30**.

CATLING, D. M. (1968). Preparing microtome sections of aquatic plants without embedding. *Proc. R. Microsc. Soc.* **3**, 126–8.

ČELAKOVSKÝ, L. J. (1896). Über den phylogenetischen Entwickelungsgang der Blüthe und über den Ursprung der Blumenkrone. I. *Sber. kön. böhm. Ges. Wiss., math.-nat. Kl.* 1–91.

CHANDLER, M. E. J. (1923). The geological history of the genus *Stratiotes. Q. J. geol. Soc., Lond.* **79**, 117–38.

CHARLTON, W. A. (1968). Studies in the Alismataceae. I. Developmental morphology of *Echinodorus tenellus. Can. J. Bot.* **46**, 1345–60.

—— (1973). Studies in the Alismataceae. II. Inflorescences of Alismataceae. *Can. J. Bot.* **51**, 775–89.

—— (1974). Studies in the Alismataceae. V. Experimental modification of phyllotaxis in pseudostolons of *Echinodorus tenellus* by means of growth inhibitors. *Can. J. Bot.* **52**, 1131–42.

—— (1976). Studies in the Alismataceae. VI. Specialized rhizome structure of *Burnatia enneandra. Can. J. Bot.* **54**, 30–8.

—— (1979a). Studies in the Alismataceae. VII. Disruption of phyllotactic and organogenetic patterns in pseudostolons of *Echinodorus tenellus* by means of growth-active substances. *Can. J. Bot.* **57**, 215–22.

—— (1979b). Studies in the Alismataceae. VIII. Experimental modification of organogenesis in *Ranalisma humile. Can. J. Bot.* **57**, 223–32.

—— and AHMED, A. (1973a). Studies in the Alismataceae. III. Floral anatomy of *Ranalisma humile. Can. J. Bot.* **51**, 891–97.

—— (1973b). Studies in the Alismataceae. IV. Developmental morphology of *Ranalisma humile* and comparisons with two members of the Butomaceae, *Hydrocleis nymphoides* and *Butomus umbellatus. Can. J. Bot.* **51**, 899–910.

CHATIN, A. (1855). *Mémoire sur le Vallisneria spiralis L.* Mallet-Bachelier, Paris. 31 pp.

CHATIN, A.-A. (1856, 1862). *Anatomie Comparée des Végétaux. Plantes Aquatiques.* Baillière, Paris. 96 pp.

CHAUVEAUD, G. (1897a). Sur la structure de la racine de l'*Hydrocharis morsus-ranae. Rev. gén. Bot.* **9**, 305–12.

—— (1897b). Recherches sur le mode de formation des tubes criblés dans la racine des monocotylédones. *Annls. Sci. nat., Bot.*, Sér. 8, **4**, 307–81.

—— (1901). Sur le passage de la disposition alterne des éléments libériens et ligneux à leur disposition superposée dans le Trocart (*Triglochin*). *Bull. Mus. Hist. nat., Paris* **7**, 124–30.

CHEADLE, V. I. (1942). The occurrence and types of vessels in the various organs of the plant in the Monocotyledoneae. *Am. J. Bot.* **29**, 441–50.

—— (1943a). The origin and certain trends of specialization of the vessel in the Monocotyledoneae. *Am. J. Bot.* **30**, 11–17.

—— (1943b). Vessel specialization in the late metaxylem of the various organs in the Monocotyledoneae. *Am. J. Bot.* **30**, 484–90.

—— (1944). Specialization of vessels within the xylem of each organ in the Monocotyledoneae. *Am. J. Bot.* **31**, 81–92.

CHEADLE, V. I. and TUCKER, J. M. (1961). Vessels and phylogeny of Monocotyledoneae. *Recent Adv. Bot.* **1**, 161–5.

—— and UHL, N. W. (1948). Types of vascular bundles in the Monocotyledoneae and their relation to the late metaxylem conducting elements. *Am. J. Bot.* **35**, 486–96.

CHRYSLER, M. A. (1907). The structure and relationships of the Potamogetonaceae and allied families. *Bot. Gaz.* **44**, 161–88.

CLAUSEN, P. (1927). Ueber das Verhalten des Antheren-Tapetums bei einigen Monokotylen und Ranales. *Bot. Arch.* **18**, 1–27.

CLAVAUD, A. (1878). Sur le véritable mode de fécondation du *Zostera marina. Actes Soc. linn. Bordeaux* **32**, 109–15.

CLIFFORD, H. T. (1970). Monocotyledon classification with special reference to the origin of the grasses (Poaceae). *Bot. J. Linn. Soc.* **63** (Suppl. 1), 25–34.

CLOS, D. (1856). Mode de propagation particulier au *Potamogeton crispus* L. *Bull. Soc. bot. Fr.* **3**, 350–2.

CÖSTER, B. F. (1875). Om *Potamogeton crispus* L. och dess groddknoppar. *Bot. Notiser*, 97–102.

COHEN, E. (1939). An account of the marine angiosperms of Inhaca, P.E.A. *S. Afr. J. Sci.* **36**, 246–56.

COLOMB, G. (1887). Recherches sur les stipules. *Annls Sci. nat., Bot.,* sér. 7, **6**, 1–76.

COOK, C. D. K. (1981). Floral biology of *Blyxa octandra* (Roxb.) Planchon ex Thwaites (Hydrocharitaceae). *Aquat. Bot.* **10**, 61–8.

—— GUT, B. J., RIX, E. M., SCHNELLER, J., and SEITZ, M. (1974). *Water plants of the world. A manual for the identification of the genera of freshwater macrophytes.* W. Junk, The Hague.

COOK, M. T. (1908). The development of the embryo-sac and embryo of *Potamogeton lucens. Bull. Torrey Bot. Club* **35**, 209–18.

CORDEMOY, C. J. DE (1862–3). Organogénie des *Triglochin. Adansonia (Baillon)* **3**, 12–14.

CORDES, W. C. (1959). Description and physiological properties of lipid-containing, non-chlorophyllous cells in *Elodea. Physiol. Plant.* **12**, 62–9.

CORMACK, R. G. H. (1937). The development of root hairs by *Elodea canadensis. New Phytol.* **36**, 19–25.

CORRELL, D. S. and CORRELL, H. B. (1975). *Aquatic and wetland plants of Southwestern United States* (2 vols.). Stanford University Press.

COSSON, E. (1860). Note sur la stipule et la préfeuille dans le genre *Potamogeton*, et quelques considérations sur ces organes dans les autres monocotylées. *Bull. Soc. bot. Fr.* **7**, 715–19.

COSTANTIN, J. (1884). Recherches sur la structure de la tige des plantes aquatiques. *Annls Sci. nat., (Bot.),* sér. 6, **19**, 287–331.

—— (1885a). Observations critiques sur l'épiderme des feuilles des végétaux aquatiques. *Bull. Soc. bot. Fr.* **32**, 83–8.

—— (1885b). Influence du milieu aquatique sur les stomates. *Bull. Soc. bot. Fr.* **32**, 259–64.

—— (1885c). Recherches sur la Sagittaire. *Bull. Soc. bot. Fr.* **32**, 218–23.

—— (1886a). Observations sur la note de M. Mer. *Bull. Soc. bot. Fr.* **33**, 192–6.

—— (1886b). Études sur les feuilles des plantes aquatiques. *Annls Sci. nat., Bot.,* sér. 7, **3**, 94–162.

CRANWELL, L. M. (1953). New Zealand pollen studies. The monocotyledons. *Bull. Auckland Inst. Mus.,* No. 3, 1–91.

CRONQUIST, A. (1968). *The evolution and classification of flowering plants.* Houghton Mifflin, New York.

CUNNINGTON, H. M. (1912). Anatomy of *Enhalus acoroides* (Linn. f.) Zoll. *Trans. Linn. Soc. Lond. (Bot.),* ser. 2, **7**, 355–71.

CURRIER, H. B. and SHIH, C. Y. (1968). Sieve tubes and callose in *Elodea* leaves. *Am. J. Bot.* **55**, 145–52.

CUTLER, D. F. (1969). *Juncales.* Vol. IV, *Anatomy of the Monocotyledons* (ed. C. R. Metcalfe). Clarendon Press, Oxford.

CUTTER, E. G. (1964). Observations on leaf and bud formation in *Hydrocharis morsus-ranae. Am. J. Bot.* **51**, 318–24.

—— (1978). *Plant Anatomy. Part 1. Cells and tissues* (2nd edn). Addison-Wesley, Reading, Mass. and Edward Arnold, London.

—— and FELDMAN, L. J. (1970*a*). Trichoblasts in *Hydrocharis.* I. Origin, differentiation, dimensions and growth. *Am. J. Bot.* **57**, 190–201.

—— —— (1970*b*). Trichoblasts in *Hydrocharis.* II. Nucleic acids, proteins, and a consideration of cell growth in relation to endopolyploidy. *Am. J. Bot.* **57**, 206–11.

—— and HUNG, C.-Y (1972). Symmetric and asymmetric mitosis and cytokinesis in the root tip of *Hydrocharis morsus-ranae* L. *J. Cell Sci.* **11**, 723–37.

CZAJA, A. T. (1930). Die photometrischen Bewegungen der Blätter von *Aponogeton ulvaceus. Ber. dt. bot. Ges.* **48**, 349–62.

DAHLGREN, K. V. O. (1939). Endosperm- und Embryobildung bei *Zostera marina. Bot. Notiser* 607–15.

DALE, H. M. (1951). Carbon dioxide and root hair development in *Anacharis* (*Elodea*). *Science, N.Y.* **114**, 438–9.

—— (1957). Developmental studies of *Elodea canadensis* Michx. I. Morphological development at the stem apex. *Can. J. Bot.* **35**, 13–24.

D'ALMEIDA, J. F. R. (1942). A contribution to the study of the biology and physiological anatomy of Indian marsh and aquatic plants. *J. Bombay nat. Hist. Soc.* **43**, 92–6.

DAMMER, U. (1888). *Beiträge zur Kenntniss der vegetativen Organe von* Limnobium stoloniferum Griseb. *nebst einigen Betrachtungen über die phylogenetische Dignität von Diclinie und Hermaphroditismus.* Diss. Freiburg, Berlin.

DANDY, J. E. and TAYLOR, G. (1946). An account of x*Potamogeton suecicus* Richt. in Yorkshire and the Tweed. *Trans. bot. Soc. Edinb.* **34**, 348–60.

DATTA, S. C. and BISWAS, K. K. (1976). Autecological studies on weeds of West Bengal. VI. *Ottelia alismoides* (L.) Pers. *Bull. bot. Soc. Bengal* **30**, 1–9.

DAUMANN, E. (1970). Das Blütennektarium der Monocotyledonen unter besonderer Berücksichtigung seiner systematischen und phylogenetischen Bedeutung. *Feddes Repert.* **80**, 463–590.

DAVIS, G. L. (1966). *Systematic embryology of the angiosperms.* John Wiley, New York.

DAVIS, J. S. and TOMLINSON, P. B. (1974). A new species of *Ruppia* in high salinity in Western Australia. *J. Arnold Arbor.* **55**, 59–66.

DELAY, C. (1941). Sur les trichoblastes des racines de *Limnobium bogotense* Rich. (Hydrocharidacées). *Bull. Soc. bot. Fr.* **88**, 480–5.

DENNY, P. (1980). Solute movement in submerged angiosperms. *Bot. Rev.* **55**, 65–92.

DIANNELIDIS, T. (1950). Zellphysiologische Beobachtungen an den Schliesszellen von *Stratiotes aloides. Protoplasma* **39**, 441–9.

DIBBERN, H. (1903). Über anatomische Differenzierungen im Bau der Inflorescenzachsen einiger diklinischen Blütenpflanzen. *Beih. bot. Zbl.* **13**, 341–60.

DONÁ DELLE ROSE, A. (1946). Sulla biologia e morfologia di una pianta marina (*Ruppia maritima* L.). *Bol. Soc. Ital. Biol. Sper.* **22**, 449–52.

DOOHAN, M. E. and NEWCOMB, E. H. (1976). Leaf ultrastructure and $\delta^{13}C$ values of three seagrasses from the Great Barrier Reef. *Aust. J. Pl. Physiol.* **3**, 9–23.

DRAWERT, H. (1937). Protoplasmatische Anatomie des fixierten *Helodea*- Blattes. *Protoplasma* **29**, 206.

—— (1938). Elektive Färbung der Hydropoten an fixierten Wasserpflanzen. Ein Beitrag zur protoplasmatischen Anatomie fixierter Gewebe. *Flora, Jena* **132**, 234–52.

DREW, M. C., CHAMEL, A., GARREC, J.-P., and FOURCY, A. (1980). Cortical air spaces (aerenchyma) in roots of corn subjected to oxygen stress. Structure and influence on uptake and translocation of ^{86}rubidium ions. *Pl. Physiol.* **65**, 506–11.

DRYSDALE, F. R. and BARBOUR, M. G. (1975). Response of the marine angiosperm *Phyllospadix torreyi* to certain environmental variables: a preliminary study. *Aquat. Bot.* **1**, 93–106.

DUCHARTRE, P. (1872). Quelques observations sur les caractères anatomiques des *Zostera* et *Cymodocea* à propos d'une plante trouvée près de Montpellier. *Bull. Soc. bot. Fr.* **19**, 289–302.

DUCKER, S. C., FOORD, N. J., and KNOX, R. B. (1977). Biology of Australian seagrasses: the genus *Amphibolis* C. Agardh (Cymodoceaceae). *Aust. J. Bot.* **25**, 67–95.

—— and KNOX, R. B. (1976). Submarine pollination in seagrasses. *Nature, Lond.* **263**, 705–6.

—— —— (1978). Alleloparasitism between a seagrass and algae. *Naturwissenschaften* **65**, 391–2.

—— PETTITT, J. M., and KNOX, R. B. (1978). Biology of Australian seagrasses: pollen development and submarine pollination in *Amphibolis antarctica* and *Thalassodendron ciliatum* (Cymodoceaceae). *Aust. J. Bot.* **26**, 265–85.

DUDLEY, W. R. (1893). The genus *Phyllospadix. Wilder Quarter-Century Book*, pp. 403–20. Comstock, Ithaca, New York.

—— (1894). *Phyllospadix*, its systematic characters and distribution. *Zoe* **4**, 381–5.

DUTAILLY, G. (1875). Observations sur l'*Aponogeton distachyum. Ass. fr. Av. Sci. Nantes* 707–24.

DUVAL-JOUVE, J. (1873*a*). Particularités des *Zostera marina* L. et *nana* Roth. *Bull. Soc. bot. Fr.* **20**, 81–91.

—— (1873*b*). Diaphragmes vasculifères des Monocotylédones aquatiques. *Mém. Acad. Sci. Montpellier* **8**, 157–76.

EAMES, A. J. (1961). *Morphology of angiosperms.* McGraw-Hill, New York.

—— and MACDANIELS, L. H. (1951). *An introduction to plant anatomy* (2nd edn). McGraw-Hill, New York.

EBER, E. (1934). Karpellbau und Plazentationsverhältnisse in der Reihe der Helobiae. Mit einem Anhang über die verwandtschaftlichen Beziehungen zwischen Ranales und Helobiae. *Flora, Jena* **127**, 273–330.

ECKARDT, T. (1964). Monocotyledonae. 1. Reihe. Helobiae. In *Syllabus der Pflanzenfamilien* edn. 12 (ed. A. Engler) Vol. 2, pp. 499–512. Borntraeger, Berlin.

EDGECUMBE, D. J. (1980). Some preliminary observations on the submerged aquatic *Zostera capensis* Setchell. *Jl. S. Afr. Bot.* **46**, 53–66.

EICHLER, A. (1875). *Blüthendiagramme* I. Wilhelm Engelmann, Leipzig.

ENGLER, A. (1879). Notiz über die Befruchtung von *Zostera marina* und das Wachsthum derselben. *Bot. Ztg.* **37**, 654–5.

—— (1887). Beiträge zur Kenntnis der Aponogetonaceae. *Bot. Jb.* **8**, 261–74.

—— (1889*a*). Aponogetonaceae. In *Die natürlichen Pflanzenfamilien* II (ed. A. Engler and K. Prantl) Vol. 1, pp. 218–22.

—— (1889*b*). Triuridaceae. In *Die natürlichen Pflanzenfamilien* II (ed. A. Engler and K. Prantl) Vol. 1, pp. 235–8.

—— (1904). *Syllabus der Pflanzenfamilien,* Edn. 4. Wilhelm Engelmann, Leipzig.

—— (1909). Eine bisher in Afrika nicht nachgewiesene Pflanzenfamilie, Triuridaceae. *Bot. Jb.* **43**, 303–7.

ERDTMAN, G. (1943). *An introduction to pollen analysis.* Chronica Botanica, Waltham, Mass.

—— (1952). *Pollen morphology and plant taxonomy. Angiosperms.* Chronica Botanica, Waltham, Mass.

ERNST, A. (1872). Über Stufengang und Entwickelung der Blätter von *Hydrocleis nymphoides* Buchenau (*Limnocharis humboldtii* C. L. Richard). *Bot. Ztg.* **30**, 518–20.

ERNST-SCHWARZENBACH, M. (1945). Zur Blütenbiologie einiger Hydrocharitaceen. *Ber. schweiz. bot. Ges.* **55**, 33–69.

—— (1953). Zur Kompatibilität von Art- und Gattungs-Bastardierung bei Hydrocharitaceen. *Öst. bot. Zeit.* **100**, 403–23.

—— (1956). Kleistogamie und Antherenbau in der Hydrocharitaceen-Gattung *Ottelia. Phytomorphology* **6**, 296–311.

ESAU, K. (1977). *Anatomy of seed plants,* 2nd edn. John Wiley, New York.

ESENBECK, E. (1914). Beiträge zur Biologie der Gattungen *Potamogeton* und *Scirpus. Flora, Jena* **107**, 151–212.

EYJÓLFSSON, R. (1970). Isolation and structure determination of triglochinin, a new cyanogenic glucoside from *Triglochin maritimum. Phytochemistry* **9**, 845–51.

FALKENBERG, P. (1876). *Vergleichende Untersuchungen über den Bau der Vegetationsorgane der Monocotyledonen.* Stuttgart, 202 pp.

FASSETT, N. C. (1955). *Echinodorus* in the American tropics. *Rhodora* **57**, 133–56; 174–88; 202–12.

—— (1957). *A manual of aquatic plants.* University of Wisconsin Press, Madison.

FELDMANN, J. (1936). Les monocotylédones marines de la Guadeloupe. *Bull. Soc. bot. Fr.* **83**, 604–13.

—— (1938). Sur la répartition du *Diplanthera wrightii* Aschers. sur la côte occidentale d'Afrique. *Bull. Soc. Hist. nat. Afr. nord* **29**, 107–12.

FELGER, R. S. and MOSER, M. B. (1973). Eelgrass (*Zostera marina* L.) in the Gulf of California: discovery of its nutritional value by the Seri Indians. *Science, N.Y.* **181**, 355–6.

—— —— and MOSER, E. W. (1980). Seagrasses in Seri Indian culture. In *Handbook of Seagrass Biology* (ed. R. C. Phillips and C. P. McRoy). Garland STPM Press, New York.

FERGUSON-WOOD, E. J. (1959). Some East Australian sea-grass communities. *Proc. Linn. Soc. N.S.W.* **84**, 218–26.

FERNALD, M. L. (1932). The linear-leaved North American species of *Potamogeton*, Section Axillares. *Mem. Amer. Acad. Arts Sci.* **17** (1), 183 pp. (Also *Mem. Gray Herb. Harvard Univ.* **3**.)

FIEBRIG, A. (1922). Fanerógamas saprófitas: *Triuris mycoides* sp. nov. *Revta. Jard. bot. Paraguay* **1**, 164–5.

FISCHER, G. (1901–4). Beiträge zur Kenntnis der bayerischen Potamogetoneen. *Mitt. bayer. bot. Ges.* Nos. 19–21, 27, 31, 32, 37.

—— (1907). Die bayerischen Potamogetonen und Zannichellien. *Ber. bayer. bot. Ges.* **11**, 20–162.

FISHER, J. B. (1974). Axillary and dichotomous branching in the palm *Chamaedorea*. *Am. J. Bot.* **61**, 1046–56.

—— (1976). Development of dichotomous branching and axillary buds in *Strelitzia* (Monocotyledoneae). *Can. J. Bot.* **54**, 578–92.

FLAHAULT, C. (1908). *Zostera* L. In *Lebensgeschichte der Blütenpflanzen Mitteleuropas.* (ed. O. von Kirchner, E. Loew, and C. Schröter) Vol. 1(1), pp. 516–29. Stuttgart.

FLEET, D. S. VAN (1942). The development and distribution of the endodermis and an associated oxidase system in monocotyledonous plants. *Am. J. Bot.* **29**, 1–15.

FONTELL, C. W. (1908–9). Beiträge zur Kenntnis des anatomischen Baues der *Potamogeton*-Arten. *Öfvers. finska Vet-Soc. Forh.* **51** (14), 91 pp.

FORTI, A. (1927). La propagazione dell' *Halophila stipulacea* (Forsk.) Aschers. anche del Mediterranea. *Nuovo G. bot. ital.* **34**, 714–16.

FRANÇOIS, L. (1908). Recherches sur les plantes aquatiques. *Annls Sci. nat., Bot.* sér. 9, **7**, 25–110.

FREIDENFELT, T. (1904). Der anatomische Bau der Wurzel in seinem Zusammenhange mit dem Wassergehalt des Bodens. *Bibl. bot.* **12** Heft. 61, 118 pp.

FRYER, A. (1886–97). Notes on pondweeds. *J. Bot., Lond.* **24**, 337–8, 378–80 (1886); **25**, 50–2, 113–15, 163–5, 306–10 (1887); **26**, 273–8, 297–9 (1888); **27**, 8–10, 33–6, 65–7 (1889); **28**, 137–9, 173–9, 225–7, 321–6 (1890); **30**, 33–7 (1892); **31**, 353–5 (1893); **32**, 97–100, 337–40 (1894); **33**, 1–3 (1895); **34**, 1–3 (1896); **35**, 355–6, 446–7 (1897).

—— and BENNETT, A. (1898–1915). *The potamogetons (pond weeds) of the British Isles.* L. Reeve, London.

GAMERRO, J. C. (1968). Observaciones sobre la biología floral y morfología de la Potamogetonácea *Ruppia cirrhosa* (Petag.) Grande (= *R. spiralis* L. ex Dum.). *Darwiniana* **14**, 575–608.

GARDNER, G. (1845). Description of *Peltophyllum*, a new genus of plants allied to *Triuris* of Miers, with remarks on their affinities. *Trans. Linn. Soc. Lond.* **19**, 155–60.

GARDNER, R. O. (1976). Binucleate pollen in *Triglochin* L. *N. Z. Jl. Bot.* **14**, 115–16.

GÉNEAU DE LAMARLIÈRE, L. (1906). Sur les membranes cutinisées des plantes aquatiques. *Rev. gén. Bot.* **18**, 289–95.

GÉRARD, R. (1881). Recherches sur le passage de la racine à la tige. *Annls Sci. nat., Bot.,* sér. 6, **11**, 279–430.

GESSNER, F. (1968). Die Zellwand mariner Phanerogamen. *Mar. biol.* (Germ.) **1**, 191–200.

GIBBS, R. D. (1974). *Chemotaxonomy of flowering plants,* Vols. I–IV. McGill-Queen's University Press, Montreal and London.

GIBSON, R. J. H. (1905). The axillary scales of aquatic monocotyledons. *J. Linn. Soc. (Bot.)* **37**, 228–37.

GIESEN, H. (1938). Triuridaceae. In *Das Pflanzenreich* IV (ed. A. Engler) Vol. 18.

GLÜCK, H. (1901). Die Stipulargebilde der Monokotyledonen. *Verh. naturhist. med. Ver., Heidelberg* (N.S.) **7**, 1–96.

—— (1905). *Biologische und morphologische Untersuchungen über Wasser- und Sumpfgewächse. Teil 1. Die Lebensgeschichte der europäischen Alismaceen.* G. Fischer, Jena.

—— (1924). *Biologische und morphologische Untersuchungen über Wasser- und Sumpfgewächse. Teil 4: Untergetauchte und Schwimmblattflora.* G. Fischer, Jena.

GODFREY, R. K. and WOOTEN, J. W. (1979). *Aquatic and wetland plants of Southeastern United States. Monocotyledons.* University of Georgia Press, Athens, Georgia.

GOEBEL, K. (1880). Beiträge zur Morphologie und Physiologie des Blattes (Schluss). *Bot. Ztg.* **38,** 833–45.

—— (1893). *Pflanzenbiologische Schilderungen,* 2/2. Marburg.

—— (1896). Ueber Jugendformen von Pflanzen und deren künstliche Wiederhervorrufung. *Sber. k. bayer. Akad. Wiss., München, math.-phys. Kl.* **26,** 447–97.

GOFFART, J. (1900). Quelques mots sur la structure et la fonction des organes de sudation chez les plantes terrestres et les plantes aquatiques. *Bull. Soc. bot. Belg.* **39,** 54–80.

GOVINDARAJALU, E. (1967). Further contributions to the anatomy of the Alismataceae: *Sagittaria guayanensis* H.B.K. ssp. *lappula* (D. Don) Bogin. *Proc. Indian Acad. Sci. B* **65,** 142–52.

GRAEBNER, P. and FLAHAULT, M. (1908). Reihe Helobiae. In *Lebensgeschichte der Blütenpflanzen Mitteleuropas* (ed. O. von Kirchner, E. Loew, and C. Schröter) Vol. 1 (1), pp. 394–584. Stuttgart.

GRAVES, A. H. (1908). The morphology of *Ruppia maritima. Trans. Connecticut Acad. Arts Sci.* **14,** 59–170.

GRAVIS, A. (1934). Théorie des traces foliaires. *Mém. Acad. roy. Belg., Cl. Sci.,* sér. 2, *12,* 50 pp. (Also in *Recueil de quelques travaux d'anatomie végétale exécutés à Liége de 1929 à 1935.* Bruxelles 1936.)

—— MONOYER, A., and FRITSCHÉ, E. (1943). Observations anatomiques sur les embryons et les plantules. *Lejeunia Mém.* **3,** 180 pp.

GREEN, P. S. and SOLBRIG, O. T. (1966). *Sciaphila dolichostyla* (Triuridaceae). *J. Arnold Arbor.* **47,** 266–9.

GRENIER, C. (1860). Recherches sur le *Posidonia caulini* Konig. *Bull. Soc. bot. Fr.* **7,** 362–7; 419–26; 448–52.

GRIER, N. M. (1920). Propagation of *Elodea* and *Ceratophyllum. Am. Botanist* **26,** 80–4.

GRÖNLAND, J. (1851). Beitrag zur Kenntniss der *Zostera marina* L. *Bot. Ztg.* **9,** 185–92.

GROOM, P. (1895). On a new saprophytic Monocotyledon. *Ann. Bot., Lond.* **9,** 45–58.

GUILLARMOD, A. J. and MARAIS, W. (1972). A new species of *Aponogeton* (Aponogetonaceae). *Kew Bull.* **27,** 563–5.

GUILLAUD, A. J. (1878). Recherches sur l'anatomie comparée et le développement des tissus de la tige des Monocotylédones. *Annls Sci. nat., Bot.* sér. 6, **5,** 5–176.

GUNNING, B. E. S. and PATE, J. S. (1969). "Transfer cells". Plant cells with wall ingrowths, specialized in relation to short distance transport of solutes—their occurrence, structure, and development. *Protoplasma* **68,** 107–33.

GUPTA, B. L. (1934). A contribution to the life history of *Potamogeton crispus* L. *J. Indian bot. Soc.* **13,** 51–65.

GUPTA, S. C., RAJESWARI, V. M. and AGARWAL, S. (1975). Stomatal complex in *Butomus umbellatus. Phytomorphology* **25,** 305–9.

GUTTENBERG, H. VON (1968). Die Wurzel der Hydro- und Hygrophyten. In *Der primäre Bau der Angiospermenwurzel.* [*Handbuch der Pflanzenanatomie,* Vol. VIII, 5] pp. 260–303 Gebrüder Borntraeger, Berlin.

—— and JAKUSZEIT, C. (1957). Die Entwicklung des Embryos und der Primärwurzel von *Galtonia candicans* Decne., nebst Untersuchungen über die Differenzierung des Wurzelvegetationspunktes von *Alisma plantago* L. *Bot. Stud.* **7,** 91–126.

HABERLANDT, G. (1887). Zur Kenntniss des Spaltöffnungsapparates. *Flora, Jena* **70,** 97–110.

HACCIUS, B. (1952). Über die Blattstellung einiger Hydrocharitaceen-Embryonen. *Planta* **40,** 333–45.

HAGSTRÖM, J. O. (1906). Holstia splendens. *Geol. Fören. Stockholm Förhandl.* **28,** 90–2.

—— (1908). New Potamogetons. *Bot. Notiser* 97–108.

—— (1911). Three species of *Ruppia. Bot. Notiser* 137–44.

—— (1916). Critical researches on the Potamogetons. *Kgl. sevenska Vet.-Akad. Handl.* (N.S.) **55** (5), 281 pp.

HALLÉ, F., OLDEMAN, R. A. A., and TOMLINSON, P. B. (1978). *Tropical trees and forests. An architectural analysis.* Springer Verlag, Berlin.

HARADA, I. (1942). Chromosomenzahlen bei der Gattung *Potamogeton. Mediz. Biol.* **1.**

HARBORNE, J. B. and WILLIAMS, C. A. (1976). Occurrence of sulphonated flavones and caffeic acid esters in members of the Fluviales. *Biochem. Syst. Ecol.* **4**, 37–41.

HARTOG, C. DEN (1957a). Alismataceae. *Flora Malesiana*, ser. 1, **5**, 317–34.

—— (1957b). Hydrocharitaceae. *Flora Malesiana*, ser. 1, **5**, 381–413.

—— (1960). New sea grasses from Pacific Central America. *Pacific Nat.* **1** (15), 1–8.

—— (1964). An approach to the taxonomy of the sea-grass genus *Halodule* Endl. (Potamogetonaceae). *Blumea* **12**, 289–312.

—— (1965). Some notes on the distribution of *Plasmodiophora diplantherae*, a parasitic fungus on species of *Halodule*. *Persoonia* **4**, 15–18.

—— (1970a). *The sea-grasses of the world. Verhandl. Kon. Ned. Akad. Wetensch. Nat.* **59** (1). North-Holland, Amsterdam.

—— (1970b). *Halodule emarginata* nov. sp., a new sea-grass from Brazil (Potamogetonaceae). *Blumea* **18**, 65–6.

—— (1975). *Althenia filiformis* Petit (Potamogetonaceae) in Turkey. *Aquat. Bot.* **1**, 75.

HARVEY, W. H. (1842). Account of a new genus of the natural order of Hydrocharideae from Southern Africa. *J. Bot., Lond.* **4**, 230–1.

HASLAM, S. M. (1978). *River plants. The macrophytic vegetation of watercourses.* Cambridge University Press.

HASMAN, M. and INANÇ, N. (1957). Investigations on the anatomical structure of certain submerged, floating and amphibious hydrophytes. *Istanb. Univ. Fen. Fak. Mecm.* B **22**, 137–53.

HAUMANN-MERCK, (1912). Cited in den Hartog (1957b).

HAYNES, R. R. (1974). A revision of North American *Potamogeton* subsection Pusilli (Potamogetonaceae). *Rhodora* **76**, 564–649.

—— (1977). The Najadaceae in the southeastern United States. *J. Arnold Arbor.* **58**, 161–70.

—— (1978). The Potamogetonaceae in the southeastern United States. *J. Arnold Arbor.* **59**, 170–91.

—— (1979). Revision of North and Central American *Najas* (Najadaceae). *Sida* **8**, 34–56.

HEGELMAIER, F. (1870). Ueber die Entwicklung der Blüthentheile von *Potamogeton*. *Bot. Ztg.* **28**, 283–9, 297–305, 313–19.

HEGNAUER, R. (1963). *Chemotaxonomie der Pflanzen.* II. *Monocotyledoneae.* Birkhäuser, Basel.

—— and RUIJGROK, H. W. L. (1971). *Lilaea scilloides* und *Juncus bulbosus,* zwei neue cyanogene Pflanzen. *Phytochemistry* **10**, 2121–4.

HELM, J. (1934). Über die Anlegung des Karpells von *Potamogeton trichoides* Cham. et Schlecht. *Planta* **22**, 443–4.

HENSLOW, G. (1911). The origin of monocotyledons from dicotyledons, through self-adaptation to a moist or aquatic habit. *Ann. Bot. Lond.* **25**, 717–44.

HERRIG, F. (1914). Beiträge zur Kenntnis der Blattentwicklung einiger phanerogamen Pflanzen. *Flora, Jena* **107**, 327–50.

HESSE, H. (1904). *Beiträge zur Morphologie und Biologie der Wurzelhaare.* Diss. Jena. Gebrüder Georgi, Greussen.

HIERONYMUS, J. (1882). Monografía de *Lilaea subulata. Act. Acad. Nac. Ciencias Córdoba.* **4**, 1–52.

HILDEBRAND, F. (1885). Über *Heteranthera zosteraefolia. Bot. Jber.* **6**, 137–45.

HILL, T. G. (1900). The structure and development of *Triglochin maritimum* L. *Ann. Bot., Lond.* **14**, 83–107.

HISINGER, E. (1887). Recherches sur les tubercules du *Ruppia rostellata* et du *Zanichellia polycarpa* provoqués par le *Tetramyxa parasitica. Medd. Soc. Fauna Flora fenn.* **14**, 53–62.

HOCHREUTINER, G. (1896). Études sur les phanérogames aquatiques du Rhône et du port de Genève. *Rev. gén. Bot.* **8**, 90–110; 158–67; 188–200; 249–65.

HOFMEISTER, W. (1852). Zur Entwickelungsgeschichte der *Zostera. Bot. Ztg.* **10**, 121–31; 137–49; 157–8.

—— (1861). Neue Beiträge zur Kenntniss der Embryobildung der Phanerogamen. 2. Monokotyledonen. *Abh. köngl. Sächs. Ges. Wiss.* **7**, 629–760.

HOLFERTY, G. M. (1901). Ovule and embryo of *Potamogeton natans. Bot. Gaz.* **31**, 339–46.

HOLM, T. (1885). Recherches anatomiques et morphologiques sur deux monocotylédones submergées. (*Halophila baillonii* Asch. et *Elodea densa* Casp.). *Bih. Kgl. Svenska Vetensk. Akad. Handl.* **9** (13), 24 pp.

HORN AF RANTZEN, H. (1961). Notes on the African species of *Triglochin*. *Svensk bot. Tidskr.* **55**, 81–117.

HOWARD-WILLIAMS, C., DAVIES, B. R., and CROSS, R. H. M. (1978). The influence of periphyton on the surface structure of a *Potamogeton pectinatus* L. leaf (an hypothesis). *Aquat. Bot.* **5**, 87–92.

HULBARY, R. L. (1944). The influence of air spaces on the three-dimensional shapes of cells in *Elodea* stems and a comparison with pith cells of *Ailanthus*. *Am. J. Bot.* **31**, 561–80.

HUTCHINSON, J. (1959). *The families of flowering plants II. Monocotyledons*, 2nd edn. Clarendon Press, Oxford.

IRMISCH, T. (1851). Ueber die Inflorescenzen der deutschen Potameen. *Flora, Jena* **34**, 81–93.

—— (1858a). Ueber das Vorkommen von schuppen- oder haarförmigen Gebilden innerhalb der Blattscheiden bei monokotylischen Gewächsen. *Bot. Ztg.* **16**, 177–9.

—— (1858b). Ueber einige Arten aus der natürlichen Pflanzenfamilie der Potameen. *Abh. naturwiss. Ver. Prov. Sachsen Thüringen Halle* **2**, 1–56.

—— (1859). Zur Naturgeschichte des *Potamogeton densus* L. *Flora* **42**, 129–39.

—— (1878). Bemerkungen über die Keimpflanzen einiger *Potamogeton*-Arten. *Z. ges. Natur.* **58**, 203–12.

ISAAC, F. M. (1969). Floral structure and germination in *Cymodocea ciliata*. *Phytomorphology* **19**, 44–51.

ISLAM, A. S. (1950). A contribution to the life history of *Ottelia alismoides*. *J. Indian bot. Soc.* **29**, 79–91.

JACOBS, D. L. (1946). Shoot segmentation in *Anacharis densa*. *Am. Midl. Nat.* **35**, 283–6.

JACOBS, S. W. L. and WILLIAMS, A. (1980). Notes on the genus *Zostera* s. lat. in New South Wales. *Telopea* **1**, 451–5.

JADIN, F. (1888). *Les organes sécréteurs des végétaux et la matière médicale.* Thèse, Montpellier.

JAGELS, R. (1973). Studies of a marine grass, *Thalassia testudinum*. I. Ultrastructure of the osmoregulatory leaf cells. *Am. J. Bot.* **60**, 1003–9.

JANCZEWSKI, E. DE (1874). Recherches sur l'accroissement terminal des racines dans les phanérogames. *Annls Sci. nat., Bot.*, sér. 5, **20**, 162–201.

JANSE, J. M. (1897). Les endophytes radicaux de quelques plantes javanaises. *Ann. Jard. Bot. Buitenz.* **14**, 53–201.

JENSEN, H. (1889). Zostera's spiring. *Bot. Tidsskr.* **17**, 162–9. [French summary.]

JENSEN, P. B. (1959). Untersuchungen über Determination und Differenzierung. 6. Über den Aufbau des Zellwandmusters des Blattes von *Helodea densa*. *Biol. Medd. Kgl. Danske Vidensk. Selsk.* **23**(10), 1–33.

JÖNSSON, B. (1881). Ytterligare bidrag till kännedomen om Angiospermernas embryosäckutveckling. *Bot. Notiser* (1881) 169–87.

—— (1883–4). Om befruktningen hos slägtet *Najas* samt hos *Callitriche autumnalis*. *Lunds Univ. Arsk.* **20**, 1–26.

JOHOW, F. (1889). Die chlorophyllfreien Humuspflanzen nach ihre biologischen und anatomischentwickelungsgeschichtlichen Verhältnissen. *Jb. wiss. Bot.* **20**, 475–525.

KADEJ, A. R. (1966). Organization and development of apical root meristem in *Elodea canadensis* (Rich.) Casp. and *Elodea densa* (Planck) Casp. *Acta Soc. bot. pol.* **35**, 143–58.

KAPLAN, D. R. (1970). Comparative foliar histogenesis in *Acorus calamus* and its bearing on the phyllode theory of monocotyledonous leaves. *Am. J. Bot.* **57**, 331–61.

KATTEIN, A. (1897). Der morphologische Werth des Centralcylinders der Wurzel. *Bot. Zbl.* **72**, 55–61; 91–7; 129–39.

KAUL, R. B. (1967a). Development and vasculature of the flowers of *Lophotocarpus calycinus* and *Sagittaria latifolia* (Alismaceae). *Am. J. Bot.* **54**, 914–20.

—— (1967b). Ontogeny and anatomy of the flower of *Limnocharis flava* (Butomaceae). *Am. J. Bot.* **54**, 1223–30.

—— (1968a). Floral development and vasculature in *Hydrocleis nymphoides* (Butomaceae). *Am. J. Bot.* **55**, 236–42.

—— (1968b). Floral morphology and phylogeny in the Hydrocharitaceae. *Phytomorphology* **18**, 13–35.

—— (1969). Morphology and development of the flowers of *Bootia cordata*, *Ottelia alismoides* and their synthetic hybrid (Hydrocharitaceae). *Am. J. Bot.* **56**, 951–9.

—— (1970). Evolution and adaptation of inflorescences in the Hydrocharitaceae. *Am. J. Bot.* **57**, 708–15.

—— (1972). Adaptive leaf architecture in emergent and floating *Sparganium*. *Am. J. Bot.* **59**, 270–8.

—— (1973). Development of foliar diaphragms in *Sparganium eurycarpum*. *Am. J. Bot.* **60**, 944–9.

—— (1976*a*). Conduplicate and specialized carpels in the Alismatales. *Am. J. Bot.* **63**, 175–82.

—— (1976*b*). Anatomical observations on floating leaves. *Aquat. Bot.* **2**, 215–34.

—— (1978). Morphology of germination and establishment of aquatic seedlings in Alismataceae and Hydrocharitaceae. *Aquat. Bot.* **5**, 139–47.

—— (1979). Inflorescence architecture and flower sex ratios in *Sagittaria brevirostra* (Alismataceae). *Am. J. Bot.* **66**, 1062–6.

KAUSHIK, S. B. (1940). Vascular anatomy of the pistillate flowers of *Enhalus acoroides* (L.) Steud. *Curr. Sci.* **2**, 182–4.

KAY, Q. O. N. (1971). Floral structure in the marine angiosperms *Cymodocea serrulata* and *Thalassodendron ciliatum* (*Cymodocea ciliata*). *Bot. J. Linn. Soc.* **64**, 423–9.

KERNER, MARILAUN, A. VON (1895). *The natural history of plants*, Vol. II (translated and edited by F. W. Oliver). Blackie, London.

KIRCHNER, O. VON, LOEW, E., and SCHRÖTER, C. (1908). *Lebensgeschichte der Blütenpflanzen Mitteleuropas*, Vol. 1, (1) Stuttgart.

KLEKOWSKI, E. J. and BEAL, E. O. (1965). A study of variation in the *Potamogeton capillaceus-diversifolius* complex (Potamogetonaceae). *Brittonia* **17**, 175–81.

KLINGE, J. (1880). Über *Sagittaria sagittifolia* L. *Sber. Dorpater Naturf. Ges.*, 32 pp.

KNY, L. (1878). Über korallenartig verzweigte Membranverdickungen an der Basis der Wurzelhaare von *Stratiotes aloides*. *Verh. Bot. Ver. Prov. Brandenburg* **20**, 48–50.

KRAUSE, K. and ENGLER, A. (1906). Aponogetonaceae. In *Das Pflanzenreich* (ed. A. Engler) Vol. IV (13), Heft 24, 23 pp.

KRISTEN, U. (1969). Licht- und elektronenmikroskopische Untersuchungen an den Hydropoten von *Nuphar lutea*, *Nymphoides peltata*, *Sagittaria macrophylla* und *Salvinia auriculata*. *Flora A* **159**, 536–58.

KROEMER, K. (1903). Wurzelhaut, Hypodermis und Endodermis der Angiospermenwurzel. *Bibl. Bot.* **12**, Heft 59.

KUDRYASHOV, L. V. (1964*a*). The origin of monocotyledony (as illustrated by the example of Helobiae). *Bot. Zh. SSSR* **49**, 473–86. [Russian; English summary.]

—— (1964*b*). The origin of monocotyledony and the importance of embryological features for constructing the system Helobiae. *Vtoroe mosk. sovesh. Filogen. Rast.* (2nd Moscow Conf. on Plant Phylogeny) 29–31. [Russian.] [Seen in: *Biol. Abstr.* **47** (1966) No. 73995.]

KULESZANKA, J. (1934). Rozwój ziarn pyłu u *Potamogeton fluitans*. (Die Entwicklung der Pollenkörner bei *Potamogeton fluitans*). *Acta Soc. bot. pol.* **11**, 457–62. [German summary.]

KUNTH, C. S. (1841). *Enumeratio plantarum*, Vol. III, pp. 115–17. Tubingen, Stuttgart.

KUO, J. (1978). Morphology, anatomy and histochemistry of the Australian seagrasses of the genus *Posidonia* König (Posidoniaceae). I. Leaf blade and leaf sheath of *Posidonia australis* Hook. f. *Aquat. Bot.* **5**, 171–90.

—— and CAMBRIDGE, M. L. (1978). Morphology, anatomy and histochemistry of the Australian seagrasses of the genus *Posidonia* König (Posidoniaceae). II. Rhizome and root of *Posidonia australis* Hook. f. *Aquat. Bot.* **5**, 191–206.

LAESSLE, A. F. (1953). The use of root characteristics to separate various ribbon-leaved species of *Sagittaria* from species of *Vallisneria*. *Turtox News* **31**, 224–25.

LAGERHEIM, G. (1913). Om 'ouvirandrano' och växternas nätblad. *Fauna och Flora*, 34–44. [See *Bot. Zbl.* **125**, 564–65, 1914.]

LAKSHMANAN, K. K. (1961). Embryological studies in the Hydrocharitaceae. 1. *Blyxa octandra* Planch. *J. Madras Univ. B* **31** (2), 133–42.

—— (1965). Note on the endosperm formation in *Zannichellia palustris* L. *Phyton (Argentina)* **22**, 13–14.

—— (1968). Preliminary note on the embryology of *Diplanthera uninervis* Aschers. *Curr. Sci.* **37**, 534–5.

LANCE-NOUGARÈDE, A. and LOISEAU, J.-E. (1960). Sur la structure et le fonctionnement du méristème végétatif de quelques angiospermes aquatiques ou semi-aquatiques dépourvues de moelle. *C.r. hebd. Séanc. Acad. Sci., Paris* **250**, 4438–40.

LANESSAN, J. L. DE (1875). Organogénie de la fleur et du fruit des *Z. marina* L. et *Z. nana* Roth. Rapports des *Zostera* avec les Graminées. *Ass. fr. Av. Sci., Nantes* 690–707.

LARSEN, K. (1963). Studies in the flora of Thailand 14. Cytological studies in vascular plants of Thailand. *Dansk bot. Arkiv* **20**, 211–75.

LEAVITT, R. G. (1904). Trichomes of the root in vascular cryptogams and angiosperms. *Proc. Boston Soc. Nat. Hist.* **31**, 273–313.

LE BLANC, (1912). Sur les diaphragmes des canaux aérifères des plantes. *Rev. gén. Bot.* **24**, 233–43.

LEBLOIS, A. (1887). Recherches sur l'origine et le développement des canaux sécréteurs et des poches sécrétrices. *Annls Sci. nat., Bot.*, sér. 7, **6**, 247–330.

LEE, C. L. and HSIN-YING, C. (1958). Morphological studies of *Sagittaria sinensis* I. The anatomy of roots. *Acta bot. sin.* **7**, 71–86. [Chinese. English summary.]

LEINFELLNER, W. (1973). Zur Lage des wahren Karpellrandes. *Öst. bot. Z.* **121**, 285–301.

LEINS, P. and STADLER, P. (1973). Entwicklungsgeschichtliche Untersuchungen am Androeceum der Alismatales. *Öst. bot. Z.* **121**, 51–63.

LEITGEB, H. (1857). Die Luftwege der Pflanzen. *S. B. Kais. Akad. Wiss. Wien* **18**, 334–63.

LEWIN, M. (1887). Bidrag till hjertbladets anatomi hos monokotyledonerna. *Bih. svenska Vet.-Akad. Handl.* **12** (3), 28 pp.

LIDFORSS, B. (1898). Ueber eigenartige Inhaltskörper bei *Potamogeton praelongus* Wulf. *Bot. Zbl.* **74**, 305–13; 337–43; 372–7.

LIEU, S. M. (1979*a*). Organogenesis in *Triglochin striata*. *Can. J. Bot.* **57**, 1418–38.

—— (1979*b*). Growth forms in the Alismatales. I. *Alimsa triviale* and species of *Sagittaria* with upright vegetative axes. *Can. J. Bot.* **57**, 2325–52.

—— (1979*c*). Growth forms in the Alismatales. II. Two rhizomatous species: *Sagittaria lancifolia* and *Butomus umbellatus*. *Can. J. Bot.* **57**, 2353–73.

LOHAMMAR, G. (1954). Bulbils in the inflorescences of *Butomus umbellatus*. *Svensk bot. Tidskr.* **48**, 485–8.

LORENZ, H. (1903). *Beiträge zur Kenntnis der Keimung den Winterknospen von* Hydrocharis morsus-ranae, Utricularia vulgaris *und* Myriophyllum verticillatum. Inaug. Diss. Kiel.

LÜPNITZ, D. (1969) [1970]. Histogenese und Anatomie von Primärwurzeln und sprossbürtigen Wurzeln einiger Potamogetonaceae L. *Beitr. Biol. Pfl.* **46**, 247–313.

LÜTTGE, U. (1964). Mikroautoradiographische Untersuchungen über die Funktion der Hydropoten von *Nymphaea*. *Protoplasma* **59**, 157–63.

—— and KRAPF, G. (1969). Die Ultrastruktur der *Nymphaea*-Hydropoten in Zusammenhang mit ihrer Funktion als Salz-transportierende Drüsen. *Cytobiologie* **1**, 121–31.

—— PALLAGHY, C. K., and WILLERT, K. VON (1971). Microautoradiographic investigations of sulfate uptake by glands and epidermal cells of water lily (*Nymphaea*) leaves with special reference to the effect of poly-L-lysine. *J. membrane Biol.* **4**, 395–407.

LUHAN, M. (1957) Das Verhalten der Rhizomgewebe einiger Wasser und Sumpfpflanzen bei Vitalfärbung. *Ber. dt. bot. Ges.* **70**, 361–70.

LUNDSTRÖM, A. N. (1888). Ueber farblose Oelplastiden und die biologische Bedeutung der Oeltropfen gewisser *Potamogeton*-Arten. *Bot. Zbl.* **35**, 177–81.

LUTHER, H. (1947). Morphologische und systematische Beobachtungen an Wasserphanerogamen. *Acta bot. fenn.* **40**, 1–28.

—— (1951). Verbreitung und Ökologie der höheren Wasserpflanzen in Brackwasser der Ekenäs-Gegend in Südfinnland. *Acta bot. fenn.* **50**, 1–370.

LYR, H. and STREITBERG, H. (1955). Die Verbreitung von Hydropoten in verschiedenen Verwandtschaftskreisen der Wasserpflanzen. *Wiss. Zeit. Martin-Luther Univ. Halle-Wittenberg Cl. math.-nat.* **4**, 471–84.

LY THI BA, CAVE, G., HENRY, M., and GUIGNARD, J.-L. (1978). Embryogénie des Potamogétonacées—étude en microscopie électronique à balayage de l'origine du cotylédon chez *Potamogeton lucens* L. *C. r. hebd. Séanc. Acad. Sci., Paris*, D **286**, 1351–3.

—— and GUIGNARD, J.-L. (1976). Embryogénie des Potamogétonacées. Développement de l'embryon chez *Potamogeton lucens* L. *C. r. hebd. Séanc. Acad. Sci. Paris*, D **283**, 151–3.

—— MESTRE, J.-C., and GUIGNARD, J. L. (1973). Embryogénie des Ruppiacées. Développement de l'embryon chez le *Ruppia maritima* L. *C. r. hebd. Séanc. Acad. Sci. Paris*, D **276**, 737–40.

McCANN, C. (1945). Notes on the genus *Ruppia* (Ruppiaceae). *J. Bombay nat. Hist. Soc.* **45**, 396–402.

—— (1978). A comparative field study of the Indian and New Zealand representatives of the genus *Ruppia* Linnaeus. *J. Bombay nat. Hist. Soc.* **75**, 600–10.

McCLURE, J. W. (1970). Secondary constituents of aquatic angiosperms. In *Phytochemical phylogeny* (ed. J. B. Harborne) pp. 238–68. Academic Press, London.

MAGNUS, P. (1869). Zur Morphologie der Gattung *Najas* L. *Bot. Ztg.* **27**, 769–73.

—— (1870). *Beiträge zur Kenntniss der Gattung Najas* L. Reimer, Berlin.

—— (1871). Anatomie der Meeresphanerogamen. *Sber. Ges. naturf. Freunde Berlin* 85–90. [See *Bot. Ztg.* **29**, 203–8, 215–16 (1871).]

—— (1872). Ueber Schlauchgefässe im Stamme von *Cymodocea nodosa, isoetifolia* und *manatorum* und Schlauchzellen in der Blatt-Epidermis dieser und anderer *Cymodocea*-Arten. *Sber. Ges. naturf. Freunde Berlin* 30–3. [See *Bot. Ztg.* **30**, 684–7 (1872).]

—— (1883). Ueber eine besondere geographische Varietät der *Najas graminea* Del. und deren Auftreten in England. *Ber. dt. bot. Ges.* **1**, 521–4.

—— (1889). Najadaceae. In *Die natürlichen Pflanzenfamilien* II (ed. A. Engler and K. Prantl) Vol. 1, pp. 214–18.

—— (1894). Ueber die Gattung *Najas. Ber. dt. bot. Ges.* **12**, 214–24.

MAHESHWARI, P. (1962). The overpowering role of morphology in taxonomy. *Bull. Bot. Surv. India* **4**, 85–94.

MAJUMDAR, G. P. (1938). A preliminary note on polystely in *Limnanthemum cristatum* and *Ottelia alismoides. Curr. Sci.* **6**, 383–5.

MALME, G. O. (1896). Ueber *Triuris lutea* (Gardn.) Benth. et Hook. *Bih. K. svenska Vet.-Akad. Handl.* III, **21** (14), 1–16.

MARIE-VICTORIN (Frère) (1931). L'*Anacharis canadensis.* Histoire et solution d'un imbroglio taxinomique. *Contrib. Lab. Bot. Univ. Montréal.* No. **18**, 1–43.

MARKGRAF, F. (1936). Blütenbau und Verwandtschaft bei den einfachsten Helobiae. *Ber. dt. bot. Ges.* **54**, 191–229.

MAROTI, M. (1950). Die Entwicklung der Wurzel von *Stratiotes aloides* L. *Acta biol. Acad. sci. hung.* **1**, 363–70.

MASON, H. L. (1957). *A flora of the marshes of California.* University of California Press, Berkeley.

MASON, R. (1967). The species of *Ruppia* in New Zealand. *N. Z. Jl. Bot.* **5**, 519–31.

MATSUBARA, M. (1931). Versuche über die Entwicklungserregung der Winterknospen von *Hydrocharis morsus-ranae* L. *Planta* **13**, 695–715.

MATZKE, E. B. (1948). The three-dimensional shape of epidermal cells of the apical meristem of *Anacharis densa* (*Elodea*). *Am. J. Bot.* **35**, 323–32.

—— (1949). Three-dimensional shape changes during cell division in the epidermis of the apical meristem of *Anacharis densa* (*Elodea*). *Am. J. Bot.* **36**, 584–95.

—— and DUFFY, R. M. (1956). Progressive three-dimensional shape changes of dividing cells within the apical meristem of *Anacharis densa. Am. J. Bot.* **43**, 205–33.

MAYER, F. L. S. JR (1971). Influence of salinity on fruit size in *Ruppia maritima* L. *Proc. Utah Acad. Sci. Arts Lett.* **46** (2), 140–3.

MAYR, F. (1915). Hydropoten an Wasser- und Sumpfpflanzen. *Beih. bot. Zbl.* **32** (1), 278–371.

—— (1943). Beiträge zur Anatomie der Alismataceen. Die Blattanatomie von *Caldesia parnassifolia* (Bassi) Parl. *Beih. bot. Zbl.* **62A**, 61–77.

MER, E. (1882). De quelques nouveaux exemples relatifs à l'influence de l'hérédité et du milieu sur la forme et la structure des plantes. *Bull. Soc. bot. Fr.* **29**, 81–7.

—— (1886). De la manière dont doit être interprétée l'influence du milieu sur la structure des plantes amphibies. *Bull. Soc. bot. Fr.* **33**, 169–78.

MESSERI, E. (1925). Ricerche sullo sviluppo del sistema vascolare in alcune Monocotiledoni. *Nuovo G. bot. ital.* (n.s.) **32**, 317–62.

METCALFE, C. R. (1960). *Gramineae.* Vol. I. In *Anatomy of the monocotyledons* (ed. C. R. Metcalfe). Clarendon Press, Oxford.

—— (1963). Comparative anatomy as a modern botanical discipline. In *Advances in botanical research* (ed. R. D. Preston), vol. 1, pp. 101–147. Academic Press, London.

—— (1967). Distribution of latex in the plant kingdom. *Econ. Bot.* **21**, 115–27.

—— (1971). *Cyperaceae*. Vol. V. In *Anatomy of the monocotyledons* (ed. C. R. Metcalfe). Clarendon Press, Oxford.

METSÄVAINIO, K. (1931). Untersuchungen über das Wurzelsystem der Moorpflanzen. *Ann. bot. Soc. zool. bot. fenn. Vanamo* **1** (1), 1–422.

MEYER, F. J. (1932*a*). Beiträge zur Anatomie der Alismataceen. I. Die Blattanatomie von *Echinodorus. Beih. bot. Zbl.* **49** (1), 309–68.

—— (1932*b*). Beiträge zur Anatomie der Alismataceen. II. Die Blattanatomie von *Rautanenia schinzii* Buchenau. *Beih. bot. Zbl.* **50** (1), 54–63.

—— (1932*c*). Anatomie und systematische Stellung der *Burnatia enneandra* Micheli. *Beih. bot. Zbl.* **49** (Suppl.), 272–91.

—— (1932*d*). Die Verwandtschaftsbeziehungen der Alismataceen zu den Ranales im Lichte der Anatomie. *Bot. Jb.* **56**, 53–9.

—— (1934). Beiträge zur Anatomie der Alismataceen. III und IV. Die Blattanatomie von *Lophotocarpus* und *Limnophyton. Beih. bot. Zbl.* **52B**, 96–111.

—— (1935*a*). Untersuchungen an den Leitbündelsystemen der Alismataceenblätter als Beitrag zur Kenntnis der Bedingtheit und der Leistungen der Leitbündelverbindungen. *Planta* **23**, 557–92.

—— (1935*b*). Zur Frage der Funktion der Hydropoten. *Ber. dt. bot. Ges.* **53**, 542–6.

—— (1935*c*). Beiträge zur Anatomie der Alismataceen. V. Die Gattungen *Damasonium* und *Alisma* im Lichte der Anatomie. *Beih. bot. Zbl.* **54A**, 156–69.

—— (1935*d*). Beiträge zur Anatomie der Alismataceen. VI. Die Blattanatomie von *Wiesneria. Beih. bot. Zbl.* **54A**, 494–506.

—— (1935*e*). Über die Anatomie und die morphologischen Natur der Bandblätter der Alismataceen. *Flora, Jena* **129**, 380–415.

—— (1935*f*). Die systematische Bedeutung der Milchsaftsgänge der Alismataceen. *Bot. Jhrb.* **59**, 98–104.

MEYER, N. R. (1966). On the development of pollen grains of Helobiae and on their relation to Nymphaeaceae. *Bot. Zh. SSSR* **51**, 1736–40. [Russian.]

MICHELI, M. (1881). Alismaceae, Butomaceae, Juncaginaceae. In *Monographiae Phanerogamarum* III (ed. A. De Candolle and C. De Candolle) pp. 7–112. G. Masson, Paris.

MIKI, S. (1932). On sea-grasses new to Japan. *Bot. Mag., Tokyo* **46**, 774–88.

—— (1933). On the sea-grasses in Japan. I. *Zostera* and *Phyllospadix*, with special reference to morphological and ecological characters. *Bot. Mag., Tokyo* **47**, 842–62.

—— (1934*a*). On the sea-grasses in Japan. II. Cymodoceaceae and marine Hydrocharitaceae. *Bot. Mag., Tokyo* **48**, 131–42.

—— (1934*b*). On fresh water plants new to Japan. *Bot. Mag., Tokyo* **48**, 326–37.

—— (1934*c*). On the Potamogetons of the Kuriles. *J. Limnol., Japan* **3**, 112–28. [Japanese.]

—— (1935*a*). New water plants in Asia Orientalis. I. *Bot. Mag., Tokyo* **49**, 687–93.

—— (1935*b*). New water plants in Asia Orientalis II. *Bot. Mag., Tokyo* **49**, 773–80.

—— (1937). The origin of *Najas* and *Potamogeton. Bot. Mag., Tokyo* **51**, 472–80.

MILANEZ, F. R. and MEIRA, E. (1943). Observações sôbre *Triuris alata* Brade. *Arq. Serv. Florestal, Rio de Janeiro* **2** (1), 51–61.

MINDEN, H. VON (1899). Beiträge zur anatomischen und physiologischen Kenntnis Wassersecernierender Organe. *Bibl. Bot.* **9**, Heft (46), 1–76.

MITRA, E. (1955). Contributions to our knowledge of Indian freshwater plants. I. On some aspects of the structure and life history of *Hydrilla verticillata* Presl. with notes on its autecology. *J. Asiat. Soc., Sci.* **21**, 1–17.

—— (1964) [1966]. On some aspects of the morphological and anatomical studies of turions of *Hydrilla verticillata* (Linn. f.) Royle. *J. Asiat. Soc., Sci.*, ser. 4, **6**, 17–27.

MITROIU, N. (1969) [1970]. Etudes morphopolliniques et des aspects embryologiques sur les "Polycarpicae" et Helobiae, avec des considérations phylogénétiques. *Lucr. Grâd. bot. Bucuresti*, 263 pp.

MONOYER, A. (1926) [1927]. Sur les stipules des *Potamogeton. C. r. Ass. fr. Av. Sci.* **50**, 327–9.

—— (1927). Contribution à l'anatomie et à l'éthologie des monocotylées aquatiques. *Mem. Cour. Acad. roy. Belg., Cl. Sci.,* Coll. 8°, **10**, 196 pp. (Also *Arch. Inst. bot. Univ. Liége* **17**. 1928.)

—— (1929). Les variations vasculaires dues à la manière d'être des feuilles et l'établissement des types de structure. *Bull. Soc. bot. Belg.* **62**, 69–72.

MONTESANTOS, N. (1912). Morphologische und biologische Untersuchungen über einige Hydrocharideen. *Flora, Jena* **105**, 1–32.

MONTFORT, C. (1918). Die Xeromorphie der Hochmoorpflanzen als Voraussetzung der "physiologischen Trockenheit" der Hochmoore. *Z. Bot.* **10**, 257–352.

MOORE, E. (1915). The Potamogetons in relation to pond culture. *Bull. Bureau Fisheries* **33**, 815.

MÜLLER, J. F. (1875). *Zur Entwicklungsgeschichte der* Vallisneria spiralis. Inaug.-Diss., Bonn.

MUENSCHER, W. C. (1936). The germination of seeds of *Potamogeton*. *Ann. Bot.* **50**, 805–21.

MURBECK, S. (1902). Über die Embryologie von *Ruppia rostellata* Koch. *K. svenska Vet.-Akad. Handl.* (N.F.) **36** (5), 1–21.

MURÉN, A. (1934). Tutkimuksia vesikasvien juurista. (Untersuchungen über die Wurzeln der Wasserpflanzen). *Ann. bot. Soc. zool.-bot. fenn. Vanamo* **5** (8), 1–56. [German summary.]

MYAEMETS, A. A. (1979). On the find of Siberian Arctic species of pondweed *Potamogeton subretusus* Hagstr. (Potamogetonaceae) in Bolshezemelskaya Tundra. *Bot. Zh. SSSR* **64**, 250–1. [Russian.]

NDONGALA-NLENDI-NTUNGA (1976). *Ramification chez les Hydrocharitaceae.* Thèse, Dijon.

NEDELCU, G. A. (1972). Contribuţie la anatomia tulpinii a două plante acvatice. *An. Univ. bucureşti, Biol. veg.* **21**, 147–50. [German summary.]

OBERMEYER, A. A. (1964). The South African species of *Lagarosiphon*. *Bothalia* **8**, 139–46.

—— (1966a). Aponogetonaceae. In *Flora of Southern Africa*, Vol. 1, A (ed. L. E. Codd, B. de Winter, and H. B. Rycroft) pp. 85–92. Department of Agricultural Technical Services, Pretoria.

—— (1966b). Zannichelliaceae. In *Flora of Southern Africa*, Vol. 1 (ed. L. E. Codd, B. de Winter, and H. B. Rycroft) pp. 73–81. Department of Agricultural Technical Services, Pretoria.

OGDEN, E. C. (1943). The broad-leaved species of *Potamogeton* of North America north of Mexico. *Rhodora* **45**, 57–105; 119–63; 171–216. [Also in *Contrib. Gray Herb. Harvard Univ.* **147**.]

—— (1974). Anatomical patterns of some aquatic vascular plants of New York. *N.Y. State Mus. Sci. Serv., Bull.* 424, 133 pp.

OHGA, I. and SINOTÔ, Y. (1924). Cytological studies on *Sciaphila japonica* I. On chromosome. *Bot. Mag., Tokyo* **38**, 202–7. (English.)

—— (1932). Cytological studies on *Sciaphila japonica* Mak. II. On pollen- and embryo sacdevelopment. III. On micorhiza. *Bot. Mag., Tokyo* **46**, 311–15. [Japanese.]

OHLENDORF, O. (1907). *Beiträge zur Anatomie und Biologie der Früchte und Samen einheimischer Wasser- und Sumpfpflanzen.* Diss. Osnabrück.

ONNIS, A. (1969). *Althenia filiformis* Petit in Puglia: nuovi dati sulla distribuzione ed ecologia. *G. bot. ital.* **103**, 47–57.

—— and MAZZANTI, M. (1971). *Althenia filiformis* Petit: azione della temperatura e dell'acqua di mare sulla germinazione. *G. bot. ital.* **105**, 131–43.

OSTENFELD, C. H. (1915) [1916]. *Ruppia anomala* sp. nov., an aberrant type of the Potamogetonaceae. *Bull. Torrey bot. Club* **42**, 659–62.

—— (1916). Contributions to West Australian Botany I. The seagrasses of Western Australia. *Dansk bot. Arkiv* **2** (6), 1–44.

OZIMEK, T., PREJS, A., and PREJS, K. (1976). Biomass and distribution of underground parts of *Potamogeton perfoliatus* L. and *P. lucens* L. in Mikołajogkie Lake, Poland. *Aquat. Bot.* **2**, 309–16.

PALAMAREV, E. H. (1979). Die Gattung *Stratiotes* L. in der Tertiärflora Bulgariens und ihre Entwicklungsgeschichte in Eurasien. *Fitologiya* **12**, 3–36.

—— and USUNOVA, K. (1969). Monokotylen aus den pliozänen Braunkohlen Südbulgariens. *Izv. bot. Inst., bulg. Akad. Nauk* **19**, 127–35.

PALIWAL, S. C. (1976). Epidermal structure and distribution of stomata in *Aponogeton natans* (L.) Engl. and Krause. *Curr. Sci.* **45**, 386–7.

—— and LAVANIA, G. S. (1978). Epidermal structure and distribution of stomata in *Sagittaria guayanensis* H.B. & K. *Curr. Sci.* **47**, 553–5.

—— and —— (1979). Epidermal structure and distribution of stomata in *Potamogeton nodosus* Poir. *Sci. Cult.* **45**, 75–7.

PALMGREN, O. (1939). Cytological studies in *Potamogeton*. Preliminary note. *Bot. Notiser*, 246–8.

PARK, J. (1931). Notes on salt marsh plants III. *Triglochin maritimum* Linn. *Trans. bot. Soc. Edinb.* **30**, 320–5.

PASCASIO, J. F. and SANTOS, J. K. (1930). A critical morphological study of *Thalassia hemprichii* (Ehrenb.) Aschers. from the Philippines. *Nat. Appl. Sci. Bull. Univ. Philipp.* **1** (1), 1–19.

PATE, J. S. and GUNNING, B. E. S. (1969). Vascular transfer cells in angiosperm leaves: a taxonomic and morphological survey. *Protoplasma* **68**, 135–56.

PATRIQUIN, D. G. (1972). The origin of nitrogen and phosphorus for growth of the marine angiosperm *Thalassia testudinum*. *Mar. Biol.* **15**, 35–46.

—— (1973). Estimation of growth rate, production and age of the marine angiosperm *Thalassia testudinum* König. *Caribb. J. Sci.* **13**, 111–23.

—— and KNOWLES, R. (1972). Nitrogen fixation in the rhizosphere of marine angiosperms. *Mar. Biol.* **16**, 49–58.

PEARSALL, W. H. (1930). Notes on *Potamogeton*. *Bot. Soc. Exch. Cl. Br. Isles* **9**, 148–56.

PEISL, P. (1957). Die Binsenform. *Ber. schweiz. bot. Ges.* **67**, 99–213.

PENDLAND, J. (1979). Ultrastructural characteristics of *Hydrilla* leaf tissue. *Tissue Cell* **11**, 79–88.

PERNER, E. and LOSADA-VILLASANTE, M. (1956). Die Zellorganelle der Wurzelhaare von *Trianea bogotensis*. *Protoplasma* **46**, 579–84.

PETER, R., WELSH, H., and DENNY, P. (1979). The translocation of lead and copper in two submerged aquatic angiosperm species. *J. exp. Bot.* **30**, 339–45.

PETTITT, J. M. (1976). Pollen wall and stigma surface in the marine angiosperms *Thalassia* and *Thalassodendron*. *Micron* **7**, 21–31.

—— (1980). Reproduction in seagrasses: nature of the pollen and receptive surface of the stigma in the Hydrocharitaceae. *Ann. Bot.* **45**, 257–71.

—— and JERMY, A. C. (1975). Pollen in hydrophilous angiosperms. *Micron* **5**, 377–405.

PFEIFFER, H. (1919). Zur Anatomie und Morphologie einiger kultivierter Elodeenspezies und über die Kälte als wachstumshemmenden Faktor. *Abh. Naturwiss. Ver. Bremen* **24**, 121–8.

PHILIP, G. (1936). An enalid plant association in the Humber estuary. *J. Ecol.* **24**, 205–19.

PHILLIPS, R. C. (1960a). Environmental effect on leaves of *Diplanthera* Du Petit-Thouars. *Bull. Marine Sci. Gulf Caribb.* **10**, 346–53.

—— (1960b). *Observations on the ecology and distribution of the Florida sea-grasses.* Florida State Board of Conservation, Prof. Pap. **2**, 1–72.

—— (1967). On species of the seagrass *Halodule*, in Florida. *Bull. Mar. Sci.* (*Miami*) **17**, 672–6.

—— MCMILLAN, C., BITTAKER, H. F., and HEISER, R. (1974). *Halodule wrightii* Ascherson in the Gulf of Mexico. *Contrib. Mar. Sci.* **18**, 257–261.

—— and MCROY, C. P. (eds.) (1980). *Handbook of seagrass biology.* Garland STPM Press, New York.

PICHON, M. (1946). Sur les Alismatacées et les Butomacées. *Notul. Syst.* **12**, 170–83.

PLANCHON, J. E. (1844). Sur le genre *Aponogeton* et sur ses affinités naturelles. *Annls. Sci. nat., Bot.*, sér. 3, **1**, 107–20.

—— (1849). Descriptions de quelques Hydrocharidées nouvelles (*Nechamandra roxburghii*). *Annls. Sci. nat., Bot.*, sér. 3, **11**, 78.

PORSCH, O. (1903). Zur Kenntnis des Spaltöffnungsapparates submerser Pflanzenteile. *Sber. Akad. Wiss. Wien, math.-nat. Kl.* **112**, 97–138.

—— (1905). *Der Spaltöffnungsapparat im Lichte der Phylogenie.* G. Fischer: Jena.

PORSILD, M. P. (1946). Stray contributions to the flora of Greenland XV. *Potamogeton groenlandicus* Hagström. *Medd. om Grønland* **134** (4). [Also in *Arb. danske arktiske Station Disko*, No. 16, 18–26.]

POSLUSZNY, U. and SATTLER, R. (1973). Floral development of *Potamogeton densus*. *Can. J. Bot.* **51**, 647–56.

—— (1974a). Floral development of *Potamogeton richardsonii*. *Am. J. Bot.* **61**, 209–16.

—— (1974b). Floral development of *Ruppia maritima* var. *maritima*. *Can. J. Bot.* **52**, 1607–12.

—— (1976a). Floral development of *Najas flexilis*. *Can. J. Bot.* **54**, 1140–51.

—— (1976b). Floral development of *Zannichellia palustris*. *Can. J. Bot.* **54**, 651–62.

—— and TOMLINSON, P. B. (1977). Morphology and development of floral shoots and organs in certain Zannichelliaceae. *Bot. J. Linn. Soc.* **75**, 21–46.

POTTIER, J. (1934). *Contribution à l'étude du développement de la racine, de la tige et de la feuille des phanérogames angiospermes. Les monocotylédones marines méditerranéennes:* Ruppia maritima *L.,* Cymodocea nodosa *(Ucria)* Ascherson *et* Posidonia oceanica *(L.) Delile de la famille des Potamogetonacées.* Jacques et Demontrond, Besançon, 125 pp.

POULSEN, V. A. (1886). Bidrag til Triuridaceernes Naturhistorie. *Vidensk. Medd. naturh. Foren. Kbn.* 1884–6, 161–79.

—— (1890). *Triuris major* sp. nov. Et bidrag til Triuridaceernes Naturhistorie. *Bot. Tidsskr.* **17**, 293–306.

—— (1906). *Sciaphila nana* Bl. Et Bidrag til støvvejens udvikling hos Triuridaceerne. *Vidensk. Medd. naturh. Foren. Kbn.* 161–76.

PRIESTLEY, J. H. and NORTH, E. E. (1922). Physiological studies in plant anatomy. III. The structure of the endodermis in relation to its function. *New Phytol.* **21**, 113–39.

PRILLIEUX, E. (1864). Recherches sur la végétation et la structure de l'*Althenia filiformis* Petit. *Annls Sci. nat., Bot.,* sér. 5, **2**, 169–90.

RAMATI, A., ESHEL, A., LIPHSCHITZ, N., and WAISEL, Y. (1973). Localization of ions in cells of *Potamogeton lucens* L. *Experientia* **29**, 497–501.

RAO, S. (1951). Cytological studies in *Ottelia. Curr. Sci.* **20**, 72.

RAUNKIAER, C. (1895–9). *De danske blomsterplanters naturhistorie.* Bd. I. Enkimbladede. Copenhagen, 724 pp.

—— (1903). Anatomical *Potamogeton*-studies and *Potamogeton fluitans. Bot. Tidsskr.* **25**, 253–80.

RAVN, F. K. (1894–5). Om flyddeevnen hos frøene af vore Vandog sumpplanter. (Sur la faculté de flotter chez les graines de nos plantes aquatiques et marécageuses.) *Bot. Tidsskr.* **19**, 143–88. [French summary 178–88.]

RAVOLOLOMANIRAKA, D. (1972). Contribution à l'étude de quelques feuilles de Monocotylédones. *Bull. Mus. natn. Hist. Nat., Paris,* sér. 3, **46**, *Bot.* **2**, 29–69.

REESE, G. (1962). Zur intragenerischen Taxonomie der Gattung *Ruppia* L. *Z. f. Bot.* **50**, 237–64.

—— (1963). Über die deutschen *Ruppia-* und *Zannichellia*-Kategorien und ihr Verbreitung in Schleswig-Holstein. *Schr. Naturw. Ver. Schlesw.-Holst.* **34**, 44–70.

—— (1967). Cytologische und taxonomische Untersuchungen an *Zannichellia palustris* L. *Biol. Zbl.* **86** (Suppl.), 277–306.

REINECKE, P. (1964). A contribution to the morphology of *Zannichellia aschersoniana* Graebn. *Jl. S. Afr. Bot.* **30**, 93–101.

REINHARDT, L. (1897). *Einige Mittheilungen über die Entwickelung der Spaltöffnungen bei den Pflanzen.* Charkow. [Russian.] (Not seen).

RENDLE, A. B. (1899). A systematic revision of the genus *Najas. Trans. Linn. Soc. Lond., Bot.* Ser. 2, **5**, 379–436.

—— (1901). Najadaceae. In *Das Pflanzenreich* 4 (12) (ed. A. Engler).

RICHARD, L. C. (1812): Memoire sur les Hydrocharidées. *Mém. Cl. Sci. math.-phys.* 1–81.

—— (1815). Proposition d'une nouvelle famille de plantes: les Butomées (Butomeae). *Mém. Mus. natn. Hist. nat., Paris,* **1**, 364–74.

RICHARDS, A. J. and BLAKEMORE, J. (1975). Factors affecting the germination of turions in *Hydrocharis morsus-ranae* L. *Watsonia* **10**, 273–5.

RIEDE, W. (1920). Untersuchungen über Wasserpflanzen. *Flora, Jena* **114**, 1–118.

ROBARDS, A. W., PAYNE, H. L., and GUNNING, B. E. S. (1976). Isolation of the endodermis using wall-degrading enzymes. *Cytobiologie* **13**, 85–92.

ROHRBACH, P. (1871). Beiträge zur Kenntniss einiger Hydrocharideen. *Abh. Naturfor. Ges. Halle* **12** (75), 64 pp.

RONTE, H. (1891). Beiträge zur Kenntniss der Blüthengestaltung einiger Tropenpflanzen. *Flora, Jena* **74**, 492–529.

ROSENBERG, O. (1901a). Ueber die Embryologie von *Zostera marina* L. *Bih. kgl. svenska Vet.-Akad. Handl.* **27**, Afd. III (6), 1–24. [Also in *Medd. Stockholms Högsk. bot. Inst.* **4** (10), 1–24, 1901.]

—— (1901b). Ueber die Pollenbildung von *Zostera. Medd. Stockholms Högsk. bot. Inst.* **4** (11), 1–21.

ROTH, I. (1961). Histogenese der Laubblätter von *Zostera nana. Bot. Jb.* **80**, 500–7.

ROUGIER, M. (1972). Etude cytochimique des squamules d'*Elodea canadensis*. Mise en évidence de leur sécrétion polysaccharidique et de leur activité phosphatasique acide. *Protoplasma* **74**, 113–31.

ROZE, E. (1887). Le mode de fécondation du *Zannichellia palustris* L. *J. Bot., Paris* **1**, 296–9.

—— (1892). Sur le mode de fécondation du *Najas major* Roth et du *Ceratophyllum demersum* L. *Bull. Soc. bot. Fr.* **39**, 361–4.

—— (1894). Recherches sur les *Ruppia. Bull. Soc. bot. Fr.* **41**, 466–80.

RÜTER, E. (1918). Über Vorblattbildung bei Monokotylen. *Flora, Jena* **110**, 193–261.

RUIJGROK, H. W. L. (1974). Cyanogenese bei *Scheuchzeria palustris. Phytochemistry* **13**, 161–2.

SACHET, M.-H. and FOSBERG, F. R. (1974). Remarks on *Halophila. Taxon* **22**, 439–43.

SAHAI, R. and SINHA, A. B. (1969). Sprouting behaviour of the 'dormant apices' of *Potamogeton crispus* Linn. *Experientia* **25**, 653.

ST. JOHN, H. (1925). A critical consideration of Hagström's work on *Potamogeton. Bull. Torrey bot. Club* **52**, 461–71.

—— (1961). Monograph of the genus *Egeria* Planchon. *Darwiniana* **12**, 293–307.

—— (1962). Monograph of the genus *Elodea* (Hydrocharitaceae). Part 1. The species found in the Great Plains, the Rocky Mountains and the Pacific States and Provinces of North America. *Research Studies, Washington State Univ.* **30**, 19–44.

—— (1963). Monograph of the genus *Elodea* (Hydrocharitaceae). Part 3. The species found in northern and eastern South America. *Darwiniana* **12**, 639–52.

—— (1964). Monograph of the genus *Elodea* (Hydrocharitaceae). Part 2. The species found in the Andes and western South America. *Caldasia* **9**, 95–113.

—— (1965a). Monograph of the genus *Elodea* (Hydrocharitaceae). Part 4. The species of eastern and central North America. *Rhodora* **67**, 1–35.

—— (1965b). Monograph of the genus *Elodea* (Hydrocharitaceae). Summary. *Rhodora* **67**, 155–80.

—— (1967). The pistillate flowers of *Egeria densa* Planch. *Darwiniana* **14**, 571–3.

SAKAI, T. and HAYASHI, K. (1973). Studies on the distribution of starchy and sugary leaves in monocotyledonous plants. *Bot. Mag., Tokyo* **86**, 13–25.

SALISBURY, E. J. (1926). Floral construction in the Helobiales. *Ann. Bot., Lond.* **40**, 419–45.

SANE, Y. K. (1939). A contribution to the embryology of the Aponogetonaceae. *J. Indian bot. Soc.* **18**, 79–91.

SATTLER, R. (1965). Perianth development of *Potamogeton richardsonii. Am. J. Bot.* **52**, 35–41.

—— (1968). A technique for the study of floral development. *Can. J. Bot.* **46**, 720–2.

—— (1973). *Organogenesis of flowers. A photographic text-atlas.* University of Toronto Press, Toronto.

—— and GIFFORD, E. M. (1967). Ontogenetic and histochemical changes in the shoot tip of *Najas guadalupensis* (Sprengel) Morong. *Phytomorphology* **17**, 419–28.

—— and SINGH, V. (1973). Floral development of *Hydrocleis nymphoides. Can. J. Bot.* **51**, 2455–8.

—— (1977). Floral organogenesis of *Limnocharis flava. Can. J. Bot.* **55**, 1076–86.

—— (1978). Floral organogenesis of *Echinodorus amazonicus* Rataj and floral construction of the Alismatales. *Bot. J. Linn. Soc.* **77**, 141–56.

SAUNDERS, E. R. (1929). On carpel polymorphism. *Ann. Bot., Lond.* **43**, 459–81.

—— (1937–9). *Floral morphology,* 2 Vols. Heffer, Cambridge.

SAUVAGEAU, C. (1887). Sur la présence de diaphragmes dans les canaux aërifères de la racine. *C. r. hebd. Séanc. Acad. Sci., Paris* **104**, 1–3.

—— (1888). Sur un cas de protoplasme intercellulaire. *J. Bot., Paris* **2**, 396–403.

—— (1889a). Sur la racine du *Najas. J. Bot., Paris* **3**, 3–11.

—— (1889b). Contribution à l'étude du système mécanique dans la racine des plantes aquatiques. Les *Zostera, Cymodocea* et *Posidonia. J. Bot., Paris* **3**, 169–81.

—— (1889c). Contribution à l'étude du système mécanique dans la racine des plantes aquatiques. Les *Potamogeton. J. Bot., Paris* **3**, 61–72.

—— (1890a). Observations sur la structure des feuilles des plantes aquatiques. *J. Bot., Paris* **4**, 41–50; 68–76; 117–26; 129–35; 173–8; 181–92; 221–9; 237–45.

—— (1890b). Sur la structure de la feuille des genres *Halodule* et *Phyllospadix. J. Bot., Paris* **4**, 321–32.

—— (1890c). Sur une particularité de structure des plantes aquatiques. *C. r. hebd. Séanc. Acad. Sci., Paris* **111**, 313–15.

—— (1890d). Sur la feuille des Hydrocharidées marines. *J. Bot., Paris* **4**, 269–75; 289–95.

—— (1891a). Sur les feuilles de quelques monocotylédones aquatiques. *Annls Sci. nat., Bot.,* sér. 7, **13**, 103–296.

—— (1891b). Sur la tige des Cymodocées Aschs. *J. Bot., Paris* **5**, 205–11; 235–43.

—— (1891c). Sur la racine des Cymodocées. *Ass. fr. Av. Sci. Congr. Marseille:* 472–7.

—— (1891d). Sur la tige des *Zostera. J. Bot., Paris* **5**, 33–45; 59–68.

—— (1893). Sur la feuille des Butomées. *Annls Sci. nat., Bot.,* Sér. 7, **17**, 295–326.

—— (1894). Notes biologiques sur les *Potamogeton. J. Bot., Paris* **8**, 1–9; 21–43; 45–58; 98–106; 112–23; 140–8; 165–72.

SAVICH, E. I. (1968). The formation of archesporium and the origin of tapetum in Helobiae. *Bot. Zh. SSSR* **53**, 514–23. [Russian; English summary.]

SCHADE, C. and GUTTENBERG, H. VON (1951). Über die Entwicklung des Wurzelvegetationspunktes der Monokotyledonen. *Planta* **40**, 170–98.

SCHALSCHA-EHRENFELD, M. VON (1940–1). Spross-vegetationspunkt und Blattanlage bei einigen monokotylen Wasserpflanzen (*Potamogeton crispus, Heteranthera dubia, Typha angustifolia*). *Planta* **31**, 448–77.

SCHEIFERS, B. (1877). *Anatomie der Laubsprosse von* Potamogeton. Inaug.-Diss. Bonn. [Not seen.]

SCHENCK, H. (1886a). *Die Biologie der Wassergewächse.* Max Cohen, Bonn.

—— (1886b). Vergleichende Anatomie der submersen Gewächse. *Bibl. Bot.* **1**, Heft. 1, 67 pp.

—— (1889). Über das Aërenchym, ein dem Kork homologes Gewebe bei Sumpfpflanzen. *Jb. wiss. Bot.* **20**, 526–74.

SCHENCKE, P. (1893). *Über* Stratiotes aloides, *zur Familie der Hydrocharideen gehörig.* Diss, Erlangen.

SCHERER, P. E. (1904). Studien über Gefässbündeltypen und Gefässformen. *Beih. bot. Zbl.* **16**, 67–110.

SCHILLING, A. J. (1894). Anatomisch-biologische Untersuchungen über die Schleimbildung der Wasserpflanzen. *Flora, Jena* **78**, 280–360.

SCHMUCKER, Th. (1959). Saprophytismus bei Kormophyten. In *Handbuch der Pflanzenphysiologie* **11** (ed. W. Ruhland) pp. 386–428.

SCHÖNHERR, J. (1976). Water permeability of isolated cuticular membranes: the effect of cuticular waxes on diffusion of water. *Planta* **131**, 159–64.

SCHONLAND, S. (1924). *Althenia* in South Africa. *Kew Bull.* 1920, 365–6.

SCHÜRHOFF, P. N. (1926). *Die Zytologie der Blütenpflanzen.* Stuttgart.

SCHUMANN, K. (1892). *Morphologische Studien,* Heft. 1. W. Engelmann, Leipzig.

SCHUSTER, W. (1910). Zur Kenntnis der Aderung des Monocotylenblattes. *Ber. dt. bot. Ges.* **28**, 268–78.

SCHWANITZ, G. (1967). Morphogenese des *Ruppia* pollen. *Pollen Spores* **9**, 9–48.

SCHWENDENER, S. (1874). *Das mechanische Princip im anatomischen Bau der Monokotylen mit vergleichenden Ausblicken auf die übrigen Pflanzenklassen.* Leipzig.

—— (1882). Die Schutzscheiden und ihre Verstärkungen. *Abhandl. königl. Akad. Wiss. Berlin,* 75 pp.

SCULTHORPE, C. D. (1967). *The biology of aquatic vascular plants.* Edward Arnold, London.

ŞERBĂNESCU, I. (1980). *Zannichellia prodanii* sp. nova. *Studii cerc. Biol.* [*Bucurest.*], *Biol. veg.* **32**, 27–30. [English summary.]

ŞERBĂNESCU-JITARIU, G. (1964). Zur Brachysynkarpie bei *Butomus umbellatus* L. *Rev. roum. Biol., sér. Bot.* **9**, 235–9.

—— (1966). Considérations sur le gynécée et le fruit de *Scheuchzeria palustris* L. *Rev. roum. Biol., ser. Bot.* **11**, 435–9.

—— (1972a). La morphologie du gynécée chez certaines espèces du genre *Potamogeton. Bull. Soc. Hist. nat. Afr. nord* **63** (3–4), 3–7.

—— (1972b). Considérations sur le gynécée, le fruit, la biologie de la dissémination et la germination chez *Zannichellia palustris* L. *Annl. Univ. Bucuresti Biol. Veg.* **21**, 63–70.

—— (1973a). Betrachtungen über das Gynözeum und die Keimung der Samen bei der Gattung *Limnocharis. Lucr. Grǎd. bot. Bucuresti* (1972–3), 41–50.

—— (1973b). Betrachtungen über Gynäzeum, Frucht und Keimung bei *Triglochin maritimum* L. und *Triglochin palustre* L. *Rev. roum. Biol., ser. Bot.* **18**, 9–20.

—— (1974). Observations sur le gynécée de *Ruppia maritima* L. et de *Zostera marina* L. *Bull. Soc. Hist. nat. Afr. nord* **65** (1–2), 215–25.

—— (1976). Contributions à l'étude du fruit et de la germination des semences du genre *Potamogeton. Bull. Soc. Hist. nat. Afr. nord* **67** (1–2), 13–8.

SERGUÉEFF M. (1907). *Contribution à la morphologie et la biologie des Aponogétonacées.* Thesis, Geneva.

SETCHELL, W. A. (1929). Morphological and phenological notes on *Zostera marina* L. *Univ. Calif. Publ. Bot.* **14**, 389–452.

—— (1933). A preliminary survey of the species of *Zostera. Proc. natn. Acad. Sci., Wash.* **19**, 810–17.

—— (1934). South American sea grasses. *Rev. sudamer. Bot.* **1** (4), 4 pp.

—— (1946). The genus *Ruppia* L. *Proc. Calif. Acad. Sci.*, ser. 4, **25**, 469–77.

SEVERIN, C. F. (1932). Origin and structure of the secondary root of *Sagittaria. Bot. Gaz.* **93**, 93–9.

SHAH, C. K. (1972). Some peculiar features in the embryogeny of *Aponogeton natans* (L.) Engl. & Krause. *Geobios* **1**, 104–7.

SHARPE, V. and DENNY, P. (1976). Electron microscope studies on the absorption and localization of lead in the leaf tissue of *Potamogeton pectinatus* L. *J. exp. Bot.* **27**, 1155–62.

SHINOBU, R. (1952). Studies on the stomata of *Potamogeton. Bot. Mag., Tokyo* **65**, 56–60.

—— (1954). Studies on the stomata of *Hydrocharis. Bot. Mag., Tokyo* **67**, 73–7.

SINGH, V. (1964). Morphological and anatomical studies in Helobiae. I. Vegetative anatomy of some members of Potamogetonaceae. *Proc. Indian Acad. Sci., B* **60**, 214–31.

—— (1965*a*). Morphological and anatomical studies in Helobiae. II. Vascular anatomy of the flower of Potamogetonaceae. *Bot. Gaz.* **126**, 137–44.

—— (1965*b*). Morphological and anatomical studies in Helobiae III. Vascular anatomy of the node and flower of Najadaceae. *Proc. Indian Acad. Sci., B* **61**, 98–108.

—— (1965*c*). Morphological and anatomical studies in Helobiae IV. Vegetative and floral anatomy of Aponogetonaceae. *Proc. Indian Acad. Sci., B* **61**, 147–59.

—— (1965*d*). Morphological and anatomical studies in Helobiae. V. Vascular anatomy of the flower of *Lilaea scilloides* (Poir.) Haum. *Proc. Indian Acad. Sci., B* **61**, 316–25.

—— (1966*a*). Morphological and anatomical studies in Helobiae. VI. Vascular anatomy of the flower of Alismaceae. *Proc. natn. Acad. Sci. India, B* **36**, 329–44.

—— (1966*b*). Morphological and anatomical studies in Helobiae. VII. Vascular anatomy of the flower of *Butomus umbellatus* Linn. *Proc. Indian Acad. Sci., B* **63**, 313–20.

—— (1966*c*). Morphological and anatomical studies in Helobiae. X. Trends of specialization in placentation in Helobiae. *Curr. Sci.* **35**, 250–1.

—— (1966*d*). Morphological and anatomical studies in Helobiae. VIII. Vascular anatomy of the flower of Hydrocharitaceae–Stratioideae and Thalassioideae. IX. Vallisnerioideae and Halophiloideae. *Agra Univ. J. Res., Sci.* **15** (2), 43–59; (3), 83–106.

—— (1973). Development of gynoecium of *Triglochin* in three dimensions. *Curr. Sci.* **42**, 813–15.

—— and SATTLER, R. (1972). Floral development of *Alisma triviale. Can. J. Bot.* **50**, 619–27.

—— (1973). Nonspiral androecium and gynoecium of *Sagittaria latifolia. Can. J. Bot.* **51**, 1093–5.

—— —— (1974). Floral development of *Butomus umbellatus. Can. J. Bot.* **52**, 223–30.

—— (1977*a*). Development of the inflorescence and flower of *Sagittaria cuneata. Can. J. Bot.* **55**, 1087–105.

—— —— (1977*b*). Floral development of *Aponogeton natans* and *A. undulatus. Can. J. Bot.* **55**, 1106–20.

SKOTTSBERG, C. (1913). Einige Beobachtungen über das Blühen bei *Potamogeton. Acta Soc. Fauna Flora fenn.* **37** (5), 1–15.

SMITH, C. W. and LEW, L.-F. (1970) [1971]. Cellular arrangement in the node of various angiosperms. *Bot. Gaz.* **131**, 269–72.

SNEATH, P. H. and SOKAL, R. R. (1962). Numerical taxonomy. *Nature, Lond.* **193**, 855–60.

—— (1973). *Numerical taxonomy. The principles and practice of numerical classification.* Freeman, San Francisco.

SOLEREDER, H. (1913). Systematisch-anatomische Untersuchung des Blattes der Hydrocharitaceen. *Beih. Bot. Zbl.* **30** (1), 24–104.

—— (1914). Zur Anatomie und Biologie der neuen *Hydrocharis*-Arten aus Neuguinea. *Meded. Rijksherb.* **21**.

—— and MEYER, F. J. (1933). Reihe 2. Helobiae. In *Systematische Anatomie der Monokotyledonen.* Heft. I, pt. 1. Gebrüder Borntraeger, Berlin. [Pp. 68–130, Potamogetonaceae (= Potamogetonaceae, Zannichelliaceae, Posidoniaceae, Cymodoceaceae, Zosteraceae of this account); pp. 130–8, Najadaceae; pp. 138–46, Aponogetonaceae; pp. 147–55, Scheuchzeriaceae (Juncaginaceae), (= Scheuchzeriaceae, Juncaginaceae, Lilaeaceae of this account.)]

SOUÈGES, R. (1940). Embryogénie des Potamogétonacées. Développement de l'embryon chez le *Potamogeton natans. C. r. hedb. Séanc. Acad. Sci., Paris* **211**, 232–3. [Not seen.]

—— (1954). L'origine du cône végétatif de la tige et la question de la "terminalité" du cotylédon des monocotylédones. *Annls Sci. nat., Bot.,* sér. 11, **15**, 1–20.

SPENCE, D. H. N., MILBURN, T. R., NDAWULA-SENYIMBA, M., and ROBERTS, E. (1971). Fruit biology and germination of two tropical *Potamogeton* species. *New Phytol.* **70**, 197–212.

STANT, M. Y. (1952). The shoot apex of some monocotyledons. I. Structure and development. *Ann. Bot., Lond. (N.S.)* **16**, 115–28.

—— (1954). The shoot apex of some monocotyledons. II. Growth organization. *Ann. Bot., Lond. (N.S.)* **18**, 441–7.

—— (1964). Anatomy of the Alismataceae. *J. Linn. Soc. (Bot.)* **59**, 1–42.

—— (1967). Anatomy of the Butomaceae. *J. Linn. Soc. (Bot.)* **60**, 31–60.

—— (1970). Anatomy of *Petrosavia stellaris* Becc., a saprophytic Monocotyledon. *Bot. J. Linn. Soc.* **63** (Suppl. 1), 147–61.

STEBBINS, G. L. and KHUSH, G. S. (1961). Variation in the organization of the stomatal complex in the leaf epidermis of monocotyledons and its bearing on their phylogeny. *Am. J. Bot.* **48**, 51–9.

STENAR, H. (1935). Embryologische Beobachtungen über *Scheuchzeria palustris* L. *Bot. Notiser* 78–86.

STERLING, C. (1978). Comparative morphology of the carpel in the Liliaceae: Hewardieae, Petrosavieae, and Tricyrteae. *Bot. J. Linn. Soc.* **77**, 95–106.

STREITBERG, H. (1954). Über die Heterophyllie bei Wasserpflanzen mit besonderer Berücksichtigung ihrer Bedeutung für die Systematik. *Flora* **141**, 567–97.

SUBRAMANYAM, K. (1962). Aquatic angiosperms. A systematic account of common Indian aquatic angiosperms. *Bot. Monograph* No. 3, CSIR, New Delhi.

SUTTON, C. S. (1919). On the growth, etc. of the sea tassel, *Ruppia maritima* Linn. *Victorian Nat.* **36**, 69–70.

SVEDELIUS, N. (1904). On the life-history of *Enhalus acoroides* (a contribution to the ecology of the hydrophilous plants). *Ann. Roy. Bot. Gard., Peradeniya* **2**, 267–97.

—— (1932). On the different types of pollination in *Vallisneria spiralis* L. and *Vallisneria americana* Michx. *Svensk bot. Tidskr.* **26**, 1–12.

SWAMY, B. G. L. (1963). The origin of cotyledon and epicotyl in *Ottelia alismoides. Beitr. Biol. Pfl.* **39**, 1–16.

—— and LAKSHMANAN, K. K. (1962a). The origin of epicotylary meristem and cotyledon in *Halophila ovata* Gaudich. *Ann. Bot., Lond. (N.S.)* **26**, 243–9.

—— and —— (1962b). Contributions to the embryology of Najadaceae. *J. Indian Bot. Soc,* **41**, 246–67.

—— and PARAMESWARAN, N. (1962a). On the origin of cotyledon and epicotyl in *Potamogeton indicus. Öst. bot. Z.* **109**, 344–9.

—— and —— (1962b). The helobial endosperm. *Biol. Rev.* **38**, 1–50.

SYMOENS, J.-J., VAN DE VELDEN, J., and BÜSCHER, P. (1979). Contribution à l'étude de la taxonomie et de la distribution de *Potamogeton nodosus* Poir. et *P. thunbergii* Chamb. & Schlechtend. en Afrique. *Bull. Soc. roy. Bot. Belg.* **112**, 79–95.

TAKADA, H. (1952). Untersuchungen über die gerbstoffführenden Idioblasten in Blattlamina von *Helodea densa. J. Inst. Polytech. Osaka City Univ.,* Ser. D **3**, 31–6.

TAKHTAJAN, A. (1959). *Die Evolution der Angiospermen.* Gustav Fischer, Jena.

—— (1966). *Systema et phylogenia magnoliophytorum.* Nauka, Moscow.

TARNAVSCHI, I. T. and NEDELCU, G. A. (1973). Contribuţii morfologice la plante de apă şi de mlaştină. *Lucr. Grăd. bot. Bucuresti* (1972–3), 9–27.

TAYLOR, A. R. A. (1957). Studies of the development of *Zostera marina* L. I. The embryo and seed. *Can. J. Bot.* **35**, 477–99. II. Germination and seedling development. *Can. J. Bot.* **35**, 681–95.

TAYLOR, G. (1949). Some observations on British Potamogetons. *South-East. Nat. Antiquary* **54**, 22–38.

TEPPER, J. G. O. (1882). Some observations on the propagation of *Cymodocea antarctica* (Endl.). *Trans. Proc. Rep. R. Soc. S. Aust.* **4**, 1–4; 47–9.

THEORIN, P. G. E. (1905). Tillägg till kännedomen om växttrichomerna. *Arkiv Bot.* **4** (18), 1–24.

TICHÁ, I. (1964). Einige quantitative anatomische Merkmale von Blättern verschiedener Insertionshöhe bei *Potamogeton perfoliatus* L. und *P. lucens* L. *Biologia Pl. Bohemoslov.* **6** (2), 108–16.

TIEGHEM, P. VAN (1870–1). Recherches sur la symétrie de structure des plantes vasculaires. *Annls Sci. nat., Bot.*, sér. 5, **13**, 5–314.

—— (1911). Place des Triuracées dans la classe des Monocotyles. *C. R. hebd'. Séanc. Acad. Sci., Paris* **152**, 1041–3.

—— and DOULIOT, H. (1888). Recherches comparatives sur l'origine des membres endogènes dans les plantes vasculaires. *Annls Sci. nat., Bot.*, sér. 7, **8**, 1–656.

TISCHLER, G. (1915). Die Periplasmodiumbildung in den Antheren der Commelinaceen und Ausblicke auf das Verhalten der Tapetenzellen bei den übrigen Monokotylen. *Jb. wiss. Bot.* **55**, 52–90.

TOMLINSON, P. B. (1961). *Palmae.* Vol. II, *Anatomy of the monocotyledons* (ed. C. R. Metcalfe). Clarendon Press, Oxford.

—— (1969*a*). On the morphology and anatomy of turtle grass, *Thalassia testudinum* (Hydrocharitaceae). II. Anatomy and development of the root in relation to function. III. Floral morphology and anatomy. *Bull. Mar. Sci. (Miami)* **19**, 57–71; 286–305.

—— (1969*b*). Commelinales-Zingiberales. Vol III, *Anatomy of the monocotyledons* (ed. C. R. Metcalfe). Clarendon Press, Oxford.

—— (1970). Monocotyledons—towards an understanding of their morphology and anatomy. *Adv. bot. Res.* **3**, 207–92.

—— (1971). The shoot apex and its dichotomous branching in the *Nypa* palm. *Ann. Bot., Lond.* (N.S.) **35**, 865–79.

—— (1972). On the morphology and anatomy of turtle grass. *Thalassia testudinum* (Hydrocharitaceae). IV. Leaf anatomy and development. *Bull. Mar. Sci. (Miami)* **22**, 75–93.

—— (1974*a*). Vegetative morphology and meristem dependence—the foundation of productivity in seagrasses. *Aquaculture* **4**, 107–30.

—— (1974*b*). Development of the stomatal complex as a taxonomic character in the monocotyledons. *Taxon* **23**, 109–28.

—— (1980). Leaf morphology and anatomy in seagrasses. In: *Handbook of seagrass biology* (ed. R. C. Phillips and R. C. McRoy) Garland STPM Press, New York.

—— and BAILEY, G. W. (1972). Vegetative branching in *Thalassia testudinum* (Hydrocharitaceae)—a correction. *Bot. Gaz.* **133**, 43–50.

—— and POSLUSZNY, U. (1976). Generic limits in the Zannichelliaceae (sensu Dumortier). *Taxon* **25**, 273–9.

—— and —— (1977). Features of dichotomizing apices in *Flagellaria indica* (Monocotyledones). *Am. J. Bot.* **64**, 1057–65.

—— and —— (1978). Aspects of floral morphology and development in the seagrass *Syringodium filiforme* (Cymodoceaceae). *Bot. Gaz.* **139**, 333–45.

—— and VARGO, G. A. (1966). On the morphology and anatomy of turtle grass, *Thalassia testudinum* (Hydrocharitaceae). I. Vegetative morphology. *Bull. Mar. Sci. (Miami)* **16**, 748–61.

TREUB, H. (1876). Le méristème primitif de la racine dans les monocotylédones. *Musée Bot. Leyden* **2**, 78 pp.

TROLL, W. (1931*a*). Botanische Mitteilungen aus den Tropen II. Zur Morphologie und Biologie von *Enhalus acoroides* (Linn. f.) Rich. *Flora, Jena* **125**, 427–56.

—— (1931*b*). Beiträge zur Morphologie des Gynaeceums. I. Über das Gynaeceum der Hydrocharitaceae. *Planta* **14**, 1–18.

—— (1932). Beiträge zur Morphologie des Gynaeceums. II. Über das Gynaeceum von *Limnocharis* Humb. et Bonpl. *Planta* **17**, 453–60.

—— (1943). *Vergleichende Morphologie der höheren Pflanzen.* Borntraeger, Berlin.

TSCHERMAK-WOESS, E. and HASITSCHKA, G. (1953). Über Musterbildung in der Rhizodermis und Exodermis bei einigen Angiospermen und einer Polypodiacee. Öst. bot. Z. 100, 646–51.

TUR, N. M. (1976). Observaciones teratológicas en el género Potamogeton L. Darwiniana 20, 257–68.

TUTAYUKH, V. KH. and ARAZOV, B. M. (1972). [Anatomical structure of the vegetative organs of Potamogeton natans L.] Izv. Akad. Nauk azerb. SSR, biol. Nauk (1), 8–15. [Azerb.; Russian Summary.]

TUTIN, T. G. (1938). The autecology of Zostera marina in relation to its wasting disease. New Phytol. 37, 50–71.

UHL, N. W. (1947). Studies in the floral morphology and anatomy of certain members of the Helobiae. Ph.D. Thesis, Cornell University.

—— (1976). Developmental studies in Ptychosperma (Palmae). II. The staminate and pistillate flowers. Am. J. Bot. 63, 97–109.

USPENSKIJ, E. E. (1913) [1914]. Zur Phylogenie und Ekologie der Gattung Potamogeton. I. Luft-, Schwimm- und Wasserblätter von Potamogeton perfoliatus L. Bull. Soc. imp. Nat. Moscou (N.S.) 27, 253–62.

VAKHMISTROV, D. B. and KURKOVA, E. B. (1979). [Symplastic connections in the rhizodermis of Trianea bogotensis Karst.] Fiziol. Rast. 26, 943–52. [Russian; English summary.]

VENKATESH, C. S. (1952). The anther and pollen grains of Zannichellia palustris. Curr. Sci. 21, 225–6.

—— (1956). Structure and dehiscence of the anther in Najas. Bot. Notiser 109, 75–82.

VERDOORN, I. C. (1922). Notes on Aponogeton distachyon. S. Afr. J. Nat. Hist. 3, 17–19.

VERES, M. (1908). Adatok a Stratiotes aloides L., ismeretéher (Beiträge zur Kenntnis der Stratiotes aloides L.) Inaug. Diss.: Budapest. [Just's Bot. Jber. 36 (1), Sect. 6, No. 67.]

VERHOEVEN, J. T. A. (1979). The ecology of Ruppia-dominated communities in western Europe. I. Distribution of Ruppia representatives in relation to their autecology. Aquat. Bot. 6, 197–267.

VIJAYARAGHAVAN, M. R. and KUMARI, A. V. (1974). Embryology and systematic position of Zannichellia palustris L. J. Indian bot. Soc. 53, 292–302.

WAGNER, R. (1918). Über den Aufbau der Limnocharis laforestii Duchars. Sber. Akad. wiss. Wien, math.-naturwiss. Kl.I, 127, 317–27.

WARMING, E. (1873). Untersuchungen über pollenbildende Phyllome und Kaulome. Hanstein's bot. Abh. 2 (2), 1–90.

—— (1890) [1891]. Botaniske Exkursioner. 1. Fra Vesterhavskystens Marskegne. Vidensk. Medd. naturh. Foren. Kbn. 206–39.

WATANABE, K. (1944). Morphologisch-biologische Studien über Miyoshia sakuraii Makino. J. Jap. Bot. 20, 85–93. [In Japanese; German Summary.]

WEBER, H. (1950). Über das Wachstum des Rhizoms von Butomus umbellatus L. Planta 38, 196–204.

—— (1956). Über das Wachstum des Rhizoms von Posidonia caulini Kön. (= P. oceanica Delile). Flora, Jena 143, 269–80.

WEIDLICH, W. H. (1976a). The organization of the vascular system in the stems of the Nymphaeaceae. I. Nymphaea subgenera Castalia and Hydrocallis. Am. J. Bot. 63, 499–509.

—— (1976b). The organization of the vascular system in the stems of the Nymphaeaceae. II. Nymphaea subgenera Anecphya, Lotos, and Brachyceras. Am. J. Bot. 63, 1365–79.

WEINROWSKY, P. (1898). Untersuchungen über die Scheitelöffnungen bei Wasserpflanzen. Diss. Berlin.

—— (1899). Untersuchungen über die Scheitelöffnungen bei Wasserpflanzen. Fünfstücks Beiträge wiss. Bot. (5), 205.

WETTSTEIN, R. von (1935). Handbuch der systematischen Botanik, edn. 3, pt.II. Leipzig.

WIEGAND, K. M. (1898). Notes on the embryology of Potamogeton. Bot. Gaz. 25, 116–17.

—— (1899). The development of the microsporangium and microspores in Convallaria and Potamogeton. Bot. Gaz. 28, 328–59.

—— (1900). The development of the embryo-sac in some monocotyledonous plants. Bot. Gaz. 30, 25–47.

WILDE, W. J. J. O. DE (1961). The morphological evaluation and taxonomic value of the spathe in Najas, with descriptions of three new Asiatic–Malaysian taxa. Acta bot. neerl. 10, 164–70.

WILDER, G. J. (1974a). Symmetry and development of *Butomus umbellatus* (Butomaceae) and *Limnocharis flava* (Limnocharitaceae). *Am. J. Bot.* **61**, 379–94.

—— (1974b). Symmetry and development of *Limnobium spongia* (Hydrocharitaceae). *Am. J. Bot.* **61**, 624–42.

—— (1974c). Symmetry and development of pistillate *Vallisneria americana* (Hydrocharitaceae). *Am. J. Bot.* **61**, 846–66.

—— (1975). Phylogenetic trends in the Alismatidae (Monocotyledoneae). *Bot. Gaz.* **136**, 159–70.

WILKINSON, H. P. (1979) [1980]. The plant surface. In: *Anatomy of the dicotyledons* 2nd edn (ed. C. R. Metcalfe and L. Chalk) Vol. 1, pp. 97–165. Clarendon Press, Oxford.

WILLE, N. (1882). Om kimens udviklingshistorie hos *Ruppia rostellata* og *Zannichellia palustris*. *Vidensk. Meddel. naturhist. Foren. Kjøbenhavn* (1882–6), 1–14.

WILLIAMS, W. T. and BARBER, D. A. (1961). The functional significance of aerenchyma in plants. *Symp. Soc. exp. Biol.* **15**, 132–44.

WILSON, K. (1936). The production of root-hairs in relation to the development of the piliferous layer. *Ann. Bot., Lond.* **50**, 121–54.

WINTERBOTTOM, D. C. (1917). Marine fibre. *Dept. Chem. S. Aust., Bull.* No. **4**, 1–36.

WIRZ, H. (1910). Beiträge zur Entwicklungsgeschichte von *Sciaphila* spec. und von *Epirrhizanthes elongata* Bl. *Flora, Jena* **101**, 395–446.

WIŚNIEWSKA, E. (1931). Rozwój ziarn pyłku u *Potamogeton perfoliatus* L. (Die Entwicklung der Pollenkörner bei *Potamogeton perfoliatus* L.). *Acta Soc. bot. pol.* **8**, 157–74. [German summary.]

WITMER, S. W. (1937). Morphology and cytology of *Vallisneria spiralis* L. *Am. Mid. Nat.* **18**, 309–33.

WOOTEN, J. W. (1971). The monoecious and dioecious conditions in *Sagittaria latifolia* L. *Evolution* **25**, 549–53.

WYLIE, R. B. (1904). The morphology of *Elodea canadensis. Bot. Gaz.* **37**, 1–22.

—— (1913). A long-stalked *Elodea* flower. *Bull. Lab. nat. Hist. State Univ. Iowa* **6**, 43–52.

—— (1917). The pollination of *Vallisneria spiralis. Bot Gaz.* **63**, 135–45.

YAMASHITA, T. (1970). Eigenartige Wurzelanlage des Embryos bei *Lilaea subulata* Humb. et Bonpl. und *Triglochin maritimum* L. *J. Fac. Sci. Univ. Tokyo, Sec. III, Bot.* **10**, 181–205.

—— (1972). Eigenartige Wurzelanlage des Embryos bei *Ruppia maritima* L. *Beitr. Biol. Pfl.* **48**, 157–70.

—— (1973). Über die Embryo- und Wurzelentwicklung bei *Zostera japonica* Aschers. et Graebn. *J. Fac. Sci. Univ. Tokyo, Sec. III, Bot.* **11**, 175–93.

—— (1976a). Über die Pollenbildung bei *Halodule pinifolia* und *H. uninervis. Beitr. Biol. Pfl.* **52**, 217–26.

—— (1976b). Über die Embryo- und Wurzelentwicklung bei *Aponogeton madagascariensis* (Mirbel) van Bruggen. *J. Fac. Sci. Univ. Tokyo, Sect. III, Bot.* **12**, 37–64.

YOSHIDA, Y. (1958). On the characteristics of the idioblast in *Elodea* leaf. *J. Fac. Sci. Niigata Univ.* ser., 2, **2**, 173–8.

ZIEGENSPECK, H. (1927). Die Lage des Zellkernes in den Wurzelhaaren von *Hydrocharis morsus-ranae* während des Wachsens. *Bot. Arch.* **20**, 475.

—— (1953–4). Phylogenie und Physiologie der Potamogetonaceae im Lichte moderner Methoden unter besonderer Berücksichtigung von Formen aus Uruguay. *Rev. sudamer. Bot.* **10**, 155–80, 197–212.

ZIMMERMANN, M. H. and BROWN, C. L. (1971). *Trees: structure and function.* Springer-Verlag, New York.

—— and TOMLINSON, P. B. (1972). The vascular system of monocotyledonous stems. *Bot. Gaz.* **133**, 141–55.

ADDENDUM TO BIBLIOGRAPHY

BARNABAS, A. D., BUTLER, V., and STEINKE, T. D. (1980). *Zostera capensis* Setchell II. Fine structure of the cavities in the wall of leaf blade epidermal cells. *Z. Pflanzenphysiol.* **99**, 95–103.

—— and GUILLARD, V. (1979). Observations on the fine structure of phloem parenchyma cells in the leaves of *Zostera capensis* Setchell. *Proc. electron microsc. Soc. S. Afr.* **9**, 63–4.

—— and NAIDOO, Y. (1979). Observations on the fine structure of the leaf epidermis of *Halophila ovalis* (R. Br.) Hook.f. *Proc. electron microsc. Soc. S. Afr.* **9**, 65–6.

CAYE, G. (1980). Analyse du polymorphisme caulinaire chez *Posidonia oceanica* (L.) Del. *Bull. Soc. bot. Fr.* **127**, *Lett. bot.* 257–62.

CHURCHILL, A. C. and RIVER, M. I. (1978). Anthesis and seed production in *Zostera marina* L. from Great South Bay, New York, U.S.A. *Aquat. Bot.* **4**, 81–94.

DE COCK, A. W. A. M. (1980). Flowering pollination and fruiting in *Zostera marina* L. *Aquat. Bot.* **9**, 201–20.

—— (1981). Development of the flowering shoot of *Zostera marina* L., under controlled conditions in comparison to the development in two different natural habitats in the Netherlands. *Aquat. Bot.* **10**, 99–113. [Includes a detailed description of flowering shoots.]

EISEMAN, N. J. and McMILLAN, C. (1980). A new species of seagrass, *Halophila johnsonii*, from the Atlantic coast of Florida. *Aquat. Bot.* **9**, 100–105.

GREENWAY, M. (1979). *Halophila tricostata* (Hydrocharitaceae), a new species of seagrass from the Great Barrier Reef region. *Aquat. Bot.* **7**, 67–70. [Placed in new section: *Tricostatae*.]

GREGOR, H.-J. (1980). *Trapa zapfei* Berger aus dem Untermiozän von Langau bei Geras (NÖ)— eine Hydrocharitaceae. *Ann. naturhist. Mus. Wien* **83**, 105–18.

HARTOG, C. DEN (1980). *Pseudalthenia* antedates *Vleisia*, a nomenclatural note. *Aquat. Bot.* **9**, 95.

HESSE, M. (1980). Entwicklungsgeschichte und Ultrastruktur von Pollenkitt und Exine bei nahe verwandten entomophilen und anemophilen Angiospermensippen der Alismataceae, Liliaceae, Juncaceae, Cyperaceae, Poaceae und Araceae. *Pl. Syst. Evol.* **134**, 229–67.

KORNATOWSKI, J. (1979). Turions and offsets of *Stratiotes aloides* L. *Acta Hydrobiol.* **21**, 185–204.

LAKSHMANAN, K. K. and RAJESHWARI, M. (1979). Sea-grasses of Kursadai Island in the Gulf of Mannar, India: 2. *Syringodium isoetifolium*. *Indian J. Bot.* **2**, 87–95.

LINDNER, C. (1978). Eine neue *Elodea* in Lunz. *Verh. zool.-bot. Ges. Öst.* **116**, 79–81.

LIPKIN, Y. (1980). *Halodule brasiliensis* sp. nov., a new seagrass from South America (Potamogetonaceae). *Revta bras. Biol.* **40**, 85–90.

LY THI BA and GUIGNARD, J.-L. (1979) [1980]. Phylogeny of Helobiae and embryogenic criteria. *Phytomorphology* **29**, 260–6.

MAAS, P.-J. M. (1979). Neotropical saprophytes. In *Tropical Botany* (ed. K. Larsen, & L. B. Holm-Nielsen). pp. 365–70. Academic Press, London, New York, San Francisco.

McMILLAN, C., ZAPATA, O. and ESCOBAR, L. (1980). Sulphated phenolic compounds in sea-grasses. *Aquat. Bot.* **8**, 267–78.

PETTITT, J. M., McCONCHIE, C. A., DUCKER, S. C. and KNOX, R. B. (1980). Unique adaptations for submarine pollination in seagrasses. *Nature* **286**, 487–9.

PHILLIPS, R. C. (1980). Ecological notes on *Phyllospadix* (Potamogetonaceae) in the northeast Pacific. *Aquat. Bot.* **6**, 159–170. [Includes habit illustrations of three species.]

POSLUSZNY, U. (1981). Unicarpellate floral development in *Potamogeton zosteriformis*. *Can. J. Bot.* **59**, 495–504.

PROBATOVA, N. S. and BUCH, T. G. (1981). *Hydrilla verticillata* (Hydrocharitaceae) in the Soviet Far East. *Bot. Zh. SSSR* **66**, 208–14 [Russ.; Eng. summ.].

SASTROUTOMO, S. S. (1981). Turion formation, dormancy and germination of curly pondweed, *Potamogeton crispus* L. *Aquat. Bot.* **10**, 161–74.

SMITH, G. W., HAYASAKA, S. S. and THAYER, G. W. (1979). Root surface area measurements of *Zostera marina* and *Halodule wrightii*. *Bot. Mar.* **22**, 347–58.

SORTINO, M., TRAPANI, S. and COLOMBO, P. (1976–7) [1978]. Influenza di alcuni parametri ambientali sulla ontogenesi di *Limnobium bosci* Rich. *Atti Accad. Sci. Lett. Palermo*, ser. 4, **36**, 41–53.

WOLFF, P. (1980). Die Hydrilleae (Hydrocharitaceae) in Europe. *Göttinger florist. Rundbr.* **14**, 33–56.

ZAURALOVA, N. O. (1980). The assimilative apparatus in some species of the heterophyllous freshwater plants. *Bot. Zh. SSSR* **65**, 1439–46 [Russ. only].

INDEX

Page numbers in **bold face** refer to illustrations; page numbers in *italics* to species cited in literature.

Potamogetonaceae Dumortier, 281; family description, 270; in key to genera, 271; synopsis, 13.
Potamogetonoideae, 281; synopsis, 271.
Potamogetonales, synopsis, 13.
Praelongus-type, of winter-bud, 287.
Pressures, negative and positive in xylem elements, 23.
Primitive status of Helobiae, 17.
Proliferative branching, 27.
Protolirion paradoxum Ridley, 474, 477.
—*sakuraii* (Makino) Dandy, 474, 477.
Protoxylem lacunae, 42.
Pseudalthenia Nakai, 336, 345.
Pseudanthia, discussion of, 44; in *Lilaea*, 18, 262.
Pseudowhorled leaves, in *Najas*, 452; in *Zannichellia*, 342.
Puncta pellucida, 63.
Pusillus-type, of winter-bud, 288.

Radicle, position of, 23.
Ranalisma Stapf, 58.
—*humile* (Kuntze) Hutch., 70, *70,* **84.**
Ranunculaceae, compared with Alismataceae, 69.
Rapateaceae, 15.
Raphide cells, 16.
Raphide crystals, 16.
Rautanenia schinzii, 71.
Regenerative branching, 27.
Retinacules, 422.
Rhizome, anatomy of, 37; in *Potamogeton,* 285.
Rhizosphere, 45; and nitrogen fixation, 43.
Root hairs, abundance of, 39.
Root traces, 39; in Zosteraceae, 39.
Roots, absorption by, 43; anatomy, general, 38; apical organization, 40; branching, 39; dimorphism, 39, 43; distribution of, 38; morphology, general, 38; phloem, 40; vascular tissue, 40; xylem, 40.
Ruppia L., compared with *Potamogeton,* 279; course of stem bundles, 275; description, 271; floral morphology, 279; Folia floralis, 276; infrageneric taxa, 278; key to species, 278; pollination, 278.
—*brachypus* J. Gay, 280, *281.*
—*cirrhosa* (Petag.) Grande, 280, *280,* **312, 314;** morphological illustration, **310.**
—*maritima* L., 279, 280, *280, 281;* leaf ultrastructure, **Pl. 1.**
—var. *brevirostris* Agardh, 279.
—var. *maritima,* 279.
—var. *rostellata* Agardh, 279.

NOTES

NOTES

NOTES

NOTES

PLATES

PLATE 1. Ultrastructure of leaf epidermal cells in leaf of blade. (Supplied by R. H. Jagels.)

A. *Zostera marina* (Zosteraceae). Young leaf, epidermal cells showing thin walls (W), plasmodesmata (P) between adjacent cells, no invagination of plasmalemma (PL), pro-mitochondria (PM) and pro-plastids (PC), portion of nucleus (N), numerous ribosomes, vacuole (V), and distinct cuticle (C) ($\times$ 9540).

B. *Zostera marina* (Zosteraceae). Mature leaf epidermal cells showing invaginated plasmalemma (PL), cell wall ingrowths (WI), mitochondria (M) with well developed cristae system, chloroplasts (CH), amorphous shaped liquid droplets (D) ($\times$ 9540).

C. *Thalassia testudinum* (Hydrocharitaceae). Mature leaf epidermal cell showing extensively invaginated plasmalemma (PL), reticulate matrix in apoplasmic region (RM), portion of chloroplast (CH), mitochondria with well developed cristae (M), Golgi (G), dense staining inner wall (IW) but no wall ingrowths or stubs (from Jagels 1973) ($\times$ 14160).

D. *Ruppia maritima* (Potamogetonaceae). Mature leaf epidermal cell showing more limited plasmalemma invagination (PL), absence of wall ingrowths, mitochondria with well developed cristae (M), chloroplast (CH) ($\times$ 9360).

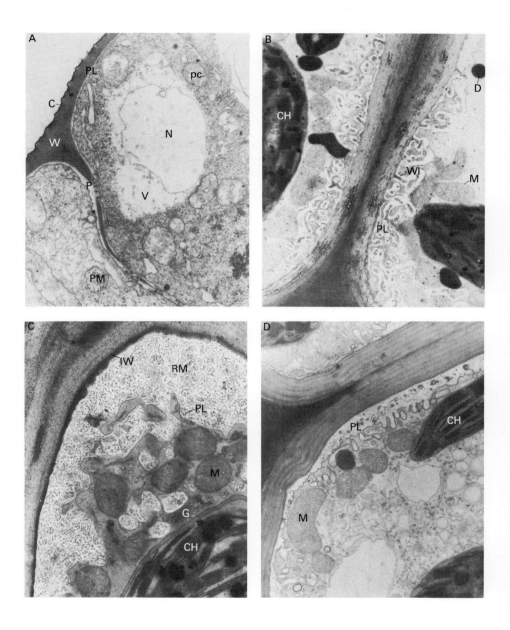

PLATE 2. Peduncle anatomy. Transverse section.

A, B. *Ottelia somaliensis* (Hydrocharitaceae).
A. (× 6), fruiting peduncle. B. (× 36), detail of single central vb.

C, D. *Cycnogeton procera* (Juncaginaceae).
C. (× 6), flower peduncle. D. (× 36), detail of peripheral tissues.

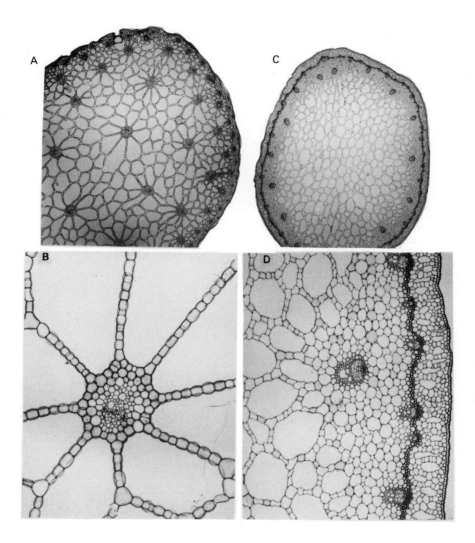

PLATE 3. Rhizome anatomy, transverse sections.

A. *Lilaea scilloides* (Lilaeaceae) ($\times$ 36), upright axis, region of the central cylinder with irregularly distributed vbs; root primordia in inner cortex are all continuous with periphery of central cylinder, out of the plane of section.

B. *Scheuchzeria palustris* (Scheuchzeriaceae) ($\times$ 36), extended rhizome, region of the central cylinder. Ground tissue of cortex and medulla aerenchymatous. Vbs of central cylinder crowded at its periphery; sections of leaf traces appear in inner cortex.

C. *Aponogeton distachyos* (Aponogetonaceae) ($\times$ 6), cormous axis; ground tissue is densely parenchymatous with abundant starch, central cylinder is broad; vbs are narrow and poorly differentiated.

D. *Ottelia cylindrica* (Hydrocharitaceae) ($\times$ 14), erect axis with narrow central cylinder, numerous roots in the cortex are continuous with the periphery of the central cylinder, out of the plane of section.

E. *Posidonia oceanica* (Posidoniaceae) ($\times$ 36), rhizome with poorly differentiated narrow central cylinder but with indistinct vbs. Cortex aerenchymatous with fibrous strands; tannin cells (black) abundant.

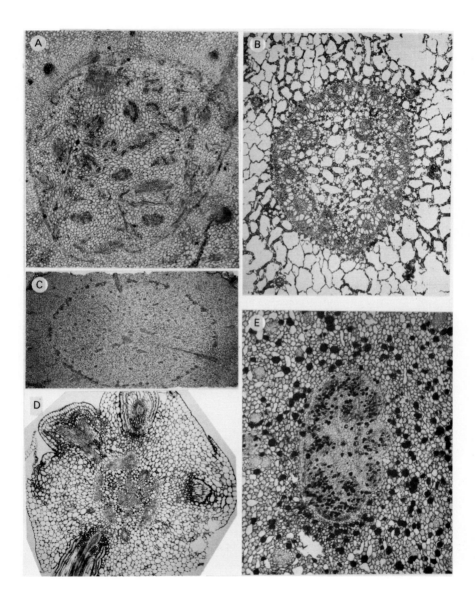

533

PLATE 4. Axis anatomy. Transverse section.

A, B. *Enhalus acoroides* (Hydrocharitaceae). Dorsiventral rhizome with indications of roots on lower surface.

A. (× 6). B. (× 30), detail of central cylinder showing poorly differentiated vascular system; tannin cells—black. Note that this section is oriented at right angles to A, of which it is an enlargement.

C, D. *Cycnogeton procera* (Juncaginaceae). Vegetative axis.

C (× 6). D (× 10), detail of central cylinder.

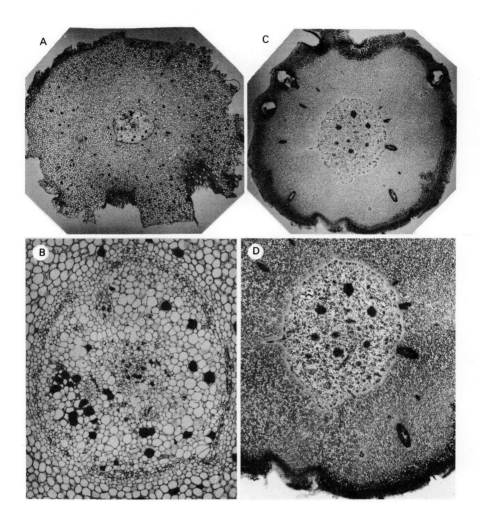

535

PLATE 5. HYDROCHARITACEAE. *Thalassia testudinum.* Axis anatomy, transverse section. Sections are bleached, tannins largely removed.

A, B. Short-shoot (leafy axis).

A. (× 14), central cylinder proportionately broad, inner cortex somewhat lacunose. B (× 40), detail of central cylinder, with median and one lateral leaf trace (right) connecting to the central vascular tissue; root trace (upper) connecting to the peripheral vascular tissue.

C, D. Long-shoot (rhizome).

C. (× 10), central cylinder proportionately narrow, inner cortex lacunose. D. (× 20), detail of central cylinder, vascular tissue poorly differentiated.

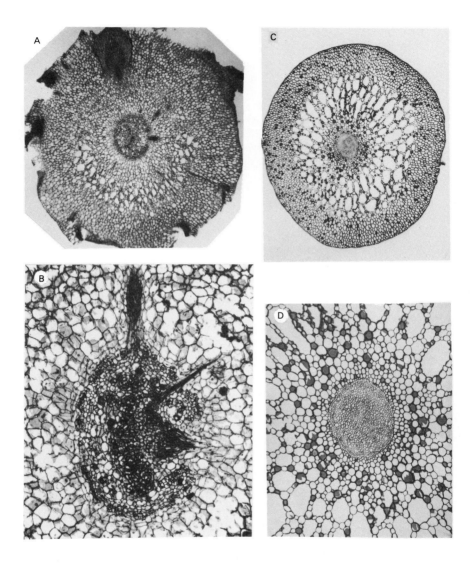

PLATE 6. ZOSTERACEAE. *Zostera marina.* Shoot development to show branch displacement. Longitudinal section.

A. Shoot apex ($\times$ 160); bud (arrow) initiated in axil of P_3, dotted line indicates approximate level of insertion of P_3.

B. Shoot apex ($\times$ 54); including parts of six leaves, with buds clearly situated in axils of P_3–P_6.

C. Extending shoot ($\times$ 27); bud (arrow) in axil of lowest leaf to left becomes associated with node above by intercalary extension of internode between it and the original node of insertion (dotted line).

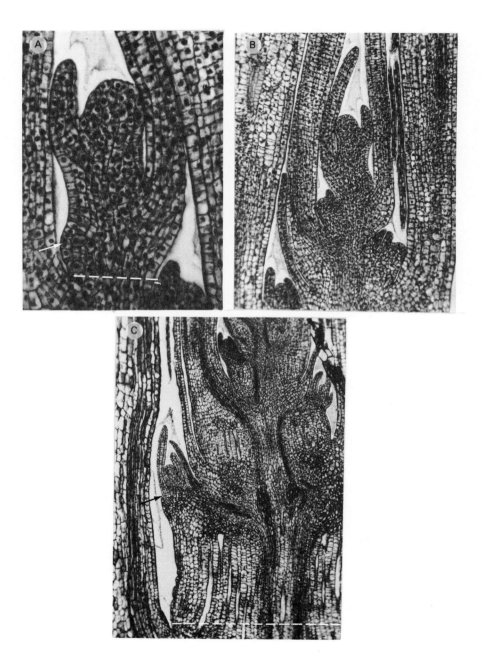

PLATE 7. ALISMATACEAE. *Echinodorus amazonicus.* Early stages in floral development to show the initiation and development of CA primordia (× 146). (After Sattler and Singh 1978.)

A. Floral bud showing three sepal primordia (K) and expanding alternisepalous areas (CA) on which petal–stamen complexes will later be formed.

B. Floral bud with two sepal primordia (K) and an expanding alternisepalous area (CA) above and between them.

C. Top view of a floral bud at the time of petal (C) inception. Above the lowest petal primordium an early indication of the first stamen pair (*arrowheads*).

D. Floral bud after the formation of three pairs of outer stamen primordia (Ao), each pair on either side of a petal primordium (C). Early indication of the inner whorl of stamens (*arrowheads*) opposite the petal primordium.

E, F. Two floral buds showing inception of three inner stamen primordia (Ai), between and above the pairs of outer stamen primordia; rK—removed sepal.

G. Floral bud after the formation of the inner stamen primordium. Carpel primordium (P) between the two groups of stamen primordia.

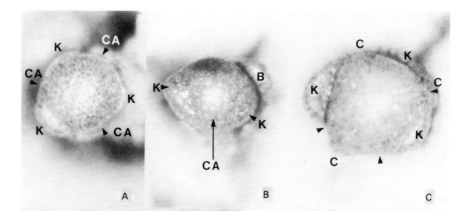

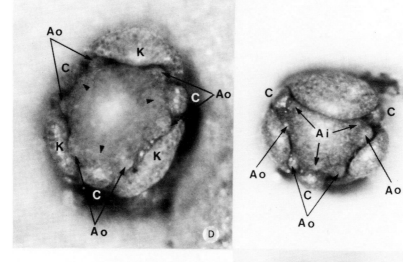

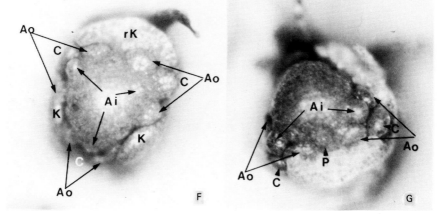

PLATE 8. ALISMATACEAE. *Sagittaria cuneata.* Early stages in development of female flower (× 120). (After Singh and Sattler 1977*a*.)

A. Three flower buds, the two uppermost (F) with their associated bract (B) just differentiating, the lower showing the inception of sepal primordia (K).

B. Side view of a flower bud with sepal primordia (K).

C. Side view of flower with sepal primordia (K) just initiated.

D. Side view of larger flower bud at a slightly later stage of development than C.

E. Side view of flower bud with a pair of staminode primordia (S) just initiated.

F. Side view of flower bud at a somewhat later stage than E, with a pair of staminode primordia (S) each to one side of the petal primordium (C).

G. Side view of flower bud with two staminode primordia of first whorl (as in F) and one staminode primordium of second whorl.

H. Top view of flower bud with staminode primordia of two whorls. In the further development of this massive apex carpel primordia are formed, beginning at its rim and progressing centripetally until the whole apex is occupied.

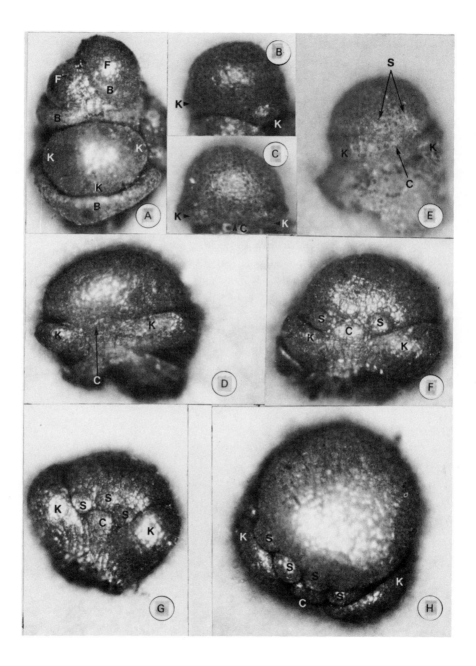

PLATE 9. BUTOMACEAE. *Butomus umbellatus.* Successive early stages (A–G) in floral development (× 120) all from above. (After Singh and Sattler 1974.)

A. Floral bud showing inception of outer tepal (sepal) primordia (K).

B. Inception of two inner tepal (petal) primordia (C) and one bulge (centre left) which can be considered a tepal–stamen primordium (CA primordium).

C. Inception of three pairs of outer stamen primordia (A) each pair on either side of an inner tepal primordium (C).

D. Inception of three inner stamen primordia, between and above a pair of outer stamen primordia.

E. Inception of antesepalous carpels (P).

F. Inception of antepetalous carpels (P).

G. Floral bud with outer appendages removed showing six carpel primordia, with conspicuous remains of floral apex (F).

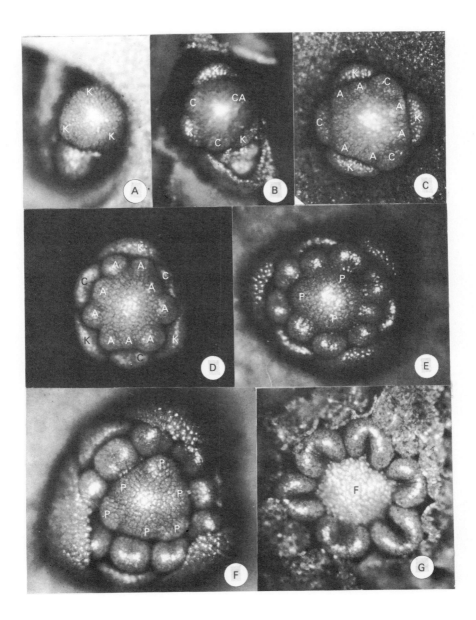

545

PLATE 10. APONOGETONACEAE. *Aponogeton undulatum.* Early stages in floral development (× 120). (After Singh and Sattler 1977.)

A. Earliest stages with successive appearance of floral appendages shown in different floral primordia on a single orthostichy; no subtending bracts are present.

B. Detail of two floral primordia, the uppermost triangular, suggesting the sites of initiation of three tepal primordia, but normally only two mature.

C. Floral primordium in centre with two tepal primordia and six stamen primordia in two whorls of three each.

D. Later stages showing initiation of carpels.

E. Latest stages, in the lowest flower the ovules are just developing.

Ai—inner stamen primordium; Ao—outer stamen primordium; G—carpel primordium; P—tepal primordium.

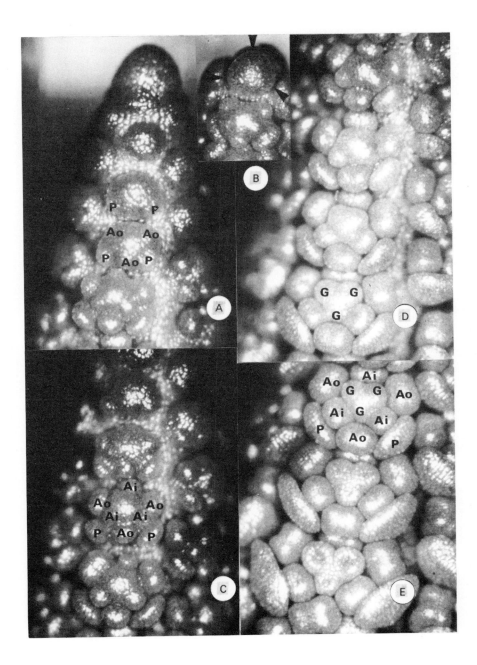

PLATE 11. SCHEUCHZERIACEAE. *Scheuchzeria palustris* L. Floral development and morphology (provided by U. Posluszny).

Side views of the inflorescence axis showing sequential stages in the development of floral apices. B—Epi-illumination, the rest SEM.

A. ($\times$ 170) Young inflorescence axis showing floral apices just prior to organ differentiation. The uppermost bract (B) envelops two floral apices; other bracts (rB) have been removed.

B. ($\times$ 103) Several of the floral apices are initiating the outer of the two trimerous whorls of tepals (P_o) and stamens (A_o). On the lowermost floral apex a tepal primordium (P_i) belonging to the inner whorl is just being initiated.

C. ($\times$ 117) Floral apices forming both inner (P_i, A_i) and outer (P_o, A_o) whorls of tepals and stamens. The terminal position of the uppermost flower is clear.

D. ($\times$ 144) Floral apices initiating carpel primordia (G).

E. ($\times$ 80) Ovules (O) being initiated on the carpel primordia.

F. ($\times$ 53) Late stage in floral development. The ovules are almost totally enclosed by the carpel wall leaving only a slit-like opening on the ventral side. Stamens removed in uppermost flower.

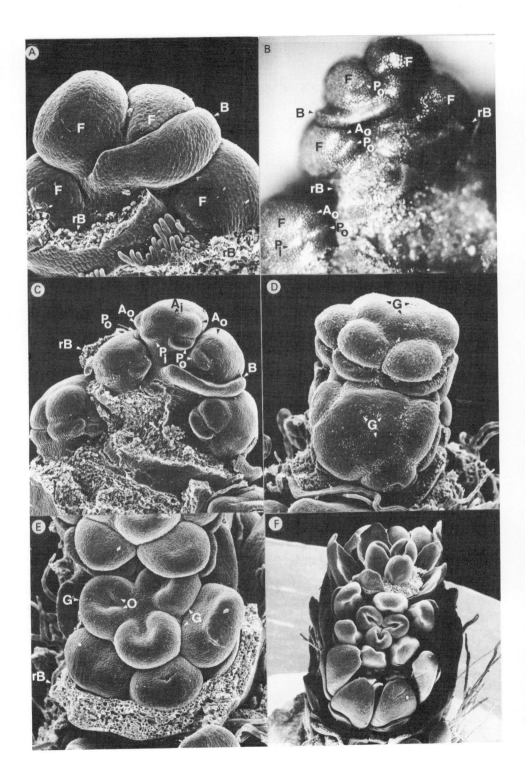

PLATE 12. LILAEACEAE. *Lilaea scilloides* (Poir.) Haum. Floral development and morphology (provided by U. Posluszny).
Side views of the inflorescence axis showing sequential stages in the development of floral apices. A, C, F—Epi-illumination; B, D, E—SEM.

A. (× 195) Young inflorescence axis (I) developing terminally. Basal pistillate flower (G) just being initiated. Renewal apex (V) has already formed a sheathing prophyll (L_s). An older sheathing prophyll (rL_s) has been removed.

B. (× 346) Several floral apices (F) being initiated on the inflorescence axis.

C. (× 245) Floral apices (F) prior to organ differentiation.

D. (× 288) Three different types of flowers developing on the inflorescence axis: (a) bisexual flower composed of a tepal (P), a stamen (A), and a carpel (G); (b) staminate flowers composed of a tepal and a stamen; (c) pistillate flowers composed of only single carpels. Squamules (S) are developing at the base of the inflorescence axis.

E. (× 240) Ovule (O) initiation in the bisexual flowers.

F. (× 110) Late stage in floral development. The basal pistillate flower is differentiating its style (Sy) and stigma (Si).

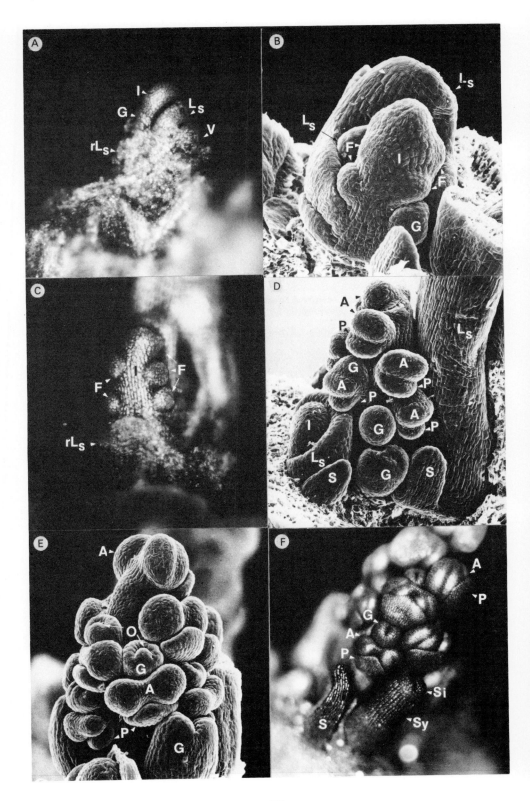

PLATE 13. ZANNICHELLIACEAE. Floral development. All magnifications, × 140.

A. *Vleisia aschersoniana.* Stages in development of male and female flowers. 1. Monopodial shoot with lateral branches. 2. Young male flowers (A). Connective outgrowth (Aco) just being initiated. 3. Sympodial male branch, terminating in a single stamen (A), with lateral branch (V) just below. Note sheathing prophyll (L_s). 4. Single stamen showing both outgrowth (Aco) and extension of tip (Ae). 5. Lateral branch below a developing stamen, ending in a dome-shaped gynoecial primordium (G). 6. Nearly mature stamen with sheathing prophyll still intact. 7. Sympodial female branch, terminating in a unicarpellate gynoecium (G). Lateral renewal branches (V) can be seen developing below the gynoecium. Membranous envelope (E) which sheaths the lower portion of the mature carpel just being initiated. 8. Carpel initiating its single ovule (O). 9. Example of the short proximal female branch at a similar stage to that in 8. 10. Nearly mature female flower with well developed stigma (Si), style (Sy) and membranous envelope (E).

B. *Althenia filiformis.* Stages in development of male and female flowers. 1. Sympodial flowering branch, terminating in a young single stamen (A) with a still younger stamen just below. 2. Three stages in flowering buds; the two uppermost buds are male flowers (A), while the lowermost bud (F) is still too young to determine its sex. Note the developing bracts (B) about the uppermost male flower. 3. Nearly mature male flower terminating sympodial floral branch; young tricarpellate female flower developing below terminal male flower. 4. Side view of two carpels showing young subtending bracts (B). 5. Nearly mature carpels with broad peltate stigmas (Si) and long style (Sy).

Key to labelling

PLATES 13 and 14.

A	=	Stamen or its primordium
Ac	=	Stamen connective
Aco	=	Connective outgrowth
Ae	=	Stamen tip or connective extension
B	=	Bract-like appendage or its primordium
E	=	Membranous envelope about pistillate flower
F	=	Floral apex
G	=	Carpel or its primordium
L	=	Vegetative appendage or its primordium
L_s	=	Sheathing prophyll or its primordium
O	=	Ovule or its primordium
r	=	Removed, e.g. rL_s = sheathing prophyll removed
S	=	Squamules
Si	=	Stigma or developing stigma
Sy	=	Style or developing style
V	=	Vegetative apex or apex of renewal shoot

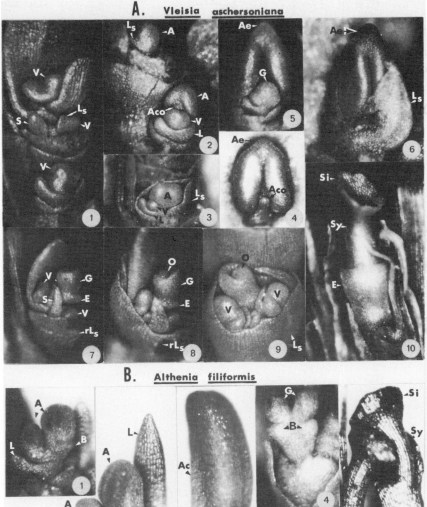

A. _Vleisia aschersoniana_

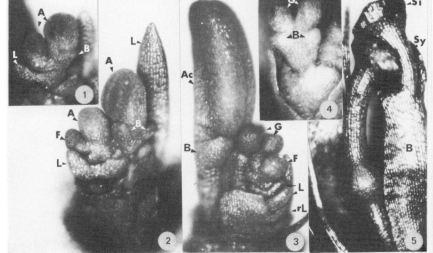

B. _Althenia filiformis_

553

PLATE 14. ZANNICHELLIACEAE. Floral development. (All × 140.)

A. *Lepilaena bilocularis.* Stages in development of male and female flowers. 1. Transitional stage in floral induction. 2. Young single stamen (A) just initiating microsporangia and bracts. 3. Early stage in development of extension of tip of stamen connective (Ae). 4. Nearly mature stage of extension of stamen connective (Ae). 5. Young tricarpellate female flower. 6. Ovule initiation (O) at adaxial portion of carpel wall. Bract primordia (B) can be seen developing below the young carpels. 7. Carpels just initiating their stigmas (Si). 8. Nearly mature carpels showing much lobed stigma (Si).

B. *L. cylindrocarpa.* Stages in development of female flowers. 1. Two stages in carpel development. Carpel (G) and bract (B) primordia just being initiated in lower portion of figure. Carpel in upper portion is at stage of stigma inception. 2. Three carpel primordia clearly distinguishable. 3. Tri-carpellate female flower in lower portion of figure, at a similar stage to that in 2. In upper portion of figure, stigmas (Si) being initiated in carpels. 4. Ovules (O) just being initiated in carpels. 5. Nearly mature carpels, in upper portion of figure, with developing styles (Sy); younger carpels (G) and undefined floral apex (F) below. 6. Nearly mature carpel showing portion of peltate stigma (Si) and subtending lobed bract.

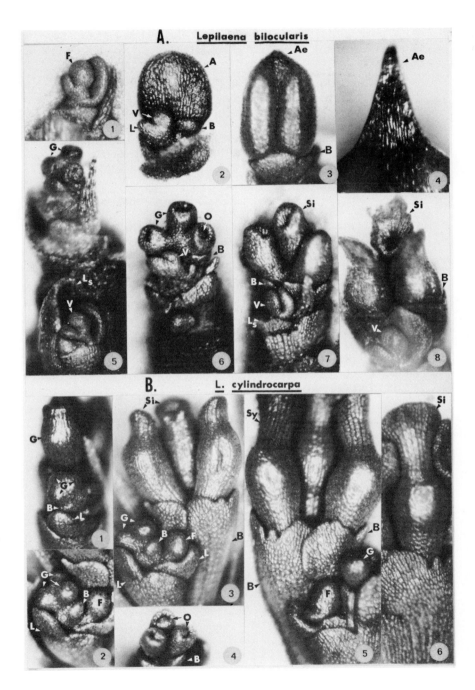

A. Lepilaena bilocularis

B. L. cylindrocarpa

PLATE 15. TRIURIDACEAE. Vegetative and floral anatomy. Except where stated, sections are all stained in safranin and Delafield's hematoxylin.

A. 'Triuridaceae' TS root. (BSIP Iromea 2535. Slide in Jodrell Laboratory, Kew.)

B. *Sciaphila africana* TS stem (Enti and Hall 36286)

C. *Andruris vitiensis* TS stem at base (Tomlinson 21.iv.69 D)

D. *Andruris vitiensis* TS stem showing scarcely differentiated stele (Tomlinson 21.iv.69 D. Section stained in phloroglucinol and concentrated HCl; mounted in glycerine).

E. *Sciaphila albescens* TS stem (HSM 22464. Slide in Jodrell Laboratory, Kew.)

F. *Sciaphila albescens* cortical cells from root, macerated and unstained to show mycelium of endophyte (Tomlinson 26.ii.72)

G. *Andruris vitiensis* TS carpellate flower showing vascular supply to numerous free carpels (Tomlinson 21.iv.69 D)

H. *Andruris vitiensis* LS staminate flower (Tomlinson 21.iv.69 D)
Scale = 200 μm; A, G, H; B and C; D–F are at the same magnification.

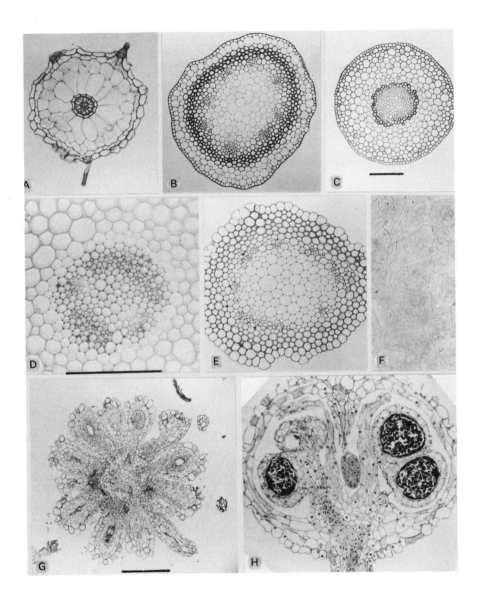

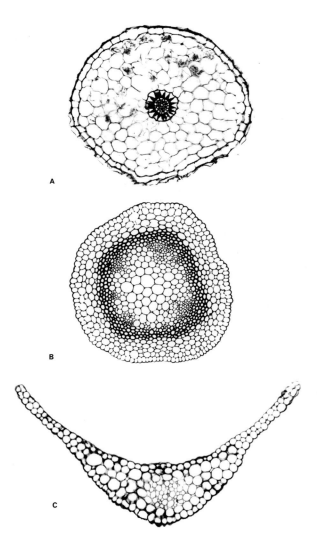

A

B

C

559